Undergraduate Topics in Computer Science

T0171856

'Undergraduate Topics in Computer Science' (UTiCS) delivers high-quality instructional content for undergraduates studying in all areas of computing and information science. From core foundational and theoretical material to final-year topics and applications, UTiCS books take a fresh, concise, and modern approach and are ideal for self-study or for a one- or two-semester course. The texts are all authored by established experts in their fields, reviewed by an international advisory board, and contain numerous examples and problems, many of which include fully worked solutions.

The UTiCS concept relies on high-quality, concise books in softback format, and generally a maximum of 275-300 pages. For undergraduate textbooks that are likely to be longer, more expository, Springer continues to offer the highly regarded Texts in Computer Science series, to which we refer potential authors.

More information about this series at http://www.springer.com/series/7592

Boris Mirkin

Core Data Analysis: Summarization, Correlation, and Visualization

Second Edition

 Springer

Boris Mirkin
Department of Data Analysis and Artificial
Intelligence, Faculty of Computer Science
National Research University Higher
School of Economics
Moscow, Russia

Professor Emeritus, Department of
Computer Science and Information Systems
Birkbeck, University of London
London, UK

ISSN 1863-7310   ISSN 2197-1781   (electronic)
Undergraduate Topics in Computer Science
ISBN 978-3-030-00270-1   ISBN 978-3-030-00271-8   (eBook)
https://doi.org/10.1007/978-3-030-00271-8

Library of Congress Control Number: 2018959271

This Springer imprint is published by the registered company Springer Nature Switzerland AG
The registered company address is: Gewerbestrasse 11, 6330 Cham, Switzerland

Acknowledgements

To 1st Edition: Too many people contributed to the material of the book to list all their names. First of all, my gratitude goes to Springer's editors who were instrumental in bringing forth the very idea of writing such a book and in channeling my efforts by providing good critical reviews. Then, of course, I thank the students at my classes in MS programs in Computer Science at Birkbeck and, more recently, in BS and MS programs in Applied Mathematics and Informatics at HSE. Here is a list of people who directly contributed to this book with advice, and sometimes with computation: I. Muchnik (Rutgers University), M. Levin (Higher School of Economics Moscow), T. Fenner (Birkbeck University of London), S. Nascimento (New University of Lisbon), T. Krauze (Hofstra University), I. Mandel (Telmar Inc), I. Mirkin (Yext), V. Sulimova (Tula Technical University), and V. Topinsky (Higher School of Economics Moscow). The HSE students J. Askarova, K. Chernyak, O. Chugunova, K. Kovaleva, and A Kramarenko helped in debugging the final version.

To 2nd Edition: My thanks go, first of all, to colleagues in the National Research University Higher School of Economics Moscow RF, for permanent support. Two of them should be mentioned here: Prof. Fuad Aleskerov, Head of Laboratory for the Analysis and Choice of Decisions, and Prof. Sergei Kuznetsov, Head of Laboratory for Intelligent Systems and Structural Analysis. Both Laboratories are supported within the framework of the Russian Academic Excellence Project "5–100". HSE students of my classes in Data Analysis contributed to the correction of mistakes and blunders.

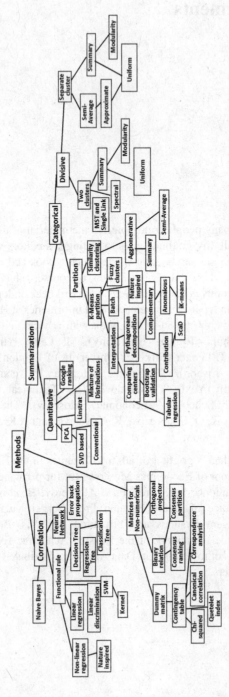

Fig. 1 T-CODA: A taxonomy of main core data analysis methods and concepts described in the text

Contents

Chapter 1
Topics in Substance of Data Analysis

Abstract This is an introductory chapter in which

(i) The goals of core data analysis as a tool helping to enhance and augment knowledge of the domain are outlined. Since knowledge is represented by the concepts and statements of relation between them, two main pathways for data analysis are summarization, for developing and augmenting concepts, and correlation, for enhancing and establishing relations.
(ii) A set of eight cases involving small datasets and related data analysis problems is presented. The datasets are taken from various fields such as monitoring market towns, computer security protocols, bioinformatics, and cognitive psychology.
(iii) An overview of data visualization, its goals and some techniques, is given.
(iv) A general view of strengths and pitfalls of data analysis is provided.
(v) An overview of the concept of classification as a soft knowledge structure widely used in theory and practice is given.

1.1 Summarization and Correlation: Main Goals of Core Data Analysis

1.1.1 The Goals for Knowledge Enhancing

The term Data Analysis has been used for quite a while, even before the advent of computer era, as an extension of mathematical statistics, starting from developments in cluster analysis and other multivariate techniques before WWII, bringing forth the concepts of "exploratory" data analysis and "confirmatory" data analysis in statistics (see, for example, Tukey 1977). The former was supposed to cover a set of techniques for finding patterns in data, and the latter to cover more conventional mathematical statistics approaches for hypothesis testing. "A possible definition of data analysis is the process of computing various summaries and derived values from the given collection of data" and, moreover, the process may become more intelligent if attempts are made to automate some of the reasoning of skilled data analysts and/or to utilize approaches developed in the Artificial Intelligence areas

© Springer Nature Switzerland AG 2019

B. Mirkin, *Core Data Analysis: Summarization, Correlation, and Visualization*, Undergraduate Topics in Computer Science, https://doi.org/10.1007/978-3-030-00271-8_1

(Berthold and Hand 2003, p. 3). Overall, the term Data Analysis is usually applied as an umbrella to cover all the various activities mentioned above, with an emphasis on mathematical statistics and its extensions.

The situation can be seen as follows. Classical statistics takes the view of data as a vehicle to fit and test mathematical models of the phenomena the data refer to. The data mining and knowledge discovery discipline claims to use data so that new knowledge is added. What is knowledge remains undefined. Knowledge of fact and knowledge of God and knowledge of regularities in nature can be distinguished though. It should be sensible then to look at those methods that relate to an intermediate level and contribute to the theoretical—rather than any—knowledge of the phenomenon. These would focus on ways for augmenting or enhancing theo- retical knowledge of a specific domain which the data being analyzed relate to. The term "knowledge" encompasses many diverse layers or forms of information, starting from individual facts to those of literary characters to major scientific laws. But when focusing on a particular domain the dataset in question comes from, its "theoretical" knowledge structure can be considered as comprised of two main types of elements: (i) concepts and (ii) statements relating concepts. *Concepts* are terms referring to aggregations of similar entities, such as apples or plums, or similar categories such as fruit comprising both apples and plums, among others. When created over data objects or features, these are referred to, in data analysis, as clusters or factors. *Statements of relation* between concepts express regular- ities relating different categories. Two features are said to correlate when a co-occurrence of specific patterns in their values is observed as, for instance, when a feature's value tends to be the square of another feature. The observance of a correlation pattern can lead sometimes to investigation of a broader structure behind the pattern, which may further lead to finding or developing a theoretical framework from which the correlation follows. It is useful to distinguish between quantitative correlations such as algebraic expressions involving data features and categorical ones expressed in a non-quantitative way, for example, as logical production rules or more complex structures such as decision trees. Correlations may be used for both understanding and prediction. In applications, the latter has been until recently by far more important. Moreover, the prediction problem is much easier to make sense of operationally, hence machine learning has concentrated on this.

What is said above suggests that there are two main pathways for data analysis to augment theoretical knowledge: (i) developing new concepts by "summarizing" data and (ii) deriving new relations between concepts by analyzing "correlation" between various aspects of the data. The quotation marks are used here to point out that each of the terms, summarization and correlation, significantly extends its conventional meaning. Indeed, while everybody would agree that the average mark does summarize the marking scores on test papers, it would be more daring to see in the same light derivation of students' hidden talent scores by approximating their test marks or finding a cluster of similarly performing students. Still, the mathe- matical structures behind each of these three activities—calculating the average, finding a hidden factor, and designing a cluster structure—are analogous, which suggests that classing them all under the "summarization" umbrella may be reasonable. Similarly, the term "correlation" which is conventionally utilized in statistics only to express the extent of linear relationship between two or more variables, is understood here in its generic sense, as a supposed affinity between two or more aspects of the same data that can be variously expressed, not necessarily by a linear equation or by a quantitative expression at all. This is what is encompassed further on as Core Data `Analysis.

It would be useful to spell out that view of the data as a subject of computational data analysis that is adhered to here. A dataset here is a table at which objects, synonymously referred to as entities, are assigned to rows. Columns contain data of the objects, usually in the form of feature values or similarity index values. Typically, in sciences and in statistics, a problem comes first, and then the investigator turns to data that might be useful in advancing towards a solution. In computational data analysis, it may also be the case sometimes. Yet the situation is reversed frequently. There is a dataset related to this or that process or phenomenon. There are many issues related to the process. Typical questions then would be: Take a look at this data set—what sense can be made out of it?—Is there any structure in the data set? Can these features help in predicting those? Can these features help in addressing this or that issue? This is more reminiscent to a traveler's view of the world rather than that of a scientist. The scientist sits at his desk, gets reproducible signals from the universe and tries to accommodate them into the great model of the universe that the science has been developing. The traveler deals with what comes on their way. Helping the traveler in making sense of data is the task of data analysis. It should be pointed out that this view much differs from the conventional scientific method in which the main goal is to identify a pre-specified model of the world, and data is but a vehicle in achieving this goal. It is that view that underlies the development of data science and its parts.

Any data set comprises two parts, data and metadata entries. Data entries are the set of measurements taken, whereas metadata is a most straightforward relation between knowledge and measurements. Metadata usually involve names for the objects and features, as well as indications of the measurement scales for the latter. Depending on the data domain, objects may be alternatively but synonymously referred to as individuals, entities, cases, instances, patterns, or observations. Data features may be synonymously referred to as variables, attributes, states, or characters. Depending on the way they are assigned to objects, features can be of an elementary structure [e.g., age, sex, or income of individual] or complex structure [e.g., an image or a statement or a cardiogram]. Metadata may involve relations between objects and other relevant information.

1.1.2   Other Approaches

The two-fold goal above clearly delineates the place of the data analysis core within the set of approaches involving various data analysis tasks. Here is a list of some popular approaches:

- *Artificial intelligence*—this term applies to any activity oriented to development of theoretical and/or practical devices, models, and methods that are supposed to perform any intelligent activity of the human brain. This began during the second half of the twentieth century, at first somewhat biased to the deductive logics side, but currently it overlaps all the directions including machine learning and data analysis.
- *Big data*—this term is used to embrace all the data analysis methods, software, and practices specifically oriented at massive datasets relating different aspects of phenomena. Big data usually come in distributed formats so that parallel computations are frequently needed.
- *Classification*—this term applies to denote either a meta-scientific area of knowledge handling via a set of separate classes to structure the phenomenon and relate different aspects of it to each other, or a discipline of supervised classification, that is, developing rules for assigning class labels to a set of entities under consideration. Data analysis can be utilized as a tool for designing the former, whereas the latter can be thought of as a problem in data analysis.
- *Cluster analysis*—is a discipline for obtaining (sets of) separate subsets of similar entities or features or both from the data, one of the most generic activities in data analysis.
- *Computational intelligence*—a discipline utilizing fuzzy sets, nature-inspired algorithms, neural nets and the like to computationally imitate human intelligence, which overlaps other areas of data analysis.
- *Data mining*—a discipline for finding interesting patterns in data stored in databases, which is considered part of the process of knowledge discovery. This has a significant overlap with computational data analysis. Yet data mining is structured somewhat differently by putting more emphasis on fast computations in large databases and finding "interesting" associations and patterns.
- *Document retrieval*—a discipline developing algorithms and criteria for query-based retrieval of as many relevant documents as possible, from a document base, which is similar to establishing a classification rule in data analysis. This area has become most popular with the development of search engines over the internet.
- *Factor analysis*—a discipline emerged in psychology for modeling and finding hidden factors in data, which can be considered part of quantitative summarization in data analysis.
- *Genetic algorithms*—an approach to globally search through the solution space in complex optimization problems by representing solutions as a population of "chromosomes" that evolves in iterations by mimicking micro-evolutionary events such as "cross-over" and "mutation". This can play a role in solving optimization problems in data analysis.
- *Knowledge discovery*—a set of techniques for deriving quantitative formulas and categorical productions to associate different features and feature sets, which hugely overlaps with the corresponding parts of data analysis.
- *Mathematical statistics*—a discipline of data analysis based on the assumption of a probabilistic model underlying the data generation and/or decision making

so that data or decision results are used for fitting or testing the models. This obviously has a lot to do with data analysis, including the idea that an adequate mathematical model is a preferred knowledge format.

- *Machine learning*—a discipline in data analysis originally oriented at producing classification rules for predicting unknown class labels at entities usually arriving one by one in a random sequence. Currently it embraces all the aspects of data analysis. The difference comes from the main goal: "Machine learning addresses the question of how to build computers that improve automatically through experience" (Jordan and Mitchell 2015), whereas data analysis is oriented at enhancing knowledge.
- *Neural networks*—a technique for modeling relations between (sets of) features utilizing structures of interconnected artificial neurons; the parameters of a neural network are learned from the data. A neural network consists of an input layer and an output layers, as well as a number of "hidden layers" Term "deep learning" refers to the use of neural networks with relatively large numbers of hidden layers.
- *Nature-inspired algorithms*—a set of contemporary techniques for optimization of complex functions such as the squared error of a data fitting model, using a population of admissible solutions evolving in iterations mimicking a natural process such as genetic recombination or ant colony search for foods.
- *Optimization*—a discipline for analyzing and solving problems in finding optima of a function such as the difference between observed values and those produced by a model whose parameters are being fitted (error).
- *Pattern recognition*—a discipline for deriving classification rules (supervised learning) and clusters (unsupervised learning) from observed data.
- *Social statistics*—a discipline for measuring social and economic indexes using observation or sampling techniques.
- *Text analysis*—a set of techniques and approaches for the analysis of unstructured text documents such as establishing similarity between texts, text categorization, deriving synopses and abstracts, etc.

1.1.3 Structure of Data Analysis Problem

This text is oriented at a description of basic models and methods for enhancing knowledge by finding in data either

(a) Correlation among features (Cor) or
(b) Summarization of entities or features (Sum),

in either of two ways, quantitative (Q) or categorical (C). Combining these two bases makes four major groups of methods: CorQ, CorC, SumQ, and SumC that form the core of data analysis. It should be pointed out that currently different categorizations of tasks related to data analysis prevail: the classical mathematical statistics focuses mostly on mathematically treatable models (see, for example, Hair et al. 2010), whereas the system of machine learning and data mining expressed in the popular account by Duda et al. (2012) concentrates on the problem of learning categories of objects, thus leaving such important problems as quantitative summarization outside.

A correlation or summarization problem typically involves the following five ingredients:

- Stock of mathematical structures sought in data
- Computational model relating the data and the mathematical structure
- Criterion to score the match between the data and structure (fitting criterion)
- Method for optimizing the criterion
- Visualization of the results.

Here is a brief outline of those that are used in this text:
Mathematical structures:

- *linear* combination of features;
- *neural network* mapping a set of input features into a set of target features;
- *decision* tree built over a set of features;
- *cluster* of entities;
- *partition* of the entity set into a number of non-overlapping clusters.

When the type of mathematical structure has been chosen, its parameters are to be learnt from the data.

A *fitting method relies on a computational model* involving a function *scoring* the adequacy of the mathematical structure underlying the rule—a criterion, and, usually, visualization aids. The data *visualization* is a way to represent the found structure to human eye. In this capacity, it is an indispensable part of the data analysis, which explains why this term is in the subtitle of this text. We briefly outline some aspects of visualization within the data analysis approach in Sect. 1.3.

A *criterion* measures either the deviation of the found structure from the target (to be minimized) or goodness of fit to the target (to be maximized).

Currently available *computational methods* to optimize the criterion encompass three major groups:

- *global* optimization, that is, finding the best possible solution, computationally feasible sometimes for linear quantitative and simple discrete structures;
- *local* improvement using such general approaches as:

  - gradient ascent and descent
  - alternating optimization
  - greedy neighborhood search (hill climbing)

- *nature-inspired approaches* involving a population of admissible solutions and its iterative evolution, an approach involving relatively recent advancements in computing capabilities, of which the following will be used in some problems:

- genetic algorithms
- evolutionary algorithms
- particle swarm optimization.

It should be pointed out that currently there is no systematic description of all possible combinations of problems, data types, mathematical structures, criteria, and fitting methods available. Here we rather focus on a very few, most generic and better explored problems in each of the four data analysis groups that can be safely claimed as being prototypical within the groups:

Summarization	Quantitative	Principal component analysis
	Categorical	Cluster analysis
Correlation	Quantitative	Regression analysis
	Categorical	Supervised classification

The four approaches on the right have emerged in different frameworks and usually are considered as unrelated. However, they are related in the context of core data analysis. They can be unified within the so-called data-driven modeling, together with the least-squares criterion, that is adopted for all main methods described in this text. In fact, the criterion is part of a unifying data-recovery perspective that has been developed in mathematical statistics for fitting proba-bilistic models and then was extended to data analysis. In data analysis, this per-spective is useful not only for supplying a nice fitting criterion but also because it involves the decomposition of the data scatter into "explained" and "unexplained" parts in all four approaches above. The data recovery approach takes in a type of mathematical structure to model the data and proceeds in three stages:

(1) fitting a model representing the structure to the data (this can be referred to as "encoding"),
(2) recovering data from the model in the format of the data used to build the model (this can be referred to as "decoding"), and
(3) analyzing discrepancies between the observed data and those recovered from the model. The smaller are the discrepancies, the better the fit—this is a principle underlying the data-driven modeling approach.

Using the data recovery approach provided the author with tools for developing and describing a number of relations, bringing together popular concepts considered sometimes as being worlds apart (Mirkin 1996, 2012). Among them:

(a) Reinterpretation and visualization of Pearson chi-square contingency coefficient as a summary association index rather than a statistical independence criterion (see Sect. 3.6);
(b) Use of anomalous patterns, an extension of the principal component analysis to clustering, for both initializing K-means and Ward clustering and setting the number of clusters (see Sect. 4.3 and Sect. 5.1);

(c) A multitude of different reformulations of the square-error clustering criterion potentially leading to different clustering strategies (see Sects. 4.3.2, 4.6.1, 5.1.4 and 5.5);

(d) Interrelation between association measures utilized for building decision trees and normalization of dummies representing categorical data (see Sect. 3.8); and

(e) A unified framework for network clustering including:

    (i)   a number of combinatorial clustering criteria (see Sects. 4.6, 4.7 and 5.5);

    (ii) spectral clustering, a recent very popular approach (see Sects. 4.3, 5.2 and 5.4);

    (iii) additive clustering, a less popular yet powerful paradigm (see Sect. 5.5).

1.1.4   Teaching Paradigm

There can be distinguished at least three different levels in studying a computational data analysis method. A reader can be interested in learning of the approach at the level of concepts only—what a concept is for, why it should be applied at all, etc. A somewhat more practically oriented tackle would be of an information system/tool that can be utilized so that no knowledge is needed beyond the input/output structure. A more technically oriented way would be studying the method involved and its properties. Each of these three levels has its comparative advantages and disadvantages, pros and cons:

	Pros	Cons
Concepts	Awareness	Superficial
Systems	Usable now	Short-term
	Simple	Stupid
Techniques	Workable	Technical
	Extendable	Boring

Many in Computer Sciences rely on the Systems approach assuming that good methods have been developed and put in there already. Although it is largely true for well-defined mathematical problems, the situation is by far different in data analysis because there are no well posed problems here—basic formulations are intuitive and rarely supported by sound theoretical results. This is why, in many aspects, intelligence of currently popular "intelligent methods" may be rather superficial potentially leading to wrong results and decisions.

Consider, for instance, a very popular concept, the power law—many say that in unconstrained social processes, such as those on the Web networks, this law, expressed with formula $y = ax^{-b}$ where x and y are some features and a and b are positive constant coefficients, dominates. Here are a few examples: the decay in the numbers of people who read a news story on the web over time; the distribution of page requests on a web-site according to their popularity; the distribution of website connections, etc. According to a very popular recipe, to fit a power law (that is, to estimate a and b from the data), one needs to fit the logarithm of the power-law equation, that is, $\log(y) = c - b * \log(x)$ where $c = \log(a)$, which is much easier to fit because it is linear. Therefore, this recipe advises: take logarithms of the x and y first and then use any popular linear regression program to find the constants. The recipe works well when the regularity is observed with no noise, which is not the case in real world social processes. With the real-world noise, this recipe may lead to big errors. For example, if x is generated randomly in the interval between 0 and 10 and y is related to x by the power law $y = 2 * x^{1.07}$, which can be interpreted as the growth with the rate of approximately 7% per time unit, with an added Gaussian noise $N(0,2)$ of the zero mean and the standard deviation equal to 2, the recipe can lead to disastrous results. With the parameter values above, in the author's computations, the linear transformation led to estimates 3.08 for a and 0.8 for b, to suggest that the process does not grow with x but rather decays. In contrast, when an evolutionary optimization method was applied to the original non-linear problem, the estimates were realistic: $a = 2.03$ and $b = 1.076$.

This is a relatively simple data analysis example, at which a correct procedure can be used. However, in more complex situations of clustering or categorization, the very idea of a correct method seems rather debatable; at least, methods in the existing systems can be of a rather poor quality.

One may compare the usage of an unsound data analysis method with that of getting services of an untrained medical doctor or car driver—the results can be as devastating indeed. This is why it is important to study not only How's but What's and Why's, which are addressed in this course by focusing on Concepts and Techniques rather than Systems. Another, perhaps even more important, reason for studying concepts and techniques is the constant emergence of new data types, such as related to internet networks or bio-medicine, that cannot be tackled by existing systems. However, concepts and methods are readily extensible to cover them.

This text is oriented towards a student in Computer Sciences or related disciplines and reflects my experiences in teaching students of this type. Most of them prefer a hands-on rather than mathematical style of presentation. This is why some parts are divided in three streams: presentation, formulation, and computation. The presentation states the problem and approach taken to tackle it, and it illustrates the solution at some data. The formulation provides a mathematical description of the problem as well as a method or two to solve it. The computation shows how to do that computationally with basic MatLab. The MatLab codes may be considered as pseudocodes as well. Each of the streams can be read independently. In this way, the reader can choose the way of using the book and adjust it to their individual style.

To help the reader to study the material actively, the text is interlaced with problems along with their solutions. Many of the problems are put as "worked examples" to show how a specific method applies to a specific dataset. More complex problems, "case studies", may involve a rule for data generation rather than a pre-specified data set or an informed way for looking at the results. Yet more complex problems may involve uncharted terrain and an investigation, however small,—these are referred to as "projects". It should be pointed out that these are not just examples to illustrate the main text but parts of main text—they do not necessarily repeat claims of the main text but rather introduce concepts and approaches via examples rather than via general statements. Therefore, it is highly advisable to work them through rather than just skip them.

There is a bias in the volumes of material devoted to correlation and summarization subjects—the latter prevails rather considerably. This can be explained by both personal and objective reasons. One of the reasons is that my main research area lies in clustering, that is, summarization. Another reason is that correlation problems are studied less thoroughly than those of summarization. The only relatively well explored part of the correlation iceberg is related to the so-called supervised learning, the case of predicting a feature using the others. Recently, this was justly called "an icing on the cake, whereas unsupervised learning is the cake itself" [see LeCun (2015); the quotation is not exact], so that summarization, as a container of the unsupervised learning requires more attention indeed. One more reason is that correlation problems, and their theoretical underpinnings, have been already subjects of a multitude of monographs and texts in statistics, data analysis, machine learning, data mining, and computational intelligence. In contrast, clustering and principal component analyses—the main constituents of summarization efforts, according to our system,—have received much less attention and usually are described as heuristics, rather than methods having a solid theoretical base. This text presents each of them as based on a model of data, which raises a number of issues that are addressed here, including that of the theoretical structure of a summarization problem. The concept of encoder–decoder is borrowed from the data processing area to draw a theoretical framework in which summarization is considered as a pair of encoding/decoding activities so that the quality of the encoding part is evaluated by the quality of decoding. Luckily, the theory of singular value decomposition of matrices (SVD) can be safely utilized as a framework for explaining the principal component analysis. A straightforward extension of the SVD equations to binary scoring vectors provides a base for K-means clustering and the like. This raises an important question of mathematical proficiency the reader should have as a prerequisite. An assumed background of the reader interested in studying formulation parts should include:

(a) basics of calculus including the concepts of function, derivative and the first-order optimality condition;
(b) basic linear algebra including vectors, inner products, Euclidean distances and matrices (these are reviewed in the Appendix),
(c) basic probability/statistics theory including the concepts of density function, conditional probability, hypothesis testing, and statistical independence, and
(d) basic set theory notation such as symbols for relations of inclusion and membership.

1.2 Case Study Problems

To be more specific, the presentation is illustrated using a number of small datasets —the sizes allow the reader to see the data by naked eye, which is always a good idea to do before engaging in analysis. The datasets and related problems are selected in such a way that methods further described could be immediately illustrated by using a relevant dataset from the collection. All the datasets are presented as data tables—this is a generic format for both theoretical thinking and practical computations in data analysis.

There will be three types of data tables presented: (I) Quantitative: those containing quantitative data only; (II) Mixed: those containing both quantitative and categorical features; (III) Similarity or network data. All the terms mentioned: feature, quantitative, categorical, similarity, network—will be explained while describing the data.

1.2.1 *Quantitative Entity-to-Feature Data Tables*

Case A. Iris
Iris is a well-known dataset collected by botanist E. Anderson and presented by R. Fisher in his founding paper on discriminant analysis (1936). The data table, Table 1.1, presents 150 Iris specimens, representing three taxa of Iris flowers, I *Iris setosa* (diploid), II *Iris versicolor* (tetraploid) and III *Iris virginica* (hexaploid), with 50 specimens from each. Each specimen is measured over four morphological variables: sepal length (w1), sepal width (w2), petal length (w3), and petal width (w4) (see Fig. 1.1).
These variables constitute the four features in Table 1.1.

The concept of data table allows for a simple formalization of the concept of feature. Indeed, entities are associated with the set I of row labels, or row indices $i = 1, 2, ..., N$, where N is the number of entities, so that $I = \{1, 2, ..., N\}$; in our case, $N = 150$.

A feature, x, then is a mapping from I to the set of its values. We say that feature x is quantitative if not only the values $x(i)$ are numbers but also that averaging them makes sense.

Let us recall that the number $x(J)$ is the average of x over a subset $J \subseteq I$ if

$$x(J) = \frac{\sum_{i \in J} x(i)}{|J|}$$

where |J| is the number of elements in J.

The taxa are defined by the genotype whereas the features are of the appearance (phenotype). The question arises whether the taxa can be described, and indeed

Table 1.1 Iris data: 150 Iris specimens measured over four features each

#	I *Iris setosa*				II *Iris versicolor*				III *Iris virginica*			
	w1	w2	w3	w4	w1	w2	w3	w4	w1	w2	w3	w4
1	5.1	3.5	1.4	0.3	6.4	3.2	4.5	1.5	6.3	3.3	6.0	2.5
2	4.4	3.2	1.3	0.2	5.5	2.4	3.8	1.1	6.7	3.3	5.7	2.1
3	4.4	3.0	1.3	0.2	5.7	2.9	4.2	1.3	7.2	3.6	6.1	2.5
4	5.0	3.5	1.6	0.6	5.7	3.0	4.2	1.2	7.7	3.8	6.7	2.2
5	5.1	3.8	1.6	0.2	5.6	2.9	3.6	1.3	7.2	3.0	5.8	1.6
6	4.9	3.1	1.5	0.2	7.0	3.2	4.7	1.4	7.4	2.8	6.1	1.9
7	5.0	3.2	1.2	0.2	6.8	2.8	4.8	1.4	7.6	3.0	6.6	2.1
8	4.6	3.2	1.4	0.2	6.1	2.8	4.7	1.2	7.7	2.8	6.7	2.0
9	5.0	3.3	1.4	0.2	4.9	2.4	3.3	1.0	6.2	3.4	5.4	2.3
10	4.8	3.4	1.9	0.2	5.8	2.7	3.9	1.2	7.7	3.0	6.1	2.3
11	4.8	3.0	1.4	0.1	5.8	2.6	4.0	1.2	6.8	3.0	5.5	2.1
12	5.0	3.5	1.3	0.3	5.5	2.4	3.7	1.0	6.4	2.7	5.3	1.9
13	5.1	3.3	1.7	0.5	6.7	3.0	5.0	1.7	5.7	2.5	5.0	2.0
14	5.0	3.4	1.5	0.2	5.7	2.8	4.1	1.3	6.9	3.1	5.1	2.3
15	5.1	3.8	1.9	0.4	6.7	3.1	4.4	1.4	5.9	3.0	5.1	1.8
16	4.9	3.0	1.4	0.2	5.5	2.3	4.0	1.3	6.3	3.4	5.6	2.4
17	5.3	3.7	1.5	0.2	5.1	2.5	3.0	1.1	5.8	2.7	5.1	1.9
18	4.3	3.0	1.1	0.1	6.6	2.9	4.6	1.3	6.3	2.7	4.9	1.8
19	5.5	3.5	1.3	0.2	5.0	2.3	3.3	1.0	6.0	3.0	4.8	1.8
20	4.8	3.4	1.6	0.2	6.9	3.1	4.9	1.5	7.2	3.2	6.0	1.8
21	5.2	3.4	1.4	0.2	5.0	2.0	3.5	1.0	6.2	2.8	4.8	1.8
22	4.8	3.1	1.6	0.2	5.6	3.0	4.5	1.5	6.9	3.1	5.4	2.1
23	4.9	3.6	1.4	0.1	5.6	3.0	4.1	1.3	6.7	3.1	5.6	2.4
24	4.6	3.1	1.5	0.2	5.8	2.7	4.1	1.0	6.4	3.1	5.5	1.8
25	5.7	4.4	1.5	0.4	6.3	2.3	4.4	1.3	5.8	2.7	5.1	1.9
26	5.7	3.8	1.7	0.3	6.1	3.0	4.6	1.4	6.1	3.0	4.9	1.8
27	4.8	3.0	1.4	0.3	5.9	3.0	4.2	1.5	6.0	2.2	5.0	1.5
28	5.2	4.1	1.5	0.1	6.0	2.7	5.1	1.6	6.4	3.2	5.3	2.3
29	4.7	3.2	1.6	0.2	5.6	2.5	3.9	1.1	5.8	2.8	5.1	2.4
30	4.5	2.3	1.3	0.3	6.7	3.1	4.7	1.5	6.9	3.2	5.7	2.3
31	5.4	3.4	1.7	0.2	6.2	2.2	4.5	1.5	6.7	3.0	5.2	2.3
32	5.0	3.0	1.6	0.2	5.9	3.2	4.8	1.8	7.7	2.6	6.9	2.3
33	4.6	3.4	1.4	0.3	6.3	2.5	4.9	1.5	6.3	2.8	5.1	1.5
34	5.4	3.9	1.3	0.4	6.0	2.9	4.5	1.5	6.5	3.0	5.2	2.0
35	5.0	3.6	1.4	0.2	5.6	2.7	4.2	1.3	7.9	3.8	6.4	2.0
36	5.4	3.9	1.7	0.4	6.2	2.9	4.3	1.3	6.1	2.6	5.6	1.4
37	4.6	3.6	1.0	0.2	6.0	3.4	4.5	1.6	6.4	2.8	5.6	2.1
38	5.1	3.8	1.5	0.3	6.5	2.8	4.6	1.5	6.3	2.5	5.0	1.9
39	5.8	4.0	1.2	0.2	5.7	2.8	4.5	1.3	4.9	2.5	4.5	1.7
40	5.4	3.7	1.5	0.2	6.1	2.9	4.7	1.4	6.8	3.2	5.9	2.3
41	5.0	3.4	1.6	0.4	5.5	2.5	4.0	1.3	7.1	3.0	5.9	2.1
42	5.4	3.4	1.5	0.4	5.5	2.6	4.4	1.2	6.7	3.3	5.7	2.5
43	5.1	3.7	1.5	0.4	5.4	3.0	4.5	1.5	6.3	2.9	5.6	1.8
44	4.4	2.9	1.4	0.2	6.3	3.3	4.7	1.6	6.5	3.0	5.5	1.8
45	5.5	4.2	1.4	0.2	5.2	2.7	3.9	1.4	6.5	3.0	5.8	2.2
46	5.1	3.4	1.5	0.2	6.4	2.9	4.3	1.3	7.3	2.9	6.3	1.8
47	4.7	3.2	1.3	0.2	6.6	3.0	4.4	1.4	6.7	2.5	5.8	1.8
48	4.9	3.1	1.5	0.1	5.7	2.6	3.5	1.0	5.6	2.8	4.9	2.0
49	5.2	3.5	1.5	0.2	6.1	2.8	4.0	1.3	6.4	2.8	5.6	2.2
50	5.1	3.5	1.4	0.2	6.0	2.2	4.0	1.0	6.5	3.2	5.1	2.0

Fig. 1.1 Sepal and petal in
an Iris flower

predicted, in terms of the features, or not. It is well known from previous studies that taxa II and III are not well separated in the variable space. Some non-linear machine learning techniques such as Neural Nets (Haykin 1999 and Sect. 4.6 further on) can tackle the problem and produce a decent decision rule involving non-linear transformation of the features. Unfortunately, rules derived with Neural Nets are typically not comprehensible to the human. The human mind needs a somewhat less artificial logic that is capable of reproducing and extending botanists' observations, such as that the petal area, roughly expressed by the product of w3 and w4, provides for much better resolution than the original linear sizes. Other problems of interest: (a) visualize the data; (b) build a predictor of sepal sizes from the petal sizes.

Case B. Market Towns

In Table 1.2 a set of Market towns in West Country, England is presented along with features characterizing population and social infrastructure according to the 1991 census. For the purposes of social planning, it would be good to monitor a smaller number of towns, each representing a cluster of similar towns. In the table, the towns are sorted according to their population size. One can see that 21 towns have less than 4000 residents. The value 4000 is taken as a divider since it is round and, more importantly, there is a gap of more than thirteen hundred residents between Kingskerswell (3672 inhabitants) and, subsequently, Looe (5022 inhabitants). The next big gap occurs after Liskeard (7044 inhabitants), separating the nine middle sized towns from two larger town groups containing six and nine towns respectively. The divider between the latter groups is taken between Tavistock

(10,222) and Bodmin (12,553). In this way, we get three or four groups of towns for the purposes of social monitoring. Is this enough, regarding the other features available? Are the groups, defined in terms of population size only, homogeneous enough for the purposes of monitoring?

As further computations will show, the numbers of services on average do follow the town sizes, but this set (as well as the complete set of about thirteen hundred England Market towns) is much better represented with seven somewhat different clusters: large towns of about 17–20,000 inhabitants, two clusters of medium sized towns (8–10,000 inhabitants), three clusters of small towns (about 5000 inhabitants), and a cluster of very small settlements with about 2500 inhabitants. Each of the three small town clusters is characterized by the presence of a facility, which is absent in two others: a Farm market or a Hospital or a Swimming pool, respectively.

One may suggest that the only difference between these seven clusters and the grouping over the town resident numbers would be just the difference in the dividing points, since both are expressed in terms of the population size only. However, one should not forget that the number of residents for the seven clusters is a posterior selection—because of our knowledge of the clusters not prior to that.

The data in Table 1.2 involve the counts of the following 12 features from those surveyed in the 1991 census:

Pop   Population resident
PS   Primary schools
D   General Practitioners
Hos   Hospitals
Ba   Banks
Sst   Superstores
Pet   Petrol stations
DIY   Do It Yourself shops
Swi   Swimming pools
Po   Post offices
CAB   Citizen Advice Bureaus
FM   Farmer markets

1.2.2   Mixed Scale Entity-to-Feature Data Tables

Case C. Company
There are eight companies and five features in Table 1.3:

(1) Income, $ Million;
(2) MarketShare—the proportion of market for the produced commodities controlled by company, %;

Table 1.2 Data of West Country England market towns 1991

Town	Pop	PS	D	Hos	Ba	Sst	Pet	DIY	Swi	Po	CAB	FM
Mullion	2040	1	0	0	2	0	1	0	0	1	0	0
So Brent	2087	1	1	0	1	1	0	0	0	1	0	0
St Just	2092	1	0	0	2	1	1	0	0	1	0	0
St Columb	2119	1	0	0	2	1	1	0	0	1	1	0
Nanpean	2230	2	1	0	0	0	0	0	0	2	0	0
Gunnislake	2236	2	1	0	1	0	1	0	0	3	0	0
Mevagissey	2272	1	1	0	1	0	0	0	0	1	0	0
Ipplepen	2275	1	1	0	0	0	1	0	0	1	0	0
Be Alston	2362	1	0	0	1	1	0	0	0	1	0	0
Lostwithiel	2452	2	1	0	2	0	1	0	0	1	0	1
St Columb	2458	1	0	0	0	1	3	0	0	2	0	0
Padstow	2460	1	0	0	3	0	0	0	0	1	1	0
Perranporth	2611	1	1	0	1	1	2	0	0	2	0	0
Bugle	2695	2	0	0	0	0	1	0	0	2	0	0
Buckfastle	2786	2	1	0	1	2	2	0	1	1	1	1
St Agnes	2899	1	1	0	2	1	1	0	0	2	0	0
Porthleven	3123	1	0	0	1	1	0	0	0	1	0	0
Callington	3511	1	1	0	3	1	1	0	1	1	0	0
Horrabridge	3609	1	1	0	2	1	1	0	0	2	0	0
Ashburton	3660	1	0	1	2	1	2	0	1	1	1	0
Kingskers	3672	1	0	0	0	1	2	0	0	1	0	0
Looe	5022	1	1	0	2	1	1	0	1	3	1	0
Kingsbridge	5258	2	1	1	7	1	2	0	0	1	1	1
Wadebridge	5291	1	1	0	5	3	1	0	1	1	1	0
Dartmouth	5676	2	0	0	4	4	1	0	0	2	1	1
Launceston	6466	4	1	0	8	4	4	0	1	3	1	0
Totnes	6929	2	1	1	7	2	1	0	1	4	0	1
Penryn	7027	3	1	0	2	4	1	0	0	3	1	0
Hayle	7034	4	0	1	2	2	2	0	0	2	1	0
Liskeard	7044	2	2	2	6	2	3	0	1	2	2	0
Torpoint	8238	2	3	0	3	2	1	0	0	2	1	0
Helston	8505	3	1	1	7	2	3	0	1	1	1	1
St Blazey	8837	5	2	0	1	1	4	0	0	4	0	0
Ivybridge	9179	5	1	0	3	1	4	0	0	1	1	0
St Ives	10092	4	3	0	7	2	2	0	0	4	1	0
Tavistock	10222	5	3	1	7	3	3	1	2	3	1	1
Bodmin	12553	5	2	1	6	3	5	1	1	2	1	0
Saltash	14139	4	2	1	4	2	3	1	1	3	1	0
Brixham	15865	7	3	1	5	5	3	0	2	5	1	0
Newquay	17390	4	4	1	12	5	4	0	1	5	1	0
Truro	18966	9	3	1	19	4	5	2	2	7	1	1
Penzance	19709	10	4	1	12	7	5	1	1	7	2	0
Falmouth	20297	6	4	1	11	3	2	0	1	9	1	0
St Austell	21622	7	4	2	14	6	4	3	1	8	1	1
Newton Abb	23801	13	4	1	13	4	7	1	1	7	2	0

Table 1.3 Company: a set of eight companies characterized by mixed scale features

Company name	Income, $ million	Market Share, %	NSup	E-C	Sector
Aversi	19.0	43.7	2	No	Utility
Antyos	29.4	36.0	3	No	Utility
Astonite	23.9	38.0	3	No	Manufacture
Bayermart	18.4	27.9	2	Yes	Utility
Breaktops	25.7	22.3	3	Yes	Manufacture
Bumchist	12.1	16.9	2	Yes	Manufacture
Civok	23.9	30.2	4	Yes	Retail
Cyberdam	27.2	58.0	5	Yes	Retail

The division of the table and company names reflect a fact not present in the data—product affinities: the first three companies mostly adhere to product group A, the next three to product group B, and the last two to product group C

(3) NSup—the number of principal suppliers;
(4) E-C—Yes or No depending on the usage of e-commerce in the company;
(5) Sector—which sector of the economy: (a) Retail, (b) Utility, and (c) Manufacture.

These features are purely illustrative and should not be taken too seriously. Examples of computational data analysis problems related to this data set:

– How to map companies to the screen with their similarity reflected in distances on the plane? (Summarization)
– Would clustering of companies reflect the product? What features would be involved then? (Summarization)
– Can rules be derived to make an attribution of the product for another company, coming outside of the table? (Correlation)
– Is there any relation between the structural features, such as NSup, and market related features, such as Income? (Correlation).

Q.1.1. Is the following statement true? "There is no data on the company products within the table".
A. Indeed: no "Product" feature is present in the table; the separating lines are not part of the data. The first letters of the names are informative of the product; but they are part of metadata, not data.

An issue related to Table 1.3 is that not all of its entries are quantitative. Specifically, there are three conventional types of feature scales in it:

– Quantitative, that is, such a feature that the averaging of its values is considered meaningful. In Table 1.3, these are: Income, ShareP and NSup;
– Binary, that is, a feature admitting one of two answers, Yes or No: this is E-C;
– Nominal, that is, a feature with a few disjoint (unordered) categories, such as Sector in Table 1.3.

This is why the table is referred to as of a mixed scale data type. Most models and methods presented in this text relate to quantitative data formats only—which does not mean that categorical data are left on their own, just the opposite. The two non-quantitative feature types, binary and nominal, can be meaningfully pre-processed into a quantitative format too, as explained further on (see also Sects. 2.3, 3.6 and 3.8, among others).

A binary feature can be recoded into 1/0 format by substituting 1 for "Yes" and 0 for "No". In the author's view, rather unconventionally, the recoded feature can be considered quantitative, because its averaging is meaningful: the average value is equal to the proportion of unities, that is, the frequency of "Yes" in the original feature.

To quantitatively recode a nominal feature in a meaningful way, the feature is first enveloped into a set of binary "Yes"/"No" features corresponding to individual categories. In Table 1.3, binary features yielded by categories of feature "Sector" are:

Is it Retail?
Is it Utility?
Is it Manufacture?

They are put as questions to make "Yes" or "No" as answers to them. These binary features now can be converted to the quantitative format advised above, by recoding 1 for "Yes" and 0 for "No". Such a 1/0 version is frequently referred to in statistics as a dummy variable.

This instruction can be applied to convert the Company data into a quantitative format (see Table 1.4).

Q.1.2. Why column 1 in Table 1.3, the company names, cannot be considered a feature, a nominal one?
A. A feature must bear information of a relation or relations on the entity set. Say, category Manu in Table 1.4 informs of two equivalence classes on the set: one class, entities 3, 5, 6 are in the Manufacture sector, the others are not. The only information supplied by the company names towards the data is that they all are different, which is analogous to the index to the list. Of course the names can bear important information regarding class or origin of the companies. This however should be counted where it belongs, in the metadata, not the data.

Case D. Student
In Table 1.5, an illustrative dataset is presented as imitating a typical set up for a group of Birkbeck University of London part-time students pursuing Master's degree in Computer Sciences.

This dataset refers to a hundred students along with six features, three of which are personal characteristics [1. Occupation (Oc): either Information Technology (IT) or Business Administration (BA) or anything else (AN); 2. Age, in years; 3. Number of children (Ch)] and three are their marks over courses in 4. Software

Table 1.4 Company data from Table 1.3 converted to the quantitative format by enveloping its categories into binary dummy features

Code	Income	Markets	NSup	E-C	Util	Manu	Retail
1	19.0	43.7	2	0	1	0	0
2	29.4	36.0	3	0	1	0	0
3	23.9	38.0	3	0	0	1	0
4	18.4	27.9	2	1	1	0	0
5	25.7	22.3	3	1	0	1	0
6	12.1	16.9	2	1	0	1	0
7	23.9	30.2	4	1	0	0	1
8	27.2	58.0	5	1	0	0	1

and Programming (SE), 5. Object-Oriented Programming (OO), and 6. Computational Intelligence (CI).

Related questions are:

– Whether the students' marks are affected by the personal features;
– Are there any patterns in marks, especially in relation to occupation?

Case E. Intrusion

With the growing range and scope of computer networks, security issues become an urgent matter. An attack on a network results in its malfunctioning. The simplest kind of attack is a denial of service (DoS). A DoS is caused by an intruder who makes some resource—in computing, memory, or I/O such as network or disk—too busy or too full to handle legitimate requests. Two of the DoS attacks are denoted as 'apache2' and 'smurf' in the data. Other types of attack include user-to-root attacks and remote-to-local attacks.

The 'apache2' attack targets a popular open source web server, the Apache HTTP Server, and results in denying services to a client by sending a request with a large number of http headers, triggering buffer overflow vulnerabilities. A 'smurf' attack works by sending forged ICMP echo messages to a host. An ICMP echo, also known as ping, is a message to a computer attached to an IP network. On receipt of this message, the receiving computer will respond with an ICMP echo reply back to the computer that sent the echo, as determined by the source IP address of the echo request. However, there are techniques that can be used to forge source IP addresses, in which case the echo reply will go to the forged source. Further, it is possible to ping multiple machines by sending an echo request to a network broadcast address. This causes every machine on that IP broadcast domain to respond to the source address (which, in the attack case, is forged). In such a way, it is possible to cause a very large number of computers across the Internet to send echo replies to a single target host, overwhelming that host's processing or network connection. A probe looking for flaws might precede an attack. One powerful probe software is SAINT—the Security Administrator's Integrated Network Tool, that uses a thorough deterministic protocol to scan

Table 1.5 Student data in two columns

Oc	Age	Ch	SE	OO	CI	Oc	Age	Ch	SE	OO	CI
IT	28	0	41	66	90	BA	51	2	75	73	57
IT	35	0	57	56	60	BA	44	3	53	43	60
IT	25	0	61	72	79	BA	49	3	86	39	62
IT	29	1	69	73	72	BA	27	2	93	58	62
IT	39	0	63	52	88	BA	30	1	75	74	70
IT	34	0	62	83	80	BA	47	0	46	36	36
IT	24	0	53	86	60	BA	38	2	86	70	47
IT	37	1	59	65	69	BA	49	1	76	36	66
IT	33	1	64	64	58	BA	45	0	80	56	47
IT	23	1	43	85	90	BA	44	2	50	43	72
IT	24	1	68	89	65	BA	36	3	66	64	62
IT	32	0	67	98	53	BA	31	2	64	45	38
IT	33	0	58	74	81	BA	31	3	53	72	38
IT	27	1	48	94	87	BA	32	3	87	40	35
IT	32	1	66	73	62	BA	38	0	87	56	44
IT	29	0	55	90	61	BA	48	1	68	71	56
IT	21	0	62	91	88	BA	39	2	93	73	53
IT	21	0	53	59	56	BA	47	1	52	48	63
IT	26	1	69	70	89	BA	39	2	88	52	58
IT	20	1	42	76	79	AN	23	0	54	50	41
IT	28	1	57	85	85	AN	34	0	46	33	25
IT	34	1	49	78	59	AN	33	0	51	38	51
IT	22	0	66	73	69	AN	31	0	59	45	35
IT	21	1	50	72	54	AN	25	0	51	41	53
IT	32	1	60	55	85	AN	40	0	41	61	22
IT	32	0	42	72	73	AN	41	0	44	43	44
IT	20	1	51	69	64	AN	42	0	40	56	58
IT	20	1	55	66	66	AN	34	0	47	69	32
IT	24	1	53	92	86	AN	37	0	45	50	56
IT	32	0	57	87	66	AN	24	0	47	68	24
IT	21	1	58	97	54	AN	34	0	50	63	23
IT	27	1	43	78	59	AN	41	0	37	67	29
IT	33	0	67	52	53	AN	47	1	43	35	57
IT	34	1	63	80	74	AN	28	0	50	62	23
IT	34	0	64	90	56	AN	28	0	39	66	31
BA	36	2	86	54	68	AN	46	0	51	36	60
BA	35	2	79	72	60	AN	27	0	41	35	28
BA	36	1	55	44	57	AN	44	0	50	61	40
BA	37	1	59	69	45	AN	47	0	48	59	32
BA	42	2	76	61	68	AN	27	0	47	56	47
BA	30	3	72	71	46	AN	27	0	49	60	58
BA	28	1	48	55	65	AN	21	0	59	57	51
BA	38	1	49	75	61	AN	22	0	44	65	47
BA	49	2	59	50	44	AN	39	0	45	41	25
BA	50	2	65	56	59	AN	26	0	43	47	24
BA	34	2	69	42	59	AN	45	1	45	39	21
BA	31	2	90	55	61	AN	25	0	42	31	32
BA	49	3	75	52	42	AN	25	0	45	33	53
BA	33	1	61	61	60	AN	50	1	48	64	59
BA	43	0	69	62	42	AN	33	0	53	44	21

various network services for possible problems. Intrusion detection systems collect information of anomalies and other patterns of communication such as compromised user accounts and unusual login behavior. The data set Intrusion consists of a hundred communication packages along with some of their features sampled from a set of artificially created data publicly available on webpage of MIT Lincoln Laboratory http://www.ll.mit.edu/mission/communications/ist/corpora/ideval/data/ intex.html. Although the value of this specific data set as a source to analyze the attacks is debatable, it does reflect the structure of the problem.

The features reflect the packet as well as activities of its source:

1. Pr, the protocol-type, which is either tcp or icmp or udp,
2. BySD, the number of data bytes from source to destination,
3. SH, the number of connections to the same host as the current one in the past two seconds,
4. SS, the number of connections to the same service as the current one in the past two seconds,
5. SE, the rate of connections (per cent in SH) that have SYN errors,
6. RE, the rate of connections (per cent in SH) that have REJ errors,
7. A, the type of attack (ap—apache, sa—saint, sm—smurf, and no attack).

Two of the seven features are nominal here, the type of attack and the type of protocol.

Out of the hundred entities in the set, the first 23 are classified as attacking the apache2 server, the 24–69 packets are normal, eleven entities 80–90 are consistent with a SAINT probe, and the last ten, 91–100, appear to be smurf attacks (Table 1.6).

These are examples of problems arising in relation to the Intrusion data:

- identify features to judge whether the system functions normally or is it under attack (Correlation);
- is there any relation between the protocol and type of attack (Correlation);
- how to visualize the data reflecting similarity of the patterns (Summarization).

Case F. Base Stations

Table 1.7 presents data of 40 base stations in a real telephone network company. Each telephone company is much concerned with the phenomenon called "customers churn"—put simply, they do not like losing customers but rather prefer keeping them happy and attracting new customers.

One of the most important factors in this is a healthy state of base stations. Blocked or dropped calls or low data throughput are main reasons for customers churn. Therefore, every station is to be subject to timely testing, maintenance, and renovation operations—still keeping the cost of these auxiliary actions under control. Therefore, they are loath to take too much care for stations that have too few customers and too little revenue.

Table 1.6 Intrusion data

Pr	BySD	SH	SS	SE	RE	A	Pr	BySD	SH	SS	SE	RE	A
Tcp	62344	16	16	0	0.94	Ap	Tcp	287	14	14	0	0	no
Tcp	60884	17	17	0.06	0.88	Ap	Tcp	308	1	1	0	0	no
Tcp	59424	18	18	0.06	0.89	Ap	Tcp	284	5	5	0	0	no
Tcp	59424	19	19	0.05	0.89	Ap	Udp	105	2	2	0	0	no
Tcp	59424	20	20	0.05	0.9	Ap	Udp	105	2	2	0	0	no
Tcp	75484	21	21	0.05	0.9	Ap	Udp	105	2	2	0	0	no
Tcp	76944	22	22	0.05	0.91	Ap	Udp	105	2	2	0	0	no
Tcp	59424	23	23	0.04	0.91	Ap	Udp	105	2	2	0	0	no
Tcp	57964	24	24	0.04	0.92	Ap	Udp	44	3	8	0	0	no
Tcp	59424	25	25	0.04	0.92	Ap	Udp	44	6	11	0	0	no
Tcp	0	40	40	1	0	Ap	Udp	42	5	8	0	0	no
Tcp	0	41	41	1	0	Ap	Udp	105	2	2	0	0	no
Tcp	0	42	42	1	0	Ap	Udp	105	2	2	0	0	no
Tcp	0	43	43	1	0	Ap	Udp	42	2	3	0	0	no
Tcp	0	44	44	1	0	Ap	Udp	105	1	1	0	0	no
Tcp	0	45	45	1	0	Ap	Udp	105	1	1	0	0	no
Tcp	0	46	46	1	0	Ap	Udp	44	2	4	0	0	no
Tcp	0	47	47	1	0	Ap	Udp	105	1	1	0	0	no
Tcp	0	48	48	1	0	Ap	Udp	105	1	1	0	0	no
Tcp	0	49	49	1	0	Ap	Udp	44	3	14	0	0	no
Tcp	0	40	40	0.62	0.35	Ap	Udp	105	1	1	0	0	no
Tcp	0	41	41	0.63	0.34	Ap	Udp	105	1	1	0	0	no
Tcp	0	42	42	0.64	0.33	Ap	Udp	45	3	6	0	0	no
Tcp	258	5	5	0	0	No	Udp	45	3	6	0	0	no
Tcp	316	13	14	0	0	No	Udp	105	1	1	0	0	no
Tcp	287	7	7	0	0	No	Udp	34	5	9	0	0	no
Tcp	380	3	3	0	0	No	Udp	105	1	1	0	0	no
Tcp	298	2	2	0	0	No	Udp	105	1	1	0	0	no
Tcp	285	10	10	0	0	No	Udp	105	1	1	0	0	no
Tcp	284	20	20	0	0	No	Tcp	0	482	1	0.05	0.95	sa
Tcp	314	8	8	0	0	No	Tcp	0	482	1	0.05	0.95	sa
Tcp	303	18	18	0	0	No	Tcp	0	482	1	0.05	0.95	sa
Tcp	325	28	28	0	0	No	Tcp	0	482	1	0.05	0.95	sa
Tcp	232	1	1	0	0	No	Tcp	0	482	1	0.05	0.95	sa
Tcp	295	4	4	0	0	No	Tcp	0	482	1	0.05	0.95	sa
Tcp	293	13	14	0	0	No	Tcp	0	482	1	0.06	0.94	sa
Tcp	305	1	8	0	0	No	Tcp	0	482	1	0.06	0.94	sa
Tcp	348	4	4	0	0	No	Tcp	0	482	1	0.06	0.94	sa
Tcp	309	6	6	0	0	No	Tcp	0	483	1	0.06	0.94	sa
Tcp	293	8	8	0	0	No	Tcp	0	510	1	0.04	0.96	sa
Tcp	277	1	8	0	0	no	Icmp	1032	509	509	0	0	sm
Tcp	296	13	14	0	0	no	Icmp	1032	510	510	0	0	sm
Tcp	286	3	6	0	0	no	Icmp	1032	510	510	0	0	sm
Tcp	311	5	5	0	0	no	Icmp	1032	511	511	0	0	sm
Tcp	305	9	15	0	0	no	Icmp	1032	511	511	0	0	sm
Tcp	295	11	25	0	0	no	Icmp	1032	494	494	0	0	sm
Tcp	511	1	4	0	0	no	Icmp	1032	509	509	0	0	sm
Tcp	239	12	14	0	0	no	Icmp	1032	509	509	0	0	sm
Tcp	5	1	1	0	0	no	Icmp	1032	510	510	0	0	sm
Tcp	288	4	4	0	0	no	Icmp	1032	511	511	0	0	sm

With no mastering of multivariate data analysis, the company makes maintenance work for about 1/3 of the base stations, those that are subject to the heaviest
load, which is measured by the numbers of active users.

However, nowadays the company wants to develop a multivariate stratification
of their base stations by adding to the customers load some financial features to
reflect the revenue—the first stratum will clearly relate to those base stations that
need application of the maintenance efforts first.

The features in Table 1.7 are as follows:

Table 1.7 Base stations data (illustrative)

#	AvNUser	AvUsRev	Inc_Meg
1	385.3	500.57	10.53
2	124.8	443.9	1.04
3	785.7	406.74	5.47
4	234.1	411.36	0.94
5	15.3	543.97	47.37
6	1580.1	525.33	19.49
7	243.9	448.36	19.39
8	1344.4	509.73	3.44
9	610.8	385.65	2.37
10	961.6	438.15	8.58
11	113.2	491.67	0.87
12	1637.2	447.04	9.08
13	81.3	418.78	0.77
14	0.4	460.32	8.11
15	82.7	424.93	2.57
16	1010.4	396.67	13.08
17	2203.4	525.37	21.26
18	175.3	578.02	10.79
19	2487.1	486.62	8.72
20	180.7	499.58	6.78
21	2284.5	375.44	4.38
22	119.1	378.25	1.58
23	2077.8	429.69	6.86
24	757	435.41	3.23
25	361.3	487.94	4.54
26	2401.5	402.6	11.64
27	1174.3	433.33	5.62
28	54.3	481.13	1.19
29	615.3	523.04	7.81
30	524.1	490.9	2.8
31	125.8	438.58	1.71
32	466.5	439.9	6.07
33	2305.1	419.05	12
34	395.2	424.58	2.97
35	655	416.37	6.43
36	674.2	442.18	16.26
37	1251.3	462.81	6.72
38	700.8	428.38	2.27
39	1112.7	609.03	17.84
40	353.4	551.38	1.76

1. AvNUser—the average nunber of unique active users a day, in thousands;
2. AvUsRev—the average number of the median values of the average revenue per customer a day;
3. IncMeg—the average income generated per a million bytes traffic.

All the measurements are taken over a year preceding the project. The income here is fairly separated over both: customers, AvUsRev, and traffic, IncMeg, to reflect the differences among the users. The average revenue per customer is rather stable; it does not much differ at different base stations; the other two features, however, may differ dramatically at different stations.

This dataset can be used to figure out which base stations should go first into maintenance work.

1.2.3 Similarity or Network Data

In contrast to the entity-to-feature type of data involving two kinds of objects, entities and features, with feature values occurring on the entities, similarity data usually involve only one kind of objects such as companies, web pages or individuals. There is only one feature under consideration, a rather distinct one. Its values are observed not on individual entities but on pairs of them. The feature scores a measure of similarity or interaction or proximity between two entities. This warrants that all the values in a similarity table are comparable across the table, which was not the case for entity-to-feature tables. The latter assume comparability only within a column related to the same feature.

Formerly, human judgement was the only supplier of similarity data; in this text, Confusion data in Table 1.8 and Eurovision song contest data in Table 1.11 are of this type. Currently, there are few more sources of similarity data:

(i) Similarity data from the analysis of complex objects such as sequences and images; instead of developing features of such objects, one may wish to

Table 1.8 Confusion data: the entries characterize the numbers of those of the participants of a psychological experiment who mistook the stimulus (row digit) for the response (column digit)

Stim	Response									
	1	2	3	4	5	6	7	8	9	0
1	877	7	7	22	4	15	60	0	4	4
2	14	782	47	4	36	47	14	29	7	18
3	29	29	681	7	18	0	40	29	152	15
4	149	22	4	732	4	11	30	7	41	0
5	14	26	43	14	669	79	7	7	126	14
6	25	14	7	11	97	663	4	155	11	43
7	269	4	21	21	7	0	667	0	4	7
8	11	28	28	18	18	70	11	577	67	172
9	25	29	111	46	82	11	21	82	550	43
0	18	4	7	11	7	18	25	71	21	818

compare them by mutually superimposing or aligning them—the alignments results will be reported as similarity values (see data of amino acid substitution rates in Table 1.9);

(ii) Data on interaction between web pages such as the number of times page i visited page j at a given time period. This is frequently referred to as network data assuming that this type of data can be treated within the framework of graph theory;

(iii) Affinity data between geometric points—these are derived with the help of so called kernel functions. A kernel function $K(x,y)$ models inner product between complex transformations of vectors x and y, $\psi(x)$ and $\psi(y)$, so that $K(x,y) = \langle \psi(x), \psi(y) \rangle$. What is nice about it—in many cases, there is no need in using the transformation $\psi(x)$ itself—the kernel $K(x,y)$ suffices. A function $K(x,y)$ is a kernel if and only if matrix of its values $(K(x_i,x_j))$ on any finite sample of points $\{x_i\}$ ($i \in I$) is positive semi-definite. Frequently, the so-called Gaussian kernel is used to derive the affinity data according to formula

$$a_{ij} = e^{-d(x_i,x_j)/s}$$

where $d(x,y) = \langle x - y, x - y \rangle$ is the squared Euclidean distance and s > 0, a constant (see Sect. 3.4.4).

Case G. Confusion

Table 1.8 presents results of a psychological experiment on errors in human judgement, specifically, on confusion of human operators between segmented numerals (drawn on Fig. 1.2). In the experiment, a digit flashes for a short time on screen before an individual (stimulus) who is to report then what digit they have seen (response): (i,j)-th entry in Table 1.8 is the proportion of response j to stimulus i (Keren and Baggen 1981). The confusion matrix is understandably not symmetric, whereas its diagonal entries contain by far the larger proportions of observations, which is typical for confusion data as well as switch data.

The problem: are there any patterns of confusion, especially if represented by clusters? If yes, can any numeral shape features be found to describe the confusion clusters more or less exclusively?

Case H. Amino Acid Substitution Rates

Table 1.9 is a symmetric table of the so-called amino acid substitution scores that are used as weight coefficients at various schemes for alignment of protein amino acid sequences. A protein amino acid sequence represents the protein prime structure (in contrast to secondary, tertiary and, frequently, quaternary structures—those relate to spatial foldings of protein molecules). This may change during the process of evolution. The main assumption for studying the evolution is that each two organisms share a common ancestry. The more similar their protein sequences are, the more recent was their common ancestor. The likelihood of the event of amino acid i substituted by amino acid j is estimated by using blocks of

Table 1.9 Amino acid substitution rates: **BLOSUM62** matrix of substitution scores between amino acids presented using 1-letter code (see Table 1.10 for decoding)

Aa	A	B	C	D	E	F	G	H	I	K	L	M	N	P	Q	R	S	T	V	W	X	Y	Z
A	4	-2	0	-2	-1	-2	0	-2	-1	-1	-1	-1	-2	-1	-1	-1	1	0	0	-3	-1	-2	-1
B	-2	6	-3	6	2	-3	-1	-1	-3	-1	-4	-3	-1	-3	0	-2	0	-1	-3	-4	-1	-3	2
C	0	-3	9	-3	-4	-2	-3	-3	-1	-3	-1	-1	-3	-3	-3	-3	-1	-1	-1	-2	-1	-2	-4
D	-2	6	-3	6	2	-3	-1	-1	-3	-1	-4	-3	1	-1	0	-2	0	-1	-3	-4	-1	-3	1
E	-1	2	-4	2	5	-3	-2	0	-3	1	-3	-2	0	-1	2	0	0	-1	-2	-3	-1	-2	5
F	-2	-3	-2	-3	-3	6	-3	-1	0	-3	0	0	-3	-4	-3	-3	-2	-2	-1	1	-1	3	-3
G	0	-1	-3	-1	-2	-3	6	-2	-4	-2	-4	-3	0	-2	-2	-2	0	-2	-3	-2	-1	-3	-2
H	-2	-1	-3	-1	0	-1	-2	8	-3	-1	-3	-2	1	-2	0	0	-1	-2	-3	-2	-1	2	0
I	-1	-3	-1	-3	-3	0	-4	-3	4	-3	2	1	-3	-3	-3	-3	-2	-1	3	-3	-1	-1	-3
K	-1	-1	-3	-1	1	-3	-2	-1	-3	5	-2	-1	0	-1	1	2	0	-1	-2	-3	-1	-2	1
L	-1	-4	-1	-4	-3	0	-4	-3	2	-2	4	2	-3	-3	-2	-2	-2	-1	1	-2	-1	-1	-3
M	-1	-3	-1	-3	-2	0	-3	-2	1	-1	2	5	-2	-2	0	-1	-1	-1	1	-1	-1	-1	-1
N	-2	-1	-3	1	0	-3	0	1	-3	0	-3	-2	6	-2	0	0	1	0	-3	-4	-1	-2	0
P	-1	-3	-3	-1	-1	-4	-2	-2	-3	-1	-3	-2	-2	7	-1	-2	-1	-1	-2	-4	-1	-3	-1
Q	-1	0	-3	0	2	-3	-2	0	-3	1	-2	0	0	-1	5	1	0	-1	-2	-2	-1	-1	3
R	-1	-2	-3	-2	0	-3	-2	0	-3	2	-2	-1	0	-2	1	5	-1	-1	-3	-3	-1	-2	0
S	1	0	-1	0	0	-2	0	-1	-2	0	-2	-1	1	-1	0	-1	4	1	-2	-3	-1	-2	0
T	0	-1	-1	-1	-1	-2	-2	-2	-1	-1	-1	-1	0	-1	-1	-1	1	5	0	-2	-1	-2	-1
V	0	-3	-1	-3	-2	-1	-3	-3	3	-2	1	1	-3	-2	-2	-3	-2	0	4	-3	-1	-1	-2
W	-3	-4	-2	-4	-3	1	-2	-2	-3	-3	-2	-1	-4	-4	-2	-3	-3	-2	-3	11	-1	2	-3
X	-1	-1	-1	-1	-1	-1	-1	-1	-1	-1	-1	-1	-1	-1	-1	-1	-1	-1	-1	-1	-1	-1	-1
Y	-2	-3	-2	-3	-2	3	-3	2	-1	-2	-1	-1	-2	-3	-1	-2	-2	-2	-1	2	-1	7	-2
Z	-1	2	-4	1	5	-3	-2	0	-3	1	-3	-1	0	-1	3	0	0	-1	-2	-3	-1	-2	5

Fig. 1.2 Simplified digit numerals over a rectangle with a line in the middle

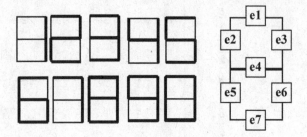

evolutionarily related protein sequences from various databases. These allow estimation of probabilities $p(i)$, $p(j)$ and $p(ij)$ of amino acids i, j and mutual substitution of i and j. Given these probabilities, the substitution scores are defined as integers proportional to logarithms of odd-ratios, $\log[p(ij)/(p(i)p(j))]$. Elements of matrix in Table 1.9 were derived by Henikoff and Henikoff (1992) using such protein sequences from database BLOCK for which pair-wise alignments involve not more than 62% of identity, which explains the name of the matrix.

This matrix leads to more reasonable results than other scoring matrices; practitioners of protein alignment have selected this matrix as a standard. We consider BLOSUM62 as a similarity matrix and are interested in finding clusters of amino acids that tend to replace each other and looking at physical and chemical properties explaining the groupings (Table 1.10).

Case H. Eurovision Song Contest Data

Table 1.11 presents the average scores given by each European country to their top 10 choices at the Eurovision song contests (up to year 2011); zero entries are presented by empty cells. The author compiled this using public data at http://www.escstats.com/. Each row of the table corresponds to one out of the nineteen selected European countries, and assigns a non-zero score to those of the other eighteen that have been among the 10 best choices of the country under consideration.

The cluster structure of the table should quantify to what extent the rumored effects of cultural and ethnical links on voting occur, because the quality of songs and performances may be considered random from year to year, so that in the ideal case when no cultural preferences are involved at evaluations, the similarity matrix should be of a random structure too. This brings forth a research question of whether any meaningful cluster structure is present in this matrix at all.

Table 1.10 Amino acids and their encoding as 3-letter and 1-letter symbolics from website http://icb.med.cornell.edu/education/courses/introtobio

1-letter	3-letter	Protein residue	Codons
A	Ala	Alanine	GCT, GCC, GCA, GCG
B	Asp, Asn	Asp. acid/asparagine	GAT, GAC, AAT, AAC
C	Cys	Cysteine	TGT, TGC
D	Asp	Aspartic acid (Aspartate)	GAT, GAC
E	Glu	Glutam. acid/glutamate	GAA, GAG
F	Phe	Phenylalanine	TTT, TTC
G	Gly	Glycine	GGT, GGC, GGA, GGG
H	His	Histidine	CAT, CAC
I	Ile	Isoleucine	ATT, ATC, ATA
K	Lys	Lysine	AAA, AAG
L	Leu	Leucine	TTG, TTA, CTT, CTC, CTA, CTG
M	Met	Methionine	ATG
N	Asn	Asparagine	AAT, AAC
P	Pro	Proline	CCT, CCC, CCA, CCG
Q	Gln	Glutamine	CAA, CAG
R	Arg	Arginine	CGT, CGC, CGA, CGG, AGA, AGG
S	Ser	Serine	TCT, TCC, TCA, TCG, AGT, AGC
T	Thr	Threonine	ACT, ACC, ACA, ACG
V	Val	Valine	GTT, GTC, GTA, GTG
W	Trp	Tryptophan	TGG
X	Xaa	Any amino acid	Any
Y	Tyr	Tyrosine	TAT, TAC
Z	Glu, Gln	Glutamic acid–Glutamine	GAA, GAG, CAA, CAG
*	STOP	Terminator	TAA, TAG, TGA

1.3 An Account of Data Visualization

1.3.1 General

Visualization can be a by-product of the model and/or method, or it can be utilized by itself. The concept of visualization usually relates to the human cognitive abilities, which are not yet well understood. Just for illustration of the role of color, a feature rarely attracting attention of both researchers and practitioners, take a look at the two double circles in Fig. 1.3.

Computationally meaningful studies of structures of visual image streams such as in a movie or video began only recently. A most recent account of the developments in information visualization can be found in Machlis (2017) (see also Mazza 2009; Ware 2012).

Table 1.11 A sample of results at the Eurovision song contests (up to year 2011)—the average scores given by each European country to their top 10 choices

	Az	Be	Bu	Es	Fr	Ge	Gr	Is	It	Ne	Pol	Por	Ro	Ru	Se	Sp	Sw	Ukr	UK
Azerbajan							6.1	4.8					5.0	6.5				9.0	
Belgium	3.8					3.9	4.0			4.7						3.4			4.2
Bulgaria	6.7						9.3							4.8	6.0			4.4	
Estonia	4.1								4.3					8.8				4.3	
France		3.7	4.3					5.6	4.7			5.4			8.0				4.1
Germany					3.4		3.7	3.5			5.5				7.0				4.2
Greece	5.4		8.0		4.1								4.0		8.0	4.4		3.8	
Israel	5.0									4.3			6.6	7.4	5.0			6.2	4.3
Italy			10.0		5.4								12.0					6.5	5.2
Netherl.	3.9	4.6				3.8		4.5							7.0				
Poland	8.4	4.3		3.9					9.0									8.2	
Portugal		3.5				4.5		4.1	8.1				5.2		5.7	4.2		7.4	4.3
Romania	5.2						8.2		6.0					4.9	8.0			3.5	
Russia	9.9						3.7	3.6							8.0			7.7	
Serbia			5.3				7.3							4.4				4.4	
Spain			7.8			5.1	4.5		7.4			4.3	7.9		4.7			4.6	
Switzerl.					4.4			4.2	4.7						10.6	4.1			4.1
Ukraine	11.1										6.0			9.8	9.0				
UK				3.6		3.9	3.8								3.7				

We are going to be concerned with presenting data as maps or diagrams or digital screen objects in such a way that relations between data rows or columns or both are reflected in distances or links, or other visual relations, between their images. Among the more or less distinct visualization goals, beyond sheer presentation that appeals to the cognitive domination of the visual over the other senses, we can distinguish between:

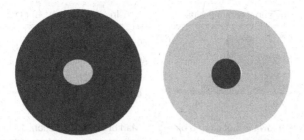

Fig. 1.3 Two colors, light-blue and red, at both parts of the figure, ideally complement each other. However, the middle circle on the left is more or less flat, as well its red complement, whereas that on the right "springs out" to the viewer's eyes, according to R. Medny (2018) The Bible of Color and Style, AST Publishers, Moscow, pp. 6–7

A. Publicizing and highlighting
B. Integrating different aspects
C. Narrating
D. Manipulating.

Of these, manipulating visual images of entities, such as in computer games, seems an interesting area yet to be developed in the framework of data analysis. There can be mentioned, though, operations of mild manipulation readily available at various sites already such as scrolling, representing an overview with possibilities of getting further details of individual fragments by zooming or windowing, and an overview that allows focusing on specific fragments by enlarging them on the same screen (Mazza 2009). The other three will be briefly discussed and illustrated in the remainder of this section.

1.3.2   Publicizing and Highlighting

Given a quantitative feature, its most comprehensive and striking to the eye visualization is the histogram. On the plane, one draws an x axis and the feature range boundaries, that is, its minimum and maximum. The range is divided then into a number of non-overlapping equal-sized sub-intervals, *bins*. Then the number of objects that fall in each bin is counted, and the counts are reflected in the heights of the bars over the bins, forming a histogram. Histograms of Population resident in Market town dataset and Petal width in Iris dataset are presented in Fig. 1.4.

The shape of the histogram may depend on the number of bins. Then it is tempting to think of an optimal number of bins. Unfortunately, no common goal for this can be specified—there is no universally good method to choose the right number of bins.

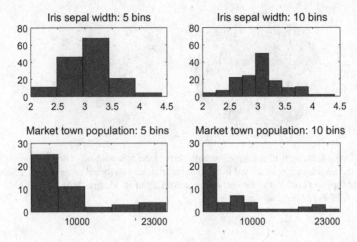

Fig. 1.4 Histograms of quantitative features in Iris and Market town data: the feature represented on the *x*-axis and the counts on the *y*-axis. The histogram shapes depend on the number of bins

Q.1.3. Why should the bins not overlap?
A. Each entity falls in only one bin if bins do not overlap, and the total of all bin counts equals the total number of entities in this case. If bins do overlap, the principle "one entity—one vote" will be broken.
Q.1.4. Why are the bar heights on the left greater than those on the right in Fig. 1.4?
A. Because bins on the right are twice as narrow as those on the left; therefore, the numbers of entities falling within them must be smaller.
Q.1.5. Is it true that when there are only two bins covering the range, the divider between them must be the midrange point?
A. Yes, because the bin sizes are equal to each other (see Fig. 1.5).

In Figs. 1.6 and 1.7 further on, two most popular types of histograms are presented. The former corresponds to the so-called power law, sometimes referred to as Pareto distribution. This type is frequent in social systems. According to numerous empirical studies, such features as wealth, group size, productivity and the like are all distributed according to a power law so that very few individuals or entities have huge amounts of wealth or members, whereas very many individuals are left with virtually nothing. However, they all are important parts of the same system with the have-nots creating the environment in which the lucky few can strive. The power law density function is $p(x) \approx a/x^{\lambda}$, where λ reflects the steepness of the frequency's fall (see Fig. 1.8 on the left). Such a law expresses what is called the Matthew's effect referring to the saying

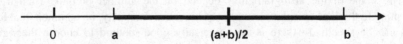

Fig. 1.5 With just two bins covering the range, the divider is mid-range

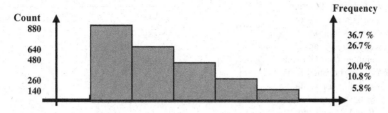

Fig. 1.6 A power law distribution

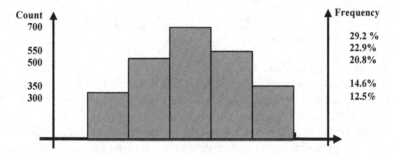

Fig. 1.7 Gaussian type distribution (bell curve)

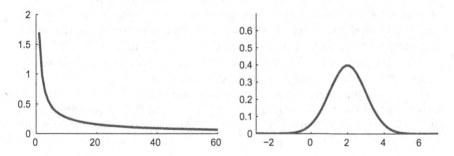

Fig. 1.8 Density functions of the power law with $\lambda = -0.8$, on the left, and normal distribution N (2,1), on the right

"He who has much, will get more; and he who has nothing, will lose even that little that he has," according to Matthew's gospel. The Matthew's effect is expressed, for example, in "the mechanism of preferential attachment": the probability that a new web surfer hits a web-site is proportional to the site's popularity, according to this mechanism.

Another type, which is frequent in physical systems, is presented in Fig. 1.7.

This type of histogram approximates the so-called normal, or Gaussian, law. Distributions of measurement errors and, in general, features being results of small random effects are thought to be Gaussian, which can be formally proven within a mathematical framework.

Fig. 1.9 A density function $p(x)$, which is a mixture of two normal distributions, $N(2,1)$ weighted 0.4, and $N(8,2)$ weighted 0.6. The area between the two dashed lines is the probability for value x to fall in the interval between 0 and 2—not too high at this $p(x)$!

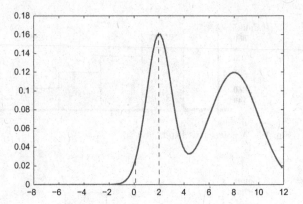

The normal, or Gaussian, law is $p(x) = C \exp[-(x - a)^2/2\sigma^2]$, where C is a constant, which is sometimes denoted as $N(a,\sigma)$. The parameters of this distribution, a and σ, have natural meaning: a expresses the expectation, or mean, and σ^2—the variance, which naturally translates in data terms (see Fig. 1.8 on the right and Sect. 2.2.3). It should be pointed out that the probability of a value x falling in the interval $a \pm \sigma$ according to the normal distribution is about 88%, and falling in the interval $a \pm 3\sigma$ about 99.7%, virtually unity, so that at modest sample sizes it is highly unlikely that a value x can fall out of this interval, which is referred to sometimes as the "three sigma rule". The Gaussian distribution can be rescaled to the standard $N(0,1)$ form, with 0 expectation and 1 the variance, by shifting the variable x to the mean, a, and normalizing it afterwards by the square root of σ^2. This transformation, sometimes referred to as z-scoring, is expressed with formula $y = (x - a)/\sigma$, where y is the transformed feature. A mixture, that is, weighted sum, of two normal density functions is presented in Fig. 1.9.

One more popular distribution is the uniform distribution, over a range $[l,r]$. Its density is a constant function equal to $p(x) = 1/(r - l)$, so that the probability of an interval (a, b) within the range is just $p = (b - a)/(r - l)$, proportional to the length of the interval.

Q.1.6. Take a look at the distributions in Fig. 1.4. Can you see which of the two types they are similar to?
A. The Population distribution is of power law type, and the Petal width is of Gaussian law type, as one would expect.

Another popular visualization of distributions is known as a pie-chart, in which the bin counts are expressed by the sizes of sectored slices of a round pie (see in the middle of Fig. 1.10).

As one can see, histograms and pie-charts cater for perception of two different aspects of the distribution; the former for the actual envelopment of the distribution along the axis x, whereas the latter caters for the relative sizes of distribution chunks falling into different bins. There are a dozen more formats of visualization of distributions, such as bubble, doughnut and radar charts, easily available in Microsoft Excel spreadsheet as well as in many other programs.

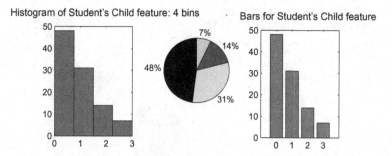

Fig. 1.10 Distribution of the number of children at the Student data. The Child feature is visualized as a 4-bin histogram on the left, pie-chart in the middle, and a bar set on the right—this seems the most appropriate of the three at the case

A categorical feature differs from a quantitative one not just because its values are strings, not numbers—they are coded by numbers anyway to be processed. The average of a quantitative feature is always meaningful, whereas the averaging of category labels, such as Occupations—BA, IT or AN—in Student data or Sector of Economy—Retail, Utility or Industry—in Company data, makes no sense even after they are coded by numbers. The applicability of the operation of averaging is indeed a characteristic property of being quantitative. For example, one may claim that a feature like the number of children, from the Student data (see Fig. 1.10), is not quantitative because its values can only be whole numbers. Still, a statement like "the average number of children per woman is 1.85" does make sense because it can be easily made legitimate by moving to counting by hundreds: there are 185 children per every hundred women.

For categorical features, there is no need to define bins: the categories themselves play the role of bins. However, their histograms typically are visualized with bars or stems, like on Fig. 1.11 that represents the distribution of categories IT, BA and AN of Occupation feature in Student data.

The distribution of the feature can be expressed in absolute numbers of entities falling in each of the categories, that is, $D = (35, 34, 31)$, or on the relative scale, by

Fig. 1.11 The distribution of categories IT, BA and AN of occupation feature in student data shown with bars on the left and stems on the right

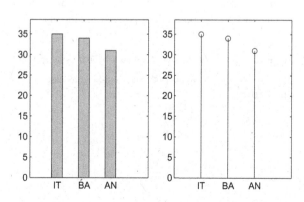

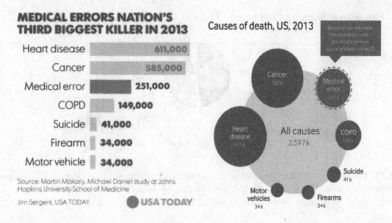

Fig. 1.12 The distribution of causes of death in the USA, 2013, according to Makary and Daniel (2016) by using strips, on the left, and bubbles (on the right). COPD is *Chronic obstructive pulmonary disease*

using proportions found by dividing frequencies over their total, 35 + 34 + 31 = 100, which leads to the relative frequency distribution d = (0.35, 0.34, 0.31).

Recently, the so-called bubble chart format, developed for multidimensional visualizations, has been applied to 1D distributions (see Fig. 1.12 at which the distribution of most frequent causes of death in the USA is visualized with strips, on the left, and bubbles, on the right). In this author's opinion, the bubble chart is more impressive. Also, it allows estimating the risk of death from a medical error, as about 10% (251 out of 2597).

Case-Study 1.1. Has There Been a Bias in S'nS' Policy?

In real life, many distributions are far from uniform. For example the distribution by race of the 878,153 stop-and-search (stop-and-frisk in the USA) cases performed by police in England and Wales was widely discussed in the media (see Table 1.12 and BBC's website http://news.bbc.co.uk/1/hi/uk/7069791.stmof29/10/07). This is far from uniform indeed: the proportion of W category is thrice greater than of the other two taken together. Yet it was a claim of racial bias because the proportion of W category in the population is even higher than that (see Table 1.12).

Table 1.12 Race distribution of stop-and-search cases in England and Wales in 2005/6

Race	Number of "Stop-and-Search" cases	Relative frequency, %
Black (B)	131,723	15
Asian (A)	70,250	8
White (W)	676,180	77
Total	878,153	100

Table 1.13 Race distribution of stop-and-search cases in England and Wales in 2016/7

Race	Total		Resulted in action	
	Number of cases	Relative frequency, %	Number of cases	Proportion, %
Black	55,140	24	17,703	32.1
Asian	28,897	11	8733	30.2
White	166,674	65	47,683	28.6
Total	256,711	100	74,119	28.9

At that time no statistics were available to rationally justify or reject the charge. This author took an indirect way by comparing the stop-and-search case ethnic distribution with that in prisons. Indeed, it is wrong to assume that police apply the Stop-and-Search policy to the population randomly. They should apply the policy only when they deem it necessary, so that the comparison should involve not all of the total population but only the criminal population. The latter is unknown, but can be likened to the population of prisons. Indeed, the distribution of subjects to Stop-and-Search policy by race has been almost identical to that of the imprisoned. Therefore, the claim of a racial bias could be declared incorrect under the assumption that the court and indictment system is not biased.

For the time being both the government and police in England and Wales have undertaken steps to much improve the Stop-and-Search policy by, first, defining rather strict rules for application of the policy and, second, by targeting the policy to a tangible action. The action may involve different measures from arrest to penalty to just warning. Application of the rules have drastically decreased the number of Stop-and-Search cases, as can be seen from Table 1.13 of application of the policy in 2016/2017 year. (The source is on https://www.gov.uk/government/statistics/police-powers-and-procedures-england-and-wales-year-ending-31-march-2017).

This Table reports of just about 250,000 Stop-and-Search cases versus almost 900,000 cases a decade ago. The former number is somewhat less than the real number (about 300,000 cases) because some police reports, about 10% of the cases, lack information of the ethnicity and about 10% of the cases report of mixed races and are not reflected in the Table either.

The data in Table 1.13 differ from those in Table 1.12 by the fact that proportions of the Stop-and-Search cases resulted in actions are reported, so that one can see how the ethnicity affects the level of disorderly behavior. It appears the differences in the proportions are not that high, 2–3% only.

To visually illustrate proportions, several formats have been proposed, of which one of the most appealing is the so-called dot plot (Cleveland 1984). A version of that is presented in Fig. 1.13. The scales chosen somehow show that the proportion of reprimands received by blacks is much greater than those by other groups, which is a bit misleading. In fact, the proportion of them, 32.1 is larger than the proportion of action for Whites, 28.6, by a mere 3.5%.

If one compares this difference with the difference between relative proportions of S-n-S cases for Blacks (24% vs. 3.3% in the population) and those for Whites

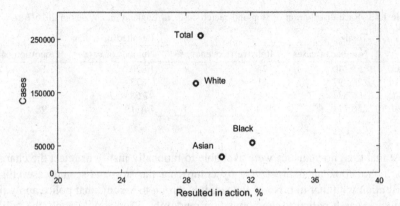

Fig. 1.13 The distribution of proportions of stop-and-search cases in England and Wales 2016/2017 resulted in a police action over ethnicity presented with a version of so-called dot plot

(65% vs. 86% in the population), one would not hesitate to conclude that the S-n-S policy indeed applies to Blacks somewhat excessively. Of course one should not take the cited proportions at the face value: relative frequencies in Table 1.12 sum to 100 whereas those for the relative proportions of ethnicities to 96.8 only (Blacks —3.3%, Asians—7.5%, Whites—86%), but the difference is quite small.

Q.1.7. What is the modal category in the distribution of Table 1.5? In Occupation on Student data?
A. These are most likely categories, W in Table 1.13 and IT in Student data, Fig. 1.11.

The fact of distortion admitted in Fig. 1.13 can be beneficial sometimes—to visually highlight a feature of an image. A local distortion of the original scale was proposed back in 1906 by H. Beck to map the London tube scheme while greatly enlarging the relative sizes of the Centre of London part. This made the center much better and more conveniently presented to the eyes of passengers but was stonewalled by the Tube authorities for about 30 years.

Such a distortion currently is a standard for metro maps worldwide (see Fig. 1.14).

In fact, this line of thinking has been worked on in geography for centuries, since the mapping of the Earth global surface to a flat sheet is impossible to do exactly. Various proxy criteria have been proposed leading to interesting highlights way beyond conventional geography maps, such as presented on Fig. 1.15 (Fullers' projection) and Fig. 1.16 (August's projection); see website http://en.wikipedia.org/wiki/World_map for more.

More recently this idea was applied by Rao and Card (1994) to table data (see Fig. 1.17); more on this can be found in Card et al. (1999) and Mazza (2009).

It should be noted that the disproportionate highlighting may lead to visual effects bordering on cheating (or being just that). This is especially apparent when relative proportions are visualized through proportions between areas, as in

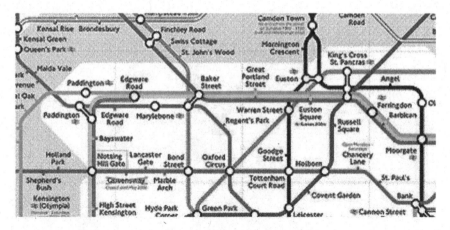

Fig. 1.14 A fragment of London tube map made after H. Beck; the central part is highlighted by a disproportionate scaling

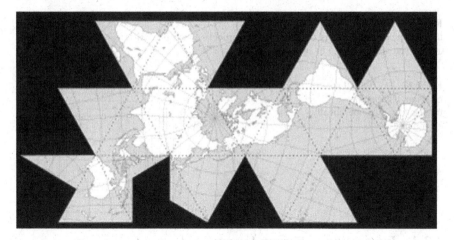

Fig. 1.15 The Fuller projection, or Dymaxion map, displays spherical data on a flat surface of a polyhedron using a low-distortion transformation. Landmasses are presented with no interruption

Fig. 1.18. An unintended effect of the picture is that the decline by half in one dimension is presented visually by the area of the doctor's body, which is just not half but one fourth of the initial size. This grossly biases the message.

Another typical case of unintentionally cheating is when the relative proportions are visualized using bars that start not at the 0 point but an arbitrary mark, as is the case presented in Fig. 1.19: a newspaper's legitimate satisfaction with its success is visualized using bars that begin at 500,000 mark rather than 0. Another mistake is that the difference between the bars' heights on the picture is much greater than the reported 220,000. Altogether, the rival's circulation bar is more than twice shorter while the real circulation is less by mere 25%.

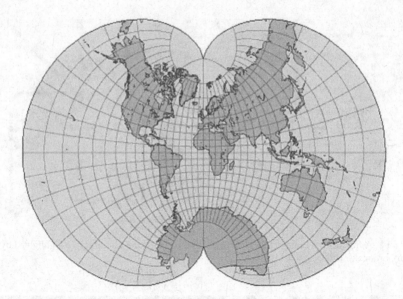

Fig. 1.16 A conformal map: the angle between *any* two lines on the sphere is the same between their projected counterparts on the map; in particular, each parallel crosses meridians at right angles; and also, the sizes at any point are the same in *all* directions

Fig. 1.17 The Table Lens machine: highlighting a fragment by disproportionally enlarging it

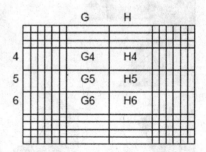

1.3.3 Integrating Different Aspects

We first describe aspects of visualization of bivariate data, that is, data with two features. These are most naturally presented via the Cartesian plots studied in high school throughout the world. Then we consider cases in which the visual image is less straightforward.

We distinguish between nominal features and quantitative features. There can be four combinations:

(i)   both features are quantitative;
(ii) the target feature is quantitative, the input feature is nominal;

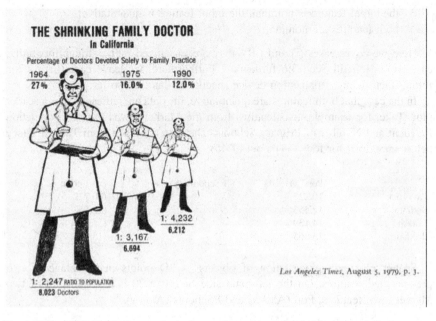

Fig. 1.18 A decline in relative numbers of general practitioner doctors in California in 70-es: the 2D image conveys a quadratic decline—not a half but a quarter of the size

Fig. 1.19 An unintended distortion: a newspaper's report (July 2005) is visualized with bars that grow from mark 500,000 rather than 0

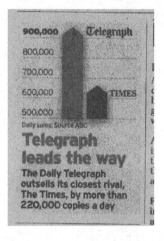

(iii)   the target feature is nominal, the input feature is quantitative;
(iv)   both features are nominal.

Here we cover cases (i) and (ii)—they are visualized rather straightforwardly; case (iv) is left till Sect. 3.6 further on. Furthermore, we leave case (iii) for the future: there is no visualization device specific to this case so far.

In the case that both features are quantitative, the data are represented as a scatter plot. Take, for example, two features from the Market towns dataset: Population Resident and Number of Primary Schools. The data are taken from Table 1.2 (see below an extract for four towns out of 45):

	Pop (x)	PSchools (y)	(x,y)-point
Tavistock	10,222	5	(10,222, 5)
Bodmin	12,553	5	(12,553, 5)
Saltash	14,139	4	(14,139, 4)
Brixham	15,865	7	(15,865, 7)

Scatter plot is a presentation of objects as 2D points in the plane of two pre-specified features. On the left-hand side of Fig. 1.20 is a scatter-plot of two Market town features, Pop (Axis x) and PSchools (Axis y).

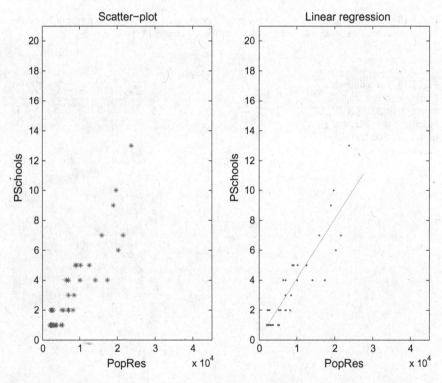

Fig. 1.20 Scatter plot of PopRes versus PSchools in Market town data. The right-hand graph includes a regression line of PSchools over PopRes

One can think that these two features are related by a linear equation $y = ax + b$, where a and b are some constant coefficients, referred to as the slope and intercept, respectively, because the number of schools should be related to the number of children, which is related to the number of residents. This equation is referred to as the linear regression of y over x. The corresponding straight line is presented on Fig. 1.20. Of course, other relations are not necessarily that simple because they may also depend on other factors such as school sizes, population's age, etc. It would be a miracle if one equation fitted well all 45 towns. The possible inconsistencies in the equation can be modeled as additive errors, or residuals. The slope a and intercept b are taken in such a way that the inconsistencies of the equation on the 45 towns are minimized.

A somewhat different outline is used to visualize interrelation between a categorical feature and a quantitative one. It is the so-called box-plot. To draw a box-plot, the x-axis is divided in bins so that each category corresponds to its own bin. Then a vertical line is drawn within each bin, and the feature's minimum and maximum within the corresponding category are shown with horizontal short lines, called "whiskers" (see Fig. 1.21). Then the upper 25% and lower 25% quantiles of the feature within the category are found. A 25% upper quantile is a cut-point within the feature range such that 25% of the values lie above the point and 75%, beneath. Similarly a 25% lower quantile is defined. Therefore, the cut-points are endpoints of an interval at which $100 - 25 - 25 = 50\%$ of all the within-category feature values fall. Then this interval is expressed with a box drawn at the end-points. A line within the box shows either the mean or median value of the feature within the category. One may specify a box for values other than 50%. For example, Fig. 1.21 illustrates the within-category range over 60%, not 50%, of its contents by removing 20% off of both its top and bottom extremes.

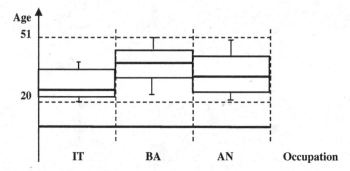

Fig. 1.21 Box-plot of the occupation-age relation with 20% quantiles; the box heights reflect the age within-category 60% ranges, whiskers show the total ranges. Within-box horizontal lines are for the within category averages

With the quantile level specified at 40%, at the category IT, Age ranges between 20 and 39, but if we sort it and remove 7 entities of maximal Age and 7 entities of minimal Age (there are 35 students in IT so that 7 makes 20% exactly), then the Age range on the remaining 60% is from 22 to 33. Similarly, Age 60% range is from 32 to 47 on BA, and from 25 to 44 on AN (see box heights on Fig. 1.21). The whiskers reflect 100% within category ranges, which are intervals [20, 39], [27, 51] and [21, 50], respectively.

The box-plot proved useful in studies of quantitative features too: one of the features is partitioned into a number of bins that are treated then as categories. Sometimes, a more precise representation of within-category distributions, called a violin plot, is utilized—this is not covered in this text.

Now consider somewhat less trivial images. Combining different features of a phenomenon into the same image can make perception easier. Figure 1.22 represents a scheme of a Smart Home in which various devices are interconnected in a wireless environment so that the home center can switch on or out each of them upon receiving a corresponding message.

The diagram in Fig. 1.23 visualizes relations between the features in Company data (Table 1.3) as a decision tree to conceptually characterize their products. For example, the left-hand branch distinctly describes Product A by combining "Not retail" and "No e-commerce" edges (See Sect. 3.8 for more on decision trees.).

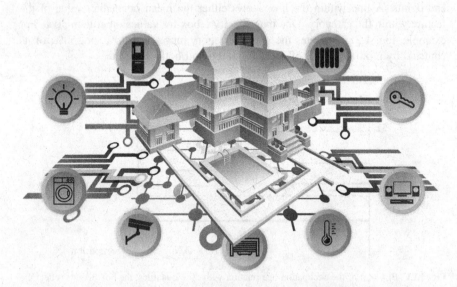

Fig. 1.22 Illustration of interrelations between home devices and gadgets, represented by conventional infographics icons, in the concept of Smart Home according to web-site http://www. communitymegaphone.com/gadgets-for-a-smart-home/ (accessed 26 January 2019)

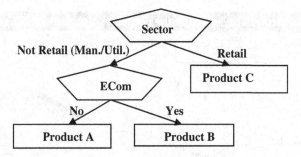

Fig. 1.23 Product decision tree for the company data in Table 1.3

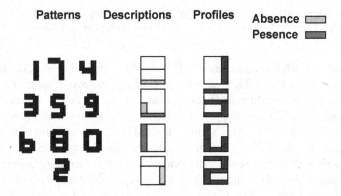

Fig. 1.24 Confusion patterns for numerals visualized from the patterns' data analysis descriptions in terms of edges being present or not. The right-hand part presents profiles of the common edges, for comparison

One more visual image, Fig. 1.24, depicts relations between the confusion patterns of decimal numerals drawn over rectangle's edges (Fig. 1.2) and their descriptions in terms of combinations of edges of the rectangle with which they are drawn. A description may combine both edges present and absent to distinctively characterize a pattern, whereas a profile comprises edges that are present in all elements of its pattern. The confusion patterns are derived from data in Table 1.8 according to clustering of numerals in Sect. 4.6 and Mirkin 2012.

1.3.4   Narrating a Story

In a situation in which data features involve a temporal and/or spatial aspects, integrating them in one image may lead to a visual narrative of a story, with its starting and ending dates, all on the same screen. Such a narrative of a military

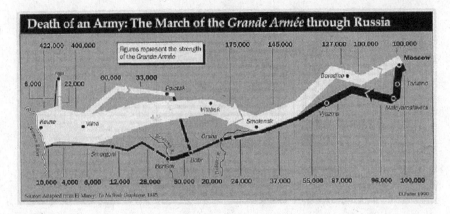

Fig. 1.25 The white band represents the trajectory of Napoleon's army moving to the east and the black band shows it moving to the west, the band width being proportional to the army's strength

company from the rich history of Europe (Napoleon's Grand army invading Russia 1812) is presented in Fig. 1.25. It shows a map of Russia, with Napoleon's army trajectory drawn forth, in white, and back, in black, so that the time is represented in this static image via the trajectory. The directions are shown with arrows. The trajectory's width shows the army's strength steadily declining in time on a dramatic scale, in the summer time and in the absence of major fighting.

Segel and Heer (2010) present a more general account for narrating a story with data visualization, possibly using multiple pictures, including a system to involve three divisions of features: genre, visual narrative tactics, and narrative structure tactics. Lee et al. (2015) present a scenario for the process of translation from data to storytelling.

All the images presented can be considered illustrations of a principle accepted further on. According to this principle, to visualize data, one needs to specify first a "ground" image, such as a map or grid or coordinate plane, which is supposed to be well known to the user. Visualization, as a computational device, can be defined as mapping data to the ground image in such a way that the analyzed properties of the data are reflected in properties of the image. Of the goals considered, integration of data will be the priority since no temporal aspect is considered in this text.

1.4   The Role of Data Analysis

1.4.1   Current Success of Data Analysis and Data Table

This author holds the view that data analysis is not just a convenient term to cover all the data processing techniques that are known as mathematical statistics or applied statistics or machine learning or data mining or, more recently, big data analysis or data analytics. The concept of data analysis, in this text, is focused on such techniques and processing tools that are oriented toward enhancing theoretical knowledge of the phenomenon to which the data refer. This may be considered akin to the concept of "data analytics", which, however, refers to techniques that lead straightforwardly to conclusions, or even decisions, regarding the phenomenon in question.

First of all, one should note that data analytics does not necessarily require a serious data processing technique. Frequently, simple comparisons of frequencies or means may bring immediate conclusions and corresponding decisions. Especially, this refers to data of big or not so big organizations in which most difficulties are related to the issues of data collection, rather than transformation. The worldwide process of digitalization brings novel opportunities, leading to major transformations in production, economics and society. Consider, for example, a story of using a computer system in a municipality of London UK (see https://www.logianalytics. com/case-study/london-borough-of-islington-saves-800k-annually-with-logi/, visited 5 June 2017). This system includes all the traffic wardens in the municipality—there are about 200 of them—and it shows their activities in real time. A few immediate results:

- It appears, the performance, that is, the numbers of penalty tickets issued, of traffic wardens go down after the lunch break (this was easily taken care of by the corresponding orders).
- It has been determined which traffic wardens wrote tickets that were successfully disputed by the motorists, and they have been retrained.
- Some places have been found to be administered so well they had no parking violations. As there is no sense in sending wardens there, the number of wardens could be reduced.

All these resulted in real economies and thus effectively increased the municipality's budget.

This example clearly shows that the effect here comes from acquiring a computer system for data collection and processing rather than from applying a data analysis method. The data analysis here pertains to specific issues and is not that complicated. This is a feature of perhaps a great majority of real-life applications.

Consider a more challenging task—analysis and prediction of car traffic in a big city. A Russian search engine, Yandex, runs such a service on-line for dozens of big cities including Moscow, the capital of Russia. This service, called "Yandex-Probki" (Yandex traffic jam), scores the speed of traffic within a city at a hundred locations over main streets and avenues. This is done by collecting the location signals from the GPS (Global Positioning System) devices used by car drivers for orientation on the streets.

All the city is divided in subareas so that calculating the difference between the time a car enters and the time it leaves a subarea, determines its average speed. The level of traffic on a route in the subarea is related to the mean speed of individual GPS devices: the smaller the speed the greater the traffic jam value. A random noise, such as the signal from a GPS device used by a pedestrian can be easily filtered out by considering related data such as the period of time at which the GPS has been switched on: pedestrians usually use GPS devices during very short time periods. Even simpler approach, the averaging can be done over 95% of the signals by removing 2.5% of both the slowest and fastest speeds—without trying to penetrate what were the signaling objects. Matching a current map of speeds and similar maps at the same weekday, at an appropriate time, under similar weather conditions, it is not difficult to give a short-term forecast of the distribution of speeds and even give recommendations for optimal routes. This is a good example of "big data" analysis. How does this happen? The main causes of the success:

– huge speed of computing,
– huge memory,
– wireless connections to electronic devices, sensors and processors.

No sophisticated methods for data analysis are used—just ubiquitous networks and very fast computation. Has someone cheated by stating in their CV that they graduated from Harvard University? No need to collect various facts and information here and there; no need for complex reasoning. Just make a search through Harvard's digital archive for the corresponding name.

This is the essence of the current movement generally referred to as "big data"— in fact, rapidly merging together various information bits—this is the root of the success. The issues of data processing and fusion are enormous, especially when considering their dynamics. Reliable distributed systems and fast parallel computations are hence in great demand. However, this is outside the scope of this book. This book is about producing meaningful statements from analysis of the data after they have been arranged as a data table.

Examples of data tables have been given in Sect. 1.2—Almost all are just entity-to-feature tables. When units of observation correspond one-to-one to entities, that is, rows of a corresponding data table, drawing a data table is not difficult since the only problem remaining is to define appropriate features, corresponding one-to-one with columns of the table. This is relatively easy in the case of a categorical feature whose categories are not disjoint. Survey questions such as "What languages can you speak?" or "List brand names of your cars" with answers such as "English, German, Dutch" and "BMW, Ford, Toyota" are examples. In these cases, each of the categories is assigned a "dummy" binary feature of its own with the value 1, if the object falls in the category, or 0, if not. Less formal considerations must be taken into account when the units of observation are protein sequences or unstructured texts. Here one should think of features useful in relation to a specific problem at hand such as "The proportion of hydrophobic residues in the first half of the sequence" and "The number of occurrences of the word 'go' and its derivatives (such as "went", "gone", etc.) in the text".

The book by Levitt and Dubner (2005) gives a number of much inventive examples of defining features related to socio-economic phenomena of common interest.

Q.1.8.   A student of a Russian School of Data Analysis, Ivan Petrovich Sidorov, took a data set of metro systems of the world according to the web-site https://en.wikipedia.org/wiki/List_of_metro_systems (visited 8 June 2017) and put a fragment of the data in a table like Table 1.14

He claimed that the table was a data table with the following features:

«System»	Subway's name (Nominal)
«City»	City in which the Subway is located (Nominal)
«State»	The country in which the Subway is located (Nominal)
«Continent»	The continent in which the Subway is located (Nominal)
«Opened»	The year in which the Subway was opened (Nominal/Ordinal)
«Lines»	The number of lines in the system (Quantitative)
«Length»	The total length of lines in kilometers (Quantitative)
«Stations»	The total number of stations (Quantitative)

Would you agree with Mr. I. P. Sidorov?

A. Not exactly. First of all, the two columns on the left, Name of the system and City, are not features because the number of categories in each of them is the same as the number of objects; so they give no information about relations between the objects except that all of them are different. Of course, one should also note that the two columns are almost identical.

Table 1.14 List of metro systems in the Former Soviet Union according to https://en.wikipedia. org/wiki/List_of_metro_systems (visited 8 June 2017) and some data on them

#	System	City	State	Continent	Opened	Lines	Length	Stations
1	Baku metro	Baku	Azerb	Asia	1967	3	36.7	25
2	Yerevan metro	Yerevan	Armenia	Asia	1981	1	12.1	10
3	Minsk metro	Minsk	Belarus	Europe	1984	2	37.3	29
4	Tbilisi metro	Tbilisi	Georgia	Asia	1966	2	26.3	22
5	Almaty metro	Almaty	Kazakh	Asia	2011	1	11.3	9
6	Moscow metro	Moscow	Russia	Europe	1935	12	338.9	203
7	Petersburg metro	Petersburg	Russia	Europe	1955	5	113.6	67
8	Nizhny Novgorod metro	N. Novgorod	Russia	Europe	1985	2	18.9	14
9	Novosibirsk metro	Novosibirsk	Russia	Asia	1985	2	15.9	13
10	Samara metro	Samara	Russia	Europe	1987	1	10.3	10

Then we have the column "State" which has different values for half the entity set and thus yields no information on that half. This column should be thrown out too or changed to a binary feature with categories "Russia" and "Not Russia"; the latter embracing all the six cases of other states. One more dubious definition is "Nominal/Ordinal" for the scale type of feature "Opened". This feature in fact refers to the timing and, therefore, should be counted as quantitative. Another thing is that it could be categorized, for example, at "Before 1970" and "After 1970" categories.

Table 1.15 Data of metro systems in the Former Soviet Union according to https://en.wikipedia.
org/wiki/List_of_metro_systems (visited 8 June 2017)—corrected

#	City	State	Continent	Opened	Lines	Length	Stations
1	Baku	Other	Asia	1967	3	36.7	25
2	Yerevan	Other	Asia	1981	1	12.1	10
3	Minsk	Other	Europe	1984	2	37.3	29
4	Tbilisi	Other	Asia	1966	2	26.3	22
5	Almaty	Other	Asia	2011	1	11.3	9
6	Moscow	Russia	Europe	1935	12	338.9	203
7	Petersburg	Russia	Europe	1955	5	113.6	67
8	N. Novgorod	Russia	Europe	1985	2	18.9	14
9	Novosibirsk	Russia	Asia	1985	2	15.9	13
10	Samara	Russia	Europe	1987	1	10.3	10

But any quantitative feature can be categorized in this manner. Therefore, a proper
data table derived from Table 1.14 would be like Table 1.15.

The features in Table 1.15 are:

«State»	The country in which the metro is located (Nominal, Russia or Other)
«Continent»	The continent at which the metro is located (Nominal, Europe or Asia)
«Opened»	The year at which the metro was opened (Quantitative)
«Lines»	The number of lines at the metro system (Quantitative)
«Length»	The total length of lines in kilometer (Quantitative)
«Stations»	The total number of stations (Quantitative)

Sometimes the unit of observation differs from that of the data entity. For
example, when collecting data of villages one may survey features of families, or
households, or the school, or surgery to consider them, in a summarized form, as
village features. These are usually taken into account by the data collectors.

In the remainder of this section we give instructive examples of data analysis,
some successful, some not, to derive useful lessons for practical applications.

1.4.2   A Great Success Story: Kepler's Planetary Motion

Here is an example of data analysis as we understand it in "big science". Johannes
Kepler (1571–1630), a great astrologist and astronomer working at the court of the
Emperor Rudolf II in Prague, then the capital of a great Empire stretching from
Portugal on the West to Balkans on the East, gets hold of data meticulously col-
lected by his boss, Danish astronomer Tycho Brahe—because Tycho suddenly dies
while at a court function. Tycho's data are of planet locations on the sky for a
40-years period. The task is, as usual, to find patterns or regularities in the data.
Some bits of knowledge which were already available at the time:

– Copernicus' suggestion that all the planets revolve around Sun in circles (1543);
– All the orbits lie in (or close to) a plane, the ecliptic, known from the time
  immemorial as the plane of the visible annual path of the Sun;

- The Pythagorean "music-of-spheres" view of the solar system, including that the average distances of the six known planets from the Sun obey the "Platonic solids law" (Kepler 1594; see also Kepler 2014).

First of all, Kepler found some constancy in planet motions including what is currently known as Kepler's 2rd Law of planetary motion: for any planet, the area of the sector between Sun and the planet in its orbit, swept out in a unit of time, is constant. Then Kepler proceeded to the analysis of deviations of the Mars orbit from circularity, which culminated in establishing of the essence of Kepler's 1st Law: the planets revolve around the Sun in ellipses rather than circles. He found these two laws in a short time. Then, after more than a dozen years, Kepler discovered his 3rd Law of planetary motion, of the constancy between the period P and the average distance D to Sun of planet orbits. It appears that it is not the ratio P/D itself which is constant but rather the ratio of their, respectively, square and cube:

$$\frac{P^2}{D^3} = const \tag{1.1}$$

How can one derive such an unintuitive formula from the data? The square of P and, especially, the cube of D are beyond intuition. Some, including this author, thought that J. Kepler did know of the law of gravitation (see further on in this section), at least in this formulation: the Sun acts on the planets with a force proportional to the inverse of squares of their distances to the Sun. The constancy of the quotient above can be mathematically derived from that—this is how J. Kepler could have come up with his third Law of planetary motion.

The others, in fact an overwhelming majority, said that they had no clue regarding the source for the Kepler's formula and they did not care. Recently one more group emerged. This group thinks that J. Kepler arrived at the formula by the way of data analysis, namely, by using the then recently invented concept of logarithm: that was proposed by John Napier in Scotland in 1614, and J. Kepler became aware of the concept in 1616.

The data of these two features, the period, P, and the average distance to Sun, D, are presented in Table 1.16. As is clearly seen on Fig. 1.26, the planet points fit no line.

But, if one takes a look at the points corresponding to the logarithms of the features, they do not need to be a rocket scientist to figure out that the points do lie on a line and, thus, do satisfy the Kepler's equation—even if they had no prior knowledge of that (Qin and Simon 1990). Since Kepler himself was a great advocate of the concept of logarithm, of course he did it!—claim enthusiasts. Figure 1.27 represents a scatter plot over logarithms of the features, log(P) and log (D)—one cannot help but see that the points exactly fit a straight line with the tangent of the angle between the line and x-axis equal to 3/2. That is exactly the Kepler's equation, with the constant = 1. The Kepler's 3rd Law of Planetary Motion says just that.

Table 1.16 Data planetary
motion: the average distance
from the Sun, in Earth
distances, and the period, in
Earth periods

Planet	Period	Distance
Mercury	**0.241**	**0.39**
Venus	**0.615**	**0.72**
Earth	**1**	**1**
Mars	**1.88**	**1.52**
Jupiter	**11.8**	**5.2**
Saturn	**29.5**	**9.54**
Uranus	84	19.18
Neptune	165	30.06
Pluto	248	39.44

The fonts are boldfaced except for the three last rows referring to
a recent addition: these celestial bodies were not known in the
17th century

Fig. 1.26 The (D,P)-
scatterplot according to data
in Table 1.16

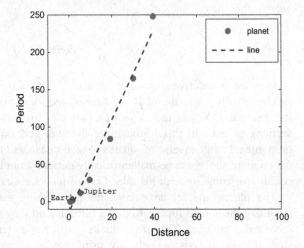

Fig. 1.27 Success: all 9
planet points fit the line
$y = 3/2x$ exactly, thus leading
to equation $P^2 = D^3$, in Earth
units

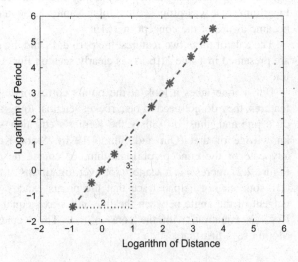

Fig. 1.28 Gravity law
illustrated

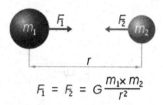

$$F_1 = F_2 = G\frac{m_1 \times m_2}{r^2}$$

What is so grand about the Kepler's Laws? Well, a simple fact is that they all can be mathematically derived from Newton's universal gravitation law presented in Fig. 1.28. The universal gravitation law is the most significant early result of modern science. It claims that there is an invisible force, called gravitation, between any two celestial bodies, that is proportional to the product of the bodies' masses and inversely proportional to the squared distance between them. Moreover, the gravitation force is immediate, however far away from each other the bodies are, with no delays. This law fits well with the classical mechanics.

One should not forget, though, that some other efforts by J. Kepler—along his favorite line, the Pythagorean perspective of the affinity between sounds and planetary motions—has the reader ever heard of "Music of Spheres"?,—even with regularities found in relations between the planet's number and its distance to Sun such as the "Bode-Titius Law" for some planets, were largely unsuccessful and remain outside of the body of science so far.

1.4.3   Failures of Data Analysis: Counter-Intuitive Cases

In many cases, patterns found with data analysis have no simple or useful explanation. In some cases they contradict a tradition or "ground truth" knowledge and, thus, fail in eyes of people. This author recalls, for example, results of the analysis of a huge—at that time—data file collected, in a 50,000-strong survey, by a Russian medical officer of health services V. Shanin, undertaken by the author and the late Peter Rostovtsev (1950–2002) in a big research center near Novosibirsk, Russia, in 1981. V. Shanin was concerned with the respiratory diseases that were abundant there at that time; some of them quite infectious. In accordance with the dominant views of the period, he suspected that there were two major risk factors: (a) smoking and (b) alcohol consumption; and he asked us to demonstrate that convincingly by analyzing his data.

To that end, we developed a classification of two dozen respiratory diseases that were present, in various combinations, in replies to the survey questionnaire. Since the dataset was real huge—it did not fit within a single computer's working memory—we used an in-house decision tree technique by Mirkin and Rostovtsev (see Mirkin 1985) for which the criterion was maximization of the association between the partition being built and the binary features representing different respiratory diseases. An account of the discovered partition is presented in Fig. 1.29; its major divisions were

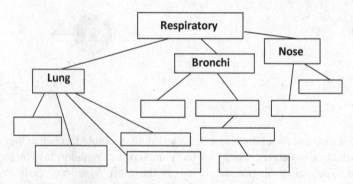

Fig. 1.29 The structure of hierarchical clusters obtained by B. Mirkin and S. Rostovtsev over two dozen respiratory diseases occurring among respondents in south-west Siberia 1981 in various combinations

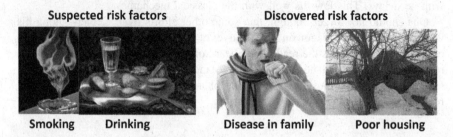

Fig. 1.30 Derived causes of respiratory diseases are on the right, wrongly suspected risk factors are on the left

related to lungs, bronchi, and the nose, respectively. Then we looked at the association tables between this partition and various features available either from data directly or by combining answers to individual questions.

The results obtained are illustrated in Fig. 1.30. The suspected risk factors were statistically independent—well, almost, in our computations, from the disease partition both on the entire dataset and various tested parts of it. The only exception, as far as I can remember, was the subset of relatively young women. It was a positive association in that case: the moderate drinking associated with better health. This, currently quite reasonable, result heavily contradicted the medical dogmas of that time and was pronounced to be nonsensical.

Among the data miners, another story—to an extent, similar—is popular; the story of the struggle by a local general practitioner doctor, John Snow, against an outbreak of cholera in Soho district, London, 1854, by using data visualization. Two weeks into the outbreak, Dr. Snow went over all houses in the vicinity and made as many tics at their locations on his map as many deaths from cholera had

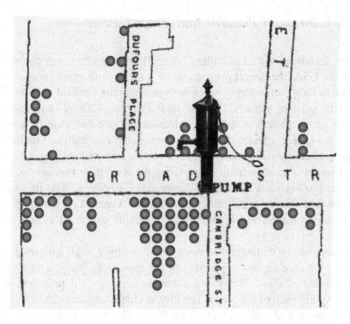

Fig. 1.31 A scheme of a fragment of Dr. Snow's map demonstrating that indeed most deaths (labeled by circles) had occurred near the water pump

occurred in that location (Figure 1.31 illustrates a fragment of Dr. Snow's map). The circles were densest near the water pump, which made Dr. Snow convinced that the pump was the cholera source.

(In fact, Dr. Snow's service in India predisposed him to the idea of contaminated water flows transmitting the disease.) He discussed his findings with the priest of the local parish, who then removed the handle of the pump, after which the deaths stopped. This all is true. But there is more to this story. The death did stop—because too few remained in the district, not because of the removal: the handle was ordered back on the very next day after it had been removed. Moreover, in a later discussion the borough council refused to accept Dr. Snow's "water pump theory" because it contradicted the then dominant theory that cholera progressed through stench in the air rather than through water flow. More people died in Soho a decade later in the next cholera outbreak. The water pump theory was not accepted until much later, when the science of microbes had become developed. Specifically, Dr. Koch in Germany 1883 discovered the cause and medium of infection—vibrio cholerae. The story is instructive in both the power of visual insight and the fact that data analysis results are not conclusive: a data based conclusion needs a reasonable explanation to get accepted [A strange coincidence: in fact, the vibrio cholera was first discovered by Filipo Pacini in Italy in the very same year, 1854, when John Snow fought for his findings in the UK; but the discovery was then ignored by the scientists as a kind of nonsense].

1.4.4   Insubstantial Patterns and Their Causes

In the first decade of the 21st century, the following pattern was discerned over schools in the USA: the largest proportions of students with excellent marks were in small schools rather than larger schools. Some parents decided then to put their children into smaller schools. Wainer and Zwerling (2006) came up with an explanation of the pattern as a feature of randomness rather than of any specific conditions at smaller schools. Their explanation also implied that smaller schools have larger percentages of students with low marks, too; so that there is no reason to want to study in a small school. Since the data of relative numbers of poor and excellent grades were not available, Wainer and Zwerling (2006) illustrated their point on the data of cancer of kidney occurrences. Figure 1.32 clearly shows that counties with the lowest and highest risks of getting cancer of kidney are in the same states.

These states can be distinguished as those prevailingly rural, with small numbers of residents, who are fervent supporters of the republican party, as they share the Christian fundamentalists' views and autarchic ideology. In principle, these are enough to explain each of the maps. The lowest kidney cancer risks? Of course, the rural life, clean air and water, simple fresh foods. The highest kidney cancer risks? That is quite clear, too: poverty, fat foods, alcohol, and low medical help. The only question remains: how can these incompatible explanations both be true?

In fact, they are both wrong. The reason is not the rural pattern of life. The reason is low population density. The smaller the population, the greater the probability of an unusual pattern. Take, for example 3 random balls from an urn with equal quantities of white and black balls. The color of a ball can be either black or white, with equal chances. Therefore, there are 8 possibilities for 3 balls, each with a probability of 1/8. The event that one ball is white, and two are black, occurs at each of the three combinations out of the eight: w-b-b, b-w-b, b-b-w. The event that all the balls are black occurs at only one of the combinations, b-b-b; therefore, its probability is 1/8. The same goes for the event at which all the three balls are white. If one takes 4 balls, then the probability of an "unusual event" that all four are black (or all four are white), is 1/16. If one samples 7 balls, then the probability

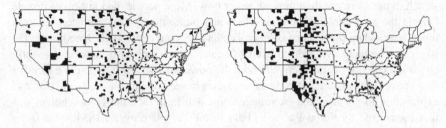

Fig. 1.32 Two copies of the map of the USA: on the left, counties at which the numbers of cancer of kidney occurrences was the lowest are shown as black; on the right, the same is for counties with the highest levels of cancer of kidney

of the unusual event is even smaller, 1/128 for all black, and 1/128 for all white. The greater the number of balls, the smaller is the probability of the unusual events. Consider counties as the sampled balls, the high level of cancer—as all black color, and the low level of cancer, as all white color. Then one may say that the probability of an unusual event is much greater in counties with smaller populations. That is the phenomenon behind the patterns presented in Fig. 1.32.

This gives an important lesson in data analytics: even if data analysis produces a pattern, one cannot immediately recommend that the user take advantage of the pattern. One must take care that the recommended behavior indeed leads to expected outcomes.

An insubstantial pattern may emerge not only by chance. There can be many other causes, of which one should mention:

    I. An unfortunate heterogeneity of the data set
   II. The dataset being non-representative or biased
 III. Correlation caused by unobserved features
 IV. Correlation caused by the research in data collection itself.

Even more fallacies may emerge when comparing different populations using sampling—but these are out of the scope of this text. It is even worse if the data values themselves have been "adjusted", "corrected" or plain changed.

Let us now give examples to each of the four fallacies above.

I. Heterogeneity of the data set

The effects of heterogeneity of datasets can be illustrated by what is called the paradox of G. U. Yule (1871–1951), with an extract from his paper (Yule 1903). The paradox concerns a sample with two binary attribute features. The sample consists of two subsamples so that the attributes are statistically independent in each of the subsamples. However, the attributes can appear highly related to an innocent user who views the set as a whole and has no knowledge of the subsamples (see Tables 1.17 and 1.18).

Consider an attribute which is inherited in neither female, nor male. Let, for example, fathers and sons have a 50% chance to have the attribute, whereas only 10% of mothers and daughters have that. Then the distribution of the attribute in each of the samples will be such as presented in Table 1.17.

Table 1.17 Illustrative data of an attribute which is inherited neither in men nor in women; however, it is present in these two subpopulations differently—only 10% in mothers and 50% in fathers

Father	Son+	Son−	Total	Mother	Daughter+	Daughter−	Total
Father+	25	25	50	Mother+	1	9	10
Father−	25	25	50	Mother−	9	81	90
Total	50	50	100	Total	10	90	100

Signs are to indicate whether the attribute is present or not

Table 1.18 Summary inheritance data

Parent	Child+	Child−	Total
Parent+	26	34	60
Parent−	34	106	140
Total	60	140	200

Signs are to indicate whether the attribute is present or not

Assuming that the values in Table 1.17 are numbers of respondents, one can build a combined table for parent-child inheritance just by adding the corresponding numbers in two gender-related tables (see Table 1.18).

Let us compare the conditional probability for a child to have the attribute, 26/60 = 0.433 and the probability for a person to have the attribute, 60/200 = 0.300. The former is greater than the latter (26/60)/(60/200) = 52/36 = 1.444 times, that is, by 44.4%. This is a rather high difference, which bears no sense.

A similar effect can be observed for a quantitative feature. At the dawn of modern era, Karl Pearson proposed a simple example of two bivariate Gaussian distributions, with zero correlation each, yet with a difference in means, so that the resulting distribution had a significant correlation between the variables.

II. **The dataset is non-representative**

Incorrect conclusions out of unrepresentative samples can be stressful. Usually no nation risks a significant error in determining its population by using surveys rather than census at least once a decade. A famous case of unrepresentative fallacy is the Literary Digest's poll of telephone subscribers and automobile owners in 1936 USA asking their voting preferences and, in this way, predicting a landslide victory of Mr. Alf Landon against the incumbent president Roosevelt on Election Day in 1936.

When the real election results, greatly contradicting those of the poll, became known, the poll organizers realized that the sample of better-off people were not representative of the opinions of the entire population, most of whom were suffering in the Great Depression. There are many other similar examples including the fact that the overwhelming majority of forecasts predicted that the president Donald Trump would lose in the US elections 2016.

III. **Correlation caused by unobserved features**

Examples of correlations manifesting a hidden explaining variable or two are currently increasing because many e-services provide data series, which can be used for matching other data series.

One of the latest services is Google Correlate. Given a keyword, it immediately finds, over a thousand-day window, the set of queries containing this keyword and computes the proportion of this set among all the queries. The series of proportions is the profile of this keyword within the universe under consideration. Then Google Correlate finds profiles of all other keywords it accumulated for the same thousand days, after which it computes correlation coefficients between the keyword profile and those other profiles. Then it outputs 10 other keywords for which the correlation coefficient is maximum, along with the values of the correlation coefficient. An example of the output is presented in Table 1.19. It is striking how different are profiles in the USA and the UK. Still, some of them seem reasonable while the others seem perfect examples of "false correlation". The case of profiles for "Data analysis" may serve as an example of the former, whereas the UK's profile for "Machine learning" is an example of the latter.

Every statistics textbook contains a few examples of "false correlation", that is, a high correlation value between features which are apparently unrelated. Among them one of the most popular is the correlation between the number of drowned swimmers and quantities of ice-cream sold. These two things are not related by themselves. However, their trends are highly affected by the weather: the hotter the Sun, the greater crowds go to the beaches and the greater crowds eat ice-cream.

Another association, frequently reported, so far has received no similar natural explanation, although popular beliefs do make it rather plausible. A high correlation is frequently mentioned between the numbers of newborn babies and the numbers of brooding pairs of storks allegedly observed in Sweden and Denmark in various periods, although with no documented evidence. However, for the post-war Germany, there is evidence of a drastic simultaneous decline in both (1965–1980), see a figure and short comment by Helmut Sies in Science, 1988, 332, p. 495. A bit less amazing but still quite valid correlation across 17 countries is noted by Robert Matthews (Storks deliver babies, Teaching Statistics, 2000, 22, no. 2, 36–38). A few more mysterious correlations:

- Social drinking and earnings—drinkers earn more money (see B. L. Peters and E. Stringham (2006) *No Booze? You May Lose: Why Drinkers Earn More Money Than Nondrinkers, Journal of Labor Research*, 27(3), 411–421).
- Chocolate consumption and the numbers of Nobel prize winners, both relative to the population size (see F. H. Messerli (2012). Chocolate consumption, cognitive function, and Nobel laureates, *The New England Journal of Medicine*, *367*(16), 1562).
- The numbers of McDonald's restaurants (per million residents) and the proportion of overweight individuals (see A. Alheritiere, S. Montois, M. Galinski, K. Tazarourte and F. Lapostolle (2013). Worldwide relation between the number of McDonald's restaurants and the prevalence of obesity. *Journal of internal medicine*, *274*(6), 610–611).

Several interesting correlations between features of countries are observed by Roberts and Winters (2013) including a positive association between the linguistic diversity of a nation and the numbers of road fatalities per 1000 habitants.

IV. Correlation caused by the data collection itself

Let us turn now to cases of correlation caused by the fact of data collection itself. This phenomenon became apparent at first in quantum physics—as the so-called

Table 1.19 Results of four queries (in columns) to Google correlate in the USA (top) and the UK (bottom)

	Data analysis	Data mining	Machine learning	Statistics
USA	0.9333functional	0.9746java code	0.9846cloudformation	0.9857education in
	0.9302secondary	0.9721c++ code	0.9834o365	0.9838education statistics
	0.9147measurementof	0.9718c code	0.9823npm package	0.9800drosophila
	0.9116linear	0.9715modulation	0.9821pandas dataframe	0.9791disorders
	0.9114criteria	0.9709c++	0.9820aws cloudformation	0.9775economics of
	0.9113data sets	0.9691statistical	0.9812read_csv	0.9767drug abuse
	0.9112and services	0.9678databases	0.9812webhook	0.9761case studies
	0.9109function	0.9662implementation	0.9810cloudwatch	0.9746critique
	0.9100to conduct	0.9654computation	0.9808node_modules	0.9742prevention of
	0.9093context	0.9652and international	0.9807okta	0.9735opinions on
UK	0.8823determine	0.9456implementing	0.9660unsplash	0.9882industry
	0.8797correlation	0.9440statistical	0.9660pixabay	0.9868diagrams
	0.8790method	0.944e-commerce	0.9650nice cks	0.9848production
	0.8729interactions	0.938strategic	0.9647digital transform.	0.9845organisations
	0.8701collaboration	0.938implementation	0.9637spring boot	0.9832advertisements
	0.8673annual report	0.937en france	0.9633giphy	0.9821research
	0.8627patients	0.937quality management	0.9617industry 4.0	0.9820statistics for
	0.8624response	0.936remote sensing	0.9615aws cli	0.9816circuits
	0.8605stress	0.936knowledgemanagment	0.9612o365	0.9807developing
	0.8594policy	0.935design of	0.9605sharepoint login	0.9800databases

The table presents keywords whose profiles are most similar to profiles of the query words principle of complementarity leading to impossibility of simultaneously measuring, say, both the location and velocity of nuclear objects. As N. Bohr has famously put it: "A complete elucidation of one and the same object may require diverse points of view which defy a unique description." (N. Bohr (1934). *Atomic theory and the description of nature* (Vol. 1). CUP Archive.). Such phenomena are readily available in economics and sociology studies. One of the most interesting is the case of the Great Depression crisis of world economics (1929–1933 or later), There are self-consistent, and of course incompatible, explanations of that: (a) in terms of the Keynesian demand-driven theory (the trigger: loss of confidence and, hence, under-consumption); (b) in terms of monetarist approaches (the trigger: a wrong policy of the Federal Reserve causing the money supply to shrink and, in this way, a usual recession to slip into an overwhelming depression); (c) Marxist class war views (trigger: a great wave of unemployment caused by the greediness of capitalists in the wake of technology revolution caused by the introduction of electricity), etc. Also, one of the well-known is the experimental study of the worker productivity (1924–1932) in the Hawthorne Works in Illinois. It appeared that the members of the control group changed their behavior because they learned of their role in the experiment rather than because of conditions of the experiment (see a review in Adair 1984). Quite famous, although short-lived, are cases of pre-election polls predicting a party to win the ballot and, therefore, triggering the opposition to come to ballot box in masses and turn the voting around to get it their way.

Summing up, the cases considered are yet another warning of caution when turning the data analysis results into the data analytic recommendations. To make a pattern to serve as a basis for recommendation, one should propose a mechanism for the pattern emergence and functioning. The mechanism should be compatible with the existing views of the phenomena in question. Otherwise, one should be prepared to wait for the times when attitudes would change. This may take years or dozens of years. The task of formulation of such a mechanism may require knowledge of specifics of the phenomena under consideration. Therefore, specialists in the specifics of the phenomenon to which the data relate should participate in the process of data analysis.

1.5 Knowledge Shaped as Classification

1.5.1 Classification as Soft Knowledge

At the time of writing the concept of knowledge is neither well understood, nor well defined, even though it traces its origins thousands years back and was decisively promoted through ages starting from two of the greatest ancient philosopher-scientists, knowledge reformers, Pythagoras (c570–c495 BC) and Aristotle (384–322 BC). The latter developed a theory of classification which remains relevant to this day.

Table 1.20 The three-fold vertical world view top/middle/bottom by Scythians (VII–II centuries BC) according to D. S. Rayevsky (1985) World Pattern in Scythian Culture, Nauka Publishers, Moscow, 256 p. (in Russian)

Top	Sky	Leaders	North	World of Gods
Middle	Earth	Priests	Midlands	World of alive
Bottom	Water	Farmers	South	World of dead

Scientific theories in sciences such as mechanics or molecular chemistry are examples of what can be called "hard" knowledge. Hard knowledge relies on reproducible experimental observations and mathematical deductions. Much more phenomena in the world are covered by what can be referred to as "soft" knowledge. Soft knowledge includes observations of unreproducible phenomena involving features whose role and associations are unclear such as in economics and politics. Moreover, soft knowledge may include beliefs and subjective judgments. Much of soft knowledge is shaped in the form of classification.

In Table 1.20, an example of soft knowledge is provided. A "vertical" three-fold classification of the world phenomena is presented along with parallel divisions in space, society, etc. It is claimed that the classification can be derived from images and artifact remains attributed to the Scythians, a nomadic people roaming 25 centuries ago over the steppes surrounding the Black Sea in East Europe. The classification, as well as our knowledge of it, is rather incomplete. Still, it can be of interest to the modern-day human as a rationale for preferring three-fold classifications in our world-views.

Figure 1.33 presents another classification, much "harder" than the vertical Scythian classification, a classification of quadratic algebraic equations $x^2 + px + q = 0$ according to the properties of their roots, that is, values x at which the equation holds indeed, expressed in terms of the coefficients p and q. This classification summarizes 25 centuries of the development of mathematics completed by the genius K. F. Gauss (1777–1855) who developed, among many things, a theory of complex numbers. A parameter, crucial for distinguishing between the case at which both roots are real and the case at which both roots are complex, is the sign

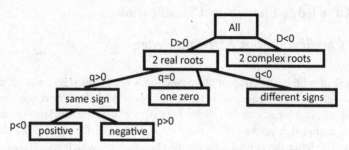

Fig. 1.33 Classification of the roots of the quadratic equations $x^2 + px + q = 0$ expressed in terms of the coefficients p and q

Fig. 1.34 An outline of the contemporary biological taxonomy involving three major divisions (Bacteria, Archae, and Eucaryota)

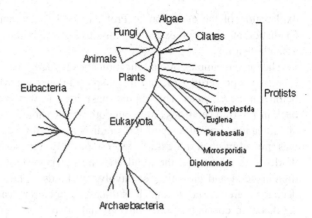

of discriminant $D = p^2 - 4q$, since the equation $x^2 + px + q = 0$ has two roots, $x1$ and $x2$, computed according to formulas $x1 = (-p + D^{1/2})/2$ and $x2 = (-p - D^{1/2})/2$ (at $D = 0$, the roots coincide, $x1 = x2 = -p$—this trivial case is omitted here from the classification). Depending on the sign of q, the case of two real roots, $D > 0$, can be distinguished at the next level of the tree on Fig. 1.33. Next layer is occupied by the sign of p—this affects the common sign of two real roots at the branch defined by conjunction $D > 0$ and $q > 0$.

Let us take a real-world classification—that of living organisms by Carl Linnaeus (1707–1778) who classified plants and animals based on similarity between their reproductive organs. In fact, he continued a long line of taxonomist efforts starting probably with Aristotle himself. Linnaeus name is attached to the biological taxonomy, in spite of the fact that its modern version (see Fig. 1.34) much differs from the Linnaeus' version, probably because both apply the so-called binomial nomenclature, another idea which can be traced back to Aristotle. The largest difference comes from two Kingdoms of bacteria invisible to the human eye, Eubacteria and Achaea, which currently have been added to the root of Linnaean taxonomy.

The binomial nomenclature assigns each biological taxon T with the name of the "parental" taxon, genus, P to which T belongs, accompanied with a T identifier. The idea of binomial nomenclature is universal and practical: a great accomplishment by itself.

However, the biological taxonomy obtained its scientific value only after Charles Darwin (1809–1882) interpreted it as a system describing the evolution of living organisms. This was a groundbreaking idea leading to a complete revamp of the picture of the world and its history, from the creationism of the Bible to a more balanced view of millions of years of natural history, in which the Sun and Earth are but ordinary exemplars of celestial objects.

The picture of biological evolution outlined in Fig. 1.34 is not yet complete (see Koonin 2011). There are many blank spots and gaps such as the issues related to the explanation of the evolution of Protista, the bacteria and the like, or the role of Cladistics, an approach explaining any divergence in the species tree by changes in a single feature.

The three examples of classification above clearly demonstrate that classification is indeed a powerful structure for shaping knowledge which does not necessarily belong in the sciences. This should make it clear, why the incipient age of science back in 17–18th centuries threw out both, the Aristotelian claims and the concept of classification, from the set of tools indispensable for science. The rigor of science was probably then necessary for establishing the scientific perspective at the worldviews. Currently, the trend goes in the opposite direction. The processes of digitalization and globalization involve all kinds of knowledge since all kinds of decisions are to be taken. It suffices to mention various relaxations of the set-theoretic concepts, from conventional crisp sets to probabilistic density functions to fuzzy sets to possibilistic sets to hesitant fuzzy sets.

1.5.2 Goals of Classification

The classification is among the most useful concepts in the emerging field of knowledge discovery, maintenance and engineering.

Generally, *classification* is considered *a form of soft knowledge* that involves a division of objects under consideration in similarity groups with any or all of the following goals:

   I. Analysis of the structure of the phenomenon under consideration;
  II. Establishing relations between different aspects of the phenomenon;
 III. Forming and shaping knowledge.

One cannot help but realize that each of the three above-mentioned classifications concerns all three of the goals above. Say, the biological taxonomy establishes the structure of the living matter, points to the relation between the structure and evolution of life, and, in this way, shapes and preserves knowledge of the phenomenon.

When discussing various sides of phenomena and processes, it is useful to distinguish between the following five facets covering their major aspects:

- Structure
- Function
- History
- Attitude
- Action.

One should note that whereas the triplet structure-function-history may relate to any phenomenon, the latter two facets, attitude and action, make sense as related to social systems only.

Here are three well-known scientific classifications relating some of the five facets to each other: structure to function, structure to history, and structure to attitude/action.

Structure-to-Function: Periodic Table of the Elements
The periodic table of elements is a celebrated example of a quite profound classification because it links together four facets of the elements: the internal structure of the atoms, their combination into molecules, their chemical interaction properties, and their physical features. The version proposed by D. Mendeleev (1834–1907) in 1869 (see https://www.slideshare.net/dianekinney/history-of-the-periodic-table-of-elements) is presented in Fig. 1.35. One can see a number of question marks—these elements were unknown at that time, but Mendeleev was able to correctly predict not only their existence, but many properties as well. The periodic table "provides a stimulus and a guide in chemical research, constantly suggesting as it does new experiments to be tried and providing a basis for critically evaluating and checking information already obtained.... The very existence of the Periodic Law as an empirical principle provided a tremendous stimulus to the development of our knowledge of atomic structure and greatly accelerated the growth of our understanding of the relationship of the structure and the properties of matter." (Sisler 1973, p. 34).

Group / Period	I	II	III	IV	V	VI	VII	VIII
1	H=1							
2	Li=7	Be=9.4	B=11	C=12	N=14	O=16	F=19	
3	Na=23	Mg=24	Al=27.3	Si=28	P=31	S=32	Cl=35.5	
4	K=39	Ca=40	?=44	Ti=48	V=51	Cr=52	Mn=55	Fe=56,Co=59 Ni=59
5	Cu=63	Zn=65	?=68	?=72	As=75	Se=78	Br=80	
6	Rb=85	Sr=87	?Yt=88	Zr=90	Nb=94	Mo=96	?=100	Ru=104,Rh=104 Pd=106
7	Ag=108	Cd=112	In=113	Sn=118	Sb=122	Te=125	J=127	
8	Cs=133	Ba=137	?Di=138	?Ce=140				
9								
10			?Er=178	?La=180	Ta=182	W=184		Os=195,Ir=197 Pt=198
11	Au=199	Hg=200	Tl=204	Pb=207	Bi=208			
12				Th=231		U=240		

Fig. 1.35 A version of the periodic table of elements proposed by D. Mendeleev in 1869 (translated from German)

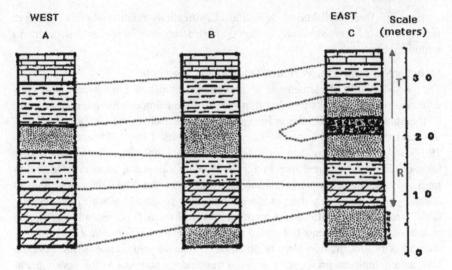

Fig. 1.36 Similarity between different coastal areas can be established according to the sequence of strata

Structure-to-History: Geological Correlation

In geology, the rocks and soils are usually organized in layers (strata) that rather clearly differ from each other. Any geological layer is naturally considered older than the layer just above because of an obvious observation that layers are formed by geological processes on top of the layers already existing. This is referred to as the "law of superposition" in geology. When originally formed, the strata were laterally continuous; the fossils found in the rock are remains of the organisms living in the time of formation of the rock or soil. A general method, "the geological, or stratigraphical, correlation", both is based on these principles and supports them (see Fig. 1.36).

This method has led to discovery of many oil or coal deposits, as well as to some theoretical breakthroughs, like periodization of the geological epochs from the times unknown about 4.5 billion years ago, to the Tertiary period (65 million years ago) to the current Quaternary period (started 1.6 million years ago), or development of the concept of continent drift.

Structure-to-Attitude-and-Action: "Class Struggle"

Karl Marx (1818–1883) proposed that the structure of any society is determined by social groups, or classes, that have different rights regarding the means of production of industrial and agricultural goods. In particular, he referred to two most important classes, the capitalists, the owners of the means of production, and the workers, the individuals who are employed to produce goods using the means of production. The workers produce surplus value, which is then distributed among all the classes. According to Marx, the lion's share of the surplus value goes to capitalists as the owners of the means of production. The greater the capitalists' share, the smaller the share of the workers. Therefore, according to Marx, the capitalist way of production necessarily leads to antagonistic class struggle between workers and capitalists. The only way to restore the justice is via a radical social action by the workers. "Let the ruling classes tremble at a Communistic revolution. The proletarians have nothing to lose but their chains. They have a world to win." (K. Marks, F. Engels, Manifesto of Communist Party, 1847). The reader of this book is well aware that out of all major nations, only two undertook a revolutionary action to achieve a fairer distribution of the national wealth, Russia (1917) and China (1949), although that was somewhat contrary to Marx's views, since both countries were overwhelmingly rural, with insignificant numbers of workers in industrial enterprises. This has happened because of fierce propaganda of the communist parties to assign farmers sometimes to the working class or sometimes to the bourgeoisie (petty capitalists) depending on political ends to achieve (and, of course, due to the help of the Soviet Army, in the case of China).

Even if Marx's views were based on any reality back in the XIX century, currently they have no substance at all since observations of the intergenerational mobility in such countries, as the USA and UK, give little evidence of stability in the social class structure from fathers to sons. Moreover, it is much clear by now that the volume of surplus value created in a society depends not only on those who are directly involved in production lines, but also the managers, banks, social and health infrastructure, etc., which make the claim of interclass antagonism rather odd. Nevertheless, throwing out the idea of class altogether currently seems a bit premature, since this concept is useful in the analysis of social inequality and stratification.

This is why the sociologists, especially Pierre Bourdieu (1930–2002) expanded the concept of capital to add social and cultural dimensions to the economic one. These, unfortunately, have not made it into the social or governmental statistics as yet. Here we describe an approach taken by Savage et al. (2013) in pursuing this idea with a survey conducted by BBC Lab UK in 2011.

To measure social capital, M. Savage et al. analyzed how well various occupations were known to respondents. Specifically, they used the mean status score of the occupations that respondents knew and the mean number of social contacts reported. To measure cultural capital, two indexes have been used: (1) the "highbrow" capital based on the extent of respondents' engagement with classical music, attending stately homes, museums, art galleries, jazz, theatre and French restaurants; (2) the 'emerging' cultural capital based on the extent of respondents' engagement with video games, social network sites, the internet, playing or watching sport, spending time with friends, etc. Measures of economic capital involved household income, household savings and house price. To structure the sample over these indexes, Savage et al. used method of "latent class structuring" invented by Paul Lazarsfeld (1901–1976), one of the first advocates of application of mathematics in social research. In contrast to clustering methods described further on, this method tries to find such clusters that are statistically homogeneous in the following sense: In each of the clusters, the features used for clustering are statistically independent. This criterion differs greatly from our clustering criteria

Table 1.21 Summary of social classes found by Savage et al. (2013); the meaning of marks: 7—highest, 6—very high, 5—high/good, 4—moderately good, 3—moderately poor, 2—low/poor, 1—lowest

Class	Percent	Capital marks			
		Econ.	Soc.	Cultural	
				High	Emergent
Elite	6	7	6	7	4
Established middle class	25	6	5	6	6
Technical middle class	6	6	6	4	4
New affluent workers	15	4	3	3	5
Traditional working class	14	3	2	2	2
Emergent service workers	19	3	3	2	5
Precariat	15	2	1	1	1

that require clusters to be homogeneous in a very different sense: All objects in them must be similar to each other. A summary of latent structure clusters found is given in Table 1.21. The nomenclature includes a rather recent term, Precariat, introduced in Standing (2011) to mean a class of people whose employment and income are insecure.

Q.1.9. What facets are related by the biological taxonomy?
A. Out of the five facets mentioned above (Structure, Function, History, Attitude, and Action), the biological taxonomy relates the structure of living organism to their history—the evolution, according to Ch. Darwn's theory. Also, the Function facet is involved, as related to reproduction of the living matter, since K. Linnaeus put that as the base of his classification.
Q.1.10. What facets are related by the Periodic Table?
A. The Periodic Table relates Structure and Function in both, Physics and Chemistry, perspectives. The structural perspective concerns both within-atomic properties and combining atoms in molecules.

Obviously, these clusters still bear a traditional industrial flavor, but they also involve features related to the types of capital and, as well, avoid traditional divisions between "manual–not-manual" and "blue-collar–white-collar". Instead, they point to an important role of services and "technical–non-technical" division. Also, two significant groups are distinguished on the extremes of the social ladder, the elite and precariat.

Another example of a classification to relate the structure and attitude is presented in Fig. 1.37. Many know—but few are serious—about a hands-on typology of fingernails, which is claimed to reflect important personality traits (see Fig. 1.37).

To complete this section, let us mention a highly universal classification relating all the five aspects from Structure to Action above, although expressed in an ancient terminology using the Chinese language and system including the so-called "yin and yang", the two parts of the whole, as shown in Fig. 1.38. This system, called I

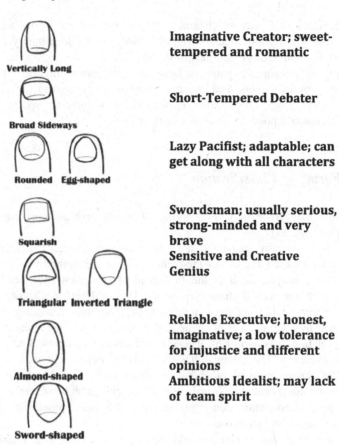

Vertically Long — **Imaginative Creator; sweet-tempered and romantic**

Broad Sideways — **Short-Tempered Debater**

Rounded Egg-shaped — **Lazy Pacifist; adaptable; can get along with all characters**

Squarish — **Swordsman; usually serious, strong-minded and very brave**
Sensitive and Creative Genius

Triangular Inverted Triangle

Almond-shaped — **Reliable Executive; honest, imaginative; a low tolerance for injustice and different opinions**
Ambitious Idealist; may lack of team spirit

Sword-shaped

Fig. 1.37 Seven types of fingernails and related personality features according to http://www.yourchineseastrology.com/palmistry/fingernail/ (accessed 29 June 2017)

Fig. 1.38 Eight trigrams of I Ching along with their meanings (from http://www.hiddentreasuresministries.org/i-ching.html)

Ching (The Book of Changes), classifies the universe and its various aspects by using a set of eight trigrams; each of trigrams being a set of three binary 1/0 digits represented by solid and broken lines, respectively.

Using the eight trigrams in pairs, the hexagrams, one can distinguish between 64 different states of the universe and its aspects, and use hexagrams cast by chance as a kind of oracle, suggesting various relations between different aspects of one's current situation and possible outcomes of their actions.

1.5.3 Forms of Classification

Classification may come in various forms, of which perhaps most popular are taxonomy and typology.

Taxonomy is usually shaped as a set of classes structured like a hierarchy represented by a rooted tree so that the higher ranks of the hierarchy correspond to more general concepts. Such is the biological classification mentioned above. Similarly, a library classification such as the Dewey Decimal Classification or Universal Decimal Classification structures book collections according to a hierarchy defined by relation "a is b". Note that, in contrast, the biological taxonomy accords to relation "a is part of b". This corresponds to the empirical nature of the biological taxonomy. The Decimal classifications are conceptual and rather speculative. The classification of quadratic equations (see Fig. 1.33 above) is also conceptual, but not speculative. It makes sense, to distinguish between two cases of the hierarchical classification: conceptual classification and empirical classification, that is taxonomy in the proper sense.

The hierarchical classification is playing a growing role as a backbone of ontology, a newest form of computationally feasible knowledge organization.

Typology is another classification structure referring to a number of types that are not hierarchical but rather of a similar level of granularity. The set of types is not required to be exhaustive; other types can be added to reflect novel analysis or novel facts. Types can be presented in different ways depending on the level of theoretical thought of the phenomenon in question. Say, in mineralogy, a type, referred to as "stratotype", is represented by an actual piece of rock or soil as an exemplar to be used for comparisons. Equally empirical, yet a bit more conceptual, are types described in words, like types of novel (see, for example, Encyclopedia Britannica, https://www.britannica.com/art/novel/Types-of-novel) or the type of "real man" among men or of "fraudster" among bank customers.

A typology is deemed more theoretical if its types are defined according to a feature forming an exhaustive system of disjoint characters over the universe under consideration. Such is the celebrated typology of social action by Max Weber (1864–1920) who claimed that there can be four different causes for social action: a goal, a value, an emotion and customs. Accordingly, Weber's typology involves four pure, or ideal, types: rationally purposeful, value rational, affective, and traditional ones.

Similarly, the ancient typology of temperament is based on the theory that there are four main fluids in the human body: blood, yellow bile, black bile, and phlegm. Normally they should be well mixed; but there can be pure cases at which one of the fluids dominates. As a theory of the day claimed, the human temper was determined by the fluids. Therefore, four ideal types of temperament are defined: "sanguine", "choleric", "melancholic", and "phlegmatic", correspondingly (see Fig. 1.39). The words relating to the four fluids are of Latin origin.

A modern typology of temperament aspiring to be known as a theoretical one is defined by two or more features in such a way that its types correspond to all possible combinations of the feature values. B. D. Nebylitsyn (1930–1972) proposed that the four types are characterized by the speed and strength of mental processes. Namely, the strong mental reaction types are choleric and phlegmatic, whereas fast mental reaction types are choleric and sanguine (see Fig. 1.39). Then the four temperament types correspond to all four combinations of values + and − of the speed and strength features. A typology involving all combinations of the feature values is referred sometimes as a faceted classification.

The classification of quadratic equations on Fig. 1.33 can be converted into a faceted classification by using all combinations of signs of the three features involved, D, p and q.

Another example of a theory-based typology: classification of snowflakes by Libbrecht (2004) who enthusiastically claims that collecting photographs of snowflakes is as interesting activity, as stamp collecting; and by far more useful!

There are dozens of snowflake classes. The shapes are determined by negative feedback processes running while snowflake grows (Libbrecht 2004). The main characteristics of the phenomenon are the levels of cold and humidity in the cloud measured by the temperature and supersaturation on the so-called Morphology Diagram in Fig. 1.40.

Fig. 1.39 Correspondence between the "fluidal" definition of temperament types and the modernized faceted one

Ancient temperament types as a faceted classification

Mental processes:		SPEED	
		+	−
STRENGTH	+	Choleric	Phlegmatic
	−	Sanguine	Melancholic

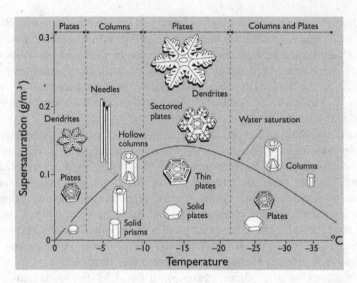

Fig. 1.40 The morphology diagram according to Libbrecht (2004): snowflake forms much depend on the relation between temperature and the water vapor and droplets in a cloud. The curve on the figure shows the boundary between the water oversaturated states and water under-saturated states. The basic shapes form within the vertical strips

Let us elaborate a bit over the typology formats mentioned above. Consider a set of features x1, x2, ..., xp, potentially applicable to objects of set 1. These can be, for example, occupation of survey respondents or an element of a computer communication protocol or speed of nervous processes (in Fig. 1.39). Assume that each of them has a finite number of values, and denote the set of values, or categories, of feature xn by Vn (n = 1,2, ..., p).

There are many ways for forming a multivariate classification here:

(i) A faceted classification is defined as the classification over the Cartesian product of sets Vn, V = V1 × V2 × ... × Vp, so that classes one-to-one correspond to p-series v = (v1,v2, ..., vp) ∈ V and, in fact, are defined by them.

(ii) A peaked classification is defined by specifying a single value vn* at each Vn (n = 1, 2, ..., p) so that its n-th class is defined by the set V(vn*) of such p-series (v1, v2, ..., vp) ∈ V at which vn = vn*, whereas vl may be any vl ∈ Vl for any l ≠ n (l = 1, 2, ..., p).

An example of the faceted classification is four temperament types defined by all combinations of ± values of speed and strength of reaction features. In contrast, the "fluid" based temperament types form a peaked classification. Indeed, the types are defined here by the over-presence of one of the four liquids which are the features in this definition.

It should be noted that the concept of faceted classification gained popularity after R. S. Ranganathan applied that in library science in 1933 (see a later edition of his seminal book Ranganathan 2006). His innovative approach involved a uniform set of five facets for any specific aspect of reality:

- Personality (the core point)
- Matter (basic elements/materials/products)
- Energy (action and interaction of persons or objects)
- Space
- Time.

An example of Ranganathan's class from Encyclopaedia Britannica: "The category of dental surgery, for example, symbolized as L 214:4:7, is created by combining the letter L for medicine, the number 214 for teeth, the number 4 for diseases, and the number 7 for surgery (https://www.britannica.com/science/Colon-Classification)."

Two other typical formats of typology:

(iii) Ranking, based on one of the features whose set of values is ordered;
(iv) Specifying a set of points in the feature space, v^1, v^2, ..., $v^K \in V = V1 \times V2 \times ... \times Vp$, to be used as explicit combinations of feature values, each expressing a type.

When talking of ranking, that can be ad hoc, as for example, for boxing weight divisions created to reduce the effect of weight differences at fist-fighting. There are several traditions; one of them, Amateur Weight Division, currently contains 11 categories such as the Lightweight (from 132 to 140 lb), Welterweight (from 152 to 164 lb) and the Super Heavyweight (over 201 lb). Or, that can be coordinated with many a human activity, such as soil classification over particle sizes, from 0 to over 200 mm, presented on Fig. 1.41. As to the type points typology, take examples such as stratotype and the literary character type. One more example comes from the health psychology. Some distinguish between contrasting type A and type B personalities. Type A relates to a personality that is competitive, highly organized, ambitious, impatient, highly aware of time management and aggressive. Type B comprises more relaxed personalities.

Table 1.22 summarizes forms of classification mentioned above.

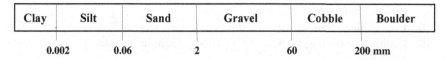

Clay	Silt	Sand	Gravel	Cobble	Boulder

0.002	0.06	2	60	200 mm

Fig. 1.41 Classification (ranking) of soils over particle sizes (according to http://environment.uwe.ac.uk/geocal/SoilMech/classification/default.html)

Table 1.22 Forms of classification as soft knowledge

Hierarchy	Typology
Taxonomy	Faceted
Conceptual system	Peaked
	Ranking
	Type pointed

Some scientists distinguish between "natural" and "artificial" classifications. According to this view, the former express laws of nature, as, for example, the Periodic Chart, whereas the latter are created for human use, as, for example, any sports classification. Unfortunately, this view is not easy to operationalize; the recent essays collected by Kendig (2015) do not lead to any reasonable computationally feasible definition. The only thing certain is that in a natural classification, having established the position of an entity within the classification, a great deal of information of principal features and tendencies of its behavior becomes available.

Overall, classification is ubiquitous and much important in all walks of life. Just a few more examples:

- In technology: the systems of technological standards which channel all the industrial manufacture and production;
- In education: the nomenclature of specialties both in subject and level; university ranking tables;
- In politics: the international blocks; political parties;
- In internet: standard menu systems; social networks.

Nevertheless, the business of designing and improving classifications has not made it to the mainstream as yet. It seems that classification research is not considered by the research community as valuable as exploring natural phenomena. Perhaps this is why such outstanding researchers as D. Mendeleev, with his Periodic Table, and C. Woese, with his groundbreaking Archaea domain in the Tree of Life (see, for example, Woese 1987), have been rejected by the Nobel Prize committees.

The contents of this section bring forth the following conclusion. Clustering, that is, activity of finding cohesive groups in data, should be considered as part of classification activities. Therefore, it should be used for any or all of the three classification goals:

I. Structuring the data to judge of the structure of the phenomenon under consideration;
II. Establishing relations between different feature sets;
III. Forming and shaping knowledge, first of all, as cluster partitions and hierarchies.

Systematic methods for doing that are yet to be developed.

1.6 Summary

This Chapter introduces four core problems in data analysis as related to either (i) summarization or correlation, in either (ii) quantitative or categorical way. The problems reflect the structure of theoretical knowledge as comprised, first of all, of concepts and statements of relation among them. Each of these four will be given a specific attention in the text further on. After covering quantitative summarization method of Principal Component Analysis and its derivatives, Chap. 2, we will move on to problems of correlation in Chap. 3, both for quantitative data (regression, Sects. 3.2–3.5) and categorical data (Sects. 3.6–3.9).

The Chapter also presents several small real-world data sets and related data analysis problems as pertained to (1) purely quantitative entity-to-feature data, (2) mixed scale data tables, and (3) square similarity data matrices.

The final part of the Chapter is devoted to three controversial subjects, which have not yet found any clear-cut or customary solution and remain challenges for the future. They are:

- Data visualization and its role as a data analysis tool and result;
- Data analysis strategic issues related to the need in theoretical substantiation of found patterns;
- Classification, its goals and structures as soft knowledge specifically related to cluster analysis;

No data science practitioner may ignore them.

References

M. Berthold, D. Hand, *Intelligent Data Analysis* (Springer, New York, 2003)

S.K. Card, J.D. Mackinlay, B. Shneiderman, *Readings in Information Visualization: Using Vision to Think* (Morgan Kaufmann Publishers, San Francisco, CA, 1999). ISBN 1-55860-533-9

R.O. Duda, P.E. Hart, D.G. Stork, *Pattern Classification*, 2nd edn. (Wiley-Interscience, 2012). ISBN 0-471-05669-3

J.F. Hair, W.C. Black, B.J. Babin, R.E. Anderson, *Multivariate Data Analysis*, 7th edn. (Prentice Hall, 2010). ISBN-10: 0-13-813263-1

S.S. Haykin, *Neural Networks*, 2nd edn. (Prentice Hall, 1999). ISBN 0132733501

C. Kendig (ed.), *Natural Kinds and Classification in Scientific Practice* (Routledge, Oxford, 2015). ISBN 9781848935402

J. Kepler, *Harmonies of the World* (originally 1619, translated to English by C.G. Wallis, 1939) (Global Grey Publisher, 2014). (E-publication http://www.24grammata.com/wp-content/uploads/2014/08/Kepler-Harmonies-Of-The-World-24grammata.pdf. Accessed 11 June 2017)

E.V. Koonin, *The Logic of Chance: the Nature and Origin of Biological Evolution* (FT Press Science, 2011)

S.D. Levitt, S.J. Dubner, *Freakonomics* (William Morrow, New York, 2005). See also a free extension in about 300 episodes in voice and print: http://freakonomics.com/archive/. Accessed 17 June 2017

K. Libbrecht, *The Snowflake: Winter's Secret Beauty* (Voyageur Press, 2004)

R. Mazza, *Introduction to Information Visualization* (Springer, New York, 2009). ISBN: 978-1-84800-218-0

B. Mirkin, *Methods for Grouping in SocioEconomic Research* (Finansy I Statistika Publishers, Moscow, 1985). (in Russian)

B. Mirkin, *Mathematical Classification and Clustering* (Kluwer Academic Press, 1996)

B. Mirkin, *Clustering: a Data Recovery Approach* (Chapman & Hall/CRC, 2012). ISBN: 1-4398-3841-9

S.R. Ranganathan, *Colon Classification* (Ess Ess Publications, 2006). ISBN-10: 8170004608

H.H. Sisler, *Electronic Structure, Properties, and the Periodic Law* (Van Nostrand Reinhold Company, 1973)

G. Standing, *The Precariat—the New Dangerous Class* (Bloomsbury Academic, London, 2011)

J.W. Tukey, *Exploratory Data Analysis* (Addison-Wesley, Reading MA, 1977)

C. Ware, *Information Visualization: Perception for Design*, 3rd edn. (Elsevier, 2012). ISBN: 978-0-12-381464-7

Articles

J.G. Adair, The Hawthorne effect: a reconsideration of the methodological artifact. J. Appl. Psychol. **69**(2), 334–345 (1984)

H. Brody, M.R. Rip, P. Vinten-Johansen, N. Paneth, S. Rachman, Map-making and myth-making in Broad Street: the London cholera epidemic, 1854. Lancet **356**(9223), 64–68 (2000)

W.S. Cleveland, Graphical methods for data presentation: full scale breaks, dot charts, and multibased logging. Am. Stat. **38**, 270–280 (1984)

R. Fisher, The use of multiple measurements in taxonomic problems. Ann. Eugenics **7**, 179–188 (1936)

S. Henikoff, J. Henikoff, Amino acid substitution matrices from protein blocks. PNAS USA **89** (22), 10915–10919 (1992)

M.I. Jordan, T.M. Mitchell, Machine learning: trends, perspectives, prospects. Science **349**(6245), 255–260 (2015)

G. Keren, S. Baggen, Recognition models of alphanumeric characters, *Perception and Psychophysics,* **29**(3), 234–246 (1981)

Y. LeCun, Obstacles on the path to AI (2015), https://drive.google.com/file/d/0BxKBnD5y2M8NbWN6XzM5UXkwNDA/view?pli=1. Accessed 6 Feb 2018

B. Lee, N.H. Riche, P. Isenberg, S. Carpendale, More than telling a story: a closer look at the process of transforming data into visually shared stories. IEEE Comput. Graph. Appl. **35**(5), 84–90 (2015)

S. Machlis, 22 free tools for data visualization and analysis (2017) https://www.computerworld.com/article/2507728/enterprise-applications/enterprise-applications-22-free-tools-for-data-visualization-and-analysis.html?page=1. Accessed 14 Jan 2018

B. Mirkin, Summary and semi-average similarity criteria for individual clusters, in *Models, Algorithms, and Technologies for Network Analysis*, ed. by B. Goldengorin, V. Kalyagin, P. Pardalos (Springer, New York, 2013), pp. 101–126

Y. Qin, H.A. Simon, Laboratory replication of scientific discovery processes. Cogn. Sci. **14**(2), 281–312 (1990)

R. Rao, S.K.Card, The table lens: merging graphical and symbolic representations in an interactive focus+ context visualization for tabular information. In *Proceedings of the ACM SIGCHI conference on Human factors in computing systems,* pp. 318–322 (1994)

S. Roberts, J. Winters, Linguistic diversity and traffic accidents: lessons from statistical studies of cultural traits. PLoS ONE **8**(8), e70902 (2013)

M. Savage, F. Devine, N. Cunningham, M. Taylor, Y. Li, J. Hjellbrekke, B. Le Roux, S. Friedman, A. Miles, A new model of social class? Findings from the BBC's Great British class survey experiment. Sociology **47**(2), 219–250 (2013)

E. Segel, J. Heer, Narrative visualization: telling stories with data. IEEE Trans. Vis. Comput. Graph. **16**(6), 1139–1148 (2010)

H. Wainer, H.L. Zwerling, Evidence that smaller schools do not improve student achievement. Phi Delta Kappan **88**(4), 300–303 (2006)

C.R. Woese, Bacterial evolution. Microbiol. Rev. **51**(2), 221 (1987)

G.U. Yule, Notes on the theory of association of attributes in statistics. Biometrika **2**(2), 121–134 (1903)

Chapter 2
Quantitative Summarization

Abstract Before going to the thick of the multivariate summarization, this chapter first considers the concept of feature and its summarizations into histograms, density functions and centers. Two perspectives are defined, the probabilistic and vector-space ones, for defining concepts of feature centers and spreads. Also, current views on the types of measurement scales are described to conclude that the binary scales are both quantitative and categorical. The core of the Chapter describes the method of principal components (PCA) as a method for fitting a data-driven data summarization model. The model proposes that the data entries, up to the errors, are (sums of) products of hidden factor scores and feature loadings. This, together with the least-squares fitting criterion, appears to be equivalent to finding what is known in mathematics as part of the singular value decomposition (SVD) of a rectangular matrix. Three applications of the method are described: (1) scoring hidden aggregate factors, (2) visualization of the data, and (3) Latent Semantic Indexing. The conventional, and equivalent, formulation of PCA via covariance matrices involving their eigenvalues is also described. The main difference between the two formulations is that the property of principal components to be linear combinations of features is postulated in the conventional approach and derived in that SVD based. The issue of interpretation of the results is discussed, too. A novel promising approach based on a postulated linear model of stratification is presented via a project. The issue of data standardization in data summarization problems, remaining unsolved, is discussed at length in the beginning. A powerful application using eigenvectors for scoring node importance in networks and pair comparison matrices, the Google PageRank approach, is described too.

2.1 Encoder–Decoder Data Summarization

2.1.1 Structure of a Summarization Problem

Summarization as a concept covers many activities from data compression to labeling a dataset in archeology with a phrase like "Archeology finds indicate no

© Springer Nature Switzerland AG 2019
B. Mirkin, *Core Data Analysis: Summarization, Correlation,*
and Visualization, Undergraduate Topics in Computer Science,
https://doi.org/10.1007/978-3-030-00271-8_2

King David Palace at the time of King David". In contrast to a correlation problem
(see further in Chap. 3), the features are not divided here into those belonging to
input or target of the phenomenon under consideration. One may think of this as
that all features available are target features so that those to be constructed for a
summary are in fact "hidden input features".

In this way, the structure of a summarization problem in Fig. 2.1a may be
likened to that of a correlation problem, on Fig. 2.1c, if a decoding rule is provided
to predict all the original data entries using the summary. That is, the original data
in the summarization problem act as the target data in the correlation problem. That
implies that there should be two rules involved in a summarization problem: one for
building the summary, that is, encoder; the other to provide a feedback from the
summary to the observed data, that is, decoder.

The issue of data summarization sometimes is considered somewhat simplisti-
cally as just deriving a summary from data without any feedback (see Fig. 2.1b).

A proper consideration of the structure of a summarization problem should rely
on the existence of a decoder to provide a feedback from the summary back to the
data and make the summarization process more or less similar to that of the correlation process (see Fig. 2.1a vs. 2.1c). More exactly, a decoder is a device that
converts the summary representation encoded in the chosen summarization rule
back into the original data format. Then one may compare the original data and
those output by the decoder: the less the difference between them, the better the

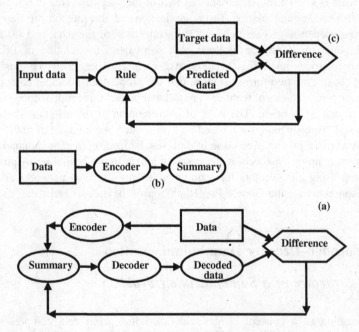

Fig. 2.1 A diagram for encoder/decoder data summarization, in **a**, versus learning input-target
correlation, in **c**, or summarization with no decoder, in **b**. Rectangles are for observed data, ovals
for computational constructions, hexagons for feedback comparisons

summary. Of course, the set of all admissible encoders and decoders should be pre-specified. This text concerns only those methods at which both the encoder and decoder are based on the so-called matrix factorization models. Matrix factorization models represent the data matrix as a product of matrices corresponding to some pre-specified "ideal" data structures.

The data recovery approach in data summarization is based on the assumption that there is a regular structure in the phenomenon of which the observed dataset informs. This regular structure C is the summary to be found. When C is determined, this can feed back to the observed data Y in the format of the decoded data $D(C)$ that should coincide with Y, up to residuals, that are due to possible flaws in any or all of the following three aspects:

(a) bias in entity sampling,
(b) selecting and measuring features,
(c) adequacy of the set of admissible C structures to the phenomenon in question.

Each of these three can greatly affect results. However, so far only the simplest of the aspects, (a) the sampling bias, has been addressed scientifically, in mathematical statistics—as a random bias, due to the probabilistic nature of the data selection process. The other two are subjects of much effort in specific domains but not in the general computational data analysis framework as yet. Rather than focusing on accounting for the causes of errors, let us consider the underlying equation in which the errors are looked at as a whole:

$$Observed\_Data\ Y = Model\_Data\ D(C) + Residuals\ E \qquad (2.1)$$

2.1.2   Least-Squares Data Summarization and Pythagorean Decomposition

Equation (2.1) brings in a natural data recovery criterion for the assessment of the quality of the model A in recovering data Y—according to the level of residuals E: the smaller the residuals, the better the model. Since any data model in data analysis involves unknown parameter values, this naturally leads to the idea of fitting these parameter values to the data so that the residuals are as small as possible.

In many cases this principle can be rather easily implemented as the least squares criterion because of an extension of the Pythagoras theorem relating the square lengths of the hypotenuse and two other sides in a right-angle triangle connecting "points" Y, $D(C)$ and 0, the space origin (see Fig. 2.2). The least squares criterion requires fitting the model C by minimizing the sum of the squared residuals. Geometrically, that often means an orthogonal projection of the data set considered as a multidimensional point onto the space of all possible models represented by the x axis on Fig. 2.2. In such a case the dataset (pentagram), its projection (rectangle) and the origin (0) form a right-angle triangle for which a multidimensional extension

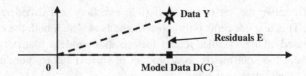

Fig. 2.2 Geometric illustration of the Pythagorean relation between the observed data (pentagram), the fitted model data (black rectangle), and the residuals (connecting line)

of the Pythagoras' theorem holds. The theorem states that the squared length of the hypotenuse is equal to the sum of squares of lengths of the two other sides. The squared hypotenuse translates into the data scatter, that is, the sum of all the data entries squared, being decomposed in two parts, the part explained by the summary model C, that is, the contribution of the line between 0 and rectangle, and the part left unexplained by C. The latter part is the contribution of the residuals E expressed as the sum of squared residuals, which is exactly the least squares criterion.

When the data can be considered as a random sample from a multivariate Gaussian distribution, the least squares principle can be derived, under some simplifying assumptions, from a major statistical principle, that of maximum likelihood. In the data analysis framework, the data do not necessarily come from a probabilistic population. Still, the least squares framework frequently provides for solutions that are both practically relevant and theoretically sound. The least squares will be the only criterion utilized in this text further on.

A decoder based linear—or, more exactly, bilinear—summarization problem can be stated as follows. Given N row-vectors $y_i = (y_{i1}, ..., y_{iV})$, $i = 1, 2, ..., N$, forming an $N \times V$ data matrix $Y = \{(y_{iv})\}$ and a set of admissible summary structures Z represented by sets of mutually orthogonal, usually normed, N-dimensional vectors $z_k = (z_{ik})$, build a summary C so that the decoded $N \times V$ data matrix $D(C)$ is as close to Y as possible. In other words, we assume an encoder to draw a summary C using Z, and a decoder $\tilde{Y} = D(C)$ such that the error, which is the difference between the decoded data $\tilde{Y} = D(C)$ and observed data Y, is minimal over the class of admissible rules Z. More explicitly, we assume that

$$Y = D(C) + E \tag{2.2}$$

where E is matrix of residual values, or errors: the smaller the errors, the better the summarization C. According to the least-squares approach, the errors are minimized by minimizing the summary squared error:

$$E^2 = \langle Y - D(C), Y - D(C) \rangle \tag{2.3}$$

with respect to all admissible summarization (encoding) rules.

In the follow-up data summarization methods, the Principal Component Analysis, K-means partitioning, and Ward hierarchical clustering (see Sects. 2.4, 4.2 and 5.1, respectively), the encoder and decoder are defined within the same bilinear structure, so that $D(C) = ZC$ where Z is an $N \times K$ matrix of mutually orthogonal N-dimensional vectors in which K is a relatively small integer. Indeed, consider Z to be either:

(a) Set of the first K normed singular vectors z_k of matrix Y, so that $Y^T z_k = \mu_k c_k$ and $Y c_k = \mu_k z_k$ where $\mu_1 > \mu_2 > \cdots > \mu_K$ and $k = 1, 2, ..., K$, or

(b) Set of the K binary 0/1 membership vectors $z_k = (z_{ik})$ for non-overlapping clusters forming a partition $S = \{S_1, S_2, ..., S_K\}$ of the set of objects, so that $z_{ik} = 1$ if $i \in S_k$ and $z_{ik} = 0$, otherwise $(k = 1, 2, ..., K)$, or

(c) Set of the K ternary 0/a/$-b$ split vectors $z_k = (z_{ik})$ for a binary up-hierarchy consisting of K binary splits $S_k = \{S_{k1}, S_{k2}\}$ where S_k is a cluster and S_{k1}, S_{k2} its non-overlapping non-empty split parts, so that $z_{ik} = 0$ if $i \notin S_k$ and $z_{ik} = a$ if $i \in S_{k1}$ and $z_{ik} = -b$ if $i \in S_{k2} (k = 1, 2, ..., K)$. Values a and b are defined as $a = (1/|S_{k1}| - 1/|S_k|)^{1/2}$ and $b = (1/|S_{k2}| - 1/|S_k|)^{1/2}$, so that all vectors z_k are centered, normed and mutually orthogonal.

Each of these should be a subject for a special consideration. Indeed, items (a) and (b) are covered in this and next chapter, respectively. As to the item (c), we felt that putting it here would be an overly technical challenge to the reader and decided to skip it, referring an interested reader to Mirkin (2012) and Kovaleva and Mirkin (2015) for more detail. Once more, let us point out that the encoder and decoder are defined within the same bilinear structure here, so that $D(C) = ZC$ where both Z and C are unknown, with an additional constraint that Z is an $N \times K$ matrix of mutually orthogonal N-dimensional vectors defined as in (a), (b), or (c) above—such a representation is usually referred to as matrix factorization. Matrix factorization has been extended to several other issues, first of all, fuzzy clustering and text analysis. Developments in data analysis with encoders and decoders set apart are yet to be developed.

Luckily, the least-squares criterion admits a rather straightforward solution. Given matrix Z, the optimal ZC minimizing the criterion $E^2 = \langle Y - ZC, Y - ZC \rangle$ in (2.3) is the orthogonal projection $ZC = P_Z Y$ of the data matrix Y, onto the span $L(Z)$ of Z, so that $P_Z = Z(Z^T Z)^{-1} Z^T$, the matrix of the orthogonal projector, where $L(Z)$ is the linear subspace $L(Z) = \{z: z = Zc \text{ for some } c\}$ referred to as the span of Z. If Z consists of normed mutually orthogonal column vectors, then $Z^T Z = I$, where I is the identity matrix, so that $ZC = ZZ^T Y$ and $C = Z^T Y$.

In this case, the data matrices Y and $D(C)$, considered as multidimensional points, form a "right-angle triangle" at the origin 0, as presented on Fig. 2.2. In such a case $\langle Y, D(C) \rangle = \langle D(C), D(C) \rangle$ and the square error (2.3) becomes part of a multivariate analogue to the Pythagorean decomposition relating the squares of the "hypotenuse", Y, and of the right-angle "sides", $D(C)$ and E:

$$\langle Y, Y \rangle = \langle D(C), D(C) \rangle + E^2 \tag{2.4}$$

or using matrix entries,

$$\sum_{i \in I} \sum_{v \in V} y_{iv}^2 = \sum_{i \in I} \sum_{v \in V} d_{iv}^2 + \sum_{i \in I} \sum_{v \in V} e_{iv}^2 \qquad (2.4')$$

The data is an $N \times V$ matrix $Y = (y_{iv})$ that can be considered as either set of rows/entities $y_i (i = 1, \ldots, N)$ or set of columns/features y_v ($v = 1, \ldots, V$) or both. The item on the left in (2.4) is usually referred to as the data scatter and denoted by $T(Y)$,

$$T(Y) = \sum_{i \in I} \sum_{v \in V} y_{iv}^2 \qquad (2.5)$$

Why is this termed "scatter"? It is not difficult to see that $T(Y)$ in (2.5) is the sum of Euclidean squared distances from all the objects to 0, thus a measure of scattering data points around 0. Of course, $T(Y)$ admits a dual interpretation. On the one hand, $T(Y)$ is the sum of row-based object contributions, the squared distances $d(y_i, 0)$ $(i = 1, \ldots, N)$. On the other hand, $T(Y)$ is the sum of column-based feature contributions $f_v = \Sigma_{i \in I} y_{iv}^2$. In the case when the average c_v has been subtracted from all values of the column v, the summary contribution f_v is N times the variance, $f_v = N\sigma_v^2$ which is proven in Q.2.1 below.

Q.2.1. Prove that the summary contribution t_v is N times the variance, $t_v = N\sigma_v^2$ if feature v is centered.

A.   Indeed, $t_v = \sum_{i \in I} y_{iv}^2 = \sum_{i \in I} (y_{iv} - c_v)^2 = N[\sum_{i \in I} (y_{iv} - c_v)^2 / N] = N\sigma_v^2$, where c_v is the mean of feature v.

2.2   Quantitative Features and Their Characteristics

The concept of feature, synonymously referred to as character (in biology), variable (in statistics), attribute (in logics), is a key concept in data analysis. So far, we have no good understanding of what feature is, which is not uncommon with prime concepts in sciences. Empirically, there must be a measurement or identification procedure that assigns every object of a specific type with a value of the characteristic which is measured or identified, be it, for example, the human height or eye color. The lack of understanding in the nature of the measurement phenomenon leads to two different co-existing mathematical formalizations of the feature concept.

One, of the data analysis proper, is very close to the empirical view. There is a finite set of objects that are characterized just by their labels, or index values. Then a feature is just a mapping from the set of objects to a set of its values, usually reals or even strings such as "good" and "bad". The values indexed by the object indexes form a vector, which leads to vector and matrix algebra as the natural habitat of the theoretical constructions and mathematical derivations, the main setting in this text.

A major shortcoming of this concept of feature is its relying on the set of objects under consideration. Whatever is observed or stated refers to this set only. This makes it quite difficult, if not impossible, to translate the conclusions to another dataset.

The other formalization, of classical statistics, is the concept of random variable, that is usually represented by a so-called density function defined over a real line or vector space. The density function allows easily assigning probabilities to real-line intervals, as well as to set-theoretic combinations of them. The density function language is well adapted for describing universal phenomena, making probabilistic statements and testing statistical hypotheses about them. Its shortcoming is exactly where the strength of the data analysis approach lies. In contrast to the latter, the probabilistic view disregards individual objects, considering them, in the best case, as results of random, usually independent, sampling from a population representing a density function. In contrast, the notion of individual object is in the focus of data analysis perspective.

When one deals with prediction of weather or harvest, the probabilistic view is adequate and useful. When one deals with the analysis of European state statistics or data of within-company messages, the probabilistic view can be useful only after the structure of inter-state or inter-department intricate inter-relations has been analyzed—this is why data analysis has emerged and become popular in the digital era.

2.2.1   Data Analysis Perspective: Scale Types, Minkowski Distance and Center

2.2.1.1   Quantitative and Categorical Feature Scales

According to the data analysis perspective, a feature x_v is a mapping from a set of objects O to the set of feature values. That is, at every $i \in O$, a feature value, $x_v(i)$, is defined. From now on, we assume the values are coded in numbers. Of course, this definition represents a very limited perspective. It does not take into account that there is a measurement method behind any feature. The method is applicable to all the objects of the sort in our dataset, so that the dataset can be easily extended to include more objects of the same kind, be it web sites or territorial units or gene products. This is missed in the definition above but is always meant in practical computations.

Conventionally, the following five feature scale types are distinguished:

(A) Absolute;
(B) Ratio;
(C) Interval;
(D) Ordinal;
(E) Nominal.

This classification is based on a conventional formalism of admissible transformations. A numeric function f is referred to as an admissible transformation for feature x if $f(x)$ is considered the same feature. The set of admissible transformations F specifies the scale type. Physics parameters can be used as examples of quantitative scale types. Say, physical mass or height is measured up to a scale factor: 5 kg is 5000 g, and 6 ft. height is 182.88 cm, as well as 20 °C is 68 °F. A measurement scale is referred to as an absolute scale if the set of admissible transformations consists of the only transformation, the identity mapping leaving values intact: $f(x) = x$. Any counting scale—the number of respondents, the number of objects—is considered as an example of the absolute scale type. A measurement scale is referred to as a ratio scale if the set of admissible transformations consists of multiplication by a positive real, called scale factor, $F = \{f: f(x) = ax$ for some $a > 0\}$. A measurement scale is referred to as an interval scale if the set of admissible transformations consists of multiplication by a positive real, called scale factor, with a follow-up summation with a constant called scale shift, $F = \{f: f(x) = ax + b$ for some real $a > 0, b\}$. Obviously, the weight and height measurements belong to the ratio scales. In this author's view, any counting scale should be considered a ratio scale as well. Indeed, one can count in dozens or hundreds rather than in unities. On the other hand, the temperature and historical time are considered as measured in interval scales. Indeed, the Celsius temperature is converted to that of Fahrenheit by multiplying it by a scale factor $a = 1.8$ with the follow up addition of the scale shift $b = 32$. The time measurements are not that smooth because of differences in the periods of the Sun and the Moon differently taken into account in different cultures. Conversion of the Gregorian calendar to the Islamic one can be done only approximately. Although the scale shift, the date of Hijri, the New Year of 622 AD, is constant, the lengths of the "Sun" year in the former is variable, 365 or 366 days, as well as that of the "Moon" year in the latter, 354 or 355 days, which prevents the scale factor to be defined as a single value. For example, 1st February 2018 is 15th Jumada Al-Awwal, 1439 Hijri.

The names of the scale types inform of the properties remaining invariant after any admissible transformation. Indeed, any ratio of measures, say $x(i)/x(j)$ remains the same under any admissible ratio scale transformation: $x'(i)/x'(j) = x(i)/x(j)$ if $x' = ax$, for any a. Similarly, any ratio of intervals, say $[x(i) - x(j)]/[x(k) - x(l)]$ remains the same under any admissible transformation: $[x'(i) - x'(j)]/[x'(k) - x'(l)]$ if $x' = ax + b$, for any scale factor a and scale shift b. What is important, though, is that the concept of average remains valid within the concept of interval scale. Specifically, assume that c and d are means of some data series $\{c_i\}$ and $\{d_j\}$, respectively. Consider an admissible transformation with the scale factor a and scale shift $b, c_i' = ac_i + b$ and $d_i' = ad_i + b$. It is easy to check that the averages satisfy equations $c' = ac + b$ and $d' = ad + b$. Since $a > 0$, $c > d$ if and only if $c' > d'$. However, the ratio is invariant for intervals only, not for the average values themselves.

A feature is said to be measured in the ordinal scale if the set of admissible transformations consists of monotonically increasing mappings so that whatever is better remains so under any admissible transformation. A feature is said to be measured in the nominal scale if the set of admissible features consists of one-to-one mappings so that whatever is different remains so under any admissible transformation. The nominal and ordinal scale types are examples of non-metric scales. In spite of seemingly loose, non-metric, character, both are abstractions, sometimes unrealistic, of the scale types in real data.

The ordinal scale type is an abstraction of ranking. Rankings come from personal preferences, ranking survey questions, and recommendation systems. In a simplest case, this may come as an answer to a question whether the respondent likes this or that event or public personality, "yes" or "no" or "indifferent". (The answer "have never heard of" is usually considered as a missing.) These three categories are usually numerically coded by 1, 0, −1 or 7, 0, −7. According to the definition of ordinal scale, equally good are numerical codes: (a) 1000000, −2, −30; (b) 1, 0, −1000000; (c) 1, −1000000, −1000001. The imbalances in the codes contradict our intuition and customs: imbalances usually are taken into account in natural language structures. For example, the coding (a) above may correspond to formulations: "like very-very much", "slightly dislike", and "dislike" rather than those in the original question.

The concept of nominal scale formalizes a type of features that divide the set of objects in unrelated non-overlapping parts. Such is a question like: (i) "What is your preferred European vacation destination?", with answers like "Greece", "Sicily", and "Malta"; or (ii) "In what region your household is located?", with answers like "West Europe", "North Europe", "Balkans", etc.; or (iii) "What is your occupation?", with answers like "Engineering", "Medicine", "Management", etc. Three main assumptions in the definition above are: (a) no relation between the feature values; (b) no values may co-occur on the same object; (c) no missings. Of these, the assumption (b) is especially difficult to provide: say, more than one destination may be preferable (question (i)) and some may have households in more than one region (question (ii)) and some may be engaged in several occupations (question (iii)).

The concept of scale type above allows one to shift the issue of a proper definition of the scale type to the issue of adequacy of a statement involving measurements. A statement is adequate if its truth modality (true or false) does not change under admissible transformations. In this sense, a numerical statement $x + y = y + x$ is adequate even if x and y are measured in a non-metrical scale. Indeed, it reflects the nature of arithmetic which depends on no concrete domain.

The issue of how this or that method of measurement and the corresponding measure values acquire a set of admissible transformations is far from being solved. Two extreme points of view: (a) the scale type is derived within that theory at which the corresponding feature is part of; (b) the scale type is derived from the properties of the relations between objects that are immanent to them. The former takes justification from physics, whereas the latter has its roots in psychology.

A good theory for data analysis should embrace all the data types occurring in real data analyses. Unfortunately, this is not the case so far, although some movement towards developing such a theory can be seen in 21st century.

In this text, three scale types are distinguished:

(1) Quantitative;
(2) Binary;
(3) Nominal;

as those that can be treated within the same formalism of linear algebra. To this end, we abandon the formalism of admissible transformations and define the scale type for a measurement, as based on common sense.

Consider the following dataset Company from Table 1.3 as an example.

Table 2.1 contains data of eight companies whose names are in the column on the left.

Features:

(1) Income, $ Million;
(2) Market Share—the proportion of market controlled by company, %;
(3) NSup—the number of principal suppliers;
(4) E-C—Yes or No depending on the usage of e-commerce in the company;
(5) Sector—sector of the economy: (a) Retail, (b) Utility, and (c) Manufacture.

A feature, as already mentioned, is a mapping of the set of objects to a set of feature values. A quantitative feature, according to our common-sense definition, is that one for which the operation of averaging of its values makes sense. In contrast, a nominal feature must have relatively few values which admit no operation over them; these values usually are referred to as categories. A nominal feature is adequately represented by the corresponding partition of the set of objects. Parts of this partition are sets of objects falling into the same category.

There are three quantitative features and two nominal features in Table 2.1. Indeed, one may average features Income, MarketS and NSup. The averages over, for example, A-companies are: $(19.0 + 29.4 + 23.9)/3 = 24.10$ (Income), $(43.7 + 36.0 + 38.0)/3 = 39.23$ (Market Share), $(2 + 3 + 3)/3 = 2.67$ (Principal

Table 2.1 Company

Company	Income, $ million	Market share, %	NSup	E-C	Sector
Aversi	19.0	43.7	2	No	Utility
Antyos	29.4	36.0	3	No	Utility
Astonite	23.9	38.0	3	No	Manufacture
Bayermart	18.4	27.9	2	Yes	Utility
Breaktops	25.7	22.3	3	Yes	Manufacture
Bumchist	12.1	16.9	2	Yes	Manufacture
Civok	23.9	30.2	4	Yes	Retail
Cyberdam	27.2	58.0	5	Yes	Retail

Suppliers). Some say that the last average, 2.67 suppliers, makes no sense, so that the operation of averaging is not admissible for such counting features. This author replies: "Yes, indeed, the number of suppliers 2.67 makes no sense because it relates to a single company. But one could think of a hundred companies like these. The hundred would have 267 suppliers, which makes a perfect sense." The operation of averaging cannot be applied to string values of features E-C and Sector. These correspond to partitions of the Company set: {{Av, An, As}, {Ba, Br, Bu, Ci, Cy}}, for E-C, and {{Av, An, Ba}, {As, Br, Bu}, {Ci, Cy}}, for Sector.

The assumption of equal importance of features currently underlies all the efforts, which makes the entire edifice of data analysis somewhat crippled—but there is nothing new in this. As the history of science clearly demonstrates, any breakthrough in the sciences starts with a rather shaky base.

To balance contributions of features to the data scatter, one conventionally applies the operation of data standardization comprising two transformations, shift of the origin and rescaling.

We will encounter the issue of standardization in Chap. 3 while studying multivariate classifiers, decision trees and neural networks. In neural networks, as well as in Support vector machine, the standardization involves the scale shift to the midrange and rescaling by normalizing the feature values by the half-range. These parameters are distribution independent.

Another, much more popular, choice is the feature's mean for the scale shift and normalizing by the standard deviation for rescaling. This standardization is a cornerstone in mathematical statistics and it works very well if observations come from a Gaussian distribution, because the distribution becomes parameter-free if standardized by subtracting the mean followed by dividing over the standard deviation. In statistics, this transformation is frequently referred to as z-scoring. In the context of data analysis, though, distributions are rarely Gaussian and rarely of any popular family at all; moreover, observations are not necessarily random or independent. In these circumstances, the choice of shifting and rescaling needs a rethink.

First of all, there is no need in linking the two operations together: shifting the origin has nothing to do with balancing feature weights. The goal of the shifting is to position the data against a backdrop of a "norm" which is put to the origin by the shift. In this way, the analysis involves the differences of the data and the norm. The experimental evidence accumulated in the ever growing body of data analysis research suggests that it is much easier to find meaningful structures if the "normal background" has been removed from data. According to the least squares criterion, it is the mean that approximates the overall "norm" the best. Since this criterion underlies all the methods considered in this text, the mean—sometimes referred to as grand mean, to point out its position over the entire entity set—will be the choice for the origin.

The normalization seems to be better if done by half-range or, equivalently, the range, indeed. On the first glance, there is no advantage in normalization by the range. Z-scoring seems a better choice, especially since z-scoring satisfies the intuitively appealing equality principle—all features contribute to the data scatter equally after they have been divided by their standard deviations.

This view is, however, overly simplistic. In fact, the feature's contribution to the data scatter is affected by two unrelated factors: (a) the feature scale range and (b) the feature distribution shape. While reducing the effect of the former, normalization should not suppress the effect of the latter because the distribution shape is an important indicator of the data structure. But the standard deviation involves both and thus mixes them up. Take a look, for example, at distributions of two features presented on Fig. 2.3. One of them has one mode only (a), whereas the other has two modes (b). Since the features have the same range, the standard deviation is greater for the distribution (b), which means that its relative contribution to the data scatter decreases under the z-scoring standardization. This means that its clear cut discrimination between two parts of the distribution will be stretched in while the unimodal structure, which is hiding the two-part structure, will be stretched out. This is not exactly what we want of data standardization. Data standardization should help in revealing the data structure rather than concealing it. Thus, normalization by the range helps in bringing forward multimodal features by assigning them relatively larger weights proportional to their variances.

Therefore, in contrast to conventional wisdom, z-scoring standardization should be avoided unless there is a strong indication that the data come from a Gaussian distribution indeed. Any index related to the scale range can be used for normalization. In this text, the range is universally accepted. If, however, there is a strong indication that the range may be subject to outlier effects and, thus, unstable and random, more stable indexes could be used for normalization such as, for example, the distance between upper and lower 1% quantiles.

2.2.2 Centers and Spreads: Minkowski Distance

Further summarization of the data leads to presenting a feature with just two numbers, one expressing the distribution's location, its "central" or other important point, and the other representing the distribution's dispersion, the spread. We

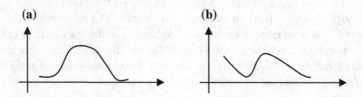

Fig. 2.3 One-modal distribution shape on **a** versus a two-modal distribution shape on **b**: the standard deviation of the latter is greater, thus making it less significant under the z-scoring standardization

review some most popular characteristics for both, the center, Table 2.2, and the spread, Table 2.3.

Worked Example 2.1. Mean

For set X = {1, 1, 5, 3, 4, 1, 2}, mean is c = (1 + 1 + 5 + 3 + 4 + 1 + 2)/7 = 17/7 = 2.42857..., or rounded up to two decimals, c = 2.43.

This is as close an approximation to the numbers as one can get, which is good. A less satisfactory property is that the mean is not stable against outliers. For example, if X in Worked Example 2.1 is supplemented with value 23, the mean becomes c = (17 + 23)/8 = 5, a much greater number. This is why it is a good idea to remove some observations on both extremes of the data range, both the minimum

Table 2.2   A review of location or central point concepts

#	Name	Explanation	Comments
1	Mean	The feature's arithmetic average	0. Minimizes the summary error squared 1. Estimates the distribution's expected value 2. Sensitive to outliers and distribution's shape
2	Median	The middle of the sorted list of feature values	1. Minimizes the summary absolute error 2. Estimates the distribution's expected value 3. Not-sensitive to outliers 4. Sensitive to distribution's shape
3	Mid-range	Middle of the range	1. Minimizes the maximum absolute error 2. Estimates the distribution's expected value 3. Very sensitive to outliers 4. Not sensitive to distribution's shape
4	P-quantile	A value dividing the entire entity set in proportion P/(1 − P) of feature values so that those with higher values constitute P proportion (upper P-quantile) or 1 − P proportion (bottom P-quantile)	1. Not-sensitive to outliers 2. Sensitive to distribution's shape
5	Mode	A maximum of the histogram	1. Depends on the bin size 2. Can be multiple

Table 2.3 A review of spread concepts

#	Name	Explanation	Comments
1	Standard deviation	The quadratic average deviation from the mean	1. Minimized by the mean 2. Estimates the square root of the variance
2	Absolute deviation	The average absolute deviation from the median	Minimized by the median
3	Half-range	The maximum deviation from the midrange	Minimized by the mid-range

and maximum, before computing the mean, which is utilized in the concept of trimmed mean in statistics.

Worked Example 2.2. Median

To compute the median of the set from the previous example, $X = \{1, 1, 5, 3, 4, 1, 2\}$, it must be sorted first: 1, 1, 1, 2, 3, 4, 5. The median is defined as the element in the middle, which is 2. This is rather far away from the mean, 2.43, which evidences that the distribution is biased towards the left end, the smaller values. With the outlier 23 added, the sorted set becomes 1, 1, 1, 2, 3, 4, 5, 23, thus leading to two elements in the middle, 2 and 3. The median in this case is the average of the two, $(2 + 3)/2 = 2.5$, which is by far lesser change than the mean of the extended set, 5.

The more symmetric a distribution, the closer its mean and median to each other. Sepal width of Iris data set (Table 1.3) has mean = 3.05 and median = 3, quite close values. In contrast, in Market town data (Table 1.4), Population resident's median, 5258, is predictably much less than the mean, 7351.4. The mean of a power law distribution is always biased towards the great values achieved by the few outliers; this is why it is a good idea to use the median as its central value. The median is very stable against outliers: the values on the extremes just do not affect the middle of the sorted set if added uniformly to both sides.

The midrange corresponds to the mean of a flat distribution, in which all bins are equally likely. In contrast to the mean and median, the midrange depends only on the range, not on the distribution. It is obviously highly sensitive to outliers, that is, changes of the maximum and/or minimum values of the sample.

The concept of P-quantile is an extension of the concept of median, which is a 50% quantile.

Worked Example 2.3. P-quantile (Percentile)

Take p = 10% and determine the upper 10% quantile of Population resident feature. This should be 5th value in its descending order, that is, 18,966. Why is the 5th value? Because 10% of the total number of entities, 45, is 4.5; therefore, the 5-th value leaves out p = 10% of the largest towns in the sample. Similarly, the lower 10% quantile of the feature is 5th value in its ascending order, 2230.

Case-Study 2.1. Mode

To determine a mode, one needs to know the feature's density function, which is available, in data analysis, as a histogram.

To draw a histogram, one first draws an x axis and the feature range boundaries, that is, its minimum and maximum. The range interval is divided then into a number of non-overlapping equal-sized sub-intervals, *bins*. Then the number of objects that fall in each bin is counted, and the counts are reflected in the heights of the bars over the bins, forming a histogram. Histograms of Population resident in Market town dataset and Petal width in Iris dataset are presented in Fig. 2.4.

Q.2.2. Why the bins are not to overlap?

A. Each entity falls in only one bin if bins do not overlap, and the total of all bin counts equals the total number of entities in this case. If bins do overlap, the principle "one entity—one vote" will be broken.

According to the histograms in the bottom of Fig. 2.4, it is the very first bin which is modal in the Population resident distribution. In the 5-bin setting, it takes one fifth of the feature range, $23,801 - 2040 = 21,761$, that is, 4352. In the 10-bin setting, it is one tenth of the feature range, that is, 2176. In the latter case, the modal bin is interval [2040, 4216], and the modal bin is as twice wider, [2040, 6392], in the former case.

Each of the characteristics of spread in Table 2.3 parallels, to an extent, a central location characteristic under the same number.

These measures intend to give an estimate of the extent of error in the corresponding centrality index. The standard deviation is the average quadratic error of the mean. Its use is related to the least-squares approach that currently prevails in data analysis and can be justified by good properties of the solutions, within the

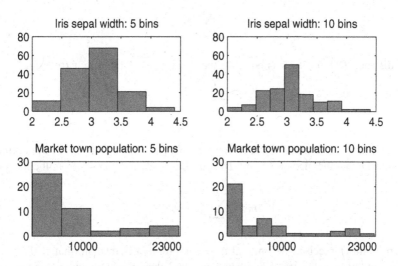

Fig. 2.4 Histograms of quantitative features in Iris and Market town data: the feature represented on x-axis and the counts on y-axis. The histogram shapes depend on the number of bins

data analysis perspective, and properties of the normal distribution, within the probabilistic perspective—these are explained further on.

The absolute deviation expresses the average absolute deviation from the median. Usually, it is calculated regarding the mean, as the average error in representing the feature values by the mean. However, it is more related to the median, because it is the median that minimizes it.

The half-range expresses the maximum deviation from the mid-range; so they should be used on par, as it is done customarily by the research community involved in building classifying rules.

There are two perspectives on data summarization and correlation that very much differ from each other. One, of the classical mathematical statistics, views the data as generated by a probabilistic mechanism and uses the data to recover the mechanism or, at least, some properties of it. The other, of data analysis, does not much care of the mechanism and tries to look for patterns in the data instead. The data analysis perspective is described here.

Given a series $X = \{x_1, ..., x_N\}$, one defines the center of X as a minimizing the average distance

$$D(X, a) = [d(x_1, a) + d(x_2, a) + \cdots + d(x_N, a)]/N \qquad (2.6)$$

At different definitions of the distance, the optimal a may have different expressions.

Consider first the least-squares formulation. According to this approach the distance is measured as the squared difference, $d(x, a) = |x - a|^2$. The minimum distance (2.6) then is reached at a equal to the mean c defined by expression

$$c = \sum_{i=1}^{N} x_i/N \qquad (2.7)$$

and distance $D(X, c)$ itself is equal to the variance s^2 defined by expression

$$s^2 = \sum_{i=1}^{N} (x_i - c)^2/N \qquad (2.8)$$

At the more traditional distance measure $d(x, a) = |x - a|$ in (2.6), the optimal a (center) is but the median, m, and $D(X, a)$ the absolute deviation from the median,

$$ms = \sum_{i=1}^{N} |x_i - m|/N \qquad (2.9)$$

To be more precise, the optimal a in this problem is median, that is the value $x_{(N+1)/2}$ in the sorted order of X, when N is odd. When N is even, any value between $x_{N/2}$ and $x_{N/2+1}$ in the sorted order of X is a solution, including the median.

If $D(X, a)$ is defined not by the sum, but by the maximum of the distances, $D(X, a) = max(d(x_1, a), d(x_2, a), \ldots, d(x_N, a))$, then the midrange mr is the solution, for $d(x, a)$ specified as both $|x - a|^2$ and $|x - a|$.

These statements explain the parallels between the centers and corresponding spread evaluations reflected in Tables 2.2 and 2.3, with each of the centers minimizing its corresponding measure of spread.

The distance minimization problem can be reformulated in the data recovery perspective. In the data recovery perspective, the observed values are assumed to be but noisy realizations of an unknown value a. This is reflected in the form of an equation expressing x_i through a:

$$x_i = a + e_i, \quad \text{for all } i = 1, 2, \ldots, N, \tag{2.10}$$

in which e_i are additive errors, or residuals, that are to be minimized.

One cannot minimize all the residuals in (2.10) simultaneously. An integral criterion is needed to embrace them all. A general family of such criteria is known as Minkowski's criterion or L_p norm. It is specified by using a positive number p as

$$L_p = (|e_1|^p + |e_2|^p + \cdots + |e_N|^p)^{1/p}$$

At a given p, minimizing L_p or, equivalently, its p-th power L_p^p, would lead to a specific solution. Most popular are values $p = 1, 2$, and ∞ (infinity) leading to:

(1) Least-squares criterion $L_2^2 = e_1^2 + e_2^2 + \cdots + e_N^2$ at $p = 2$.

Its minimization over unknown a is equivalent to the task of minimizing the average squared distance, thus leading to the mean as the optimal a.

(2) Least-modules criterion $L_1 = |e_1| + |e_2| + \cdots + |e_N|$ at $p = 1$.

Its minimization over unknown a is equivalent to the task of minimizing the average absolute deviation, thus leading to the median, optimal $a = m$.

(3) Least-maximum criterion $L_\infty = max(|e_1|, |e_2|, \ldots |e_N|)$ at $p = \infty$. Minimization of L_∞ with respect to a is equivalent to the task of minimizing the maximum deviation leading to the midrange, optimal $a = mr$.

The Minkowski's criteria (1)–(3) may look just as trivial reformulations of the distance approximation criterion (2.6). This, however, is not exactly so. Equation (2.10) adds to the solution one more equation. It allows for a decomposition of the data scatter involving the corresponding data recovery criterion.

This is rather straightforward for the least-squares criterion L_2 whose minimal value, at a equal to the mean c (2.7) is $L_2^2 = (x_1 - c)^2 + (x_2 - c)^2 + \cdots + (x_N - c)^2$. With little algebra, this becomes

$$L_2^2 = (x_1 - c)^2 + (x_2 - c)^2 + \cdots + (x_N - c)^2 + Nc^2$$
$$= x_1^2 + x_2^2 + \cdots + x_N^2 - Nc^2 = T(X) - Nc^2$$

where $T(X)$ is the quadratic data scatter defined as $T(X) = x_1^2 + x_2^2 + \cdots + x_N^2$.

This leads to equation $T(X) = Nc^2 + L_2^2$ decomposing the data scatter in two parts: that explained by the model (2.10), Nc^2, and that unexplained, L_2^2. Since the data scatter is constant, minimizing L_2^2 is equivalent to maximizing Nc^2. The decomposition of the data scatter allows measuring the adequacy of model (2.10) not by just the averaged square criterion, the variance, but by the relative value of the explained part $L_2^2/T(X)$. A similar decomposition can be derived for the least modules L_1 (see Mirkin 1996).

Q.2.3. Consider a multiplicative model for the error, $x_i = a(1 + e_i)$, assuming that errors are proportional to the values. What center a fits the data with the least-squares criterion?

A. According to the least squares approach, the fit should minimize the summary errors squared. Every error can be expressed, from the model, as $e_i = x_i/a - 1 = (x_i - a)/a$. Thus the criterion can be expressed as $L_2^2 = e_1^2 + e_i^2 + \cdots + e_N^2 = (x_1/a - 1)^2 + (x_2/a - 1)^2 + \cdots (x_N/a - 1)^2$. Applying the first order optimality condition, let us take the derivative of L_2^2 over a and equate it to zero. The derivative is $L_2^{2\prime} = -(2/a^3) \sum_i (x_i - a) x_i$. Assuming the optimal value of a is not zero, the first order condition can be expressed as $\Sigma_i(x_i - a)x_i = 0$, so that $a = \sum_i x_i^2 / \sum_i x_i = (\sum_i x_i^2/N)/(\sum_i x_i/N)$. The denominator here is but the mean, c, whereas the numerator can be expressed through the variance s^2 because of equation $s^2 = \sum_i x_i^2/N - \sum_i x_i/N$ which is not difficult to prove. With little algebraic manipulation, the least-squares fit can be expressed as $a = s^2/c + 1$. The variance to mean ratio s^2/c, equal to $a - 1$ according to the model, emerges also in statistics as a good relative estimate of the spread.

It seems rather natural that both, the standard deviation and absolute deviation, are not greater than half the range, which can be proven mathematically (see Sect. 2.3.2).

Q.2.4. Prove that Minkowski's center is not sensitive with respect to changing the scale factor.

Q.2.5. Prove that Minkowski's center grows whenever power p grows.

Q.2.6. For the Population resident feature in Market town data compute Minkowski center at $p = 0.5, 1, 2, 3, 4, 5$.

A. See solutions found using the cm.m code developed in Project 2.1 (and confirmed, at $p > 1$, with the steepest descent AG-MC method) in Table 2.4.

Table 2.4 Minkowski's metric centers of the Population resident in Market town dataset for different power values p

p	Minkowski's center	Data scatter unexplained
0.5	2611.0	0.7143
1	5258.0 (median)	0.6173
2	7351.4 (mean)	0.4097
3	8894.9	0.2318
4	9758.8	0.0584
5	10,294.5	0.1186

2.2.3   Probabilistic Perspective: Distributions, Centers and Spreads

2.2.3.1   Distribution and Density Function

The concept of probabilistic distribution is a far-reaching way for summarization of the data. There are several popular probabilistic distributions, of which one should be acquainted with at least these three:

(a) Gaussian, or normal;
(b) Power law; and
(c) Uniform distribution.

They can be represented by different mathematical objects, of which perhaps the most popular is representation via density function. The density functions of these three, in respect, are

$$f(u) = Ce^{-(u-\mu)^2/2\sigma^2}, \tag{2.11}$$

where C stands for a constant term equal to $C = (2\pi\sigma^2)^{-1/2}$, μ for the mean, σ^2 for the variance;

$$p(x) = Cx^{-\alpha} = C/x^\alpha, \tag{2.12}$$

for $x \geq x_{min}$ where $C = (\alpha - 1)x_{min}^{\alpha-1}$ and $\alpha > 1$, and

$$f(x) = \begin{cases} \frac{1}{b-a}, & x \in [a,b] \\ 0, x < a & \text{or } x > b \end{cases} \tag{2.13}$$

as illustrated in Figs. 2.5, 2.6, and 2.7. The area between the x-axis and a density function, that is, the corresponding integral, shows the probabilities. The total probability, that is, the total area is 1; and whatever interval one would take, its probability would be defined as the area between that interval and corresponding part of the density function. For example, for the Gaussian (or normal) density function, the probability of interval between μ and $\mu - \sigma$ is 0.341, between μ and $\mu - 2\sigma$,

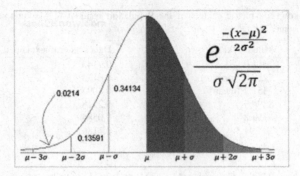

Fig. 2.5 Gaussian density function

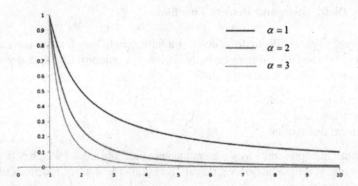

Fig. 2.6 Power law density function at different values of alpha

Fig. 2.7 Density function of
the uniform distribution in
interval [a, b]

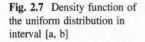
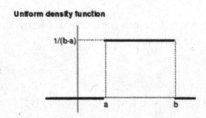

$0.341 + 0.136 = 0.477$, and between μ and $\mu - 3\sigma, 0.341 + 0.136 + 0.021 = 0.499$, respectively. The square root of the variance, σ, is referred to as the standard deviation. It represents the average quadratic deviation of the feature from the mean μ and is more intuitive than the variance because measured in the same units of scale.

The Gaussian distribution is called normal not without a reason. Sums of density functions of independent from each other distributions with a limited variation converge to a Gaussian density function. The average density function of any distribution of a limited variability converges to the density function of a Gaussian distribution. Distributions of measurement errors and, in general, features being results of small random effects are thought to be Gaussian. One more property of the Gaussian distribution: a linear transformation of feature x with a Gaussian distribution $N(\mu, \sigma)$, $y = ax + b$, also has a Gaussian distribution with accordingly transformed parameters, $\mu_y = a\mu + b$, and $\sigma_y = a\sigma$. This last property allows comparing features having different Gaussian distributions by using the so-called z-transformation $x \Rightarrow x' = (x - \mu)/\sigma$. Clearly, x' has the standardized Gaussian distribution N(0,1) with 0 mean and 1 variance.

The parameter α in the power law density function (2.12) reflects the steepness of the frequency's fall. Such a law expresses what is called the Matthew's effect referring to the saying "He who has much, will get more; and he who has nothing, will lose even that little that he has," according to Matthew's gospel (as well as that by Luke). The Matthew's effect was first observed in the distribution of household income (Pareto's Law): 80% of wealth belongs to 20% of families, whereas the rest 80% possess the remaining 20% of wealth. Similar 80-20 effects were later observed in many other areas of human activity: sciences, languages, city sizes. Currently, many phenomena over the web can be approximately described as distributed according to a power law, notably, the popularity of web-sites, that is described with the so-called "mechanism of preferential attachment". This mechanism proposes that the probability that a new web surfer hits a web-site is proportional to the site's popularity. A nice property of the power law is that it is "scale-free", meaning that change of the feature scale unit does not affect the value of parameter α: α does not change at all! At $\alpha < 2$, the mathematics says that a power law has an infinite mean. Therefore, practitioners usually do not compute averages for variables distributed according to such a power law preferring order statistics such as the median or percentile. Rather than comparing the average incomes, statistics compares incomes of the 10% of families, those with lowest income with those of the 10% top income families. Indeed, what is the value in computing the average income of a group of university professors to which a billionaire is added?

The next distribution to be considered is the uniform distribution, over an interval [a, b]. Its density is constant over this interval and equal to $p(x) = 1/(b - a)$, so that the total area of the box in Fig. 2.7 is 1. The probability of any interval (α, β) within the range is proportional to the length of the interval, $p = (\beta - \alpha)/(b - a)$.

2.2.3.2 Center and Spread

In classical mathematical statistics, a set of real numbers $X = \{x_1, x_2, ..., x_N\}$ is usually considered a random sample from a population defined by a probabilistic distribution with density $f(x)$, in which each element x_i is sampled independently from the others. This involves an assumption that each observation x_i is modeled by the distribution $f(x_i)$ so that the mean's density function is the average of density functions $f(x_i)$. The population analogues to the mean and variance are defined over function $f(x)$ so that the mean, median and the midrange are unbiased estimates of the population mean. Moreover, the variance of the mean is N times less than the population variance, so that the standard deviation tends to decrease by $\sqrt{N}$ when N grows.

Then c in (2.7) is an estimate of μ, and s in (2.8) of σ in (2.11). [These parameters amount to the population analogues of the mean and variance defined, for any density function $f(u)$, as $\mu = \int u f(u) du$ and $\sigma^2 = \int (u - \mu)^2 f(u) du$ where the integral is taken over the entire u axis.] We are going to prove this.

Consider the set X as a random independent sample from a population with a Gaussian, for the sake of simplicity, probabilistic density function $f(x) = Cexp\{-(x - \mu)^2/2\sigma^2\}$ (2.11) where μ and σ^2 are unknown parameters. The likelihood of randomly obtaining x_i then will be $Cexp\left\{-(x_i - \mu)^2/2\sigma^2\right\}$. The likelihood of the entire sample X will be the product of these values, because of the independence assumption. Therefore, the likelihood of the sample is $P(X) = \Pi_{i \in I} Cexp\{-(x_i - \mu)^2/2\sigma^2\} = C^N \exp\{-\sum_{i \in I} (x_i - \mu)^2/2\sigma^2\}$. One may even go further and express $P(X)$ as $P(X) = \exp\{N \ln(C) - \sum_{i \in I} (x_i - \mu)^2/2\sigma^2\}$ where ln is the natural logarithm (over base e). A well-established approach in mathematical statistics, the principle of maximum likelihood, claims that the values of μ and σ^2 best fitting the data X are those at which the likelihood $P(X)$ or, equivalently its logarithm, $\ln(P(X))$, reaches its maximum. The maximum $\ln(P) = N \ln(C) - \sum_{i \in I} (x_i - \mu)^2/2\sigma^2$ is reached at μ minimizing the expression in the exponent, $E = \sum_{i \in I} (x_i - \mu)^2$, which is in fact the summary quadratic distance (2.6), that is, the least-squares criterion, which thus can be derived from the assumption that the sample is randomly drawn from a Gaussian population. This, however, does not mean that the least-squares criterion is only meaningful under the normality assumption: the least-squares criterion has a meaning of its own within the data analysis paradigm.

Likewise, the optimal σ^2 maximizes the part of $\ln(P)$ depending on it, $g(\sigma^2) = N \ln(\sigma^2)/2 - \sum_{i \in I} (x_i - \mu)^2/2\sigma^2$. It is not difficult to find the optimal σ^2 from the first-order optimality condition for $g(\sigma^2)$. Let us take the derivative of the function over σ^2 and equate it to 0: $dg/d(\sigma^2) = -N/(2\sigma^2) + \sum_{i \in I} (x_i - \mu)^2/2(\sigma^2)^2 = 0$. This equation leads to $\sigma^2 = \sum_{i \in I} (x_i - \mu)^2/N$, which means that indeed the variance s^2 is the maximum likelihood estimate of the parameter σ^2 in the Gaussian distribution.

However, when μ is not known beforehand but rather found from the sample according to formula (2.7) for the mean, s^2 in (2.8) appears to be a slightly biased estimate of σ^2 and must be corrected by taking the denominator equal to $N - 1$ rather than N, which is the case in many statistical packages. The intuition behind the correction is that Eq. (2.7) is a relation imposed by the user on the N observed values, which effectively decreases the "degrees of freedom" in the number of observations from N to $N - 1$.

In situations in which the data entities can be plausibly assumed to have been randomly and independently drawn from a Gaussian distribution, the derivation above justifies the use of the mean and variance as the only theoretically valid estimates of the data center and spread. The Gaussian distribution has been proven to approximate well situations in which there are many small independent random effects adding to each other. However, in many cases the assumption of normality is highly unrealistic, which does not necessarily lead to rejection of the concepts of the mean and variance—they still may be utilized within the general data analysis perspective.

In some real-life situations, the assumption that X is an independent random sample from the same distribution seems rather adequate; current mathematical statistics developments cover the issue of centrality and spread quite well even at non-Gaussian distributions. However, in most real-world databases and multivariate samplings this assumption is far from being realistic. Nevertheless, in many real-life situations, conclusions drawn from such untrue assumptions work quite well.

2.2.3.3 Computational Validation of the Mean: Bootstrap

Having a dataset and a feature x with its mean computed over the dataset, one may wonder how representative the mean is. Mathematical statistics proposes to assign the feature with a probabilistic distribution that would allow drawing likely boundaries to the mean. Luckily, the computation power proposes a simple way to extend the dataset available to many copies of it randomly sampled from the dataset. These copies give rise to as many copies of the estimated value, the mean in the case—quite enough to draw a convincing density function for the value.

Consider, for example, the data file short.dat in Appendix A6, a 50×3 array whose columns are samples of three data types described in Table 2.5.

The normal data is in fact a sample from a Gaussian N(10, 2), that has 10 as its mean and 2 as its standard deviation. The other two are Two-modal and Power law samples. Their histograms are on the left-hand sides of Figs. 2.8, 2.9, and 2.10. Even with the aggregate data in Table 2.5 one can see that the average of Power law does not make much sense, because its standard deviation is more than three times greater than the average itself.

Many statisticians would argue the validity of characteristics in Table 2.5 not because of the distribution shapes—which would be a justifiable source of concern for at least two of the three distributions—but because of small sizes of the samples. Is the 50 entities available a good representation of the entire population indeed? To address these concerns, the Mathematical Statistics have worked out principles

Table 2.5 Aggregate characteristics of columns for short.dat array

Data type		Normal	Two-modal	Power law
Mean		10.27	16.92	289.74
Standard deviation	Real value	1.76	4.97	914.50
	Related to $\sqrt{N}$	0.25	0.70	129.33

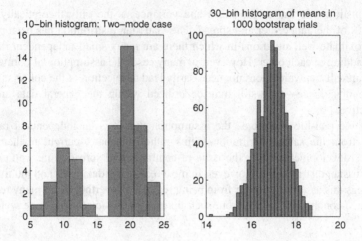

Fig. 2.8 The histograms of a 50 strong sample from a two-mode distribution (on the left) and its mean's bootstrap values (on the right)

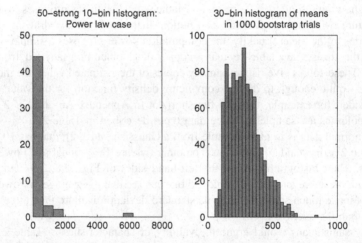

Fig. 2.9 The histograms of a 50 strong sample from a Power law distribution (on the left) and its mean's bootstrap values (on the right)

based on the assumption that the sampled entities come randomly and independently from a—possibly unknown but stationary—probabilistic distribution. The mathematical thinking would allow then, in reasonably well-defined situations, to arrive at a theoretical distribution of an aggregate index such as the mean, so that the distribution may lead to some confidence boundaries for the index. Typically, one would obtain the boundaries of an interval at which 95% of the population fall, according to the derived distribution. For instance, when the distribution is normal, the 95% confidence interval is defined by its mean plus/minus 1.96 times the

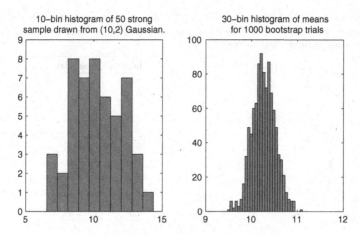

Fig. 2.10 The histograms of a 50 strong sample from a Gaussian distribution (on the left) and its mean's bootstrap values (on the right): all falling between 9.7 and 11.1

standard deviation related to the square root of the number of observations, which is 7.07 at N = 50. Thus, for the first column data, the theoretically derived 95% confidence interval will be 10 ± 1.96 * 2/7.07 = 10 ± 0.55, that is, (9.45, 10.55) (if the true parameters of the distribution are known) or 10.27 ± 1.96 * 1.76/ 7.07 = 10.27 ± 0.49, that is, (9.78, 10.76) (at the observed parameters in Table 2.5).

The difference is rather minor, especially if one recalls that the 95% confidence is a rather arbitrary notion. In probabilistic statistics, the so-called Student's distribution is used to make up for the fact that the sample-estimated standard deviation value is used instead of the exact one, but that distribution differs little from the Gaussian distribution when there are more than several hundred entities.

In many real-life applications, the shape of the underlying distribution is unknown and, moreover, the distribution is not necessarily stationary. The theoretically defined confidence boundaries are of little value then. This is why a question arises whether any confidence boundaries can be derived computationally by re-sampling the data at hand rather than by imposing some debatable assumptions. A popular approach to this is bootstrapping, which will be considered here in its two basic formats defined usually as those "pivotal" and "non-pivotal" (see Carpenter and Bithell 2000).

Bootstrapping is based on a pre-specified number, say 1000 or 5000 or 10,000 (nobody knows any rule on this), of random trials. A *trial* involves randomly drawn N entities, with replacement, from the entity set. Note that N is the size of the entity set. Since the sampling goes with replacement, some entities may be drawn two or more times so that some others are bound to be left behind. For instance, in a bootstrap trial of 10 entities, the following numbers have been drawn: 4, 5, 5, 10, 6, 1, 3, 4, 4, 3 (see the middle column in Table 2.6), so that four out of 10 indices have

Table 2.6 An illustrative bootstrap trial: the original feature x, in the left column; a sample of 10 indices randomly drawn out of 10 in the middle; and corresponding x-values, in the column on the right; the means are in the bottom line

Feature x	Trial r indices	Feature x over trial, x(r)
5	4	−3
6	5	4
−2	5	4
−3	10	1
4	6	2
2	1	5
1	3	−2
3	4	−3
−1	4	−3
1	3	−2
1.6		0.3

been left out of the trial while several multiple copies have got in. These multiples are 4, occurred three times, as well as 3 and 5, that occurred twice each. Is there any order in the random missing items?

Yes, there is. Recalling that e = 2.7182818... is the natural logarithm base, it is not difficult to prove that, on average, only approximately (e − 1)/e = 63.2% entities get selected into a trial sample. Indeed, at each random drawing from a set of N, the probability of an entity being not drawn is $1 − 1/N$, so that the approximate proportion of entities never selected in N draws is equal to $(1 − 1/N)^N \approx 1/e = 1/2.71828 \approx 36.8\%$ of the total number of entities. The approximate equation above comes from the so-called special limit, defining the e, $(1 + x)^{1/x} \Rightarrow e$ at x tending to 0.

A trial set of N randomly drawn entity indices (some of them, as explained, would coincide) is assigned with the corresponding row data values from the original data table so that coinciding entities get identical values. Then a method under consideration, currently "computing the mean", applies to this trial data to produce the trial result. After a number of trials, the user gets enough results to represent them with a histogram and derive confidence boundaries from that.

Consider a bootstrap trial for feature x in the left column of Table 2.6. at which its mean, 1.6, is presented too. The column in the middle is a list of indices randomly drawn between 1 and 10. The column on the right contains a resampled version of feature x—the values of x at the indices in the second column. One cannot help but noticing how biased the trial version of x is, just because of the bias in the trial sample; and its mean value is 0.3 now, far from the original 1.6. It is not an individual trial, but rather the set of all of them that creates a representative density function.

The bootstrap distributions for each of the three types of data generation mechanism, after 1000 trials, are presented in Figs. 2.8, 2.9 and 2.10 (Table 2.7).

The pivotal validation method is based on the assumption that the bootstrap distribution of means is Gaussian, so that having estimated its average m_b and standard deviation s_b, the 95% confidence interval is estimated as usual, with

Table 2.7 Aggregate characteristics of the results of 1000 bootstrap trials over short.dat array

Data type		Normal	Two-mode	Power law
Mean		10.24	16.94	287.54
Standard deviation	Original sample	0.25	0.70	129.33
	Bootsrap value	0.24	0.69	124.38
	Mean, %	2.46	4.05	43.26

formula $m_b \pm 1.96 * s_b = 10.24 \pm 1.96 * 0.24 = 10.24 \pm 0.47$, the interval between 9.77 and 10.71—which is very similar to that obtained under the hypothesis of Gaussian distribution—this is no wonder here because the hypothesis is true. However, the Gaussian assumption is not bad even in the case if the underlying distribution is not Gaussian because of the central limit theorem. According to this theorem, the distribution of the mean approximates the normal distribution under rather mild assumptions. Specifically, the approximating Gaussian distribution has its mean coinciding with the sample mean while the variance is equal to the sample variance divided by N.

There is theoretical evidence, presented by Efron and Tibshirani (1994), to support the view that the bootstrap can produce somewhat tighter estimate of the deviation than the estimate based on the original sample. In our case, we can see in Table 2.7 that indeed, with the means almost unchanged, the standard deviations have been slightly reduced.

The non-pivotal method makes no assumption of the distribution of bootstrap means and uses the empirical bootstrap found distribution to cut it at its 2.5% upper and bottom quantiles. To do this, we can sort values of the vector of bootstrap means and find the values at its 26th and 975th components that cut out exactly 2.5% of the set each. This action produces interval between 9.78 and 10.70, which is very close to the previously found boundaries for the 95% confidence interval for the mean value of the first sample.

Unfortunately, the bootstrap results are not that helpful in analyzing the other two distributions: as can be seen in our example, both of the means, the Two-modal and Power law ones, are assigned rather decent boundaries while, in most applications, the mean of either of these two distributions may be considered meaningless. It is a matter of applying other data analysis methods such as clustering to produce more homogeneous sub-samples whose distributions would be more similar to that of a Gaussian.

The reader is requested to provide pivotal and not-pivotal estimates of 95% confidence interval for the other two samples in short.dat dataset (Two-modal and Power law).

Worked Example 2.4 Bootstrap Confidence Interval for the Mean Sepal Width

Consider the Iris dataset and its feature w2, the Sepal Width, whose mean is m = 3.057. Let us draw its 5000-strong bootstrap resampling duplicate. Here are corresponding MatLab commands:

Fig. 2.11 A histogram of the bootstrap 5000-strong set of means for the Sepal width feature of the Iris dataset

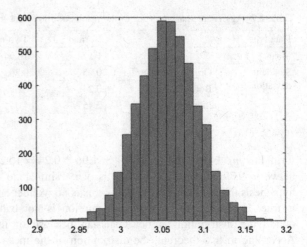

```
>> n=150; r=ceil(n*rand(n,5000));
```

The command "rand" produces a 150×5000 matrix of pseudo-random reals between 0 and 1. After multiplication by n = 150 the reals become between 0 and 150, and command "ceil" converts them into next larger integers that are between 1 and 150. Columns of matrix r are individual bootstrap trials, so that matrix xr = w2(r) is a set of bootstrap trial values of the feature w2 at the 5000 random samples. Now we are ready to draw the 5000-strong set of the corresponding means,

```
>> mx=mean(xr);
```

Let us derive the 95% confidence interval for the mean; a 25-bin histogram of vector mx is presented in Fig. 2.11 looking quite Gaussian. To use the pivotal method, let us find the mean and standard deviation of mx:

```
>> ma=mean(mx)= 3.0583
>> sa=std(mx)= 0.0360
```

Therefore, according to the pivotal method, the 95% confidence interval for the mean has its left boundary at lb = ma − 1.96 * sa = 2.988 and its right boundary at rb = ma + 1.96 * sa = 3.129.

To apply the non-pivotal method, let us first sort mx in the ascending order, making it sm, and take its 2.5% percentiles from the bottom and the top:

```
>> lb=sm(126)= 2.989
>> rb=sm(4875)= 3.131
```

2.3   Categorical Features and Mixed Scale Data

2.3.1   Distribution and Its Characteristics

A categorical feature differs from a quantitative one not just because its values are strings, not numbers—they are coded by numbers anyway to be processed. The average of a quantitative feature is always meaningful, whereas the averaging of categories, such as Occupations—BA, IT or AN—in Student data or Sector of Economy—Retail, Utility or Manufacture—in Company data, makes no sense even after they are coded by numbers. The applicability of the operation of averaging is indeed a characteristic property of being quantitative. For example, one may claim that a feature like the number of children in Student data is not quantitative because its values can only be whole numbers. Still, a statement like "the average number of children per woman is 1.85" does make sense because it can be easily made legitimate by moving to counting by hundreds: there are 185 children per every hundred women.

For categorical features, there is no need to define bins: the categories themselves play the role of bins. However, their histograms typically are visualized with bars or stems, like on Fig. 2.12 that represents the distribution of categories IT, BA and AN of Occupation feature in Student data.

The distribution of the feature can be expressed in absolute numbers of entities falling in each of the categories, that is, D = (35, 34, 31), or on the relative scale, by using proportions found by dividing frequencies over their total, 35 + 34 + 31 = 100, which leads to the relative frequency distribution d = (0.35, 0.34, 0.31).

This distribution is close to the uniform one in which all frequencies are equal to each other. In real life, many distributions are far from that. For example the

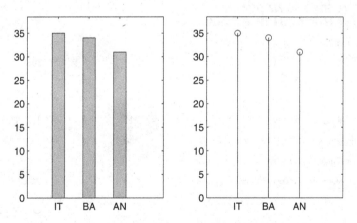

Fig. 2.12 The distribution of categories IT, BA and AN of Occupation feature in Student data shown with bars on the left and stems on the right

distribution by race of the 878,153 stop-and-search cases performed by police in England and Wales was widely discussed in the media (see Table 2.8 and BBC's website http://news.bbc.co.uk/1/hi/uk/7069791.stm of 29/10/07.) This is far from uniform indeed: the proportion of W category is thrice greater than of the other two taken together. Yet it was a claim of racial bias because the proportion of W category in the population is even higher than that (for further analysis, see Sect. 2.3).

Q.2.7. What is the modal category in the distribution of Table 2.8? In Occupation on Student data?
A. These are most likely categories, W in Table 2.8 and IT in Student data.

Recently, the so-called bubble chart format, developed for multidimensional visualizations, has been applied to 1D distributions (see Fig. 2.13) at which the distribution of most frequent causes of death in the USA is visualized with strips, on the left, and bubbles, on the right. At the current author's opinion, the bubble chart is more impressive. Also, it allows for visually estimating the risk of death from a medical error, about 10% (251 out of 2597).

Table 2.8 Race distribution of stop-and-search cases in England and Wales in 2005/6

Race	"Stop-and-search" case number	Relative frequency, %
Black (B)	131,723	15
Asian (A)	70,250	8
White (W)	676,180	77
Total	878,153	100

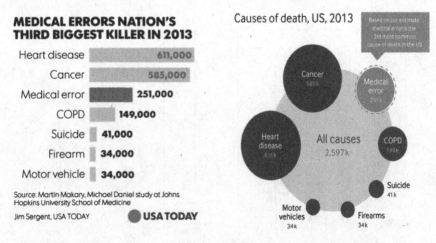

Fig. 2.13 The distribution of causes of death in the USA, 2013, according to Makary and Daniel (2016) by using strips, on the left, and bubbles (on the right). COPD is *Chronic obstructive pulmonary disease*

A number of coefficients have been proposed to evaluate how much a distribution differs from the uniform distribution. The most popular are the entropy and Gini index. The latter also is referred to as the categorical variance.

The entropy measures the information in signals being transferred over a communication channel. Intuitively, a rare signal bears more information than a more frequent one. Additionally, the levels of information in independent signals are to be summed to estimate the total information. These two requirements lead to the choice of logarithm of $1/p$, that is, $-\log(p)$, for scoring the level of information in a signal which appears with the probability (frequency) p. The logarithm's base is taken to be 2, because all the digital coding uses the binary number system. The entropy is defined as the averaged level of information in categories of a categorical feature. The unit of entropy has been chosen to be the bit, which is the entropy of a uniformly distributed binary feature, also referred to as a binary digit with two equally likely states. Intuitively, one bit is the level of information given in an answer to a Yes-or-No question in which no prior knowledge of the possible answer is assumed. The maximum entropy of a feature with m categories, $H = \log(m)$, is reached when their distribution is uniform. The maximum Gini index, $(m - 1)/m$, is reached at the uniform distribution too. Gini index measures the average level of error of the proportional classifier. Given a sequence of entities with unknown values of a categorical feature, the proportional classifier assigns entities with values chosen randomly, each with a probability proportional to its frequency. The average error of a category whose frequency is p is equal to $p(1 - p) = p - p^2$. If, for example, p = 20%, then the average error is $0.2 - 0.2 * 0.2 = 16\%$.

Worked Example 2.5. Entropy and Gini Index of a Distribution
Table 2.9 presents all the steps to compute the value of entropy, the summary $-p\log(p)$ value, and Gini index, the summary $p(1 - p)$ value where p are probabilities (relative frequencies) of categories.

Entropy is the averaged quantity of information in the three categories, $H = -p_1 \log(p_1) - p_2 \log(p_2) - p_3 \log(p_3)$. The entropy in Table 2.9 relative to the maximum is $0.99/1.585 = 0.625$ because at m = 3 the maximum entropy is $H = \log(3) = 1.585$. Gini index is defined as the average error of the proportional

Table 2.9 Entropy and Gini index for race distribution in Table 2.8

Distribution		Entropy		Categorical variance	
Category	Relative frequency p	Information $-\log(p)$	Weighted $-p\log(p)$	Error $1-p$	Variance $p(1-p)$
B	0.15	2.74	0.41	0.85	0.128
A	0.08	3.64	0.29	0.92	0.074
W	0.77	0.38	0.29	0.23	0.177
Total	1.00		0.99		0.378

prediction. The proportional prediction mechanism is defined over a stream of entities of which nothing is known beforehand except for the distribution of categories $\{p_l\}$. This mechanism predicts category l at an entity in p_l proportion of all instances. In our case, $G = p_1(1 - p_1) + p_2(1 - p_2) + p_3(1 - p_3) = 0.378$. The maximum Gini index value, $(m - 1)/m$, is reached at the uniform distribution, that is, $G = 2/3$. The relative Gini index, thus, is $0.378/(2/3) = 0.567$, which is not that different from the relative entropy.

A categorical feature such as Occupation in Students data or Protocol in Intrusion data, partitions the entity set in such a way that each entity falls in one and only one category. Categorical features of this type are referred to as *nominal* ones.

If a nominal feature has L categories $l = 1,..., L$, its distribution is characterized by quantities $N_1, N_2, ..., N_L$ of entities that fall in each of the categories. Because of the partitioning property these numbers sum to the total number of entities, $N_1 + N_2$ .... $N_L = N$. The relative frequencies, defined as $p_l = N_l/N$, sum to the unity ($l = 1, 2, ..., L$).

Since categories of a nominal feature are not ordered, their distributions are better visualized by pie-charts than by histograms.

The concepts of centrality, except for the mode, are not applicable to categorical feature distributions. Spread here is also not quite applicable. However, the variation—or diversity—of the distribution ($p_1, p_2, ..., p_L$) can be measured. There are two rather popular indexes that evaluate dispersion of the distribution, Gini index, or categorical variance, and entropy.

Gini index G is the average error of the proportional prediction rule. According to the proportional prediction rule, each category l, $l = 1,2, ..., L$, is predicted randomly with the distribution (p_l), so that l is predicted at Np_l cases out of N. The average error rate of predictions of l in this case is equal to $1 - p_l$, which makes the average error rate to be equal to:

$$G = \sum_{l=1}^{L} p_l(1 - p_l) = 1 - \sum_{l=1}^{L} p_l^2 \tag{2.14}$$

Entropy averages the quantity of information in category l as measured by $\log(1/p_l) = -\log(p_l)$. over all l. The entropy is defined as

$$H = -\sum_{l=1}^{L} p_l \log p_l \tag{2.15}$$

Fig. 2.14 Graphs of functions of the error $f(p) = 1 - p$ involved in Gini index (dashed line) and the information $f(p) = -log(p)$

There is an affinity between H and G indexes, because at small p, $-\log(1 - p)$ and $1 - p$ coincide, up to a very minor difference, as is well known from calculus (see Fig. 2.14).

If the distribution of a feature is in vector df, then a command like

>> bar(df, .4);h=axis;axis(1.1*h);

will produce its bar drawing. The parameters here are: 0.4 the width of bars, 1.1 the rescaling to allow some air between the histogram and the border in the drawing frame (see Fig. 2.12 on the left).

Computation of the entropy and Gini index for the distribution presented in vector df can be done with commands:

>> df=df/sum(df); h=-sum(df .*\log2(df)); % h is entropy
>> df=df/sum(df); g=-sum(df .*(1-df)); % h is Gini

Q.2.8. Take nominal features from the Intrusion data set and generate category-based binary features, after which compute their individual means and variances. Compare the variances with Gini index for the original features.

2.3.2 Binary Features

A very important class of nominal features consists of features with only two categories—binary features. A feature admitting only two, either "Yes" or "No", values is conventionally considered Boolean in Computer Sciences, thus relating it to Boolean algebra with its "True" and "False" statement evaluations. We do not adhere to this strict logic approach but rather engage the numbers and arithmetic, thus referring to two-value nominal features as Binary. The values are coded by numerals 1, for "Yes", and 0, for "No". This coding is conventionally referred to, in statistics, as dummy. In the context of data analysis, the coding is meaningful because it makes the operation of averaging meaningful. Indeed, the average of a dummy feature over any set of objects is the relative frequency of the category coded by 1, on the set.

Curiously, binary features can be meaningfully treated within the perspective of scale types defined with admissible transformations. Indeed, the scale type of a binary feature is twofold: nominal and interval simultaneously. To prove that, consider a binary feature with categories x1 and x2. Then a one-to-one mapping φ can transform them into x1′ and x2′ which are not equal to each other. Therefore, equations x1′ = ax1 + b and x2′ = ax2 + b with two unknown reals, a and b, have one and only one solution, $a = (x1 - x2)/(x1' - x2')$ and $b = x1' - a$x1, thus defining an interval scale transformation with scale factor a and scale shift b. Obviously, this transformation coincides with the mapping φ.

Moreover, a binary feature may be considered even as belonging in the ratio scale format. Just take $b = 0$ in the above discussion, and you obtain any $0/a$ coding in which 0 is assigned to the "No" category and a, to the "Yes" category. This is especially convenient when using linear algebra transformations of the data matrix. Consider, say, multiplication of the data matrix Y by a column-vector c on the right, $y = Yc$. If Y contains a dummy column v, then its multiplication by the corresponding component c_v corresponds to the $0/a$ ratio scale coding where $a = c_v$. A question emerges: can matrix algebra be used to maintain the original interval scale type for a binary feature? Yes, it can! The only thing needed for that is doubling the dummy feature. A binary feature can be represented by two dummy columns, v and v', one for the "Yes" category and the other for the "No" category, so that they complement each other and sum to a vector of all unities. Then, in Yc, the categories will be represented by two reals, c_v and $c_{v'}$, thus corresponding to b and $a + b$ where a is the scale factor and b, the scale shift.

These combine properties of both categorical and quantitative features. Indeed, an important difference between categorical and quantitative features is in their admissible coding sets. An admissible numerical recoding of values of a feature changes them consistently, in such a way that the relations between entities according to the feature remain intact. For example, the human heights in centimeters can be recoded in millimeters, by multiplying them by 10, or temperatures at various locations expressed in Fahrenheit can be recoded in Celsius, by subtracting 32 and dividing the result by 1.8. Such a recoding would not change the relations between locations that have been put in effect when Fahrenheit temperatures had been recorded. If, however, we assign arbitrary values to the temperatures, the new set will be inconsistent with the previous one and give a very different information. This is the borderline between quantitative and nominal features: the nominal feature can only compare if the categories are the same or not, thus admitting any one-to-one recoding as admissible, whereas the quantitative feature can only admit shifts of the origin of the scale and changes of the scale factor. This borderline however is not quite hard. Binary features, as nominal ones, admit any numerical recoding. But the recoding, in this case, can always be expressed as a shift of the origin and change of the scale factor. Indeed, for any two numbers, α and β, a conversion of the feature values from 0 to α and from 1 to β can be achieved in a conventional quantitative fashion by using two rescaling parameters: the shift of the origin (α) and scaling factor ($\beta - \alpha$).

Thus, a binary feature can be always considered as coded into a quantitative 1/0 format, 1 for Yes and 0 for No. Thus coded, a binary feature sometimes is referred to as a dummy variable.

The mean of a 1/0 coded binary feature is the proportion of its "Yes" values, which is rather meaningful. The other popular central values bear much less information. The median is 1 only if the proportion of ones is 0.5 or greater; otherwise, it is 0. In a rare event when the number of entities is even and the proportion of ones is exactly one half, the median is one half too. The mode is either 1 or 0, the same as the median.

To compute the variance s^2 of a binary feature whose mean $c = p$, sum Np items $(1 - p)^2$ and $N(1 - p)$ items p^2, and divide the result by N, which altogether leads to $s^2 = p(1 - p) = p - p^2$. Accordingly, the standard deviation is the square root of the variance, $s = \sqrt{p(1 - p)}$. Obviously, this is maximum when $p = 0.5$, that is, both binary values are equally likely. The range is always 1. The absolute deviation, in the case when $p < 0.5$ so that median $m = 0$, comprises Np items that are 1 and N $(1 - p)$ items that are 0, so that $sm = p$. When $p > 0.5$, $m = 1$ and the number of unity distances is $N(1 - p)$ leading to $sm = 1 - p$. That means that, in general, $sm = \min(p, 1 - p)$, which is less than or equal to the standard deviation. Indeed, if $p \leq 0.5$, then $p \leq 1 - p$ and, thus, $p^2 \leq p(1 - p)$, so that $ms \leq s$. Analogously, if $p > 0.5$ then $p > 1 - p$ and, thus, $p(1 - p) > (1 - p)^2$, so that again $sm < s$, which proves the statement.

When a categorical feature is converted into a set of binary features corresponding to its categories, the total variance of the L binary variables is equal to the Gini index, or categorical variance, of the original feature.

There are some probabilistic underpinnings to binary features. Two models are popular, one by Bernoulli and another by Poisson. Given p, $0 \leq p \leq 1$, Bernoulli model assumes that every x_i is either 1, with probability p, or 0, with probability $1 - p$. Poisson model suggests that, among the N binary numerals, random pN are unities, and $(1 - p)N$ zeros. Both models yield the same mathematical expectation, p. However, their variances differ: the Bernoulli distribution's variance is $p(1 - p)$, whereas the Poisson distribution's variance is p, which is obviously greater for all positive p, because the factor at Bernoulli standard deviation, $1 - p$, is less than 1 under this condition. Similar models can be considered for nominal features with more than two categories.

2.3.3   Dummy Matrix; Spanning Subspace and Equivalence Relation

Given a nominal feature x on the set $I = \{1, 2, ..., N\}$ with values v_k ($k = 1, 2, ..., K$), each value v_k defines a set $R_k \subseteq I$ consisting of such entities i that $x(i) = v_k$. Roughly speaking, R_k is the set of entities falling in the category v_k. For the mathematical simplicity, it is assumed that no R_k is empty, that is, that for every category, there are some objects falling in it. Of course, in reality, the set I may be a subset of a larger sample so that some categories could be absent from I. For example, if one takes only enterprises from the Manufacturing sector, then no object will have the Retail category of the Sector feature. Those categories that are absent from the set, must be removed from the set of feature values before applying a data analysis method. The R_k ($k = 1, 2, ..., K$) sets form a partition $R = \{R_1, R_2, ..., R_K\}$ which may be considered a portrayal of the feature over set I. To apply linear algebra operations, any K-valued nominal feature x can be represented by an $N \times K$ 1/0-matrix $X = (x_{ik})$ where $x_{ik} = 1$ if $x(i) = v_k$, or, equivalently, if $i \in R_k$, and $x_{ik} = 0$, otherwise

$(k = 1,2,\ldots, K)$. This matrix X will be referred to as the dummy matrix of the nominal feature x.

It is not difficult to prove that the dummy matrix of any nominal feature satisfies the following two properties:

- Each its row i, $i = 1$, 2, $\ldots$, N, contains one and only one unity, that one corresponding to the category k such that $x(i) = k$;
- Each its column contains at least one unity.

Reciprocally, any binary matrix satisfying these properties is the dummy matrix of a nominal feature.

It is not difficult to prove then that the sum of X columns is an N-dimensional unity vector. Let us denote 1_n an n-dimensional column-vector whose all components are unities, so that $I_n^T = (1, 1, \ldots, 1)$. Then this property can be written as a linear algebraic equation $X1_K = 1_N$.

Let us consider the span of X, $L(X) = \{y = Xa$ for some K-dimensional $a\}$. This linear subspace $L(X)$ consists of vectors $y = Xa$, each of which represents the feature x by a quantitative encoding its categories. More precisely, every category k is quantitatively coded in y as a_k $(k = 1,2,\ldots, K)$. $L(X)$ corresponds to all possible numerical encodings of the categories including those at which different categories k and l may be coded by the same values $a_k = a_l$. That is, the space $L(X)$ represents the nominal scale x in an extended sense, at which admissible transformations are not only one-to-one mappings but any mappings admitting merging categories as well.

Unfortunately, all the spaces $L(X)$ have a purely formal part in them, the bisector, the one-dimensional subspace consisting of all the N-dimensional vectors $\alpha 1_N$ consisting of equal values α where α is any real, because of the property of dummy matrices that their columns sum to the 1_N-vector. Therefore, it would be more productive to consider the space $L(X)$ without the bisector, that is, an orthogonal complement $L^-(X)$ to the bisector consisting of all centered vectors. Indeed, $\langle z, 1_N \rangle = 0$ if and only if z is a centered vector so that $z_1 + z_2 + \cdots + z_N = 0$.

Consider the orthogonal projector $P_{L^-(X)}$ onto the space $L^-(X)$. It is not difficult to prove that $P_{L^-(X)} = P_X - P_1$ where $P_X = X(X^T X)^{-1} X^T$ is the orthogonal projector onto $L(X)$ and $P_1 = 1_N (1_N^T 1_N)^{-1} 1_N^T$ is the orthogonal projector onto the bisector.

To see what is P_X, let us first find $X^T X$, that is a diagonal matrix because columns of X are mutually orthogonal, and the diagonal consists of frequencies $N_1 = |R_1|, N_2 = |R_2|, \ldots, N_K = |R_K|$ of the categories. Therefore, $(X^T X)^{-1}$ is a diagonal matrix with diagonal elements equal to $1/N_1, 1/N_2, \ldots, 1/N_K$. Matrix XX^T is a 1/0 binary matrix of a block structure. For every k, it has a unity for every pair $(i,j) \in R_k \times R_k, k = 1, 2, \ldots, K$, while 0 stands at all other pairs (i, j). Matrix $P_X = X(X^T X)^{-1} X^T$ has a similar structure: For every k, its (i, j)-th element is $1/N_k$ for all $(i,j) \in R_k \times R_k, k = 1, 2, \ldots, K$, while 0 stands at all other pairs (i, j).

We can similarly find that $1_N^T 1_N = N$, so that $(1_N^T 1_N)^{-1} = 1/N$, so that $P_1 = 1_N (1_N^T 1_N)^{-1} 1_N^T$ has all its elements equal to $1/N$.

This gives us the structure of the $N \times N$ orthogonal projector matrix $P_{L^-(X)}$: For every k, its (i, j)-th element is $1/N_k - 1/N = (1 - N_k/N)/N_k$ for all $(i, j) \in R_k \times R_k$, $k = 1,2,\ldots, K$, while $-1/N$ stands at all other (i, j) places.

It should be noted that the auxiliary 1/0 binary matrix XX^T of a block structure has an independent meaning as the matrix representing a binary equivalence relation. A binary relation is a logics concept that can be expressed in a set-theoretic way. In this case, the relation can be formulated as follows: Any objects $i, j \in I$ are ε_x-equivalent if and only if they fall in the same x-category, that is, $x(i) = x(j)$. This relation can be expressed in a set-theoretic way as the set of ε_x-equivalent pairs (i, j), that is, $\varepsilon_x = R_1 \times R_1 \cup R_2 \times R_2 \cup \ldots \cup R_K \times R_K$. Matrix XX^T is the indicator of ε_x. Indeed, as we have seen, for every k, its elements are unities for all pairs $(i, j) \in R_k \times R_k$, $k = 1,2,\ldots, K$, while 0 stands at all other pairs (i, j).

A binary relation ρ is said to be an equivalence relation if it is: (i) reflexive, so that $(i, i) \in \rho$ for any $i \in I$; (ii) symmetric, so that $(i, j) \in \rho$ implies $(j, i) \in \rho$ for all $i, j \in I$; and (iii) transitive, so that $(i, j) \in \rho$ and $(j, k) \in \rho$ implies $(i, k) \in \rho$ for any triplet $i, j, k \in I$. An equivalence relation one-to-one corresponds to a partition of the set of objects in equivalence classes. An example: ε_x-equivalence and partition $R = \{R_1, R_2, \ldots, R_K\}$.

There is a rather natural, though somewhat less recognized, relation between quantitative and binary features: the variance of a quantitative feature is always smaller than that of the corresponding binary feature. To explicate this according to Mirkin (2012), assume the interval [0, 1] to be the range of data $X = \{x_1,\ldots,x_N\}$. Assume that the mean c divides the interval in such a way that a proportion p of the data is greater than or equal to c, whereas proportion of those smaller than c is $1 - p$. The question then is this: given p, at what distribution of X the variance is maximized. To address the question, assume that X is any given distribution within interval [0, 1] with its mean at some interior point c. According to the assumption, there are Np observations between 0 and c. Obviously, the variance can only increase if we move each of these points to the boundary, 0. Similarly, the variance will only increase if we push each of $N(1 - p)$ points between c and 1, into the opposite boundary 1. That means that the variance $p(1 - p)$ of a binary variable with Np zero and $N(1 - p)$ unity values is the maximum, at any p. The following is proven. Given a $p > 0$, a binary variable, whose distribution is $(p, 1 - p)$, has the maximum variance, and the standard deviation, among all quantitative variables of the same range and the proportion p of entries below its average.

This implies that no variable over the range [0, 1] has its variance greater than the maximum ¼ reached by a binary variable at $p = 0.5$. The standard deviation of this binary variable is ½, which is just half the range. Therefore, the standard deviation of any variable cannot be greater than its half-range.

The binary variables also have the maximum absolute deviation among the variables of the same range, which can be proven similarly.

Worked Example 2.6. Standardizing the Iris Dataset
Consider Iris dataset in Table 1.3. Its grand mean and midrange are presented in Table 2.10, along with its range and standard deviations.

Table 2.10 Characteristics of Iris dataset

Characteristics	Features			
	w1	w2	w3	w4
Mean, m	5.84	3.06	3.76	1.20
Midrange, mr	6.10	3.20	3.95	1.30
Standard deviation, s	0.83	0.44	1.77	0.76
Range, ra	3.60	2.40	5.90	2.40

These have been found by using the following MatLab commands:

```
>> iris=load('Data\iris.dat');
>> m=mean(iris); % grand mean
>> ma=max(iris);% maximum
>> ma=min(iris);% minimum
>> mr=(ma+mi)/2; % midrange
>> s=std(iris);% standard deviation
>> ra=ma-mi;% range
```

In Table 2.10, midrange more or less follows the grand mean, but there are some discrepancies between the range and standard deviation. For example, ranges of w2 and w4 are the same, whereas standard deviations differ by almost 100%.

Let us take three different standardizations:

A—range related, $y \Leftarrow (x - mr)/ra$;
B—mean/range standardization, $y \Leftarrow (x - m)/ra$;
C—z-scoring, $y \Leftarrow (x - m)/s$;

and evaluate feature contributions to the data scatter after each of them (Table 2.11).

Feature contributions under A and B are similar, because both involve division by the range. According to these standardizations features w3 and w4 contribute most, because they are bimodal (Sect. 2.3.2) and, thus play important role in further summarization methods, both Principal component analysis and cluster analysis. This concurs with the botanists' view that it is these sizes that determine the belongingness of an Iris specimen to a specific taxon (see references in Mirkin 2012). Moreover, at building a classification tree over Iris dataset, in Sect. 3.8, feature w4 will be involved in the splits according to three goodness criteria. In contrast, the first line assigns contributions according to feature values so that the lengths w1 and w3 get much larger contributions than the widths w2 and w4. And z-scoring (standardization C) makes all features contribute similarly, even in spite of the fact that two of them are bimodal.

Table 2.11 Iris feature contributions to data scatter at different standardizations, percent to the data scatter value

Standardization	Feature contributions, %			
	w1	w2	w3	w4
No standardization	54.76	15.00	27.07	3.17
A: midrange/range	20.16	12.70	31.48	35.66
B: mean/range	19.15	11.94	32.40	36.51
C: mean/std	25.00	25.00	25.00	25.00

The problem of standardization can be addressed by the user if they know the type of the distribution behind the observed data—the parameters of the distribution typically lead to a reasonable standardization. For example, the data should be standardized by z-scoring if the data is generated by independent one-dimensional Gaussian distributions. According to the formula for Gaussian density, a z-scored feature column would then fit the conventional $N(0, 1)$ distribution making all features comparable to each other. A similar strategy applies if the data is generated from a multivariate Gaussian density, just the data first need to be transformed into mutually orthogonal singular vectors or, equivalently, principal components. Then z-standardization applies.

If no reasonable distribution can be assumed in the data, then there is no universal advice on standardization. However, with the summarization problems that we are going to address, the principal component analysis and clustering, some advice can be given in terms of the data scatter.

The data transformation effected by the standardization can be expressed as

$$y_{iv} = (x_{iv} - a_v)/b_v \tag{2.16}$$

where $X = (x_{iv})$ stands for the original and $Y = (y_{iv})$ for standardized data, whereas $i \in I$ denotes an entity and $v \in V$ a feature. Parameter a_v stands for the shift of the origin and b_v for normalizing factor at each feature $v \in V$. In other words, one may say that the transformation (2.16), first, shifts the data origin into the point $a = (a_v)$, after which each feature v is rescaled separately by dividing its values over b_v.

The position of the space's origin, zero point $0 = (0,0, \ldots, 0)$, at the standardized data Y is unique because any linear transformation of the data, that is, matrix product AY, for any A, can be expressed as a set of rotations of the coordinate axes around the origin, so that the origin itself is invariant. The principal component analysis can be expressed mathematically as a set of twisted linear transformations of the data features as becomes clear in Sect. 2.4, which means that all the action in this method occurs around the origin. Metaphorically, the origin can be likened to the eye through which data points are looked at by the methods below. Therefore, for the purposes of data analysis, the origin should be put somewhere in the center of the data set, for which the gravity center, the point of all within-feature averages, is a best candidate. Let us recall that the data Y scatter is defined as the sum of the squares of all its entries, $T(Y) = \sum_{i,v} y_{iv}^2$. What is nice about locating the origin to the center of gravity of the data, is that the feature contributions to the scatter of the center-of-gravity standardized data are equal to $t_v = \sum_{i \in I} y_{iv}^2$ $(v \in V)$, which means that they are proportional to the feature variances. Indeed, after the average c_v has been subtracted from all values of the column v, the summary contribution satisfies equation $t_v = N\sigma_v^2$ so that t_v is N times the variance. Even nicer properties of the gravity center as the origin have been derived in the framework of the simultaneous analysis of categorical and quantitative data, see in Sects. 3.6, 3.8 and 4.4.

As to the normalizing coefficients, b_v, their choice is underlied by the idea of balancing the features weights. A most straightforward expression of the principle of feature equal importance is the use of the standard deviations as the normalizing coefficients, $b_v = \sigma_v$. This standardization makes the variances of all the variables $v \in V$ equal to 1 so that all the feature contributions become equal to $t_v = N$, which is seen at Table 2.11.

A very popular way to take into account the relative importance of different features is by using weight coefficients of features in computing the distances. This, in fact, is equivalent to and can be achieved with a proper standardization. Take, for instance, the weighted squared Euclidean distance between arbitrary entities $x = (x_1, x_2,..., x_V)$ and $y = (y_1, y_2,..., y_V)$ which is defined as

$$D_w(x, y) = w_1(x_1 - y_1)^2 + w_2(x_2 - y_2)^2 + \cdots + w_V(x_V - y_V)^2$$

where w_v are pre-specified weights of features $v \in V$. Let us define (additional) normalizing parameters $b_v = 1/\sqrt{w_v}$ $(v \in V)$ to transform x and y into $x'_v = x_v/b_v$ and $y'_v = y_v/b_v$. It is rather obvious that

$$D_w(x, y) = d(x', y')$$

where d is the unweighted Euclidean squared distance.

That is, the following fact holds: for the Euclidean squared distance, the feature weighting is equivalent to an appropriate normalization as described above.

Q.2.9. Is it true that the sum of feature values standardized by subtracting the mean is zero?

A. Yes, because the sum is proportional to the mean which is zero after centering.

Q.2.10. Consider a reversal of the operations in standardizing data: the scaling to be followed by the scale shift. Is it that different from the conventional standardization?
A. Denote the scale shift and rescaling factor by a and b. Then the conventional standardization produces $y = (x - a)/b = x/b - a/b$ from x, whereas that suggested gives $z = x/b - a$. These differ at $a \neq 0$. To make them equal, the scale shift in the latter case must be a/b.

2.3.4 Quantification and Standardization of Mixed Scale Data

Many data analysis specialists prefer treating different scale type data differently, by converting the non-numerical part of data into an object-to-object similarity matrix while keeping the quantitative part as is. This author, however, prefers keeping the mixed data scales all together by quantifying the entire data table in a most non-destructive way. Given a data table containing mixed scale features, quantitative, rank, binary, and nominal, among them, the user first should take care of rank features. As there is no theory to embrace the rank scale type into a unified format, rank features should be converted into either quantitative or nominal format. In this author and his colleagues experiences, following C. Spearman by coding all the ranks by consecutive numbers (say, $-2, -1, 0, 1, 2$ or $1, 2, 3, 4, 5$, which does not matter if a unified centering applies as a pre-processing option) works quite well in practice. Next, all the binary features are to be considered quantitative by assigning 1 to their "Yes" value and 0, to the "No" value. Then, every nominal feature should be quantified by substituting its dummy matrix instead of the feature's column. This is examplified at the Company dataset in the Project 2.1 below. If the number of values of a given nominal feature is m, then in this way, m columns are inserted instead of just one column in the original data table.

This creates a disbalance in the data scatter, an important characteristic of the data, according to data-driven summarization models which will be considered in further sections. The data scatter is the sum of squares of all the data table entries. Instead of N items in the sum from the orginal feature, the dummy matrix adds mN items, thus making the contribution of the feature, in general, m times greater. To make up for such a growth, the dummy matrix should be divided by $\sqrt{m}$. Why the square root? Because all the values are squared in the data scatter. After this, thus quantified data matrix should be standardized by subtracting a background central value from every column and dividing the result by a factor to balance contributions of different features. This is another twilight zone of data analysis at which the customs prevail rather than theory (see Project 2.1 for illustration).

The utilized method of quantification of categoricak values via dummy variables is not without a merit. First of all, the binary representation unifies the concept of the average and the frequency: the latter is the average of a dummy variable. Second, in clustering problems, geometric characteristics of binary data over partitions being built are mathematically equivalent to popular association measures defined for cross-classifications, such as Pearson's chi-squared. This bridges the gaps between the geometric and statistical perspectives on the analysis of categorical data.

Table 2.12 Data of eight companies producing goods A, B, or C according to the initial symbol of company's name

Company	Income	MShare	NSup	EC	Sector
Aversi	19.0	43.7	2	No	Utility
Antyos	29.4	36.0	3	No	Utility
Astonite	23.9	38.0	3	No	Manufacture
Bayermart	18.4	27.9	2	Yes	Utility
Breaktops	25.7	22.3	3	Yes	Manufacture
Bumchist	12.1	16.9	2	Yes	Manufacture
Civok	23.9	30.2	4	Yes	Retail
Cyberdam	27.2	58.0	5	Yes	Retail

Project 2.1. Standardization of Mixed Scale Data and Its Effect

Pr2.1.A Data Table and Its Quantification

Consider the Company dataset described in Sect. 1.2 and copied here in Table 2.12. Let us convert it into a quantitative format. The table contains two categorical variables, EC, with categories Yes/No, and Sector, with categories Utility, Manufacture and Retail. The former feature, EC, in fact represents just one category, "Using E-Commerce" and can be recoded as such by substituting 1 for Yes and 0 for No. The other feature, Sector, has three categories. To be able to treat them in a quantitative way, one should substitute each by a dummy variable.

Specifically, the three dummy features are:

(i)   Utility: Is it Utility sector?
(ii) Manufacture: Is it Manufacture sector?
(iii) Retail: Is it Retail sector?

Each of them admits Yes or No values, respectively substituted by 1 and 0. In this way, the original heterogeneous table will be transformed into a quantitative matrix in Table 2.13.

The first two features, Income and MShare, dominate the data in Table 2.13, especially with regard to the data scatter, that is, the sum of all the data entries squared, equal to 14,833. As shown in Table 2.14, the two of them contribute more

Table 2.13 Quantitatively recoded Company data table, along with summary characteristics

Company	Income	Market Share	NSup	EC	Util	Manu	Reta
Aversi	19.0	43.7	2	0	1	0	0
Antyos	29.4	36.0	3	0	1	0	0
Astonite	23.9	38.0	3	0	0	1	0
Bayermart	18.4	27.9	2	1	1	0	0
Breaktops	25.7	22.3	3	1	0	1	0
Bumchist	12.1	16.9	2	1	0	1	0
Civok	23.9	30.2	4	1	0	0	1
Cyberdam	27.2	58.0	5	1	0	0	1
Average	22.45	34.12	3.0	5/8	3/8	3/8	1/4
St deviation	5.26	12.10	1.0	0.48	0.48	0.48	0.43
Midrange	20.75	37.45	3.5	0.5	0.5	0.5	0.5
Range	17.3	41.1	3.0	1.0	1.0	1.0	1.0

Table 2.14 Within-column sums of the squared entries in Table 2.13

Contribution	Income	MShare	NSup	EC	Util	Manu	Retail	Data scatter
Absolute	4253	10,487	80	5	3	3	2	14,833
Percent	28.67	70.70	0.54	0.03	0.02	0.02	0.01	100.00

than 99% to the data scatter. To balance the contributions, features should be rescaled. Another important transformation of the data is the shift of the origin, because it affects the value of the data scatter and the decomposition of it in the explained and unexplained parts, as can be seen on Fig. 2.2.

Pr2.1.B Visualization of the data at different normalizations

One can take a look at the effects of different standardization options. Table 2.15 contains data of Table 2.12 standardized by the scale shifting only: in each column, the within-column average has been subtracted from the column entries. Such standardization is referred to as centering.

The relative configuration of the 7-dimensional row-vectors in Table 2.14 can be captured by projecting them onto a plane, which is two-dimensional, in an optimal way; this is provided by the two first singular values and corresponding singular vectors, as will be explained later in Sect. 2.5. This visualization is presented on Fig. 2.15 at which different product companies are shown with different shapes: squares (for A), triangles (for B) and circles (for C). As expected, this bears too much on features 2 and 1, that contribute 83.2 and 15.7%, respectively; a slight change from the original 70.7 and 23.7% according to Table 2.14. The features seem not related to products at all—the products are randomly intermingled with each other in the picture.

Pr2.1.C Standardization by z-scoring

Consider now a more balanced standardization involving not only feature centering but also feature normalization over the standard deviations—z-scoring, as presented in Table 2.16.

Table 2.15 The data in Table 2.13 standardized by the shift scale only, with the within-column averages subtracted

Ave	−3.45	9.58	−1.00	−0.62	0.62	−0.38	−0.25
Ant	6.95	1.88	0	−0.62	0.62	−0.38	−0.25
Ast	1.45	3.88	0	−0.62	−0.38	0.62	−0.25
Bay	−4.05	−6.22	−1.00	0.38	0.62	−0.38	−0.25
Bre	3.25	−11.82	0	0.38	−0.38	0.62	−0.25
Bum	−10.4	−17.22	−1.00	0.38	−0.38	0.62	−0.25
Civ	1.45	−3.92	1.00	0.38	−0.38	−0.38	0.75
Cyb	4.75	23.88	2.0	0.38	−0.38	−0.38	0.75
Cnt	221.1	1170.9	8.0	1.9	1.9	1.9	1.5
Cnt %	15.7	83.2	0.6	0.1	0.1	0.1	0.1

The values are rounded to the nearest two-digit decimal part, choosing the even number when two are the nearest. The bottom rows represent contributions of the columns to the data scatter as they are and percent

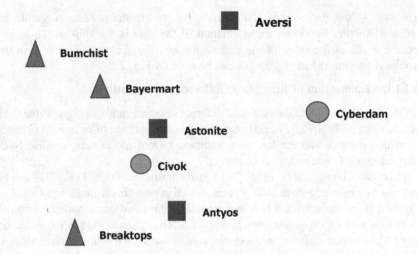

Fig. 2.15 Visualization of the entities in Companies data, centered only

Table 2.16 The data in Table 2.13 standardized by z-scoring

Ave	−0.66	0.79	−1.00	−1.29	1.29	−0.77	−0.58
Ant	1.32	0.15	0	−1.29	1.29	−0.77	−0.58
Ast	0.28	0.32	0	−1.29	−0.77	1.29	−0.58
Bay	−0.77	−0.51	−1.00	0.77	1.29	−0.77	−0.58
Bre	0.62	−0.98	0	0.77	−0.77	1.29	−0.58
Bum	−1.97	−1.42	−1.00	0.77	−0.77	1.29	−0.58
Civ	0.28	−0.32	1.00	0.77	−0.77	−0.77	1.73
Cyb	0.90	1.97	2.00	0.77	−0.77	−0.77	1.73
Cnt	8	8	8	8	8	8	8
Cnt, %	14.3	14.3	14.3	14.3	14.3	14.3	14.3

The bottom rows represent contributions of the columns to the data scatter as they are and percent

An interesting property of this standardization is that contributions of all features to the data scatter are equal to each other, and moreover, they total to the number of entities, 8! This is not a coincidence but a property of the z-scoring standardization.

The data in Table 2.16 projected onto the plane of two first singular vectors better reflect the products—on Fig. 2.16, C companies are clear-cut separated from the others; yet A and B are still intertwined.

Pr2.1.D Range Normalization and Rescaling of Dummy Features

We would like now to standardize the data in a mixed way by: (a) shifting the scales to the averages, as in z-scoring, but (b) dividing the results not by the feature's standard deviations but rather their ranges. However, when using both categorical and quantitative features, there is a catch here: each of the categories represented by dummy binary variables will have a greater variance than any of the

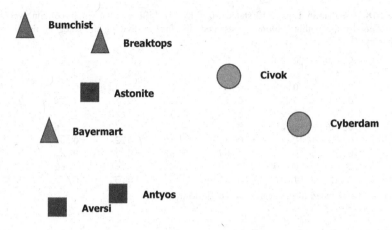

Fig. 2.16 Visualization of the entities in Companies data after z-scoring (Table 2.16)

Table 2.17 Within-column sums of the entries squared in the data of Table 2.13 standardized by subtracting the averages and dividing the results by the ranges

Contribution	Income	MShar	NSup	EC	Util	Manu	Retail	Data scatter
Absolute	0.739	0.693	0.889	1.875	1.875	1.875	1.500	9.446
Percent	7.82	7.34	9.41	19.85	19.85	19.85	15.88	100.00

quantitative counterparts after dividing by the ranges. Table 2.17 represents contributions of the range-standardized columns of Table 2.13.

Binary variables contribute much greater than the quantitative variables according to this standardization. The total contribution of the three categories of the original variable Sector looks especially odd—it is more than 55% of the data scatter, by far greater than should be assigned to one of the five original variables. This is partly because the variable Sector in the original table has been enveloped into three variables corresponding to its categories, thus blowing up the contribution accordingly. To make up for this, the summary contribution of the three dummies should be decreased back three times. This can be done by making further normalization of them by dividing the normalized values by the square root of their number—3 in our case. Why the square root is used, not just 3? Because contribution to the data scatter involves not the entries themselves but their squared values.

The data table after additionally dividing entries in the three right-most columns over $\sqrt{3}$ is presented in Table 2.18. One can see that the contributions of the last three features did decrease threefold from those in Table 2.17, though the relative contributions changed less. Now the most contributing feature is the binary EC that divides the sample along the product-based lines. This probably has contributed to the structure visualized in Fig. 2.17. The product defined clusters, much blurred on the previous figures, are clearly seen here, which shows that the original features

Table 2.18 The data in Table 2.13 standardized by: (i) shifting to the within-column averages, (ii) dividing by the within-column ranges, and (iii) further dividing the category based three columns by $\sqrt{3}$

Av	−0.20	0.23	−0.33	−0.63	0.36	−0.22	−0.14
An	0.40	0.05	0	−0.63	0.36	−0.22	−0.14
As	0.08	0.09	0	−0.63	−0.22	0.36	−0.14
Ba	−0.23	−0.15	−0.33	0.38	0.36	−0.22	−0.14
Br	0.19	−0.29	0	0.38	−0.22	0.36	−0.14
Bu	−0.60	−0.42	−0.33	0.38	−0.22	0.36	−0.14
Ci	0.08	−0.10	0.33	0.38	−0.22	−0.22	0.43
Cy	0.27	0.58	0.67	0.38	−0.22	−0.22	0.43
Cnt	0.74	0.69	0.89	1.88	0.62	0.62	0.50
Cnt %	12.42	11.66	15.95	31.54	10.51	10.51	8.41

The values are rounded to the nearest two-digit decimal part

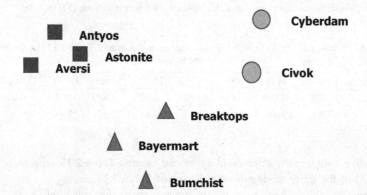

Fig. 2.17 Visualization of data in Table 2.18 standardized by the ranges with further subdividing the binary category features by the square root of 3

indeed are informative of the products—just a proper standardization has to be carried out.

Note: only two different values stand in each of the four columns on the right—why? Moreover, the entries within every column sum to 0 (see Q.2.9 and Table 2.18).

Q.2.11. How to do z-scoring in MatLab?

A. Take [n, v] = size(X) where X is the data matrix. Then define Y = (X-repmat (mean(X, n, 1)))./repmat(std(X, n, 1).

Q.2.12. What are the feature contributions after z-scoring?
A. They all are equal to the same value, the data scatter related to V, the number of features.

Q.2.13. How distances are affected if a different set of scale shifts is applied?

A. Since the coordinates of two points are subtracted from each other in the distance, the scale shifts cancel each other and have no effect on the distances: the distances do not depend on the location of the origin of the space.

2.4 Principal Component Analysis (PCA): Model, Method, Usage

The method of principal component analysis (PCA) has emerged in the research of "inherited talent" undertaken on the verge of 19th and 20th centuries by F. Galton (1822–1911) and K. Pearson (1857–1936). For the time being, it has become one of the most popular methods for data summarization and visualization. The mathematical structure and properties of the method are based on the so-called singular value decomposition of data matrices (SVD); this is why in some publications terms PCA and SVD are used as synonymous. In the UK and USA, though, the term PCA frequently refers only to a technique for the analysis of inter-feature covariance or correlation matrix by extracting most contributing linear combinations of features, which utilizes no specific data models and is considered as purely heuristic. However, this method can be related to a genuine encoder–decoder based data summarization model that is underlied by the SVD equations—in the case when the data matrix has been centered beforehand. But the centering can hardly make a big difference to the method as such; this is why I refer to the method, even when the data matrix is not centered, as PCA.

There are many motivations for this method, of which the following will be considered further on:

1. **Scoring a hidden factor** (Sect. 2.6.1)
2. **Data visualization** (Sect. 2.5)
3. **Latent semantic analysis** (Sect. 2.5.2)

Q.2.14. What could be a purpose to aggregate the features in the Market towns' data?
A. Since all the features relate to the extent of development of a town, the aggregate feature perhaps would express the extent of the town's development.

2.4.1 A Multiplicative Decoder

Let us consider a data matrix X with entries x_{iv} and standardize it into $Y = (y_{iv})$ $(i = 1,2,..., N; v = 1,2, ...,V)$. The PCA model assumes hidden factor scores z_i^* and feature loadings $c_v{}^*$ such that their product $z_i{}^* c_v{}^*$ is the decoder for y_{iv}, which can be explicated, by using additive residuals e_{iv}, as

$$y_{iv} = c_v{}^* z_i{}^* + e_{iv} \tag{2.17}$$

where the residuals are to be minimized using the least squares criterion

$$L^2 = \sum_{i \in I} \sum_{v \in V} e_{iv}^2 = \sum_{i \in I} \sum_{v \in V} (y_{iv} - c_v{}^* z_i{}^*)^2 \tag{2.18}$$

The decoder in (2.17), as a mathematical model for deriving $z_i{}^*$ and $c_v{}^*$, has a flaw from the technical point of view: its solution cannot be defined uniquely! Indeed, assume that we have obtained a factor score $z_i{}^*$ for object i and the loading $c_v{}^*$ at feature v, to produce $z_i{}^* c_v{}^*$ as the estimate for the value of the feature at the object. However, the same estimate will be produced if we halve the factor score vector and simultaneously double the loading vector: $z_i{}^* c_v{}^* = (z_i{}^*/2)(2c_v{}^*)$. Any other real taken as the divisor/multiplier would, obviously, do the same.

A conventional remedy to this is the following: specify the norms of vectors z^*, c^* to unity, and treat the multiplicative effect of the two as a real $\mu > 0$. Then put the product $\mu z_i c_v$ in (2.17) and (2.18) instead of $z_i{}^* c_v{}^*$ where z and c are normed versions of z^* and c^*, and μ is their multiplicative effect, the product of norms of z^* and c^*. The (Euclidean) norm $||x||$ of vector $x = (x_1, ..., x_N)$ is defined as its length, that is, the square root of $||x||^2 = x^T x = x_1^2 + x_2^2 + \cdots + x_N^2$. Thus, a vector is referred to as normed if its length is 1, $||x|| = 1$. After μ, z and c minimizing (2.18), thus adjusted, are determined, return to the talent score vector z^* and loading vector c^* with formulas: $z^* = \mu^{1/2} z$, $c^* = \mu^{1/2} c$. Any asymmetric version of the recovery rules, say, $z^* = \mu^a z$, $c^* = \mu^{1-a} c$, at $0 < a < 1$, would not work unless $a = 1/2$, as proven below, see Property 1. It should be pointed out that a different norming condition such as say $|x_1| + |x_2| + \cdots |x_N| = 1$ would lead to a different solution, which is not popular, since the Euclidean normalization leads to an elegant SVD-based solution described below.

The first-order optimality conditions for a triplet (μ, z, c) to be the least-squares solution to (2.17) and (2.18) imply that $\mu = z^T Y c$ is the maximum value satisfying equations

$$Y^T z = \mu c \quad \text{and} \quad Y c = \mu z \tag{2.19}$$

These equations for the optimal scores give the transformation of the data leading to the summaries z^* and c^*. The transformation, denoted by $D(C)$ in (2.1), appears to be linear, and combines optimal c and z so that each determines the other. It appears, this type of summarization is well known in linear algebra.

A triplet (μ, z, c) consisting of a non-negative μ and two vectors, c (size $V \times 1$) and z (size $N \times 1$) is referred to as to a singular triplet for Y if it satisfies (2.19); μ is referred to as a singular value and z, c the corresponding singular vectors. What can be proven immediately is the following:

Property 1 For any singular triplet (μ, z, c) satisfying (2.19) at $\mu \neq 0$, vectors z and c have the same norm.

Indeed, by multiplying the left-side equation in (2.19) by c^T, and the right-side equation by z^T, both from the left, one arrives at equations $c^T Y^T z = \mu c^T c$ and $z^T Y c = \mu z^T z$. Since $c^T Y^T z = (z^T Y c)^T$ and both are just real numbers, the equation $c^T c = z^T z$ holds because $\mu \neq 0$. Typically, the norms of c and z are taken to be unities. However, at the Principal components in (2.17), they are equal to the square root of the singular value μ, which proves the statement.

Any matrix Y can have only a finite number of nonzero singular values, and the number is equal to the rank of Y. Singular vectors z corresponding to different singular values are necessarily mutually orthogonal, as well as singular vectors c. When two or more singular values coincide, their singular vectors form a linear subspace and can be chosen to be orthogonal, which is the case in computational packages such as MatLab.

Therefore, $z* = \mu^{\frac{1}{2}} z$ and $c* = \mu^{\frac{1}{2}} c$ is a solution to the model (2.17) minimizing (2.18) defined by the maximum singular value of matrix Y and the corresponding normed singular vectors. Vectors $z*$ and $c*$ obviously also satisfy (2.19). This leads to other nice mathematical properties.

Property 2 The score vector $z*$ is a linear combination of columns of Y weighted by $c*$'s components: $c*$'s components are feature weights in the score $z*$.

Equations (2.19) allow mapping additional features or entities onto the other part of the hidden factor model. Consider, for example, an additional N-dimensional feature vector y standardized same way as Y. Its loading $c*(y)$ is determined as $c*(y) = \langle z*, y \rangle / \mu$ for the talent score $z*$. Similarly, an additional standardized V-dimensional entity point h has its hidden factor score defined according to the other part of (2.19), $z*(h) = \langle c*, h \rangle / \mu$.

Property 3 Pythagorean decomposition of the data scatter T(Y) relating the least squares criterion (2.18) and the singular value holds as follows:

$$T(Y) = \mu^2 + L^2 \tag{2.20}$$

This implies that the squared singular value μ^2 expresses the proportion of the data scatter explained by the principal component $z*$.

2.4.2 Extension of the PC Encoder–Decoder to the Case of Many Factors

It is rather naïve to expect that a single hidden factor can explain the structure of observed data. Assume a relatively small number K of different hidden factors $z*_k$ and corresponding feature loading vectors $c*_k$ ($k = 1, 2, \ldots, K; K < V$), with objects and features differently scored over them so that the observed data, after standardization, are sums of scores provided by different factors, as expressed by Eq. (2.21).

This is again a decoder that can be used for deriving a summary from the

$$y_{iv} = \sum_{k=1}^{K} c_{kv}^* z_{ik}^* + e_{iv}, \qquad (2.21)$$

standardized matrix $Y = (y_{iv})$ so that the hidden score and loading vectors z^*_k and c^*_k are found by minimizing residuals, e_{iv}. To eliminate the mathematical ambiguity, we again assume that $z^*_k = \mu^{\frac{1}{2}} z_k$ and $c^*_k = \mu^{\frac{1}{2}} c_k$, where z_k and c_k are normed vectors.

Assume that the rank of Y is r and $K < r$. Assume that the singular values of Y are sorted so that $\mu_1 \geq \mu_2 \geq \cdots \geq \mu_r$. It can be proven that the least-squares solution to (2.21) is provided by the maximal singular values μ_k and corresponding normed singular vectors z_k and c_k ($k = 1, 2, \ldots, K$).

The underlying mathematical property is that any matrix Y can be decomposed

$$y_{iv} = \sum_{k=1}^{r} \mu_k c_{kv} z_{ik}, \qquad (2.22)$$

over its singular values and vectors, as expressed in (2.22), which is referred to as the singular value decomposition (SVD). In matrix terms, SVD (2.22) can be expressed as

$$Y = \sum_{k=1}^{r} \mu_k z_k c_k^T = ZMC^T \qquad (2.22')$$

where the right-hand item Z is $N \times r$ matrix with columns z_k and C is $V \times r$ matrix with columns c_k and M is an $r \times r$ diagonal matrix with the (k, k)-th diagonal entry equal to μ_k.

Since the singular vectors are mutually orthogonal, Eq. (2.22 implies, that the scatter of matrix Y is decomposed into the sum of the squared singular values:

$$T(Y) = \mu_1^2 + \mu_2^2 + \cdots + \mu_r^2 \qquad (2.23)$$

This implies that the least-squares fitting of the PCA model in Eq. (2.21) decomposes the data scatter into the sum of contributions of individual singular triplets and the least-squares criterion $L^2 = \sum_{i,v} e_{iv}^2$:

$$T(Y) = \mu_1^2 + \mu_2^2 + \cdots + \mu_K^2 + L^2 \qquad (2.24)$$

This provides for scoring the relative contribution of the model (2.21) to the data scatter as $\left(\mu_1^2 + \mu_2^2 + \cdots + \mu_K^2 \right) / T(Y)$.

In particular, this part of decomposition (2.22) is used at 2D visualization:

$$y_{iv}{}^* \approx z_{ii}{}^* c_{1v}{}^* + z_{i2}{}^* c_{2v}{}^* \tag{2.22''}$$

where the elements on the right come from the two first principal components. This equation holds not 100% of the data scatter but $100 * (\mu_1^2 + \mu_2^2)/T(Y)$ percent. Every entity $i \in I$ is represented on a 2D Cartesian plane by the coordinate pair $(z_{ii}{}^*, z_{i2}{}^*)$. Moreover, because of the symmetry, every feature v can be represented, on the same plot, by the pair $(c_{1v}{}^*, c_{2v}{}^*)$. Such a simultaneous representation of both entities and features is referred to as a joint display or a biplot. As a matter of fact, features are presented on a biplot by not just the corresponding points, but by lines joining them to 0. This reflects the fact, that projections of points, representing the entities (entity markers), to these lines are meaningful. For a variable v, the length and direction of the projection of an entity marker to the corresponding line reflects the value of v on the entity.

It should be added that the decomposition of the data scatter (2.24) in the explained part and unexplained part, as well as its special case (2.20), can be considered a multivariate analogue to the popular Pythagoras theorem. Pythagoras, an Ancient Greece philosopher, whose life spanned almost all of the 6th century BC, is credited for the development of the concept of mathematical proof, seeing parallels between the numbers, music and outer space phenomena and, in this way, establishing the classical system of education. The Pythagoras theorem states that in any right-angled triangle, the square of the hypotenuse is equal to the sum of squares of the other two sides. The data scatter then serves as an analogue to the squared hypothenuse, whereas the other two items, the other two sides, the singular values squared and L^2, the PCA encoding model and the decoder's error (see Fig. 2.2 on p. 80 for an illustration of this).

2.4.3 Conventional Formulation of PCA Using Covariance Matrix

In the English-language literature, PCA is conventionally introduced in a different way: not via the encoder–decoder based model (2.17) or (2.21), but rather as a heuristic technique to build most contributing linear combinations of features with the help of the data covariance matrix.

The covariance matrix is defined as $V \times V$ matrix $C = Y^T Y/N$, where Y is a centered version of the data matrix X, so that all its columns are centered. The (v', v'')-entry in the covariance matrix is the covariance coefficient between features v' and v''; and the diagonal elements are variances of the corresponding features. The covariance matrix is referred to as the correlation matrix if Y has been z-score standardized, that is, if, after shifting each column to its mean, it was further normalized by dividing by its standard deviation. In this case, elements of C are correlation coefficients between corresponding variables. (Note how a bivariate concept is carried through to multivariate data by using matrix multiplication.)

One may express the conventional PCA formulation as follows. Given a centered $N \times V$ data matrix Y, find a normed V-dimensional vector $c = (c_v)$ such that the sum of Y columns weighted by c, $f = Yc$, has the largest variance possible. This vector is the principal component, termed so because it shows the direction of the maximum variance in data. Vector f is centered for any c, since Y is centered. Therefore, its variance is $s^2 = \langle f, f \rangle / N = f^T f / N$. The last equation comes under the convention that a V-dimensional vector is a $V \times 1$ matrix, that is, a column. By substituting Yc for f, this leads to equation $s^2 = c^T Y^T Yc/N$. Maximizing this with respect to all vectors c that are normed, that is, satisfy condition $c^T c = 1$, is equivalent to unconditionally maximizing the quotient

$$q(c) = \frac{c^T Y^T Yc}{c^T c} \tag{2.25}$$

over all V-dimensional vectors c. Expression (2.25) is well known in linear algebra as the Rayleigh quotient for matrix $Y^T Y$ which is proportional to the covariance matrix $A = Y^T Y/N$. The maximum of Rayleigh quotient is reached at c being an eigenvector of matrix A, corresponding to its maximum eigenvalue $q(c)$ (2.25).

Vector a is referred to as an eigenvector for a square matrix B if $Ba = \lambda a$ for some, possibly complex, number λ which is referred to as the eigenvalue corresponding to a. In the case of a covariance matrix all eigenvalues are not only real but non-negative as well. The number of eigenvalues of B is equal to the rank of B, and eigenvectors corresponding to different eigenvalues are orthogonal to each other. In data analysis, the eigenvalues are always assumed to be different because the probability that two or more of them are equal to each other is negligible.

Therefore, the first principal component, in the conventional definition, is vector $f = Yc$ defined by the eigenvector of the covariance matrix A corresponding to its maximum eigenvalue. The second principal component is conventionally defined as another linear combination of columns of Y, which maximizes its variance under the condition that it is orthogonal to the first principal component. It is defined, of course, by the second eigenvalue and corresponding eigenvector. Other principal components are defined similarly, in a recurrent manner, under the condition that they are orthogonal to all preceding principal components; which implies that they correspond to other eigenvalues, in the descending order, and the corresponding eigenvectors.

This construction seems rather remote from how the principal components are introduced above. However, it is not difficult to prove that the two definitions are computationally equivalent. Indeed, take equation $Yc = \mu z$ from (2.19), express z from this as $z = Yc/\mu$, and substitute this z into the other equation in (2.19): $Y^T z = \mu c$, leading to $Y^T Yc/\mu = \mu c$. This implies that μ^2 and c, defined by the multiplicative encoder–decoder model, satisfy equation

$$Y^T Y c = \mu^2 c, \tag{2.26}$$

that is, c is an eigenvector of square matrix $Y^T Y$, corresponding to its maximum eigenvalue $\lambda = \mu^2$. Matrix $Y^T Y$, in the case when Y is centred, is the covariance matrix A up to the constant factor $1/N$. Therefore, c in (2.26) is an eigenvector of A corresponding to its maximum eigenvalue. This proves that the two definitions are equivalent when the data matrix Y is centered. The given proof also establishes a simple relation between the eigenvalues λ of A and singular values μ of Y: $\lambda = \mu^2/N$.

In spite of the computational equivalence, there are some conceptual differences between the two definitions. In contrast to the definition based on the multiplicative encoder–decoder model, the conventional definition is purely heuristic, assuming no underlying model whatsoever. It makes sense only for centered data because of its reliance on the concept of covariance. Moreover, the fact that the principal components are linear combinations of features is postulated in the conventional definition, whereas this is a derived property of the optimal solution to the multiplicative decoder model which involves no assumptions on a linear or nonlinear relation between features and hidden factors.

Q.2.15. Can you write equations defining μ^2 and z as an eigenvalue and corresponding eigenvector of matrix YY^T. Does this square matrix have any meaning of its own?
A. By multiplying the left-side equation in (2.19) by Y on the left, we obtain $YY^T z = \mu Y c = \mu^2 z$, the latter equation following from the right-hand equation in (2.19). Elements of matrix YY^T are inner products of rows of matrix Y to express similarities between corresponding entities.

2.4.4 Computing Principal Components

The SVD decomposition is found with MatLab's svd.m function

$$[Z, M, C] = \text{svd}(Y);$$

where Y is $N \times V$ data matrix after standardization whose rank is r. Typically, if all data entries come from observation, the rank $r = \min(N, V)$.

The output consists of three matrices:

Z—$N \times N$ matrix of which only r columns are meaningful, r factor score normed columns;
C—$V \times V$ matrix of corresponding feature loading columns (normed) of which only r are meaningful;
M—$N \times V$ matrix with $r \times r$ diagonal submatrix of corresponding singular values sorted in the descending order on the top left, the part below and to the right is all zeros.

Q.2.16. Assume that a category covers subset S of entities and $y(S)$ represents the feature mean vector over S. Prove that the supplementary introduction of $y(S)$ onto the plane of singular vectors z via equation $z* = \sqrt{\mu}z = Y * y(S)/\sqrt{\mu}$ from (2.19) onto the 2D PCA display is equivalent to representing the category by the averages of the 2D points $z_{1i}*$ and $z_{2i}*$ over $i \in S$.

A. Indeed, the operation of averaging involves but addition and dividing by a constant, which are not affected by the linear operation of matrix multiplication.

2.4.5 Interpretation of Principal Components

We consider here issues of interpretation of solutions of the Principal Component Analysis model. These are not quite straightforward. The fact, that the concept of interpretation is quite vague itself, does not help either. We consider two aspects of interpretation, the geometry and the substance, in the next two sections, respectively.

2.4.5.1 Geometric Interpretation

Take all talent score points $z = (z_1, \ldots, z_N)$ that are normed, that is, satisfy equation $\langle z, z \rangle = 1$ or $z^T z = 1$ or $z_1^2 + \cdots + z_N^2 = 1$: they form a sphere of radius 1 in the N-dimensional "entity" space (Fig. 2.18a). The image of these points in the feature space, $Y^T z$, forms a skewed sphere, an ellipsoid in the feature space, consisting of points μc where c is normed. The longest axis of this ellipsoid corresponds to the maximum μ, that is the first singular value of Y [Indeed, the first singular value μ_1 and corresponding normed singular vectors c_1, z_1 satisfy equation $Yc_1 = \mu z_1$ and, thus, its transpose, $c_1^T Y^T = \mu_1 z^T$. Multiplying the latter by the former, one gets equation $c_1^T Y^T Y c_1 = \mu_1^2$, because $z^T z = 1$.] (Fig. 2.18).

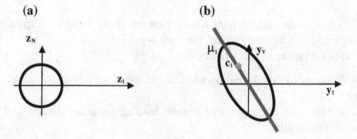

Fig. 2.18 Sphere $z^T z = 1$ in the entity space (a) and its image, ellipsoid $c^T Y^T Y c = \mu_1^2$, in the feature space (b). The first component, c_1, corresponds to the maximal axis of the ellipsoid with its length equal to $2\mu_1$

What the longest axis has to do with the data? This is exactly the direction which is looked for in the conventional definition of PCA. The direction of the longest axis of the data ellipsoid makes minimum of the summary distances (Euclidean squared) from data points to their projections on the line (see Fig. 2.19), because of the least-squares optimality of the decoder in (2.17) so that this axis is the best possible 1D representation of the data.

This property extends to all subspaces generated in the order of extraction of principal components: the first two PCs make a plane that is the best two-dimensional approximation of the dataset; the first three make a 3D space best approximating the dataset, etc.

One more illustration concerns the difference between SVD for the raw data and that after centering; see Fig. 2.20. One can see that at the raw data, Fig. 2.20a, the longest axis follows the direction between the origin and the points cloud, whereas after centering it follows the structure of the dataset, Fig. 2.20b.

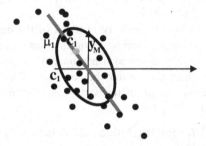

Fig. 2.19 The direction of the longest axis of the data ellipsoid makes minimum the summary distances (Euclidean squared) from data points to their projections onto the line

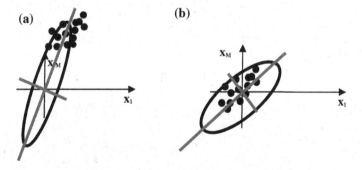

Fig. 2.20 **a** PCA at data before centering, **b** PCA at the data after centering

2.4.5.2   Substantive Interpretation

The principal components provide for the best possible least-squares approximation of the data in a low dimension space. The quality of such a data compression is usually judged over (i) the proportion of the data scatter taken into account by the reduced dimension space and (ii) interpretability of the hidden factors supplied by the PCs.

Contribution of the PCA model to the data scatter is reflected in the sum of squared singular values corresponding to the principal components in the reduced data table. This sum should be related to the data scatter or, equivalently, to the total of all squared singular values to observe the impact to the data scatter.

For example, the 2D representation of 4D student data on Fig. 2.21 in the further down Worked Example 2.7 contributes 72.11% to the data scatter.

The Fig. 2.2 on p. 80 shows that the contribution, in spite of its firm mathematical footing, can be rather shaky an argument because it much depends on the data standardization options. The criterion of interpretability gives a different, and frequently more reliable, perspective.

To interpret PCA results substantively in terms of the data feature space, one should use the feature loadings according to the singular vectors related to features. These straightforwardly show how much a principal component is affected by a feature: the larger the value, the greater the correlation. Any PCA component is interpreted by using features with relatively high positive or negative coefficients, as illustrated in the example below.

Worked Example 2.7. Interpretation of Principal Components at the Standardized Student Data

Consider four features in the Students dataset—the Age and marks for SEn, OOP and CI subjects. Let us center it by subtracting the mean vector $a = (33.68, 58.39, 61.65, 55.35)$ from all the rows, and normalize the features by their ranges

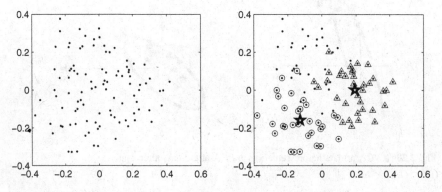

Fig. 2.21 Scatter plot of the Student data (age and marks over SP, OO, CI) row points on the plane of two first principal components, centered and rescaled. Pentagrams on the right represent the mean points of the occupation categories AN and IT

Table 2.19 Components of the normed loading parts of principal components for the standardized part of Student data set; corresponding singular values, along with their squares expressed both percent to their total, the data scatter, and in real

Singular value	Singular value squared	Contribution, percent	Singular vector			
			Age	SEn	OOP	CI
3.33	11.12	42.34	−0.59	0.03	0.59	0.55
2.80	7.82	29.77	0.53	0.73	0.10	0.42
2.03	4.11	15.67	−0.51	0.68	−0.08	−0.51
1.79	3.21	12.22	−0.32	0.05	−0.80	0.51

$r = (31, 56, 67, 69)$. The latter operation seems a necessity because the Age, expressed in years, and subject marks, percent, are not exactly comparable. Characteristics of all the four singular vectors of these data for feature loadings are presented in Table 2.19.

The summary contribution of the two first principal components (PCs) to the data scatter is $42.34 + 29.77 = 72.11\%$, which is not that bad for educational data; it warrants a close representation of the entities on the principal components plane. The two principal components are found by multiplication of each of the corresponding left singular vectors z_1 and z_2 by the square roots of the corresponding singular values.

Take a look at the first singular vector in Table 2.19 corresponding to the maximum singular value 3.33 at the standardized Students data. One can see that the first component positively relates to marks over all subjects, perhaps except SEn at which the loading is almost zero, and negatively to the Age. That means that on average, the first factor is greater when a student gets better marks and is younger. Thus, the first component can be interpreted as the "Age-related Computer Science proficiency". The second component (the second line in Table 2.19) is positively related to all of the features, especially SEn marks, which can be interpreted as "Age defying inclination towards Software Engineering". Further insights into interpretation can be found by looking at a visualization of the data on the plane of two first principal components (see Worked Example 2.8 further on).

In spite of the fact that the occupation has not been involved in building the PC space, its categories appear to occupy different parts of the plane, which will be shown later, in Worked Example 2.8. Then the triangle and circle patterns on the right of Fig. 2.21 show that AN laborers are on the minimum side of the age-related CS proficiency, whereas IT occupations are high on that—all of which seem rather reasonable. Both are rather low on the second component, though, in contrast to students represented by dots, thus belonging to BA occupation category, that get the maximum values on it.

In the early days of the development of factor analysis, yet within the psychology community, researchers were trying to explore the possibility of achieving a more interpretable solution by rotating the axes of the found subspace. The goal was to find a simple structure of the loadings, in which most of the loading elements are zero with a few non-zero values that should be as close to either 1 or −1 as possible. This goal, however, is subject to too much of arbitrariness and remains an open issue, although using the so-called regularizers, relatively simple functions of the solution being sought, to the optimized penalty function, can make a difference (see, for example, Tibshirani et al. 2015). Keeping singular vectors as they are, not rotated, has the advantage that each of them contributes to the data scatter as much as possible. This relates to frequently occurring real world situations in which factors underlying the phenomenon of interest contribute to it differently. The PCA factors express such a structure formally: that one contributing the most is followed by the second best contributing, then by the third best contributing, etc.

2.5 SVD Based Data Visualization

2.5.1 PCA Data Visualization

For the purposes of *visualization* of the data entities on a 2D plane, the data set is usually first centered to put it against the backdrop of the center—we mentioned already that more structure in the dataset can be seen when looking at it from the center of gravity, that is, the grand mean location. Solutions to the multiple factor model, that is, the hidden factor scoring vectors which are singular vectors of the data matrix, in this case, are referred to as principal components (PCs). Two principal components corresponding to the maximal singular values are needed for a 2D representation.

What is warranted in this arrangement is that the PC plane approximates the data, as well as the between-feature covariances and between-entity similarities, in the best possible way. The coordinates provided by the singular vectors/principal components are not unique, though, and can be changed by rotating the axes, but they do hold a unique property that each of the components maximally contributes to the data scatter.

Worked Example 2.8. Visualization of a Fragment of Students Dataset
Let us return to the Worked Example 2.7, regarding four features in the Students dataset—the Age and marks for SEn, OOP and CI. We visualize the data on the plane formed by the first two singular vectors as axes after multiplying each of them by the square root of the corresponding singular value.

Each entity $i = 1, 2, ..., N$ is represented on the PC plane by the pair of the first and second PC values (z_{1i}^*, z_{2i}^*) (see Fig. 2.21). The left part is just the data with no labels. On the right part, two occupational categories are visualized using triangles (IT) and circles (AN); remaining dots relate to category BA.

To produce the scatter-plot of Fig. 2.21 with 100×4 Student data matrix Y, the following MatLab commands can be used:

```
≫ subplot(1, 2, 1); plot(z1*,z2*,'k.'); %Fig. 2.21, picture on the left
≫ subplot(1,2,2);
```

≫ plot(z1*, z2*,'k.', z1*(1:35),z2*(1:35),'k^',...z1*(70:100),z2*(70:100),'ko', ad1,ad2,'kp');

In the last command, there are several items to be shown on the same plot:

(i) z1*, z2*, 'k.'—these are black dot markers for all 100 entities in the plot on the left;

(ii) z1*(1:35), z2*(1:35), 'k^'—these are triangles to represent entities 1–35— those in category IT;

(iii) z1*(70:100), z2*(70:100), 'ko'—these are circles to represent entities 70– 100—those in category AN;

(iv) ad1, ad2,'kp'—ad1 is a 2 × 1 vector of the averages of z1* over entities 1 to 35 (category IT) and 70–100 (category AN), and ad2 a similar vector of within-category averages of z2*. These are represented by pentagrams.

Q.2.17. Why we recommend multiply the singular vectors by the square roots of the corresponding singular values for a 2D data visualization?
A. Because of Eq. (2.22″) on p. 127 underlying the recommendations. The singular values are taken into account in this way in the equation.

Worked Example 2.9. Evaluation of the Quality of Visualization of the Standardized Student Data

To evaluate how well the data are approximated by the PC plane such as that on Fig. 2.21, according to Eq. (2.21), one needs to assess the summary contribution of the first two squared singular values in the total data scatter. To get the squares one can multiply matrix mu below by itself and then see the proportion of the first two values in the total:

≫ mu=m(1:4,:); %no need in the MatLab's 4 × 100 matrix output, have a square size 4 × 4
≫ la=diag(mu*mu);% make squares and put them as a vector
≫ lar=la*100/sum(sum(la)) % vector of the relative contributions of the PCs
≫ lar(1)+lar(2) % relative contribution of the 2 first components
The last two lines print to the screen:

lar =
    42.3426
    29.7719
    15.6664
    12.2191

ans =
    72.1145

The latter is the sum of two first elements of the former—the proportion of the data scatter, percent, taken into account by the 2D visualization on Fig. 2.21.

2.5.2 Latent Semantic Analysis

The number of papers applying PCA to various problems—image analysis, information retrieval, gene expression interpretation, complex data storage, etc.—makes many hundreds published annually. Some of the applications are well established techniques of their own. We present one such technique: Latent semantic indexing (analysis), see Deerwester et al. (1990), Landauer (2006). Another application, to categorical features, Correspondence analysis, is left to Sect. 3.6.3.

Latent semantic analysis is an application of PCA to document analysis—information retrieval, first of all, using document-to-keyword data.

Information retrieval is an application that no computational data analysis may skip: given a set of records or documents stored, find out those related to a specific query expressed by a set of keywords. Initially, at the dawn of computer era, when all the documents were stored in the same database, the problem was treated in a hard manner—only documents containing the query words were to be given to the user. Currently, this is a much softer problem, that is being constantly and efficiently solved by various search engines such as Google, for millions of World Wide Web users, see Manning et al. (2008).

In its generic format, the problem can be illustrated with data in Table 2.20. It refers to a number of newspaper articles related to subjects such as entertainment, feminism and households, conveniently coded with letters E, F and H, respectively. Columns correspond to keywords, or terms, listed in the first line of the table, and

Table 2.20 Database of 12 newspaper articles along with 10 keywords and the conventional coding of term frequencies

Article	Keyword									
	Drink	Equal	Fuel	Play	Popular	Price	Relief	Talent	Tax	Woman
F1	1	2	0	1	2	0	0	0	0	2
F2	0	0	0	1	0	1	0	2	0	2
F3	0	2	0	0	0	0	0	1	0	2
F4	2	1	0	0	0	2	0	2	0	1
E1	2	0	1	2	2	0	0	1	0	0
E2	0	1	0	3	2	1	2	0	0	0
E3	1	0	2	0	1	1	0	3	1	1
E4	0	1	0	1	1	0	1	1	0	0
H1	0	0	2	0	1	2	0	0	2	0
H2	1	0	2	2	0	2	2	0	0	0
H3	0	0	1	1	2	1	1	0	2	0
H4	0	0	1	0	0	2	2	0	2	0
Df	5	5	6	7	7	8	5	6	4	5
Idf	0.88	0.88	0.69	0.54	0.54	0.41	0.88	0.69	1.1	0.88

The articles are labeled F for Feminism, E for Entertainment and H for Household. Two bottom lines hold document frequencies of terms (df) and the other, inverse document frequency weights (idf)

entries refer to term frequency in the articles, according to a conventional coding scheme:

0—no occurrence,
1—occurs once,
2—occurs twice or more.

The user may wish to retrieve all the articles on the subject of households, but they have to inquire by using the listed keywords only. For example, query "fuel" will retrieve all four of the household related articles, and, in fact more than that— E1 and E3 will show up too; query "tax" will get four items, three—H1, H3, and H4—on the subjects of household and one—E3—on the subject of entertainment. No combination of these two can improve the result.

This is very much a problem of description of a class, that is, of finding a decision rule covering the required set of documents as tight as possible. Just the decision rules must be queries, combinations of keywords. The error of such a query is characterized by two characteristics, precision and recall (see Sect. 3.10). For example, the "fuel" query's precision is $4/6 = 2/3$ since only four out of six are relevant and recall is 1 because all of the relevant documents have been returned. Similarly, for "tax" query both the precision and recall are ¾.

The rigidity of the query format does not fit well into the polysemy of natural language—such words as "fuel" or "play" have more than one meaning—thus leading to impossibility of exact information retrieval in many cases.

The method of latent semantic indexing (LSI) utilizes the SVD decomposition of the document-to-term data to soften and thus improve the query system by embedding both documents and terms into a subspace of singular vectors of the data matrix.

Before proceeding to SVD, the data table sometimes is pre-processed, typically, with what is referred to Term-Frequency-Inverse-Document-Frequency (tf-idf) normalization. This procedure gives a different weight to any keyword according to the number of documents it occurs at (document frequency df). The intuition is that the greater the document frequency, the more common and thus less informative is the word. The idf weighting assigns each keyword with a weight inversely proportional to the logarithm of its document frequency. For Table 2.20, df and idf weights are in the bottom lines.

The term frequency, tf, value is that of the corresponding data matrix entry referring to the frequency of the occurrence of the column word in the row document related to the document size.

After the SVD of the data matrix is obtained, the documents are considered points of the subspace of a few first singular vectors. The dimension of the space is not very important here, though it still should be much smaller than the original dimension. Good practical results have been reported at the dimension of about 100–200 when the number of documents in tens and hundred thousands and the number of keywords in thousands. A query is also represented as a point in the same space. The principal components, in general, are considered as "orthogonal" concepts underlying the meaning of terms. This however, should not be taken too literally, as the singular vectors can be quite difficult to interpret. Also, the representation of documents and queries as points in a Euclidean space is referred to sometimes as the vector space model in information retrieval.

The Euclidean space format allows to measure similarity between items using the inner product or even what is called cosine—the inner product between rows that have been pre-normalized. Then a query would return the set of documents whose similarity to the query point is greater than a threshold. This tool may provide for a better resolution in the problem of information retrieval, because it well separates different meanings of synonyms.

Worked Example 2.10. Latent Semantic Space for the Article-to-term Data
Let us illustrate the LSI at the data in Table 2.20. To apply tf-idf normalization, we assume that all the documents have the same length so that the absolute term frequencies in Table 2.20 can be used as tf estimates. Then the tf-idf coding of any entry is equal to the entry value multiplied by the corresponding idf value (the last line in Table 2.20).

Consider the combination fuel-price-relief-tax as a query Q that should relate to Household category. Table 2.21 contains data that are necessary for computing coordinates of the query Q in the concept space. The query is represented by 1/0 vector in the bottom line of Table 2.21. The first coordinate of Q image on the map is computed by summing all the corresponding components of the first left singular vector and dividing the result by the square root of the first singular value: $u1 = (-0.34 - 0.42 - 0.29 - 0.24)/8.6^{\frac{1}{2}} = -0.44$. The second coordinate is computed similarly from the second singular vector and value: $u2 = (-0.25 - 0.22 - 0.35 - 0.33)/5.3^{\frac{1}{2}} = -0.48$. These correspond to the pentagram on the left part of Fig. 2.22. One can think that query Q corresponds to an additional row in the data table which is equal to the last line in Table 2.21, so that the visualization algorithm of PCA applies to that line.

Figure 2.22 represents the documents in the space of two first principal components; left part corresponds to the original term frequency codes and the right part to the data tf-idf normalized.

As one can see, both representations separate the three subjects, F, E and H, more or less similarly, and provide the query Q with a rather good resolution. Taking into account the position of the origin of the concept space—the circle in the middle of the right boundary, the four H items are indeed have very good angular similarity to the pentagram representing the query Q.

The SVD representation of documents is utilized in other applications such as text mining, web search, text categorization, software engineering, etc.

The full SVD of data matrix F leads to equation $F = ZMC^T$ where Z and C are matrices whose columns are right and left normed singular vectors of F and M is a diagonal matrix with the corresponding singular values of F on the diagonal. By leaving only K columns in these matrices, we substitute matrix F, by matrix $F_K = Z_K M_K C_K^T$ so that the entities are represented in the K-dimensional concept space by the rows of matrix $Z_K M_K^{\frac{1}{2}}$.

Table 2.21 Two first singular vectors of term frequency data in Table 2.20

	Ei.value	Contrib., %	Left singular vectors normed									
1st comp.	8.6	46.9	−0.25	−0.19	−0.34	−0.40	−0.39	−0.42	−0.29	−0.32	−0.24	−0.22
2nd comp.	5.3	17.8	0.22	0.34	−0.25	−0.07	0.01	−0.22	−0.35	0.48	−0.33	0.51
Query Q			0	0	1	0	0	1	1	0	1	0

To translate a query presented as a vector q in the V-dimensional space into the corresponding point u in the K-dimensional concept space, one needs to take the product $g = C_K^T q$, which is equal to $g = z M_K^{1/2}$ according to the definition of singular values, after which z is found as $z = g M_K^{-1/2}$. Specifically, k-th coordinate of vector z is calculated as $z_k = \langle c_k, q \rangle / \mu_k^{1/2} (k = 1, 2, \ldots, K)$.

The similarities between rows (documents), corresponding to row-to-row inner products in the concept space are computed as $Z_K M_K^2 Z_K^T$ and, similarly, the similarities between columns (keywords) are computed according to the dual formula $C_K M_K^2 C_K^T$. Applying this to the case of the K-dimensional point z representing the original V-dimensional vector q, its similarities to N original entities are computed as $z M_K^2 Z_K^T$.

Let X be $N \times V$ array representing the original frequency data. To convert that to the conventional coding, in which all the entries larger than 1 are coded by 2, one can use this operation:

```
>> Y=min(X,2*ones(N,V));
```

Computing vector df of document frequencies over matrix Y can be done with this line:

```
>> df=zeros(1,V); for k=1:V;df(k)=length(find(Y(:,k)>0));end;
```

and converting df to the inverse-document-frequency weights, with this:

```
>> idf=log(N./df);
```

After that, it-idf normalization can be made by using command

```
>>YI=Y.*repmat(idf, N,1);
```

Given term frequency matrix Y, its K-dimensional concept space is created with commands:

```
>> [z,m,c]=svd(Y);
>>zK=z(:, [1:K]); cK=c(:, [1:K]); mK=m([1:K], [1:K]);
```

Worked Example 2.11. Drawing Fig. 2.22
For MatLab coding, consider that z is the matrix of normed document score singular vectors, c the matrix of normed keyword loading vectors, and m the matrix of singular values of the data in Table 2.20 as they are.

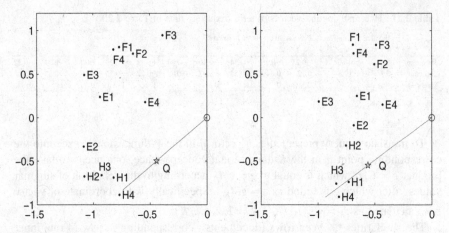

Fig. 2.22 Two first principal components plane for the data in Table 2.21, both in the original format (left) and after the tf-idf normalization (right). Query Q combining fuel-price–relief-tax keywords is mapped to the pentagram; the line connects it with the origin 0

To draw the left part of Fig. 2.22, one can define the coordinates with vectors z1 and z2:

```
≫ z1=z(:,1)*sqrt(m(1,1)); %first coordinates of N entities in the concept space
≫ z2=z(:,2)*sqrt(m(2,2)); %second coordinates of N entities in the concept space
```

Then prepare the query vector and its translation to the concept space:

```
≫ q=[0 0 1 0 0 1 1 0 1 0]; % "fuel, price, relief, tax" query vector
≫ d1=q*c(:,1)/sqrt(m(1,1)); %first coordinate of query q in the concept space
≫ d2=q*c(:,2)/sqrt(m(2,2)); %second coordinate of query q in the concept space
```

After this, an auxiliary text data should be put according to MatLab requirements:

```
≫ tt={'E1','E2', …, 'H4'}; % cell of 12 names of the items in data matrix
≫ ll=[0:.04:1.5]; zd1=d1*ll; zd2 = d2*ll;
% pair zd1, zd2 will draw a line through the origin and point (d1, d2)
```

Now we are ready for plotting the left drawing on Fig. 2.22:

```
≫ subplot(1,2,1);
≫ plot(u1,u2,'k.',d1,d2,'kp',0,0,'ko',ud1,ud2);text(u1,u2,tt);
≫ text(d1,d2, 'Q');
≫ axis([-1.5 0 -1 1.2]);
```

The arguments of the plot drawing command here are:

u1, u2, 'k.'—black dots corresponding to the original entities;
d1,d2,'kp'—black pentagram corresponding to query q;
0,0,'ko'—black circle corresponding to the space origin;

ud1,ud2—coordinates of the line through the query and the origin.

Command *text* provides for string labels at the corresponding points. Command *axis* specifies the boundaries of the Cartesian plane box on the figure, which can be used for making different plot boxes uniform.

The plot on the right of Fig. 2.22 is coded similarly by using SVD of tf-idf matrix YI rather than Y.

Q.2.18. In many situations, (a) the first singular vector is all positive and (b) the second singular vector is half negative. Why can be that? **A.** A typical situation: (a) All features are positively correlated which implies that the first eigenvector is positive; (b) the second must be orthogonal to the first, to make 0 their inner product.

Q.2.19. Is matrix

$$\begin{matrix} 1 & 2 \\ 2 & 1 \end{matrix}$$

of rank 1 or not?
A. The rows are not proportional to each other, thus not.

Q.2.20. Prove that if matrix Y is symmetric then its eigenvalues and vectors (λ,z) are simultaneously its singular triplets $(\lambda,\ z,\ z)$.
Q.2.21. Find a matrix of rank 1 that is the nearest to matrix in Q.2.19 according to the least-squares criterion.
A. The solution is given by the first singular value and corresponding singular vectors which are the same as the first eigenvalue and corresponding eigenvector, $\lambda = 3$ and $z = (1/\sqrt{2}, 1/\sqrt{2})$, thus leading to matrix

$$\begin{matrix} 3/2 & 3/2 \\ 3/2 & 3/2 \end{matrix}$$

as the solution.

Q.2.22. For positive a and b, inequality $a < b$ can be equivalently expressed as $a/b < 1$. The difference between a and b does not change if $c > 0$ is added to both of them, but the ratio does. Prove that for any $c > 0$, $(a + c)/(b + c) > a/b$—this would illustrate that the further away the positive data are from zero, the greater the contribution of the first principal component.
Q.2.23. There is another representation of a singular value problem as an eigenvalue problem. Given a rectangular $N \times V$ matrix Y, consider a square

$(N + V) \times (N + V)$ matrix Y^* that consists of four blocks, two of which, the diagonal $N \times N$ and $V \times V$ blocks, are all zeros, while the others are Y and Y^T:

$$Y^* = \begin{pmatrix} 0 & Y \\ Y^T & 0 \end{pmatrix}.$$

Prove that a triplet (μ, z, c) is singular for Y if and only if μ is eigenvalue for Y^* corresponding to eigenvector $y = (z,c)$ in which first N components are taken by z and the remaining V components, by c.

A. Consider an arbitrary eigenvalue μ and corresponding eigenvector y of matrix Y^* and denote the vector of its first N component by z, and the rest by c so that $y = (z, c)$. The product $Y^* y$ will have its first N components equal to $0z + Yc = Yc$ and the next V components equal to $Y^T z + 0c = Y^T z$. Since $Y^* y = \mu y$, that means that $Yc = \mu z$ and $Y^T z = \mu c$, which proves the statement.

Q.2.24. Prove that if an eigenvector $y = (z,c)$ of Y^* is normed, then its components z and c both have norms of $0.5^{1/2}$.

A. Indeed, $||y||^2 = ||z||^2 + ||c||^2$ because these are just sums of the squared components. On the other hand, as proven in Q.2.23, z and c are singular vectors of Y so that they must have equal norms because of Property 1 on p. 125. This leads to equation $||z||^2 = ||c||^2 = 1/2$, which proves the statement.

Q.2.25. Prove that if μ is an eigenvalue of Y^* corresponding to its eigenvector $y = (z, c)$ then so is its negation $-\mu$ corresponding to eigenvector $y = (-z, c)$.

A. Indeed, equations $Yc = \mu z$ and $Y^T z = \mu c$ hold if and only if $Yc = (-\mu)(-z)$ and $Y(-z) = (-\mu)c$.

2.6   Ranking in Feature Space and Networks

2.6.1   Scoring a Hidden Factor

2.6.1.1   Encoder–Decoder Multiplicative Model

Consider the following problem. Given student's marks at different subjects, can we derive from this their score at a hidden factor of talent that is supposedly reflected in the marks? Take a look, for example, at the first six students' marks over the three subjects in Table 2.22 extracted from Students data, Table 1.5.

To judge of the relative strength of a student, the average mark is used in practice. This ignores the relative work load that different subjects may impose on a student—can you see that CI mark is greater than SEn mark for each of the six students?—and in fact, is purely empirical and does not allow much theoretical speculation. Let us assume that there is a hidden factor, not measurable straightforwardly, the talent, that is manifested in the marks. Suppose that another factor manifested in the marks is subject load, and, most importantly, assume that these factors multiply to make a mark, so that a student's mark over a subject is the product of the subject's loading and the student's talent:

$$\text{Mark}(\text{Student}, \text{Subject}) = \text{Talent\_Score}(\text{Student}) * \text{Loading}(\text{Subject}) \quad (*)$$

One may point out two issues related to this model—one internal, the other external.

The external issue is that the mark, as observed, depends on many other factors differently affecting different students—the weather, a sleepless night or malady, level of interest in the subject, etc., which make the model as is overly simplistic and prone to errors. Well, a proponent would say, sure the model is simplistic—it takes on only most important factors. The others will cause errors indeed, but these can be tackled by minimizing them: the idea is that the hidden talent and loading factors can be found by minimizing the differences between the real marks and those derived from the model. The PCA method is based on the least-squares approach so that it is the sum of squared differences between the observed and computed marks that is minimized in PCA.

The internal issue is that the model as is admits no unique solution because it is the product of mark by loading that matters, not their individual values—if one multiples all the talent scores by a number, say $\text{Talent\_Score}(\text{Student}) * 5$, and simultaneously divides all the subject loadings by the same number, Loading (Subject)/5, the product will not change. How one is supposed to compute something which admits no definite representation? To make a solution unique, conventionally, a constant norm of one or both of the items is assumed so that one more item into the product is admitted—that expressing the product's magnitude. Then, there is a unique solution indeed, with the magnitude expressed by the so-called maximum singular value of the data matrix with the score and load factors being its corresponding normed singular vectors.

Specifically, the maximum singular value of matrix in Table 2.22 is 291.4, and the corresponding normed singular vectors are $z = (0.40, 0.34, 0.42, 0.42, 0.41, 0.45)$, for the talent score, and $c = (0.50, 0.57, 0.66)$, for the loadings. That means that every mark in the matrix is the product of three items. For example, to compute

Table 2.22 Marks at three subjects for six students from Students data Table 1.5

#	SEn	OOP	CI	Average
1	41	66	90	65.7
2	57	56	60	57.7
3	61	72	79	70.7
4	69	73	72	71.3
5	63	52	88	67.7
6	62	83	80	75.0

the model SEn value for student 6, one takes 291.4 * 0.45 * 0.50 = 65.6, which is not that far from the observed mark of 62. Yet the model involves the product of two items only. To get back to our model, we need to distribute the singular value between the vectors. There is only one way to do it complying with the singular value Eqs. (2.22) and (2.22″)—by multiplying each of the vectors by the same value, the square root of the singular value which is 17.1. Thus, the denormalized talent score and subject loading vectors will be $z' = (6.85, 5.83, 7.21, 7.20, 6.95, 7.64)$ and $c' = (8.45, 9.67, 11.25)$. According to the model, the score of student 3 over subject SEn is the product of the talent score, 7.21, and the loading, 8.45, which is 60.9, quite close to the observed mark 61. Similarly, product 5.83 * 9.67 = 56.4 is close to 56, student 2's mark over OOP. The differences can be greater though: product 5.83 * 8.45 is 49.3, which is rather far away from the observed mark 57 for student 2 over SEn.

In matrix terms, the model can be represented by the following equation

$$
\begin{bmatrix} 6.85 \\ 5.83 \\ 7.21 \\ 7.20 \\ 6.95 \\ 7.64 \end{bmatrix} * \begin{bmatrix} 8.45 & 9.67 & 11.25 \end{bmatrix} = \begin{bmatrix} 57.88 & 66.29 & 77.07 \\ 49.22 & 56.37 & 65.53 \\ 60.88 & 69.72 & 81.05 \\ 60.83 & 69.67 & 81.00 \\ 58.69 & 67.22 & 78.15 \\ 64.53 & 73.91 & 85.92 \end{bmatrix}
\tag{2.27}
$$

whereas its relation to the observed data matrix, by equation

$$
\begin{bmatrix} 41 & 66 & 90 \\ 57 & 56 & 60 \\ 61 & 72 & 79 \\ 69 & 73 & 72 \\ 63 & 52 & 88 \\ 62 & 83 & 80 \end{bmatrix} = \begin{bmatrix} 57.88 & 66.29 & 77.07 \\ 49.22 & 56.37 & 65.53 \\ 60.88 & 69.72 & 81.05 \\ 60.83 & 69.67 & 81.00 \\ 58.69 & 67.22 & 78.15 \\ 64.53 & 73.91 & 85.92 \end{bmatrix} + \begin{bmatrix} -16.88 & -0.29 & 12.93 \\ 7.78 & -0.37 & -5.53 \\ 0.12 & 2.28 & -2.05 \\ 8.17 & 3.33 & -9.00 \\ 4.31 & -15.22 & 9.85 \\ -2.53 & 9.09 & -5.92 \end{bmatrix}
\tag{2.28}
$$

where the left-hand item is the observed mark matrix; that in the middle, the model-computed evaluations of the marks; and the right-hand item comprises the differences between the real and decoded marks.

2.6.1.2 Error of the Model

Among questions that arise with respect to the matrix equation such as that in (2.2) are the following:

(i)   Why are the errors appearing at all?
(ii) How can the overall level of errors be assessed?
(iii) Can any better fitting estimates for the talent be found?

We address them in turn.

(i) **Differences between real and model-derived marks**

The differences emerge because the model imposes significant constraints on the model-derived estimates of marks. They are generated as products of components of just two vectors, the talent score and the subject loadings. This means that every row in the model-based matrix (*) is proportional to the vector of subject loadings, and every column, to the vector of talent scores. Therefore, the rows are mutually proportional as well as the columns. Real marks, generally speaking, do not satisfy such a property: mark rows or columns are typically not proportional to each other. More formally, this can be expressed in the following way: 6 talent scores and 3 subject loadings together can generate not more than $6 + 3 = 9$ independent estimates. (One more degree of freedom may go because the norms of these two vectors are the same.) The number of marks however, is the product of these, $6 * 3 = 18$. The greater the size of the data matrix, $M \times V$, the smaller the proportion of the independent values, $M + V$, that can be generated from the model.

In other words, matrix (*) is one-dimensional. It is well recognized in mathematics in the concept of matrix rank which corresponds to the "inner" dimension of a matrix—matrices that are products of two vectors are referred to as matrices of rank 1.

Q.2.26. For the data in Table 2.22, as well as many others, svd function in MatLab produces first singular vectors z and c negative, which contradicts the meaning of them as talent scores and subject loadings. Can anything be done about that?

A. Yes, they can be changed to $-z$ and $-c$ without compromising their singular vector status.

(ii) **Assessment of the level of differences**

A conventional measure of the level of error of the model is the ratio of the scatters of the model derived matrix and the observed data matrix in (2.28). The scatter of matrix A, $T(A)$, is the sum of the squares of all of A-entries or, which is the same, the sum of the diagonal entries in matrix $A * A^T$, the $trace(A * A^T)$.

Worked Example 2.12. Explained Proportion of the Data Scatter in Eq. (2.28)
Consider scatters of three matrices in (2.28) in Table 2.23. The residual data scatter is rather small and accounts for only $\varepsilon^2 = 1183.2/86,092 = 0.0137$, that is, 1.37%, of the original data scatter. Its complement to unity, 98.63%, is the proportion of the data scatter explained by the multiplicative model. This also can be straightforwardly

Table 2.23 Scatters of matrices in Eq. (2.28)

	Scatter of		
	Data matrix	Model matrix	Residual matrix
Absolute	86,092	84,908.80	1183.20
Proportion	100	98.63	1.37

derived from the singular value, 291.4: its square shows the part of the data scatter explained by the model, $291.4^2/86{,}092 = 0.9863$ (see Eq. (2.20) on p. 125).

Q.2.27. In spite of the fact that some errors in (2.24) are rather high, the overall squared error is quite small, just about 1% of the data scatter. Why is that?
A. Because the data values are far away from 0—see Q.2.22 explaining the effect in general.

(iii) The singular vector estimates are the best

The squared error is the criterion optimized by the estimates of talent scores and subject loadings. No other estimates can give a smaller value to the square error for data matrix in Table 2.22 than $\varepsilon^2 = 1.37\%$.

2.6.1.3 Formulaic Expression of the Hidden Factor Through the Data

The relations between singular vectors (see Eq. (2.19) in Sect. 2.4.1) provide us with a conventional expression of the talent score as a weighted average of marks at different subjects. The weights are proportional to the subject loadings $c' = (8.45, 9.67, 11.25)$: weight vector w is the result of dividing of all entries in c' by the singular value, $w = c'/291.4 = (0.029, 0.033, 0.039)$. For example, the talent score for student 1 is the w weighted average of their marks, $0.029 * 41 + 0.033 * 66 + 0.039 * 90 = 6.88$.

The model-derived averaging allows one also to score the talent of other students, those not belonging to the sample being analyzed. If marks of a student over the three subjects are $(50, 50, 70)$, their talent score will be the w-weighted average: $0.029 * 50 + 0.033 * 50 + 0.039 * 70 = 5.83$.

A final touch to the hidden factor scoring can be given by rescaling it in a way conforming to the application domain. Specifically, one may wish to express the talent scores in a 0–100 scale resembling that of the original mark scales. That means that the score vector z' has to be transformed into $z'' = \alpha * z' + \beta$, where α and β are the scaling factor and shift coefficients, that can be found from two natural conditions: (a) z'' is 0 when all the marks are 0 and (b) z'' is 100 when all the marks are 100. Condition (a) means that $\beta = 0$, and condition (b) calls for calculation of the talent score of a student with all top marks. Summing the three 100 marks with weights from w leads to the value $zM = 0.029 * 100 + 0.033 * 100 + 0.039 * 100 = 10.10$ which implies that the rescaling coefficient α must be $100/zM = 9.92$ or, equivalently, weights must be rescaled as $w' = 9.92 * w = (0.29, 0.33, 0.38)$. Talent scores found with these weights are presented in the right column of Table 2.24—hardly a great difference from the average scores, except that the talent scores are slightly higher, due to a greater weight assigned to the mark-earning CI subject.

Worked Example 2.13. SVD for Six Students Dataset
For the centered version of the 6×3 matrix in Table 2.22 the SVD matrices are as follows:

Table 2.24 Marks and talent scores for six students

#	SEn	OOP	CI	Average	PC talent
1	41	66	90	65.7	68.0
2	57	56	60	57.7	57.8
3	61	72	79	70.7	71.5
4	69	73	72	71.3	71. 5
5	63	52	88	67.7	69.0
6	62	83	80	75.0	75.8

$$
Z = \begin{array}{cccccc}
-0.7086 & 0.1783 & 0.4534 & 0.4659 & 0.1888 & 0.0888 \\
0.2836 & -0.6934 & 0.4706 & 0.0552 & 0.2786 & 0.3697 \\
0.0935 & 0.1841 & -0.0486 & -0.1870 & 0.9048 & -0.3184 \\
0.4629 & 0.0931 & -0.1916 & 0.8513 & 0.0604 & -0.1092 \\
\\
-0.3705 & -0.3374 & -0.7293 & 0.1083 & 0.2279 & 0.3916 \\
0.2391 & 0.5753 & 0.0455 & -0.0922 & 0.1116 & 0.7673
\end{array}
$$

$$
C = \begin{array}{ccc}
-0.6566 & 0.0846 & 0.7495 \\
-0.2514 & 0.9123 & -0.3232 \\
0.7111 & 0.4006 & 0.5778
\end{array}
\qquad
M = \begin{array}{ccc}
27.37 & 0 & 0 \\
0 & 26.13 & 0 \\
0 & 0 & 17.26
\end{array}
$$

In spite of the fact that the original model does not assume any averaging of the marks, the optimal scoring is a form of averaging indeed. However, one should note that it is the model that provides us with both the weights, which are the optimal subject loadings, and the error—these are entirely out of the picture at the empirical averaging.

This line of thinking can be applied to any other hidden performance measures such as quality of life in different cities using scorings over its different aspects (housing, transportation, catering, pollution, etc.) or performance of different sections of management in a big company or government.

2.6.1.4 Sensitivity of the Hidden Factor to Data Standardization

One big issue related to the multiplicative hidden factor model is its instability with respect to data standardization that has been clearly seen at different data normalization options in Project 2.1. Here is another example.

Worked Example 2.14. Principal Components After Feature Centering
Consider the data set in Table 2.22 analyzed above. Take the means of marks over different disciplines in this table, 58.8 for SEn, 67.0 for OOP, and 78.2 for CI, and subtract them from the marks, to shift the data to the mean point (see Table 2.25). This would not much change the average scores presented in Tables 2.19 and 2.20—just shifting them back by the average of the means, $(58.8 + 67.0 + 78.2)/3 = 68$.

Table 2.25 Centered marks for six students and corresponding talent scores, first, as found as explained in A.1, and, second, that rescaled to produce extreme values 0 and 100 if all subject marks are 0 or 100, respectively

#	SEn	OOP	CI	Average	Talent score	Talent rescaled
1	−17.8	−1.0	11.8	−2.3	−3.71	13.69
2	−1.8	−11.0	−18.2	−10.3	1.48	17.60
3	2.2	5.0	0.8	2.7	0.49	16.85
4	10.2	6.0	−6.2	3.3	2.42	18.31
5	4.2	−15.0	9.8	−0.3	−1.94	15.02
6	3.2	16.0	1.8	7.0	1.25	17.42

Everything changes, though, in the multiplicative model, starting from the data scatter, which is now 1729.7—a 50 times reduction from the case of uncentered data. The maximum singular value of the feature centered matrix in Table 2.25, is 27.37 so that the multiplicative model now accounts for only $27.37^2/$ $1729.7 = 0.433 = 43.3\%$ of the data scatter. This goes in line with the idea that much of the data structure can be seen from the "grand" mean (see Fig. 2.20 in Sect. 2.4.5.1 illustrating the point), however, this also greatly increases the error. In fact, the relative order of errors does not change that much, as can be seen in formula (2.29) decomposing the centered data (in the box on the left) in the model-based item, the first on the right, and the residual errors in the right-hand item. What changes is the denominator. The model-based estimates have been calculated in the same way as those in formula (2.27)—by multiplying every entry of the new talent score vector $z* = (−3.71, 1.48, 0.49, 2.42, −1.94, 1.25)$ over every entry of the new subject loading vector $c* = (3.10, 1.95, −3.73)$.

$$
\begin{bmatrix}
-17.8 & -1.0 & 11.8 \\
-1.8 & -11.0 & -18.2 \\
2.2 & 5.0 & 0.8 \\
10.2 & 6.0 & -6.2 \\
4.2 & -15.0 & 9.8 \\
3.2 & 16.0 & 1.8
\end{bmatrix}
=
\begin{bmatrix}
-11.51 & -7.24 & 13.83 \\
4.60 & 2.90 & -5.53 \\
1.52 & 0.96 & -1.82 \\
7.52 & 4.73 & -9.04 \\
-6.02 & -3.79 & 7.23 \\
3.88 & 2.44 & -4.67
\end{bmatrix}
+
\begin{bmatrix}
-6.33 & 6.24 & -2.00 \\
-6.44 & -13.90 & -12.63 \\
0.65 & 4.05 & 2.66 \\
2.65 & 1.27 & 2.87 \\
10.18 & -11.21 & 2.60 \\
-0.72 & 13.56 & 6.50
\end{bmatrix}
\quad (2.29)
$$

Worked Example 2.15. Rescaling the Talent Score from Worked Example 2.14
Let us determine rescaling parameters α and β that should be applied to $z*$, or to the weights $c*$, in Worked Example 2.14 so that at 0 marks over all three subjects the talent score would be 0 and at all 100 marks the talent score would be 100.

As in the previous section, we first determine what scores correspond to these situations in the current setting. All-zero marks, after centering, become minus the average marks, −58.8 for SEn, −67.0 for OOP, and −78.2 for CI. Averaged according to the loadings $c*$ from Worked Example 2.14, they produce 3.10 * (−58.8) + 1.95 * (−67) − 3.73 * (−78.2) = −21.24. Analogously, all-hundred marks, after centering, become 41.2 for SEn, 33.0 for OOP, and 21.8 for CI to produce the score 3.10 * (41.2) + 1.95 * (33) − 3.73 * (21.8) = 110.8. The difference between these,

110.8 − (−21.2) = 132.0 divides 100 to produce the rescaling coefficient a = 100/ 132 = 0.75, after which shift value is determined from the all-0 score as b = −a* (−21.24) = 16.48. Thus rescaled talent scores are in the column on the right of Table 2.25. These are much less related to the average scoring than it was the case at the original data. One can see some drastic changes such as, for instance, the formerly worst student 2 becoming second best, since their deficiency over CI has been converted to an advantage because of the negative loading at CI.

For a student with marks (50, 50, 70) that becomes (−8.8, −17.0, −8.2) after centering, the rescaled talent score comes from the adjusted weighting vector w = a * c = 0.75 * (3.10, 1.95, −3.73) = (2.34, 1.47, −2.81) as the weighted average 2.34 * (−8.8) + 1.47 * (−17) − 2.81 * (−8.2) = −22.73 plus the shift value b = 16.48 so that the result is, paradoxically, −6.25—less than at all zeros! This is again a result of the negative loading at CI.

This example illustrates not only the idea of a great sensitivity of the multiplicative model, but, also, that there should be no mark centering when evaluating performances.

Project 2.2. Stratification: The PCA and an Alternative Scoring Method

Consider the following illustrative data of 8 projects evaluated over two criteria (Table 2.26).

These data are illustrated on Fig. 2.23 which is just a data scatter-plot on the plane of the two criteria, x_1 and x_2. Additional information, the parallel dash-lines A, B, and C show three strata, A, B, and C, to which the projects belong.

The term "stratum (singular)—strata (plural)" denotes here a line at which an aggregate weighted criterion $f = w_1x_1 + w_2x_2$ is constant. It is not difficult to check that the strata A, B, and C on Fig. 2.16 are defined by the equation $f = 1/3x_1 + 2/3x_2 = \alpha$ where α = 2.67, 2.00, and 0.67, respectively. The weights form a vector, $w = (w_1, w_2) = (1/3, 2/3)$ which shows the relative contribution of the criteria in the aggregate criterion. This vector is orthogonal to the strata lines, that is, normal vector, and is represented by the dash-axis on Fig. 2.23.

Table 2.26 Eight projects evaluated over two criteria

Project	Criterion x_1	Criterion x_2	LinStrat	First singular vector
C1	2	0	0.67	1.54
C2	0	1	0.67	0.23
B1	6	0	2.00	4.63
B2	5	0.5	2.00	3.97
B3	3	1.5	2.00	2.66
B4	1	2.5	2.00	1.34
A1	4	2	2.67	3.54
A2	2	3	2.67	2.23

The columns on the right contain scores assigned to the projects by two alternative criteria combining methods

In this regard, a question emerges—have the strata A, B, C and the weight vector w anything to do with the SVD decomposition? Specifically—does this solution coincide with that provided by the first singular vector of the data matrix in Table 2.26? On the first glance, it should. Indeed, the aggregate criterion exactly represents the data, thus making the error of the one dimensional representation zero. It is the error, in its squared form, which is minimized by the method of principal components, is not it? It appears not!

Let us first compute the first principal component, with no preliminary data centering. To this end, let as compute the squared matrix $A = X^T X$:

$$A = \begin{pmatrix} 95.00 & 23.50 \\ 23.50 & 22.75 \end{pmatrix}$$

Operation eig in MatLab produces the maximum eigenvalue of A, 101.971 and the corresponding eigenvector $c = (-0.9587, -0.2844)$. The negative sign, at the positive A, makes no sense. Thus, we multiply it by -1 to equivalently arrive at $c = (0.9587, 0.2844)$. Normalizing this to weights summing to unity, we obtain a weight vector $w1 = (0.7712, 0.2289)$. This shifts the main weight from x_2 to x_1! Moreover, the quadratic error of the component is not zero, but 13.4% of the data scatter (Fig. 2.23).

What is the matter? Well, let us recall that the PCA encoder–decoder model assumes no linear relation between the hidden factor sought and the criteria. The model in (2.17) assumes that data matrix elements are products of the factor scores and feature loadings, $x_{iv} \approx z_i c_v$, up to a residual. The fact that the principal component is a linear combination of features derives from their optimality regarding that model, the least-squares criterion (2.18), from nothing else.

Indeed, if we compare the values of the Rayleigh quotient, $\lambda(c) = c^T A c/(c^T c)$, whose maximum is the maximum eigenvalue of A, at w and $w1$, we will see that $\lambda(w) = 56$, whereas $\lambda(w1) = 101.971$, which shows how far away is the right weight vector $w = (1/3, 2/3)$ from the optimum.

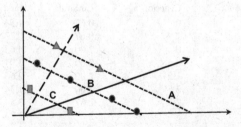

Fig. 2.23 Parallel strata on the plane of criterion x_1, x-axis, and criterion 2, x_2, y-axis, labeled by letters A, B, and C, according to groups of project points located on them. Projects of these groups are presented by rectangles (C), circles (B), and triangles (A). The dash line axis represents the normal vector to the strata, the solid line axis, the first principal component

This example motivates us to define a data analysis model which would be more appropriate to the issue of stratification than that of the PCA. According to such a model, the user has to pre-specify the number of strata, $K > 1$. The problem is to find them by minimizing their "thickness". Given M criteria $x_1, x_2,..., x_M$, over N objects, an aggregate criterion $f = \sum_{m=1}^{M} w_m x_m$ is to determine K strata and their aggregate f-values, c_k ($k = 1, 2, ..., K$), so that the objects in each of the strata would have their aggregate f-values as close to the strata value c_k, as possible (see Orlov and Mirkin 2014; Murtagh et al. 2018). Let us denote the strata to be found by S_k ($k = 1, 2,..., K$); they form a partition, so that every object belongs to one and only one of the strata. Then the criterion can be formulated as

$$L(w, S_k, c_k) = \sum_{i=1}^{N} \sum_{k=1}^{K} \sum_{i \in S_k} \left(c_k - \sum_{m=1}^{M} w_m x_{im}\right)^2 \qquad (2.30)$$

to be minimized with respect to an unknown weight vector $w = (w_m)$ such that $0 \leq w_m \leq 1$ and $w_1 + w_2 + \cdots + w_M = 1$, strata S_k and their values c_k ($k = 1,..., K$).

A method, Linstrat, proposed by Orlov and Mirkin (2014) starts from a vector of arbitrary admissible weights w and works in iterations. Each of the iterations consists of two steps.

The first step takes in vector w and forms the current aggregate criterion $f = Xw$ where X is the object-to-criterion data matrix. The criterion in (2.30) can be reformulated by using f as

$$l(S_k, c_k) = \sum_{i=1}^{N} \sum_{k=1}^{K} \sum_{i \in S_k} (c_k - f_i)^2 \qquad (2.31)$$

It is not difficult to prove two following properties of the criterion:

(1) Given strata S_k, the optimal values c_k are the within-stratum centers of f, $c_k = \frac{1}{N_k} \sum_{i \in S_k} f_i$.
(2) Given centers c_k, the optimal strata are intervals of the f range.

The statement (1) is proven by taking the derivative of (2.31) with respect to c_k and equalizing it to zero. To prove the statement (2), one should assume that on the contrary a stratum contains such i' and i'' that a j from a different stratum exists such, that $f_{i'} < f_j < f_{i''}$; in this case, moving one of them to an appropriate stratum should decrease the value of the criterion, which would prove that the initial assumption is wrong.

This warrants that the number of admissible partitions can be constrained to a binomial number, C_N^{K-1}, of sets of $K - 1$ boundary points between the strata; a method for finding an optimal partition is known for quite a while already (see Fisher 1958), but an ordinary K-means works even better, as our experiments show. An ordinary K-means algorithm begins with a set of values c_k. Then all the objects are assigned to their nearest centers, forming intervals of f-values around them. After this, the within-interval averages are computed, and the process is repeated till convergence. More on K-means can be found in Chap. 4.

The second step takes in the strata S_k found at the previous step and finds a weight vector minimizing (2.31) at given S_k and their centers c_k automatically found at every current $f = Xw$ as the within-strata f-averages. To put it more formally, let us introduce the stratification incidence $N \times K$ matrix $S = (s_{ik})$ where $s_{ik} = 1$ if $i \in S_k$, and $s_{ik} = 0$, otherwise. Define $N \times N$ projector matrix $P_S = S^T(S^TS)^{-1}S^T$, so that at any given f, its projection $f_S = P_S f$ onto the linear subspace $L(S)$ span by the columns of S consists of the within-stratum averages c_k ($k = 1, ..., K$). Therefore, the squared norm of the difference between f and f_S, which is the minimized criterion (2.30) can be equivalently reformulated as $L = (f - f_S)^T(f - f_S) = (f^T - f^T P_S)(f - P_S f)$. After a little algebra, we obtain $L = f^T f + f^T P_S P_S f - 2f^T P_S f$. But $P_S P_S = P_S$ because of the property of idempotence of orthogonal projectors. Therefore $L = f^T f - f^T P_S f = f^T(I - P_S)f$ where I is the unity matrix at which all entries are zeros except those in the main diagonal, that are unities. Now we recall that $f = Xw$, so that $L = w^T X^T(I - P_S)Xw$. Let us denote $Y = (I - P_S)X$. Then $L = w^T Y^T Yw$, since matrix $(I - P_S)$ is also idempotent. Then the problem of minimization of (2.30), given S, is to find an admissible weight vector w minimizing $L = w^T Y^T Yw$. This is a canonical quadratic programming problem, which can be solved by using a standard method such as the parametric active set algorithm coded in C++ (see Ferreau et al. 2014; there are references there to other programs). After obtaining an optimal weight vector w, the next iteration of Linstrat may begin. The computation stops upon convergence or reaching a prespecified number of steps. The convergence is warranted by the fact that each step decreases the value of criterion L in (2.30) because the number of possible strata is finite.

The results of application of the method Linstrat to the base stations data in Table 1.7 (Chap. 1) are in Table 2.26. The Table contains the raw data (the first four columns) and the resulting 3 strata stratification. The number of strata is taken to be the same as it was in the original stratification over the first feature (the so-called ABC-classification). Let us recall the criteria in this Table:

1. AvNUser—the average number of unique active users a day, in thousands;
2. AvUsRev—the average number of the median values of the average revenue per customer a day;
3. IncMeg—the average income generated by the traffic per a million bytes.

The features are measured in quite different scales so that the data are to be pre-processed using a data standardization.

Any standardization of feature x_v is defined by two reals, the origin shift a_v and the scaling factor b_v: so that the standardized feature x'_v is defined as

$$x'_v = (x_v - a_v)/b_v$$

We apply two different standardizations to the data. One standardization, popular in Operations Research, transforms all the feature values into a 0–100 scale so that

Table 2.27 The data and strata for base stations

#	AvNUser	AvUsRev	Inc_Meg	Strata
1	385.30	500.57	10.53	2
2	124.80	443.90	1.04	3
3	785.70	406.74	5.47	3
4	234.10	411.36	0.94	3
5	15.30	543.97	47.37	1
6	1580.10	525.33	19.49	1
7	243.90	448.36	19.39	2
8	1344.40	509.73	3.44	2
9	610.80	385.65	2.37	3
10	961.60	438.15	8.58	2
11	113.20	491.67	0.87	3
12	1637.20	447.04	9.08	2
13	81.30	418.78	0.77	3
14	0.40	460.32	8.11	3
15	82.70	424.93	2.57	3
16	1010.40	396.67	13.08	2
17	2203.40	525.37	21.26	1
18	175.30	578.02	10.79	2
19	2487.10	486.62	8.72	1
20	180.70	499.58	6.78	3
21	2284.50	375.44	4.38	2
22	119.10	378.25	1.58	3
23	2077.80	429.69	6.86	2
24	757.00	435.41	3.23	3
25	361.30	487.94	4.54	3
26	2401.50	402.60	11.64	1
27	1174.30	433.33	5.62	2
28	54.30	481.13	1.19	3
29	615.30	523.04	7.81	2
30	524.10	490.90	2.80	3
31	125.80	438.58	1.71	3
32	466.50	439.90	6.07	3
33	2305.10	419.05	12.00	1
34	395.20	424.58	2.97	3
35	655.00	416.37	6.43	3
36	674.20	442.18	16.26	2
37	1251.30	462.81	6.72	2
38	700.80	428.38	2.27	3
39	1112.70	609.03	17.84	1
40	353.40	551.38	1.76	3

Table 2.28 Criteria characteristics and weights for the base stations data

Characteristics	Criteria		
	AvNUser	AvUsRev	Inc_Meg
Range	2486.7	233.6	46.6
Standard deviation	763.2	55.2	8.5
Range/Std	3.3	4.2	5.5
Standardization	Criteria weights		
Statistics	0.46	0.18	0.36
Operations research	0.34	0.17	0.49

the minimum feature value is shifted to 0 and the maximum value, to 100. To achieve this, the parameters are defined by $a_v = \min(x_v)$ and $b_v = (\max(x_v)-\min(x_v))/100$. The other standardization, popular in Statistics, is the conventional z-scoring at which a_v is taken to be the mean of x_v, and b_v, the standard deviation of x_v'.

The stratification in Table 2.27 has been obtained under each of the two standardizations. One can see that the stratum 1 referring to the heaviest load contains only 7 base stations, by far less than 1/3 of the stations, 13, normally taken for the maintenance work. This would give the company a cheer as the result hints that the maintenance work could be somewhat reduced.

However, the criteria weight coefficients obtained under these standardizations somewhat differ (see Table 2.28 which also contains spread characteristics of the criteria, as well as their ratios). Although the ratios of the ranges and standard deviations in Table 2.28 do show a degree of differences between their actions as the divisors in the data standardization formula, the weights obtained do not show that much of differences. Both standardization give a lesser weight to the Revenue per User criterion while giving almost equal weighting to the two other criteria.

2.6.2   PCA and Ranking Network Nodes: PageRank

Take a look at the graph in Fig. 2.24 representing a small fragment of a large social network.

This is a directed graph with 4 nodes and 5 asymmetric arcs. Each node has inward and outward links according to the arc directions. For example, node 1 has 3 outward links, to nodes 2, 3, and 4, and 0 inward links, whereas node 2 has 1 outward link, to node 3, and 2 inward links, from nodes 1 and 3 (Fig. 2.24).

To score importance of a node, assume that each node has a unit of influence which is uniformly distributed across its outward links, so that node 2 sends 1/3 to each of the nodes 2, 3, and 4. This can be represented by a 4×4 matrix Q:

Fig. 2.24 An illustrative
network fragment

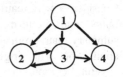

$$Q = \begin{pmatrix} \begin{array}{cccc} 1 & 2 & 3 & 4 \\ \hline 0 & 1/3 & 1/3 & 1/3 \\ 0 & 0 & 0 & 0 \\ 0 & 1/2 & 0 & 1/2 \\ 0 & 0 & 0 & 0 \end{array} \end{pmatrix} \begin{array}{c} 1 \\ 2 \\ 3 \\ 4 \end{array}$$

The (i, j)-th entry of Q, q_{ij} is the share of influence sent by i to j. What is nice about Q, its mathematical property—that the rows sum to unity. Well, not all of them: the row 4 cannot sum to unity because it is all zero; node 4 is terminal, sending no influence to other nodes. There is a simple remedy: assume that any terminal node randomly distributes its influence across the nodes by sending $1/N$ to each of the N nodes in the network, including herself. Matrix Q becomes Q':

$$Q' = \begin{pmatrix} \begin{array}{cccc} 1 & 2 & 3 & 4 \\ \hline 0 & 1/3 & 1/3 & 1/3 \\ 0 & 0 & 1 & 0 \\ 0 & 1/2 & 0 & 1/2 \\ 1/4 & 1/4 & 1/4 & 1/4 \end{array} \end{pmatrix} \begin{array}{c} 1 \\ 2 \\ 3 \\ 4 \end{array}$$

In general, this transformation can be described as follows. Given a network of N nodes, introduce a binary vector, $t = (t_i)$, where $t_i = 1$ if i is a terminal node, and $t_i = 0$, otherwise. Define $e = (1,1,\dots,1)^T$, a $N \times 1$ vector consisting of all unities. Then Q' is defined by equation $Q' = Q + te/N$.

A matrix P is referred to as a stochastic one if all its rows are non-negative and sum to unity. Therefore, Q' is a stochastic matrix. A stochastic $N \times N$ matrix P corresponds to a probabilistic process, the so-called Markov chain with N states. In the process, a particle wanders between the states jumping from i to j with probability p_{ij} in P. Therefore, our analysis of the distribution of influences in the networks will be valid for Markov chains as well.

Before we begin, let us introduce one more transformation of matrix Q to make it positive so that the Perron-Frobenius theorem works straightforwardly, with no need in addressing the intricacies of the structure of the network graph. Such an intricacy exists in the structure of the network in Fig. 2.24: the subset of nodes 2, 3, and 4 forms a closed subset, so that any particle visiting either of them can never go to node 1 outside because of the zero probabilities.

Let us introduce the possibility of "teleportation" between any two nodes with a small probability which does not depend on the structure of the network.

Consider $N \times N$ matrix $\beta ee^T/N$ with the same number β/N, where β is assumed rather small, in all its entries and add it to Q' or, better, to $(1 - \beta)Q'$ so that the summary matrix remains stochastic. This is what makes PageRank a most powerful application. The original influence matrix Q, with many zeros and potential structural complexities, is transformed into a positive influence matrix

$$P = \alpha(Q + te^T) + (1 - \alpha)ee^T/N \tag{2.32}$$

The damping factor $\alpha = 1 - \beta$ is taken at the level 0.85 in Page et al. (1999); to this author's knowledge, nobody has challenged this value so far.

In our example, at $\alpha = 0.85$, the matrix P (2.32) is:

	1	2	3	4	
	0.0375	0.3208	0.3208	0.3208	1
$P =$	0.0375	0.0375	0.8875	0.0375	2
	0.0375	0.4625	0.0375	0.4625	3
	0.2500	0.2500	0.2500	0.2500	4

Now we can turn to the essence of the PageRank method. Let us associate an importance score, r_i, with each node $i = 1,2,\dots, N$. It is natural to assume that the importance score of node i is the sum of the scores of all the network nodes weighted by the influence the node i exercises over them:

$$r_i = p_{i1}r_1 + p_{i2}r_{12} + \cdots + p_{iN}r_N \tag{2.33}$$

In matrix form, Eq. (2.33) can be rewritten as $r = P^T r$ where r is an N-dimensional column-vector $r = (r_i)$. This equation means that r is an eigenvector of matrix P^T corresponding to its eigenvalue 1. Indeed, it is proven that a positive stochastic matrix has 1 as its unique maximum eigenvalue. The PageRank method, basically, is the power method of one-by-one application of P^T to an arbitrary initial vector r_0, so that each next vector $r_{s+1} = P^T r_s$ ($s = 1,2,\dots$).

Table 2.29 presents results of PageRank iterations at r_0 with all the entries equal to $1/N$ at $N = 4$.

Computationally, the results are fine. The triumph cannot be abated by the difficulties in bringing the matrix P^s to the stationary form; it stabilizes, up to 4 decimals, only at $s = 20$. The difficulties relate to the presence of a "switching" element $p_{23} = 0.8875$ in matrix P. It should be pointed out that the converging eigenvectors are in rows of matrix P (see Q.2.28).

Q.2.28. Why the converging eigenvectors are in rows of matrix P, whereas the power method works with right eigenvectors?
A. Because in the very definition (2.33), r is a left eigenvector of P, or, equivalently, right eigenvector of P^T.

Table 2.29 Results of s-th iterations of the PageRank method for matrix P^T and r_s at $s = 0, 1, 5,$ 10, 14, 15 are in the columns of the table

Node	r_0	r_1	r_5	r_{10}	r_{14}	r_{15}
1	0.25	0.0906	0.0961	0.0957	0.0958	0.0958
2	0.25	0.2677	0.2732	0.2742	0.2742	0.2742
3	0.25	0.3740	0.3576	0.3558	0.3559	0.3559
4	0.25	0.2677	0.2732	0.2742	0.2742	0.2742

The convergence, up to 4 decimals, is observed at 14th iteration

However, the scores found seem at odds with the structure of the graph in Fig. 2.24. The most striking is the fact that the node 1 is dominant in that structure, whereas its final score is the minimum, a humble 0.0958, according to the Table 2.29. The maximum importance is assigned to node 3, dominated by node 1 in Fig. 2.24. Moreover, the most subordinate, according to the graph, node 4, shares 2–3 places in importance. These controversial results are obtained because of the very small size of the network. The random elements introduced in Eq. (2.32) are absolutely minor at any more or less realistic network, with its size N of the order of a thousand or, in a real-world search engine, a hundred million. But in this case the line 4, originally zero in Q, gets numbers as high as line 1 in the modified matrix P.

Therefore, PageRank is a two-step method to be applied to a network, both weighted and flat. A network is referred to as weighted if its arcs are assigned with reals expressing the degree of similarity or interaction between nodes. Arcs in a flat network are assigned with the same, unity, weight. At the first step, the network matrix is transformed into a positive stochastic matrix according to formula (2.32). At the second step, a power method is applied to find an output importance scoring for the nodes. This is one of the most powerful research ideas applied to an industrial application. A history of the power method can be found in Franceschet (2011). Some approaches to extending the power method to big data can be found in Andersson and Eckström (2004); its exposition is broadly followed to in the current account.

Worked Example 2.16. Pagerank Scoring

Let us find the PageRank importance scores of nodes in the graph in Fig. 2.25.

First, we are to develop an interaction matrix P; then apply the power method to P. The network adjacency matrix Q will contain a zero row corresponding to node 1. Matrix of the network, with probabilities 1/6 inserted as entries in row 1, is below

$$Q' = \begin{pmatrix} \frac{1}{6} & \frac{1}{6} & \frac{1}{6} & \frac{1}{6} & \frac{1}{6} & \frac{1}{6} \\ \frac{1}{2} & 0 & \frac{1}{2} & 0 & 0 & 0 \\ \frac{1}{3} & \frac{1}{3} & 0 & \frac{1}{3} & 0 & 0 \\ 0 & 0 & 0 & 0 & \frac{1}{2} & \frac{1}{2} \\ 0 & 0 & 0 & \frac{1}{2} & \frac{1}{2} & 0 \\ 0 & 0 & 0 & 1 & 0 & 0 \end{pmatrix}$$

Fig. 2.25 A six node
network from Andersson and
Ekström (2004)

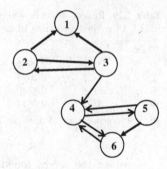

Now we add teleportation probabilities $(1 - \alpha)/N = 0.025$, where $\alpha = 0.85$ and $N = 6$, to each entry of Q' multiplied by α. This produces matrix P:

$$
P = \begin{pmatrix}
\frac{1}{6} & \frac{1}{6} & \frac{1}{6} & \frac{1}{6} & \frac{1}{6} & \frac{1}{6} \\
\frac{9}{20} & \frac{9}{40} & \frac{9}{20} & \frac{1}{40} & \frac{1}{40} & \frac{1}{40} \\
\frac{37}{120} & \frac{37}{120} & \frac{1}{40} & \frac{1}{3} & \frac{1}{40} & \frac{1}{40} \\
\frac{1}{40} & \frac{1}{40} & \frac{1}{40} & \frac{1}{40} & \frac{9}{20} & \frac{1}{9} \\
\frac{1}{40} & \frac{1}{40} & \frac{1}{40} & \frac{9}{20} & \frac{1}{9} & \frac{1}{20} \\
\frac{1}{40} & \frac{1}{40} & \frac{1}{40} & \frac{1}{8} & \frac{1}{40} & \frac{1}{40}
\end{pmatrix}
$$

Applying the power method starting from $e/6$. the iterations of the process converge by $s = 22$, as can be seen in Table 2.30. The matrix P^s to converge takes a bit longer, about 30 iterations.

An issue of theoretical nature should be mentioned regarding the one-rank approximation matrix which is sought by the power method. In contrast to symmetric similarity matrices that are approximated by one-rank matrices with (i, j)-th entry $\lambda x_i x_j$, there is no unified view of an ideal one-rank matrix $\lambda x_i y_j$ where $x = (x_i)$ is a right eigenvector and $y = (y_i)$ is a left eigenvector of an arbitrary P corresponding to its maximum eigenvalue. An issue is that entries p_{ij} in P may have different meaning depending on the nature of pair comparisons they relate to.

A radical view assumes that p_{ij} expresses how many times i is preferred to j, so that the symmetric comparison should be its reciprocal, $p_{ji} = 1/p_{ij}$. Another, less radical view, considers that p_{ij} is a share of the total preference assigned to i against j, so that $p_{ij} + p_{ji} = 1$. Still, both of these keep to an assumption, "the independence of irrelevant alternatives", that the comparison between i and j may not involve any other object. This is much convenient in the theoretical perspective, although rather naïve practically. In particular, the radical view above leads to the concept of super-transitivity of matrix P as a form of consistency in the preferences. This concept reformulates the so-called Luce axiom in psychology as follows (Mirkin 1979): matrix P is super-transitive if the equation $p_{ij}p_{jk} = p_{ik}$ holds for every triplet of objects. i, j, k. It is not difficult to prove that matrix P is super-transitive if and only if there exists a vector $r = (r_1, r_2, \ldots, r_n)$ such that $p_{ij} = r_i/r_j$ $(i, j = 1, \ldots, n)$, see Mirkin 1979 and Q.2.29. For a super-transitive P, this vector r is obviously right

Table 2.30 Results of s-th iterations of the PageRank method for 6×6 matrix P^T and $s = 0, 1, 5,$ 10, 20, 22 are in the columns of the table

Node	r_0	r_1	r_5	r_{10}	r_{20}	r_{22}
1	0.1667	0.1667	0.0833	0.0743	0.0737	0.0737
2	0.1667	0.0958	0.0572	0.0521	0.0517	0.0517
3	0.1667	0.1194	0.0639	0.0578	0.0574	0.0574
4	0.1667	0.3083	0.3351	0.3308	0.3327	0.3327
5	0.1667	0.1903	0.2926	0.3074	0.3076	0.3076
6	0.1667	0.1194	0.1679	0.1776	0.1769	0.1769

The convergence, up to 4 decimals, is observed at 22nd iteration

eigenvector of matrix P^T corresponding to its maximum eigenvalue 1, whereas vector of reciprocals $r' = (1/r_1, 1/r_2, \ldots, 1/r_n)$ is a corresponding left eigenvector of P^T. Of course, the power method for a super-transitive matrix converges in one iteration.

Q.2.29. A square matrix P is super-transitive if the equation $p_{ij}p_{jk} = p_{ik}$ holds for every triplet of indices. i, j, k. Prove that matrix P is super-transitive if and only if there exists a vector $r = (r_1, r_2, \ldots, r_n)$ such that $p_{ij} = r_i/r_j$ ($i, j = 1, \ldots, n$).

A. If $p_{ij} = r_i/r_j$, then the super-transitivity is obvious. Conversely, assume P to be super-transitive. Take any positive number r_1 and define $r_i = r_1/p_{1i}$ for all $i > 1$. Then, for any i, j, $p_{ij} = p_{i1}p_{1j}$ because of super-transitivity. But $p_{1j} = r_1/r_j$ because of definition. Similarly, $p_{i1} = 1/p_{1i} = r_i/r_1$. Therefore, $p_{ij} = r_i/r_j$, q.e.d.

The concept of super-transitive pair comparisons lies at the heart of another popular ranking method, the so-called Analytical Hierarchy Process, which also uses the Perron-Frobenius theorem and power method (Saaty 1990). Podinovski and Podinovskaya (2014) point to some shortcomings of the ranking approach based on the first eigenvector. For further reading, see Luce (2014), Cavallo and D'Apuzzo (2009).

2.7 Summary

This Chapter introduces the concept of data summarization as an encoder-decoder pair and describes the method of principal components (PCA) as a data-driven model in this framework. The data, even if of mixed scale types, is first converted into a quantitative format by using popular dummy representations of categories. This gives a chance to discuss, however brief, the concept of feature as a mathematical object. Luckily, the PCA model is based on a well developed mathematical theory of singular value decomposition (SVD) for rectangular matrices. Unlike the conventional formulation of PCA, this model does not require to postulate that the principal components are to be linear combinations of features. This property is rather derived from the model. The PCA model itself is rather simplistic and suggests that further thinking on better data summarization models should be undertaken. An extension of PCA to scoring nodes of a network, the celebrated Google PageRank approach, is described too.

Three applications of PCA—scoring hidden factors, data visualization, and Latent semantic analysis are illustrated with further instructions and Worked Examples. This is extended to network and similarity data including the currently very popular Google PageRank algorithm.

It should be noted that the PCA model is an early example of what is currently called matrix factorization approach. According to this approach the data matrix is encoded as the product of two other, unknown, matrices, $X = CZ$ which represent a hidden structure of the data and satisfy some conditions required by the underlying considerations. Such is the model of PCA above and of the partitional cluster analysis described further, in Part 4. Also much popular are the so-called non-negative matrix factorization and probabilistic topic modeling; see, in respect, Lee and Seung (2001) and Wang and Blei (2011) and references therein.

References

B. Efron, R.J. Tibshirani, *An Introduction to the Bootstrap* (CRC Press, 1994)

T.K. Landauer, *Latent Semantic Analysis* (Wiley, Hoboken, 2006)

R.D. Luce, *Utility of Gains and Losses: Measurement-theoretical and Experimental Approaches* (Psychology Press, 2014)

C.D. Manning, P. Raghavan, H. Schütze, *Introduction to Information Retrieval* (Cambridge University Press, Cambridge, 2008)

B. Mirkin, (1979) *Group Choice* (Winston and Sons, 1979). *A division of Scripta Technica* (English translation from Russian, Group Choice Problems, 1974)

B. Mirkin, *Mathematical Classification and Clustering* (Kluwer Academic Press, 1996)

B. Mirkin, *Clustering: A Data Recovery Approach* (Chapman & Hall/CRC, Boca Raton, 2012)

R. Tibshirani, M. Wainwright, T. Hastie, *Statistical Learning with Sparsity: The Lasso and Generalizations* (Chapman and Hall/CRC, Boca Raton, 2015)

Articles

E. Andersson, P.A. Ekström, Investigating Google's pagerank algorithm. A Tech. Rep. Sci. Comput. (2004)

J. Carpenter, J. Bithell, Bootstrap confidence intervals: when, which, what? A practical guide for medical statisticians. Stat. Med. **19**(9), 1141–1164 (2000)

B. Cavallo, L. D'Apuzzo, A general unified framework for pairwise comparison matrices in multicriterial methods. Int. J. Intell. Syst. **24**(4), 377–398 (2009)

S. Deerwester, S. Dumais, G.W. Furnas, T.K. Landauer, R. Harshman, Indexing by latent semantic analysis. J. Am. Soc. Inf. Sci. **41**(6), 391–407 (1990)

H.J. Ferreau, C. Kirches, A. Potschka, H.G. Bock, M. Diehl, qpOASES: A parametric active-set algorithm for quadratic programming. Math. Program. Comput. **6**(4), 327–363 (2014)

W.D. Fisher, On grouping for maximum homogeneity. J. Am. Stat. Assoc. **53**(284), 789–798 (1958)

M. Franceschet, PageRank: Standing on the shoulders of giants. Commun. ACM **54**(6), 92–101 (2011)

E.V. Kovaleva, B.G. Mirkin, Bisecting K-means and 1D projection divisive clustering: a unified framework and experimental comparison. J. Classif. **32**(3), 414–442 (2015)

D.D. Lee, H.S. Seung, Algorithms for non-negative matrix factorization. Adv. Neural Inf. Process. Syst. 556–562 (2001)

M.A. Makary, M. Daniel, Medical error—the third leading cause of death in the US. BMJ **353**, i2139 (2016)

F. Murtagh, M. Orlov, B. Mirkin, Qualitative judgement of research impact: Domain taxonomy as a fundamental framework for judgement of the quality of research. J. Classif. **35**(1), 5–28 (2018)

M. Orlov, B. Mirkin, A concept of multicriteria stratification: a definition and solution. Procedia Comput. Sci. **31**, 273–280 (2014)

L. Page, S. Brin, R. Motwani, T. Winograd, The PageRank citation ranking: bringing order to the web. Stanford InfoLab Technical Report (1999)

V. Podinovski, O.V. Podinovskaya, Criteria importance theory for decision making problems with a hierarchical criterion structure, Moscow. HSE Working Paper WP7/2014/04 (2014)

T.L. Saaty, How to make a decision: the analytic hierarchy process. Eur. J. Oper. Res. **48**(1), 9–26 (1990)

C. Wang, D.M. Blei, Collaborative topic modeling for recommending scientific articles, in *Proceedings of the 17th ACM SIGKDD International Conference on Knowledge Discovery and Data Mining* (2011), 448–456

Chapter 3
Learning Correlations

Abstract After a short introduction of the general concept of decision rule to relate input and target features, this chapter describes some generic and most popular methods for learning correlations over two or more features. Four of them pertain to quantitative targets (linear regression, canonical correlation, neural network, and regression tree), and seven to categorical ones (linear discrimination, support vector machine, naïve Bayes classifier, classification tree, contingency table, distance between partition and ranking relations, and the correspondence analysis). Of these, classification trees are treated in a most detailed way including a number of theoretical results that are not well known. These establish firm relations between popular scoring functions and bivariate measures—Quetelet indexes in contingency tables and, rather unexpectedly, normalization options for dummy variables representing target categories. Some related concepts such as Bayesian decision rules, bag-of-word model in text analysis, VC-dimension and kernel for non-linear classification are introduced too. The Chapter outlines several important characteristics of summarization and correlation between two features, and displays some of the properties of those. They are:

- linear regression and correlation coefficient for two quantitative variables (Sect. 3.2);
- tabular regression and correlation ratio for the mixed scale case (Sect. 3.8.3); and
- contingency table, Quetelet index, statistical independence, and Pearson's chi-squared for two nominal variables; the latter is treated as a summary correlation measure, in contrast to the conventional view of it as just a criterion of statistical independence (Sect. 3.6.1); moreover, a few less known least-squares based concepts are outlined, including canonical correlation and correspondence analysis.

3.1 General: Decision Rules, Fitting Criteria, and Learning Protocols

To specify a problem of learning correlation in a data table, one has to distinguish between two parts in the feature set: *predictor*, or *input*, features and *target*, or *output*, features. Typically, the number of target features is small, and in generic tasks, there is just one target feature. Target features are usually difficult to measure or impossible to know beforehand. This is why one would want to derive a decision rule relating predictors and targets so that prediction of targets can be made after measuring predictors only. Examples of learning problems include:

(a) chemical compounds: input features are of the molecular structure, whereas target features are activities such as toxicity or healing effects;
(b) types of grain in agriculture: input features are those of the seeds, soil and weather, and target features are of productivity and protein contents,
(c) industrial enterprises: input features refer to technology, investment and labor policies, whereas target features are of sales and profits;
(d) postcode districts in marketing research: input features refer to demographic, social and economic characteristics of the district residents, target features—to their purchasing behavior;
(e) bank loan customers: input features characterize demographic and income, whereas output features are of (potentially) bad debt;
(f) gene expression data: input features relate to levels of expression of DNA materials in the earlier stages of an illness, and output features to those at later stages.

A *decision rule* predicts values of target features from values of input features. A rule is referred to as a classifier if the target is categorical and as a regression if the target is quantitative. A generic categorical target problem is defined by specifying just a subset of entities labeled as belonging to the class of interest—the correlation problem in this case would be of building such a decision rule that would recognize, for each of the entities, whether it belongs to the labeled class or not. A generic regression problem—the bivariate linear regression—is considered in Sect. 3.1; its extension to the multivariate case is described in Sect. 3.3.

A decision rule is learnt over a dataset in which values of the targets are available. These data are frequently referred to as the training data. The idea underlying the process of learning is to look at the difference between predicted and observed target feature values on the training data set and to minimize them over a class of admissible rules. The structure of such a process is presented on the upper part of Fig. 3.1.

The notion that it ought to be a class of admissible rules pre-specified emerges because the training data is finite and, therefore, can be fit exactly by using a sufficient number of parameters. However, this would be valid on the training set only, because the fit would capture all the errors and noise inevitable in data collecting processes. Take a look, for example, at the 2D regression problem on Fig. 3.2 depicting seven points on (x,u)-plane corresponding to observed combinations of input feature x and target feature u.

The seven points on Fig. 3.2 can be exactly fitted by a polynomial of 6th order $u = p(x) = a_0 + a_1x + a_2x^2 + a_3x^3 + a_4x^4 + a_5x^5 + a_6x^6$. Indeed, they would lead to 7 equations $u_i = p(x_i)$ ($i = 1,...,7$), so that, in a typical case, the 7 coefficients a_k of the polynomial can be exactly determined. Having N points observed would require an $(N-1)$-th degree polynomial to exactly fit them.

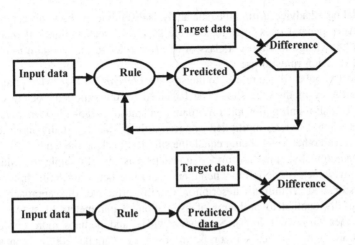

Fig. 3.1 Structure of a training/testing problem: In training, on the top, the decision rule is fitted to minimize the difference between the predicted and observed target data. In testing, the bottom part, the rule is used to predict so that no feedback to the rule is utilized

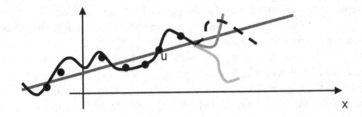

Fig. 3.2 Possible graphs of interrelation between x and u according to observed data points (black circles)

However, the polynomial, on which graph all the observations lie, has no predictive power both within and beyond the range. The curve may go either course (like those shown) depending on small changes in the data. The power of a theory —and a regression line is a theory in this case—rests on its generalization power, which, in this case, can be cast down as the relation between the number of observations and the number of parameters: the greater the better. When this ratio is relatively small, statisticians would refer to this as an *over-fitted* rule. The over-fitting normally produce very poor predictions on newly added observations. The blue straight line fits none of the points, but it expresses a simple and very robust tendency and should be preferred because it summarizes the data much deeper: the seven observations are summarized here in just two parameters, slope and intercept, whereas the polynomial line provides no summary: it involves as many parameters as the data entities. This is why, in learning decision rules problems, a class of admissible rules should be selected first. Unfortunately, as of this moment, there is no model based advice, within the data analysis discipline, on how this can be done, except very general ones like "look at the shapes of scatter plots". If there is no domain knowledge to choose a class of decision rules to fit, it is hard to tell what class of decision rules to use.

A most popular advice relates to the so-called *Occam's razor*, which means that the complexity of the data should be balanced by the simplicity of the decision rule. A British monk philosopher William Ockham (*c.* 1285–1349) made a claim: "Entities should not be multiplied unnecessarily." This is usually interpreted as saying that all other things being equal, the simplest explanation tends to be the best one. Operationally, this is further translated as the Principle of Maximum Parsimony, which is referred to when there is nothing better available. In the format of the so-called "Minimum description length" principle, this approach can be meaningfully applied to problems of estimation of parameters of statistic distributions (see Grünwald 2007). Somewhat wider, and perhaps more appropriate, explication of the Occam's razor is proposed by Vapnik (2006). In a slightly modified form, to avoid mixing different terminologies, it can be put as follows: "Find an admissible decision rule with the smallest number of free parameters to explain the observed facts" (Vapnik 2006, p. 448). However, even in this format, the principle gives no guidance about how to choose an adequate functional form. For example, which of two functions, the power function $f(x) = ax^b$ or logarithmic one, $g(x) = b\log(x) + a$, both having just two parameters a and b, should be preferred as a summarization tool for graphs on Fig. 3.3?

Another set of advices, not incompatible with those above, relates to the so-called falsifiability principle by Popper (1902–1994), which can be expressed as follows: "Explain the facts by using such an admissible decision rule which is easiest to falsify" (Vapnik 2006, p. 451). In principle, to falsify a theory one needs to give an example contradicting to it. Falsifiability of a decision rule can be

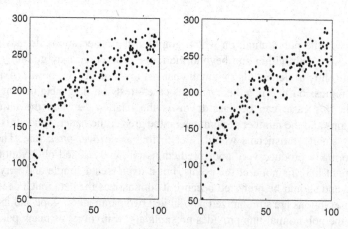

Fig. 3.3 A graph of one out of two functions, $f(x) = 65x^{0.3}$ and $g(x) = 50\log(x) + 30$, both with an added normal noise $N(0,15)$, is presented on each of the plots. Can the reader give an educated guess of which is which? (Answer: $f(x)$ is on the right and $g(x)$ on the left)

formulated in terms of the so-called *VC*-dimension, a measure of complexity of classes of decision rules: the smaller VC- dimension, the greater the falsifiability.

Let us explain the concept of VC-dimension for the case of a categorical target, so that a decision rule to be would be a classifier. However many categorical target features are specified, different combinations of target categories can be assigned different labels, so that a classifier is bound to predict a label. A set of classifiers Φ is said to shatter the training sample if for any possible assignment of the labels, a classifier exactly reproducing the labels can be found in Φ. Given a set of admissible classifiers Φ, the VC-dimension of a classifying problem is the maximum number of entities that can be shattered by classifiers from Φ. For example, 2D points have VC complexity 3 in the class of linear decision rules. Indeed, any three points, not lying on a line, can be shattered by a line; yet not all four-point sets can be shattered by lines, as shown on Fig. 3.4, the left and right parts, respectively.

The VC complexity is an important characteristic of a correlation problem especially within the probabilistic machine learning paradigm. Under the conventional conditions of the independent random sampling of the data, a reliable classifier "with probability $a\%$ will be $b\%$ accurate, where b depends not only on a, but also on the sample size and VC-dimension" (Vapnik 2006).

The problem of learning correlation in a data table can be stated, in general terms, as follows. Given N pairs (x_i, u_i), $i = 1, ..., N$, in which x_i are predictor/input p-dimensional vectors $x_i = (x_{i1},...,x_{ip})$ and $u_i = (u_{i1},...,u_{iq})$ are target/output q-dimensional vectors (usually $q = 1$), find a decision rule

$$\hat{u} = F(x) \qquad (3.1)$$

such that the difference between computed $\hat{u}$ and observed u is minimal over a pre-specified class Φ of admissible rules F.

To specify a correlation learning problem one should specify assumptions regarding a number of constituents including:

(i)   Type of target
      Two types of target features are considered usually: quantitative and categorical. In the former case, Eq. (3.1) is usually referred to as regression; in the latter case, decision rule, and the learning problem is referred to as that of "classification" or "pattern recognition".

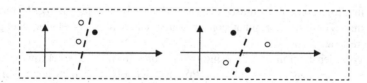

Fig. 3.4 Any two-part split of three points (not on one line) can be made by a line, but the presented case of four points on the right cannot be split by a line

(ii) Type of rule

A rule involves a postulated mathematical structure whose parameters are to be learnt from the data. The mathematical structures considered further on are:

- *linear* combination of features
- *neural network* mapping a set of input features into a set of target features
- *decision tree* built over a set of features
- *partition* of the entity set into a number of non-overlapping clusters

(iii) Criterion

Criterion of the quality of fitting depends on the framework in which the learning task is formulated. Most popular criteria are: maximum likelihood (in a probabilistic model of data generation), least-squares (data recovery approach) and relative errors. According to the least-squares criterion, the difference between u and $\hat{u}$ is measured with the average squared error,

$$E = \langle u - \hat{u}, u - \hat{u} \rangle / N = \langle u - F(x), u - F(x) \rangle / N \qquad (3.2)$$

which is to be minimized over all admissible F.

(iv) Training protocol

The rule F is to be learnt from a training dataset. The way the data become available can be referred to as the learning protocol. Three popular training protocols are: batch, random and on-line. The batch mode is the case when all the training set is available and used at once, the other two refer to cases when data entities come one by one so that the learning goes incrementally. In the random protocol, the data are available at once, yet the learning process is organized incrementally by picking up entities randomly one-by-one, possibly many times each. In contrast, in an on-line protocol each entity comes from an external source and can be seen only once.

3.2 Two-D Linear Regression and Special Cases

3.2.1 Case of Two Features

Let us first focus on a most illustrative case when only two quantitative features are considered. Three most popular concepts are: scatter plot, correlation coefficient, and regression.

We consider them in turn by using two features from the Market towns dataset, Population Resident and Number of Primary Schools. The data are taken from Table 1.4 (see below an extract for four towns out of 45):

	PopRes (x)	PSchools (y)	(x,y)-point
Tavistock	10,222	5	(10,222.5)
Bodmin	12,553	5	(12,553.5)
Saltash	14,139	4	(14,139.4)
Brixham	15,865	7	(15,865.7)

Scatter plot is a presentation of entities as 2D points in the plane of two pre-specified features. On the left-hand side of Fig. 3.5, a scatter-plot of Market town features PopRes (Axis x) and PSchools (Axis y) is presented.

One can think that these two features are approximately related by a linear equation $y = ax + b$ where a and b are some constant coefficients, referred to as the slope and intercept, respectively, because the number of schools should be related to the number of children which is related to the number of residents. This equation is referred to as the linear regression of y over x. Obviously, most relations are not necessarily that simple because they also depend on other factors such as school sizes, population's age, etc. It would be a miracle if one equation fitted well all 45 towns. The possible inconsistencies in the equation can be modeled as additive errors, or residuals. The slope a and intercept b are taken in such a way that the inconsistencies of the equation on the 45 towns are minimized.

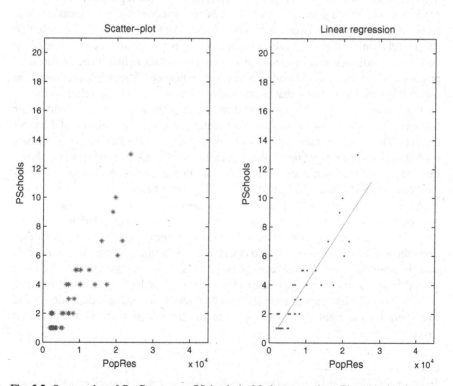

Fig. 3.5 Scatter plot of PopRes versus PSchools in Market town data. The right hand graph includes a regression line of PSchools over PopRes

When a linear regression equation is fitted, its validity should be checked. A valid equation can be used for both (i) prediction and (ii) description.

The term "regression" relates to an episode in the long struggle by Francis Galton for the recognition of his obsession with "inherited talent". He thought, in about 1888, like this: "Currently, I cannot measure the talent—why cannot I take a feature, that I can measure, say, the height? The son's height should be related to the father's height, adjusted by that of the mother, of course." This thought was possibly supported by the idea, rather relevant at that time, that the human's height may have an evolutionary dimension so that, in the UK, the taller people might have had better chances of survival. The relation appeared to exist. However, unexpectedly, it turned out to be rather counter-intuitive. The taller father's son was, on average, not as tall as his father, whereas, in contrast, the shorter fathers' sons were relatively taller than their fathers. Galton considered that as a phenomenon of regress to the mediocrity. And it took some time, to figure out that the regress did not contradict the law of natural selection established by his cousin, Charles Darwin.[1]

The Galton-Pearson theory of linear regression involves a useful and very popular parameter, the correlation coefficient, that shows the extent of linearity in the relation between two features. Its square, referred to as the determinacy coefficient, can be used for a quick check of the validity of the regression: it shows the proportion of the variance of y that is taken into account by the regression. The correlation coefficient between the two features, PopRes and PSchools, is 0.909. The correlation coefficient, in general, ranges between −1 and 1, and a value close to 1 or −1 indicates a high extent of the linear relation between the features. In physics or chemistry, a high value of the correlation coefficient is rather usual; in social sciences, rather not—that is, the current features are highly related indeed.

Most other features in Market town data—such as the numbers of Post offices or Doctors—are also highly related to Pop feature, but not the number of Farmers markets. This latter feature appears to be binary here: a town either has a farmers market or not. The low value of the correlation coefficient, just below 0.15, shows that the size of the town does not much matter in this part of the world: a farmers market is as likely in a small town as it is in a larger town.

A low or even zero value of the correlation coefficient does not necessarily mean "no relation at all", but rather just "no *linear* relation". A zero correlation coefficient may hide a different type of functional relation, as shown on Fig. 3.6, which presents three different cases of the zero correlation. Only one of these, that on the left, case is genuine—there is no relation between x and y according to the picture indeed. Each of the other two cases relates to a rather high association between x and y. Specifically, the figure in the middle refers to a quadratic dependence and the figure on the right, to a split between two subsamples of highly linear but inverse relations.

---

[1]Both, Francis Galton and Charles Darwin, were grandsons of a celebrated medical doctor and philosopher, Erasmus Darwin.

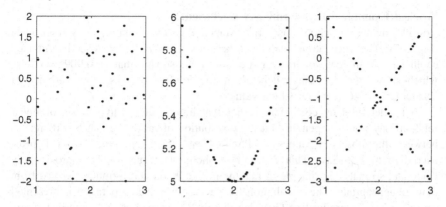

Fig. 3.6 Three scatter-plots corresponding to zero or almost zero correlation coefficient ρ; the case on the left: no relation between x and y; the case in the middle: a non-random quadratic relation $y = (x - 2)^2 + 5$; the case on the right: two symmetric linear relations, $y = 2x - 5$ and $y = -2x + 3$, each holding at a half of the entities

Then the regression equation, estimated according to formulas (3.6)–(3.8) in Sect. 3.2.3, is this:

$$PSchool = 0.401 * PopRes + 0.072 \qquad (3.3)$$

where Population resident (PopRes) is expressed in thousands to make the slope the thousand times greater than it would be if population is expressed in the absolute numbers. The slope expresses how much target changes when the input changes by 1. Because the target's values are integers, the value of slope can be rephrased as follows: the growth of population in a town by 2.5 thousand would lead, on average, to building one more primary school.

3.2.2 Validity of the Regression

A regression function built over a data set should be validated. Three types of validity checks can be considered:

(a) The proportion of the variance of target variable taken into account by the regression, the determinacy coefficient: the greater the determinacy the better the fit.
(b) The confidence intervals of regression parameters—their ranges can give an idea of how stable the regression is.
(c) The direct testing of the accuracy of prediction both on data used for building the regression and data not used for that.

Worked Example 3.1. Determinacy Coefficient

Consider feature PSchools as target versus PopRes as input, in Market Data (Fig. 3.5). The correlation coefficient between them is 0.909. The determinacy coefficient, in the case of linear regression, is its square, that is, $0.909^2 = 0.826$, which shows that the linear dependence on PopRes decreases the variance of PSchools by 83.6%, a rather high value.

If the determinacy coefficient is not that high, still the hypothesis of linear relation may hold—depending on the distribution of residuals, that is, differences between the observed values of PSchool and those computed from PopRes according to Eq. (3.3). This distribution should be Gaussian or approximately Gaussian, so that the principle of maximum likelihood and formulas derived from that are appropriate. The distribution for the case under consideration is presented on Fig. 3.7. It is similar to a Gaussian distribution indeed, at the 5 bin histogram. The histogram with 10 bins is less so because it is somewhat dented—probably the sample is too small for this level of granularity: on average, only 4–5 entities fall in each of the bins.

A more straightforward validity test can be performed without any statistic theory at all—by purely computational means using the so-called bootstrapping which is a procedure for obtaining a multitude of random estimates of the parameters of interest by using random samples from the dataset as illustrated in the Worked Example 3.2.

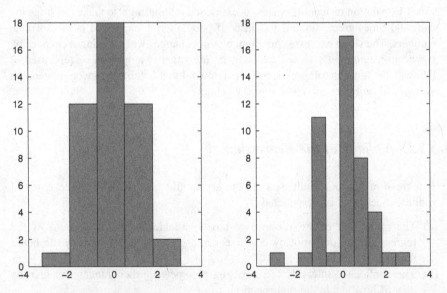

Fig. 3.7 Histograms of the residuals, the differences between values of PSchool as observed and those computed from Pop by using Eq. (3.1), with 5 bins (on the left) and 10 bins (on the right). The dents in the finer histogram can be attributed to the fact that the sample of 45 instances is too small to use 10 bins

Worked Example 3.2. Bootstrap Validity Testing
Consider the linear regression of PSchools over PopRes in Eq. (3.3) in the previous
section. How stable are its slope and intercept regarding change of the sample? This
can be tested by using an approach referred to as the bootstrap. One bootstrap trial
involves three stages:

1. Randomly choose, with replacement, as many entities as there are in the
   sample—45 in this case. Here is the sequence of indices of the entities randomly
   drawn with replacement while writing this text: r = {26,17,36,11,29,39,32,
   25,27,26, 29,4,4,33,10,1,5,45,17,16, 13,5, 42,43,28, 26,35,2,37,44,6,39,33,
   21,15, 11,33,1,44,30,26,25,5,37,24}. Some indices made it into the sample
   more than once, most notably 26—four times, whereas many others did not
   make it into the sample at all—altogether, 16 objects such as 3,7,8 are absent
   from the sample. The proportion of the absent indices is 16/45 = 0.356, which is
   rather close to the theoretic estimate 1/e = 0.3679 derived in Sect. 2.2.3.3, p. 102.
2. Take "resampled" versions of PopRes and PSchools as their values on the
   elements drawn on step 1.
3. Find values of the slope and intercept for the resampled PopRes and PSchools
   and store them.

The MatLab computation steps are similar to those in Sect. 2.2.3.3. After 400
trials the stored slopes and intercepts form distributions presented as 20 bin histograms on Fig. 3.8 a and b, respectively. After 4000 trials, the respective histograms are c and d. One can easily see the smoothing effect of the increased
number of trials on the histogram shapes—at 4000 trials they do look Gaussian.

The bootstrapping trials give a diversity needed for estimating the average values
of the slope and intercept. Moreover, one can draw confidence boundaries for the
values.

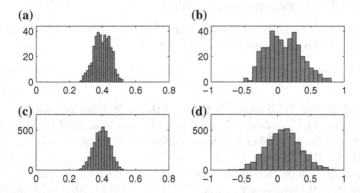

Fig. 3.8 Histograms of the distributions of the slope, on the left, and intercept, on the right, found
at 400 (on top) and 4000 (bottom) bootstrapping trials on PopRes, expressed in thousands, and
PSchool features in Market town data

Table 3.1 Parameters of the linear regression of PopRes over PSchool found on the original set, as well as on the bootstrap 400 and 4000 trials

Regression parameters	Set	400 trials			4000 trials		
		Mean	2.5%	97.5%	Mean	2.5%	97.5%
Slope	0.401	0.399	0.296	0.486	0.398	0.303	0.488
Intercept	0.072	0.089	−0.343	0.623	0.092	−0.400	0.594

The latter involves the average values, as well as the lower and upper 2.5% quantiles

How can one obtain, say, 95% confidence boundaries? According to the non-pivotal method, lower and upper 2.5% quantiles are cut out from the distribution in a symmetric way: 95% of the observations fall between the quantiles. For the case of 400 trials, 2.5% equals 10, so that the lower quantile corresponds to 11th and the upper quantile to 390th elements in the sorted set of values. For the case of 4000 trials, 2.5% equals 100: the quantiles correspond to 101st and 3900th elements of the sorted sets. They are shown in Table 3.1 at both of the cases, 400 and 4000 trials. One can see that these provide consistent and rather tight boundaries for the slope: it is between 0.303 and 0.488 in 95% of all trials, according to the 4000-trial data, and more or less the same at the 400-trial data. The values of intercept are distributed with a greater dispersion and provide for a worsened accuracy. Symmetric 95% confidence intervals for the intercept are $[−0.343, 0.623]$ at 400 trials and $[−0.400, 0.594]$ at 4000 trials.

How a pivotal bootstrapping rule can be applied here? This would provide more stable evaluations than empirical distributions. The standard deviations of the slope and intercept are 0.0493 and 0.2606, respectively, at 400 bootstrapping trials; they are somewhat smaller, 0.0477 and 0.2529, at the 4000 trials. Can one derive from this a symmetric 95% confidence interval for the slope or intercept? Tip: in a Gaussian distribution, 95% of all values fall within interval mean $\pm 1.96 *$ std. This is the so-called pivotal bootstrapping method.

Q.3.1. Can you give an estimate of the level of variance of the differences between PSchool observed and computed values?

A final validity test of the regression equation is probably the toughest one—by the prediction error (see Worked Example 3.3).

Worked Example 3.3. Prediction Error of the Regression Equation
Compare the observed values of PSchool with those computed through PopRes according to Eq. (3.3). Table 3.2 presents a few examples taken from both ends of the sorted PopRes feature.

On average, the predictions are close, but, in some cases, are less so. One can easily estimate the relative error, which is $[(1 − 0.89)/1] * 100 = 11\%$ on the first element, $[(2 − 0.97)/2] * 100 = 51.5\%$ on the second element, etc. The average relative error of Eq. (3.3) on the set of all 45 towns is equal to 30.7%. Can it be made smaller? On the first glance, no, it cannot, because Eq. (3.1) minimizes the error.

Table 3.2 Observed numbers of primary schools versus those predicted from the population resident data on some market towns	PS obs.	PS comp.	Pop	PS obse.	PS comp.	Pop
	1	0.89	2040	2	3.35	5676
	2	0.97	2230	2	3.9	7044
	2	1.06	2452	4	3.12	10,092
	2	1.19	2786	7	6.44	15,865
	1	1.54	3660	4	7.05	17,390

But, the error minimized by Eq. (3.1) is the average quadratic error, not the relative error under consideration. The two errors do differ, and Eq. (3.1) is not necessarily optimal with regard to the relative error.

The classical optimization theory has virtually nothing to propose for the minimization of the relative error—this criterion is neither linear, nor quadratic, nor convex. Instead, the evolutionary optimization approach can be applied to the task. This approach uses a population of solutions randomly evolving, iteration after iteration, in the search for better solutions as explained in Project 3.2. Applying the algorithm from that project to minimize the criterion of relative error, one can find a different solution, in fact, a set of solutions each leading to the average relative error of 26.4%, a reduction of 3.3 points, one seventh of the relative error of the Eq. (3.3).

The new solution is PSchool = 0.28 * PopRes + 0.33 expressing a smaller rate of increase in school numbers at the growth of population.

3.2.3 Fitting the Equation of Linear Regression

Let us derive parameters of linear regression. Given target feature y and predictor x at N entities (x_1,y_1), (x_2,y_2),..., (x_N,y_N), we are interested at finding a linear equation relating them so that

$$y = ax + b \qquad (3.4)$$

The exact fit can occur only if all pairs (x_i,y_i) belong to the same straight line on (x,y)-plane, which is rather unlikely on real-world data. Therefore, Eq. (3.4) will have an error at each pair (x_i,y_i) so that the equation should be rewritten as

$$y_i = ax_i + b + e_i \quad (i = 1, 2, \ldots, N) \qquad (3.4')$$

where e_i are referred to as errors or residuals. The problem is of determining the two parameters, a and b, in such a way that the residuals are least-squares minimized, that is, the average square error

$$L(a,b) = \Sigma_i e_i^2/N = \Sigma_i(y_i - ax_i - b)^2/N, \qquad (3.5)$$

reaches its minimum over all possible a and b, given x_i and y_i ($i = 1, 2, ...,N$). This minimization problem is easy to solve with the elementary calculus tools.

Indeed $L(a, b)$ is a "bottom down" parabolic function of a and b, so that its minimum corresponds to the point at which both partial derivatives of $L(a, b)$ are zero (the first-order optimality condition):

$$\partial L/\partial a = 0 \quad \text{and} \quad \partial L/\partial b = 0.$$

Leaving the task of actually finding the derivatives to the reader as an exercise, let us focus on the unique solution to the first-order optimality equations defined by the following formulas (3.6), for a, and (3.8), for b:

$$a = \rho\sigma(y)/\sigma(x) \qquad (3.6)$$

where

$$\rho = \left[\Sigma_i(x_i - m_x)(y_i - m_y)\right]/[N\sigma(x)\sigma(y)] \qquad (3.7)$$

is the so-called correlation coefficient and m_x, m_y are means of x_i, y_i, respectively;

$$b = m_y - am_x \qquad (3.8)$$

By putting these optimal a and b into (3.5), one can express the minimum criterion value as

$$L_m(a,b) = \sigma^2(y)(1 - \rho^2) \qquad (3.9)$$

The Eq. (3.4) is referred to as the linear regression of y over x, index ρ in (3.6) and (3.7) as the correlation coefficient, its square ρ^2 in (3.9) as the determinacy coefficient, and the minimum criterion value L_m in (3.9) is referred to as the unexplained, or residual, variance.

3.2.4 Correlation Coefficient and Its Properties

The meaning of the coefficients of correlation and determinacy, in the data recovery framework of data analysis, is provided by Eqs. (3.6)–(3.9). Here are some formulations.

Property 1 Determinacy coefficient ρ^2 shows the rate of decrease of the variance of y after its linear relation to x has been taken into account by the regression (follows from (3.9)).

Property 2 Correlation coefficient ρ ranges between -1 and 1, because ρ^2 is between 0 and 1, as follows from the fact that value L_m in (3.9) cannot be negative because the items in its expression (3.5) are all squares. The closer ρ to either 1 or -1, the smaller are the residuals in the regression equation. For example, $\rho = 0.9$ implies that y's unexplained variance L_m is $1 - \rho^2 = 19\%$ of the original value.

Property 3 The slope a is proportional to ρ according to (3.6); a is positive or negative depending on the sign of ρ. If $\rho = 0$, the slope is 0: in this case, y and x are referred to as not correlated.

Property 4 The correlation coefficient ρ does not change under shifting and rescaling of x and/or y, which can be seen from Eq. (3.7). Its formula (3.7) becomes especially simple if the so-called z-scoring has been applied to standardize both x and y.

To perform z-scoring over a feature, its mean m is subtracted from all the values and the results are divided by the standard deviation σ:

$$x_i' = (x_i - m_x)/\sigma(x) \quad \text{and} \quad y_i' = (y_i - m_y)/\sigma(y), \quad i = 1, 2, \ldots, N$$

Using the z-score standardization, formula (3.7) can be rewritten as

$$\rho = \Sigma_i x_i' y_i' / N = \langle x', y' \rangle / N \tag{3.7'}$$

where $\langle x', y' \rangle$ denotes the inner product of vectors $x' = (x_i')$ and $y' = (y_i')$.

The next property refers to one of the fundamental discoveries by K. Pearson, interpretation of the correlation coefficient in terms of the bivariate Gaussian distribution. A generic formula for the density function of this distribution, in the case in which the features have been pre-processed by using z-score standardization described above, is

$$f(u, \Sigma) = C \exp\{-u^T \Sigma^{-1} u / 2\} \tag{3.10}$$

where $u = (x, y)$ is a two-dimensional vector of two variables, x and y, under consideration and Σ is the so-called correlation matrix

$$\Sigma = \begin{pmatrix} 1 & \rho \\ \rho & 1 \end{pmatrix}$$

In formula (3.10), ρ is a parameter with a very clear geometric meaning. Consider a set of points $u = (x, y)$ on (x, y)–plane making function $f(u, \Sigma)$ in (3.10) equal to a pre-specified constant. Such a set makes the values of $u^T \Sigma^{-1} u$ constant too. That means that a constant density set of points $u = (x, y)$ must satisfy equation $x^2 - 2\rho xy + y^2 = \text{const}$. This equation defines a well-known quadratic curve, the ellipsis. At $\rho = 0$ the equation becomes that of a circle, $x^2 + y^2 = const$, and the greater the difference between ρ and 0, the more skewed is the ellipsis, so that at $\rho = \pm 1$ the ellipsis becomes a bisector line $y = \pm x + b$ because the left part of the equation makes a full square, in this case, $x^2 \pm 2xy + y^2 = const$, that is, $(y \pm x)^2 = const$. The size of the ellipsis is proportional to the constant: the greater the constant the greater the size.

Property 5 The correlation coefficient (3.7) is a sample based estimate of the parameter ρ in the Gaussian density function (3.10) under the conventional assumption that the sample points (y_i, x_i) are drawn from a Gaussian population randomly and independently.

This striking fact is behind a longstanding controversy. Some say that the usage of the correlation coefficient is justified only when the sample is taken from a Gaussian distribution, because the coefficient has a clear-cut meaning only in this model. This logic seems somewhat overly restrictive. True, the usage of the coefficient for esti-mating the density function is justified only when the function is Gaussian. However, when trying to linearly represent one variable through the other, the coefficient has a very different meaning in the approximation context, which has nothing to do with the Gaussian distribution, as expressed above with Eqs. (3.6)–(3.9).

3.2.5 Linearization of Non-linear Regression

Non-linear dependencies also can be fit by using the same criterion of minimizing the square error. Consider a popular case of exponential regression, that is, repre-senting correlation between target y and predictor x as $y = ae^{bx}$ where a and b are unknown constants and e the base of natural logarithm. Given some a and b, the average square error is calculated as

$$E = \left([y_1 - a\exp(bx_1)]^2 + \ldots + [y_N - a\exp(bx_N)]^2 \right)/N = \Sigma_i[y_i - a\exp(bx_i)]^2/N$$

$$(3.11)$$

There is no method that would straightforwardly lead to a globally optimal solution of the problem of minimization of E in (3.11) because it is too complex function of the unknown values. This is why conventionally the exponential regression is fit by what should be referred to as its linearization: transforming the original problem to that of linear regression.

Indeed, let us take the logarithm of both parts of the equation that we want to fit, $y = ae^{bx}$. The resulting equation is $\ln(y) = \ln(a) + bx$. This equation has the format of linear equation, $z = \alpha x + \beta$, where $z = \ln(y)$, $\alpha = b$ and $\beta = \ln(a)$. This leads to the following idea. Let us take the target be $z = \ln(y)$ with its values $z_i = \ln(y_i)$. By fitting the linear regression equation with data x_i and z_i, one finds optimal α and β, so that the original exponential parameters are found as $a = \exp(\beta)$ and $b = \alpha$. These values do not necessarily minimize (3.11), but the hope is that they are close to the optimum anyway. Unfortunately, this may be very wrong sometimes as the material in Project 3.2. clearly demonstrates.

Q.3.2. Find the derivatives of L over a and b and solve the first-order optimality conditions.

Q.3.3. Derive the optimal value of L in (3.9) for the optimal a and b.

Q.3.4. Prove or find a proof in the literature that any linear equation $y = ax + b$ corresponds to a straight line on Cartesian xy plane, for which a is the slope and b intercept.

Q.3.5. Find the inverse matrix Σ^{-1} for $\Sigma = \begin{pmatrix} 1 & \rho \\ \rho & 1 \end{pmatrix}$.

A. $\Sigma^{-1} = \begin{pmatrix} 1 & -\rho \\ -\rho & 1 \end{pmatrix} / (1 - \rho^2)$.

3.2.6 Linear Regression: Computation

Regression is a technique for representing the correlation between x and y as a linear function (that is, a straight line on the plot), $y = $ *slope* $* x + $ *intercept* where *slope* and *intercept* are constants, the former expressing the change in y when x is added by 1 and the latter the level of y at $x = 0$. The best possible values of slope and intercept (that is, those minimizing the average square difference between real y's and those found as *slope* $* x + $ *intercept*) are expressed in MatLab, according to formulas (3.6) and (3.8), as follows:

```
≫slope=corr(x,y)*std(y)/std(x);
≫intercept=mean(y) - slope*mean(x);
```

Here corr(x,y) is the MatLab command for computing Pearson correlation coefficient between x and y (3.7). There is another MatLab operation "corrcoef" which leads to an estimate of the matrix Σ above.

Project 3.1. 2D Analysis, Linear Regression and Bootstrapping
Let us take the Students data table as a 100×8 array a in MatLab, pick any two features of interest and plot entities as points on the Cartesian plane formed by the features. For instance, take Age as x and Computational Intelligence mark as y:

```
≫x=a(:,4); % Age is 4-th column of array "a"
≫y=a(:,8); % CI score is in 8-th column of "a"
```

Then student 1 (first row) will be presented by point with coordinates x = 28 and y = 90 corresponding to the student's age and CI mark, respectively. To plot them all, use command:

≫plot(x,y,'k')
% k refers to black colour, "." dot graphics; 'mp' stands for magenta pentagram;
% see others by using "help plot"

Unfortunately, this gives a very tight presentation: some points are on the borders of the drawing. To make the borders stretched out, one needs to change the axis, for example, as follows:

≫d=axis; axis(1.2*d-10);

This transformation is presented on the right part of Fig. 3.9. To make both plots presented on the same figure, use "subplot" command of MatLab:

≫subplot(1,2,1)
≫plot(x,y,'k.');
≫subplot(1,2,2)
≫plot(x,y,'k.');
≫d=axis; axis(1.2*d-10);

Whichever presentation is taken, no regularity can be seen on Fig. 3.9 at all. Let's try then whether anything better can be seen for different occupations. To do this, one needs to handle entity sets for each occupation separately:

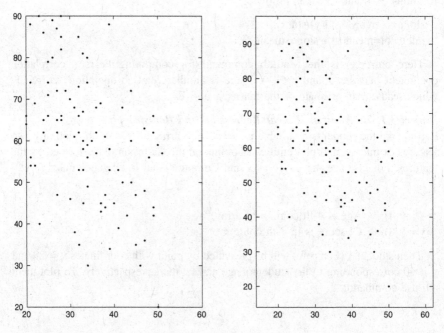

Fig. 3.9 Scatter plot of features "Age" and "CI score"; the display on the right is a rescaled version of that on the left

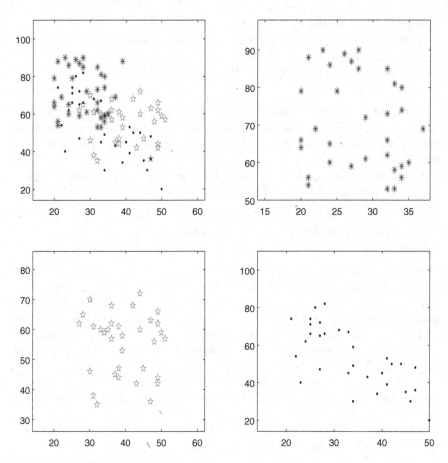

Fig. 3.10 Joint and individual displays of the scatter-plots at the occupation categories (IT star, BA pentagrams, AN dots)

```
≫o1=find(a(:,1)==1); % set of indices for IT
≫o2=find(a(:,2)==1); % set of indices for BA
≫o3=find(a(:,3)==1); % set of indices for AN
≫x1=x(o1);y1=y(o1); % the features x and y at IT students
≫x2=x(o2);y2=y(o2); % the features at BA students
≫x3=x(o3);y3=y(o3); % the features at AN students
```

Now we are in a position to put, first, all the three together, and then each of these three separately (again with the command "subplot", but this time with four windows organized in a two-by-two format, see Fig. 3.10).

```
≫subplot(2,2,1); plot(x1,y1, '*b',x2,y2,'pm',x3,y3,'.k');% all three
≫d=axis; axis(1.2*d-10);
≫subplot(2,2,2); plot(x1,y1, '*b'); % IT plotted with blue stars
```

```
≫d=axis; axis(1.2*d-10);
≫subplot(2,2,3); plot(x2,y2,'pm'); %BA plotted with magenta pentagrams
≫d=axis; axis(1.2*d-10);
≫subplot(2,2,4); plot(x3,y3,'.k'); % AN plotted with black dots
≫d=axis; axis(1.2*d-10);
```

Of the three occupation groups, some potential relation can be seen only in the AN group: it is likely that "the greater the age the lower the mark" regularity holds in this group (black dots in the Fig. 3.10's bottom right). To check this, let us utilize the linear regression.

Linear regression equation, $y = slope * x + intercept$, is estimated by using MatLab, according to formulas (3.4)–(3.6), as follows:

```
≫cc=corrcoef(x3,y3);rho=c(1,2);% producing rho = -0.7082
≫slope=rho*std(y3)/std(x3); % this produces slope=−1.33;
≫intercept=mean(y3) - slope*mean(x3); % this produces intercept=98.2;
```

Since we are interested in group AN only, we apply these commands at AN-related values x3 and y3 to produce the linear regression as $y3 = 98.2 − 1.33 * x3$. The slope value suggests that every year added to the age, in general decreases the mark by 1.33, so that aging by 3 years would lead to the loss of 4 mark points. Obviously, care should be taken to draw realistic conclusions.

Altogether, the regression equation explains $rho^2 = 0.50 = 50\%$ of the total variance of y3—not too much, as is usual in social and human sciences.

Let us take a look at the reliability of the regression equation with bootstrapping, the popular computational experiment technique for validating data analysis results (see Sect. 2.2.3.3).

Bootstrapping is based on a pre-specified number of random trials, for instance, 5000. Each trial consists of the following steps:

(i)   randomly selecting an entity N times, with replacement, so that the same entity can be selected several times whereas some other entities may be never selected in a trial. (As shown in Sect. 2.2.3.3, on average only 63% entities get selected into the sample.) A sample consists of N entities because this is the number of entities in the set under consideration. In our case, N = 31. One can use the following MatLab command:

```
≫N=31;ra=ceil(N*rand(N,1));
```
% rand(N,1) produces a column of N random real numbers, between 0 and 1 each.
% Multiplying this by N stretches them to (0,N) interval; the operation ceil rounds the numbers up to integers.

(ii) the sample ra is assigned with their data values according to the original data table:

```
≫xt=xx(ra);yt=yy(ra);
```
% here xx and yy represent the predictor and target, respectively;

% they are x3 and y3, respectively, which can be taken into account with assignments
% xx=x3; and yy=y3.

so that coinciding entities get identical feature values.

(iii) a data analysis method under consideration, currently "linear regression", that basically computes the rho, the slope and the intercept, applies to this data sample to produce the trial result.

To do a number of trials (5000, in this case), one should run (i)–(iii) in a loop:

≫for k=1:5000; ra=ceil(N*rand(N,1));

```
xt=xx(ra);yt=yy(ra);
cc=corrcoef(xt,yt);
rh(k)=cc(1,2);
sl(k)=rh(k)*std(yt)/std(xt); inte(k)=mean(yt)-sl(k)*mean(xt);
```

end
% the results are 5000-strong set of columns rh (correlations), sl (slopes)
% and inte (intercepts)

Now we can check the mean and standard deviation of the obtained distributions. Commands

≫mean(sl); std(sl)

produce values: −1.33 and 0.23 for the mean and standard deviation, respectively. That means that the original value of slope = −1.33 is confirmed with the bootstrapping, but now we have obtained its standard deviation, 0.23, as well. Similarly mean/std values for the intercept and rho are computed. They are, respectively, 98.2/9.0 and −0.704/0.095.

We can plot the 5000 values found as 30-bin histograms (see Fig. 3.11):

≫subplot(1,2,1); hist(sl,30)
≫subplot(1,2,2); hist(in,30)

The command subplot (1,2,1) creates a pane with one row consisting of two windows for plots and puts the follow-up plot into the first window (that on the left). Command subplot (1,2,2) moves the action into the second window which is on the right of Fig. 3.11.

To derive the 95% confidence boundaries for the slope, intercept and correlation coefficient, one may use both pivotal and non-pivotal methods.

The pivotal method uses the hypothesis that the bootstrap sample is indeed a random sample from a Gaussian distribution. Parameters of this distribution for slope are determined with the following commands:

≫msl=mean(sl); ssl=std(sl);

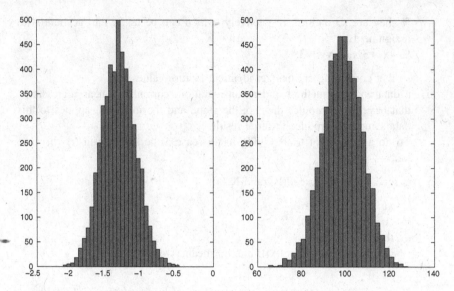

Fig. 3.11 30-bin histograms of the slope (left) and intercept (right) after 5000 bootstrapping trials

Since 95% of the Gaussian distribution fall within interval of $\pm 1.96*$std, the 95% confidence boundaries are derived, for the slope, as follows:

≫lbsl=msl − 1.96*ssl; rbsl=msl+1.96*ssl

The non-pivotal estimates require no such a hypothesis and are based on the bootstrap distribution as is. One just sorts all the values and takes 2.5% quantiles on both extremes of the range:

≫ssl=sort(sl); lbn=ssl(126);rbn=ssl(4875);

Indeed, we need to cut out 5% items from the sample, to make a 95% confidence interval. Since 5% of 5000 is 250, conventionally divided in two halves, this requires cutting off first 125 observations as well as the last 125 observations of the presorted list of the bootstrap values, which brings us to ssl(126) and ssl(4875) as the non-pivotal boundaries for the slope value.

All these estimates are presented in Table 3.3. The pivotal and non-pivotal estimates do not fall too far apart. Either can be taken as parameters of the boundary regressions.

Table 3.3 Parameters of the bootstrap distributions, as well as pivotal and non-pivotal boundaries

	Mean	St. dev.	Pivotal boundaries		Non-pivotal boundaries	
			Left	Right	Left	Right
Slope	−1.337	0.241	−1.809	−0.865	−1.800	−0.850
Intercept	98.510	9.048	80.776	116.244	80.411	116.041
Corr. coef.	−0.707	0.094	−0.891	−0.523	−0.861	−0.493

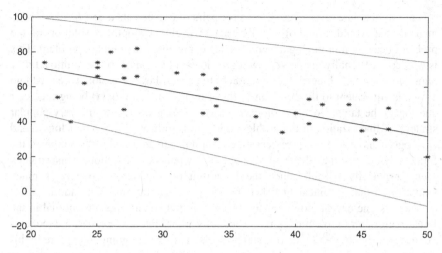

Fig. 3.12 Regression of CI score over Age (black line) within occupation category AN with boundaries covering 95% of potential biases due to sample fluctuations

This all can be visualized by, first, defining the three regression lines, the regular one and two corresponding to the lower and upper estimate boundaries, respectively, with

≫y3reg=slope*x3+intercept;
≫y3regleft=lbsl*x3+lbintercept;
≫y3regright=rbsl*x3+rbintercept;
and then plotting the four sets onto the same Fig. 3.12.

≫plot(x3,y3,'*k',x3,y3reg,'k',x3,y3regleft,'r',x3,y3regright,'r')
% x3,y3,'*k' presents student data as black stars; x3,y3reg,'k' presents the
% real regression line in black
% x3,y3regleft,'g' and x3,y3regright,'g' for boundary regressions in green

The lines on Fig. 3.12 show the boundaries of the regression line for 95% of trials.

Project 3.2. Non-linear and Linearized Regression: A Nature-Inspired Algorithm

In many domains the correlation between features is not necessarily linear. For example, in economics, processes related to the inflation over time are modeled by using the exponential function. A similar way of thinking applies to the processes of growth in biology. Variables describing climatic conditions obviously have a cyclic character; etc. The power law in social systems is nonlinear too.

Consider, for example, a power law function $y = ax^b$ where x is predictor and y target variables whereas a and b are unknown constant coefficients. Given the values of x_i and y_i on a number of observed entities $i = 1,..., N$, the power law regression problem can be formulated as the problem of minimizing the summary squared or absolute error over all possible pairs of coefficients a and b. There is no method that would straightforwardly lead to a globally optimal solution of the problem because minimizing a sum of many exponents is a complex problem. This is why conventionally the power law regression is fit by transforming it into a linear regression problem. Indeed, the equation of the power law regression, taken with no errors, is equivalent to the equation of linear regression with $\log(x)$ being predictor and $\log(y)$ the target: $\log(y) = b\log(x) + \log(a)$. This gives rise to the very popular strategy of linearization of the problem. First, transform x_i and y_i to $v_i = \log(x_i)$ and $z_i = \log(y_i)$ and fit the linear regression equation for given v_i and z_i; then convert the found coefficients into those of the original exponential function. This strategy seems especially suitable since the logarithm of a variable typically is much smoother so that the linear fit is better under the logarithm transformation.

There is one caveat, however: the fact that found coefficients are optimal in the linear regression problem does not necessarily imply that the converted exponents are necessarily optimal in the original problem. This we are going to explore in this project.

Nature-inspired optimization is a computational intelligence approach to minimize a non-linear function. Rather than look and polish a single solution to the optimization problem under consideration, this approach utilizes a population of solutions iteratively evolving from generation to generation, according to rules imitating a real-world evolutionary process and survival of the fittest. The rules typically include: (a) random changes from generation to generation such as "mutations" and "crossovers" in earlier, so-called "genetic", algorithms, and (b) policies for selecting and maintaining the best found solutions, the "elite". After a pre-specified number of iterations, the best solution among those observed in computations is reported as the outcome.

To start the evolutionary optimization process, one should first define a restricted area of admissible solutions so that no member of the population may leave the area. This warrants that the population will not explode by moving solutions to the infinity. There can be different ways for determining such an area. Let us follow this. Under the hypothesis of a power law relation $y = ab^x$, for any two entities i and j, the following equations should hold: $z_i = b * v_i + c$ and $z_j = b * v_j + c$ where $c = \log(a)$, $z_i = \log(y_i)$ and $v_i = \log(x_i)$. From these, b and c can be expressed as follows: $b = (z_i - z_j)/(v_i - v_j)$, $c = (v_i * z_j - v_j * z_i)/(v_i - v_j)$, which may lead to different values of b and c at different i and j. Denote bm and bM the minimum and the maximum of $(z_i - z_j)/(v_i - v_j)$, and cm and cM the minimum and maximum of $(v_i * z_j - v_j * z_i)/(v_i - v_j)$ over those i and j for which $v_i - v_j \neq 0$. One would expect that the admissible b and c should be within these boundaries, which means that the area of admissible solutions should be defined by the inequalities $(bm,cm) \leq (b,c) \leq (bM,cM)$. Since the optimal values of (b,c) should be around the averages of the ratios above, that is, lie deep inside the area between their maxima and minima, it helps to speed up the computation if one takes only those pairs (i,j) at which the values of v_i, v_j and z_i, z_j are not too close to 0 so that their logarithms are not that far away from 0, and, similarly, the differences between them should be neither that small nor that high. This approach is implemented in MatLab code ddr.m in Appendix A.3.

For the step of producing the next generation, let us denote the population's $p \times 2$ array by f, at the current iteration, and by f', at the next iteration. The transition from f to f' is done in three steps. First, take the row of mean values within the columns of f and repeat it p times in a $p \times 2$ array mf. Then make a Gaussian random move:

$$\mathrm{fn} = \mathrm{f} + \mathrm{randn}(\mathrm{p}, 2). * \mathrm{mf}/20$$

Here randn(p,2) is a $p \times 2$ array of (pseudo) random numbers generated according to Gaussian distribution $N(0,1)$ with 0 expectation and 1 variance. The symbol .* denotes the operation of multiplication of corresponding elements in matrices, so that (aij).*(bij) is a matrix whose (i,j)-th elements are products aij * bij. This random matrix is scaled down by $mf/20$ so that the move accounts for about 5% (one twentieth) of the average f values.

Since the move is to be restricted within the admissibility area, any a-element (first column of fn) which is greater than aM, is to be changed for aM, and any a-element smaller than am is to be changed for am. Similar trimming applies to b-elements. Denote the result by fr.

At the next step, take a $p \times 2$ array el whose rows are the same stored elite solution and arrive at the next generation f' by using the following "elite mix":

$$f' = 0.7fr + 0.3el$$

The elite mix moves all population members in the direction of the best solution found so far by 30%, which has been found work well in the examples of our interest.

This procedure is implemented in MatLab code nlr.m that relies on ddr.m at step 1 and a subroutine, delta, for evaluating the fitness (see A.4 in Appendix).

Consider now this experiment. Generate predictor x as a 50-long vector of random positive entries between 0 and 10, x = 10 * rand(1,50), and define $y = 2 * x^{1.07}$ with the normal additive noise 2 * $N(0,1)$ where 0 is the mean and 1 the variance, which is suppressed when overly negative, according to the Matlab code line

≫for ii=1:50;yy=2*x(ii)^1.07 +2*randn;y(ii)=max(yy,1.01);end;

When using the conventional linearized regression model by linearly mapping $\log(x)$ to $\log(y)$, to extract b and a (as the exponent of the found c) from this, the program llr.m implementing this approach produces $a = 3.0843$ and $b = 0.8011$ leading to the averaged squared error $y - ax^b$ equal to 3.41, so that the standard error is 3.10, about 20% of the mean y value, 10.1168. It is not only that the error is high, but also a wrong law is identified. The generated function y stretches x out ($b > 1$), whereas the found function stretches x in ($b < 1$).

Here are the results. Minimization of the averaged squared error $y - ax^b$ of the original model directly by using the code nlr.m, that implements the nature-inspired algorithm, the values are $a = 3.0293$ and $b = 1.0760$ leading to the average squared error of 0.0003 and the standard error of 0.0180. In contrast to the values found at the linearized scheme, the parameters a and b here are very close to those generated.

This obviously considerably outperforms the conventional procedure. Similar results can be found at different values of the noise variance.

Case-Study 3.1. Growth of Investment

Let us apply a similar approach to the following example involving variables x and y defined over a period of 20 time moments as presented in Table 3.4.

Variable x can be thought of as related to the time periods whereas y may represent the value of a fund. In fact, the components of x are numbers from 1 to 20 divided by 10, and y is obtained from them in MatLab according to formula $y = 2 * \exp(1.04 * x) + 0.6 * \text{randn}$ where randn is the normal (Gaussian) random variable with the mathematical expectation 0 and variance 1.

Let us, first, try a conventional approach of finding the average growth of the fund during all the period.

The average growth of the investment according to these data is conventionally expressed as the root 19, or power 1/19, of the ratio y_{20}/y_{01}, that is, 1.13. This estimates the average growth as 14% per period—which is by far greater than 4% in the data generating model.

Let us now try to make sense of the relation between x and y by applying the conventional linearization strategy to this data.

The strategy of linearization of the exponential equation outlined in Sect. 3.2.5 leads to values 1.1969 and 0.4986 for b and c, respectively, to produce $a = e^c = 1.6465$ and $b = 1.1969$ according to formulas there. As one can see, these differ from the original $a = 2$ and $b = 1.04$ by the order of 15–20%. The value of the squared error here is E = 13.90. See Fig. 3.13 representing the data.

Let us now apply the nature inspired approach to the original non-linear least-squares problem.

The program nlrm.m implementing the evolutionary approach described in Project 3.2 found a = 1.9908 and b = 1.0573. These are within 1–3% of the error from the original values a = 2 and b = 1.03. The summary squared error here is E = 7.45, which is by far smaller than that found with the linearization strategy.

The two found solutions can be represented on the scatter-plot graph, see Fig. 3.14. One can see that the linearized version has a much steeper exponent, which becomes visible at later periods.

Table 3.4 Data of investment y at time moments x from 0.10–3.00

x	0.1	0.2	0.3	0.4	0.5	0.6	0.7	0.8	0.9	1
y	1.3	1.82	3.03	3.29	3.3	3.9	3.84	3.24	3.23	6.5
x	1.1	1.2	1.3	1.4	1.5	1.6	1.7	1.8	1.9	3
y	6.93	7.23	7.91	9.27	9.45	11.18	13.48	13.51	15.4	15.91

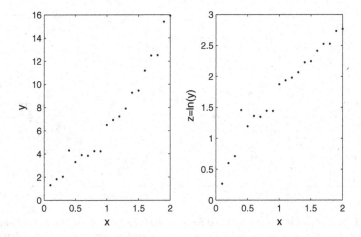

Fig. 3.13 Plot of the original pair (x,y) in which y is a noisy exponential function of x (on the left) and plot of the pair (x,z) in which $z = \ln(y)$. The plot on the right looks somewhat straighter indeed, though the correlation coefficients are rather similar, 0.970 for the plot on the left and 0.973 for the plot on the right

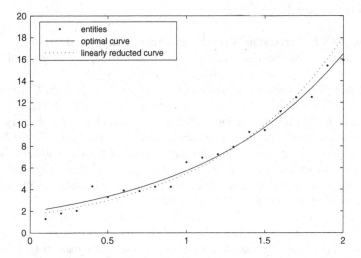

Fig. 3.14 Two fitting exponents are shown, with stars and dots, for the data in Case-Study 3.1

Q.3.6. Consider a binary feature defined on seven entities so that it is category A on the first three of them, and category B on the next four. Let us draw two dummy 1/0 variables, xA and xB, corresponding to each so that xA = 1 on the first three entities and xA = 0 on the rest, whereas xB = 0 on the first three entities and xB = 1 on the rest. What can be said of the correlation coefficient between xA and xB?
A. The correlation coefficient between xA and xB is -1 because xA + xB = 1 for all entities so that xA = −xB + 1.

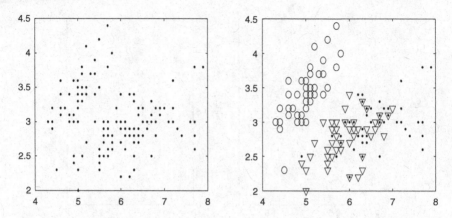

Fig. 3.15 Scatter plot of Sepal length and Sepal width from Iris data set (Table 1.1), as a whole on the left and taxon-wise on the right. Taxon 1 is presented by circles, taxon 2 by triangles, and taxon 3 by dots

Q.3.7. Extend the nature-inspired approach to the problem of fitting a linear regression with a nonconventional criterion such as the average relative error defined by formula $1/N \sum_{i=1}^{N} |e_i/y_i|$.

Case-Study 3.2. Correlation Between Iris Sepal Length and Width
Take x and y from the Iris set in Table 1.1 as the Sepal's length and width, respectively.

A scatter plot of x and y is presented on the left part of Fig. 3.15. This is a loose cloud of points which looks similar to that on the left part of Fig. 3.6, of no correlation. Indeed the correlation coefficient value here is not only very small, -0.12, but also negative, which is somewhat odd, because intuitively the features should be positively correlated as reflecting the size of the same flower.

To see a particular reason for the low, and negative, correlation, one should take into account that the sample is not homogeneous: the Iris set consists of 50 specimens of each of three different taxa. When the taxa are separated (see Fig. 3.15 on the right), the positive correlation is restored. The correlation coefficients are 0.74, 0.53 and 0.46 in taxon one, two and three, respectively. Here we see a nice example of the negative effect of the non-homogeneity of the sample on the data analysis results.

3.3 Multivariate Linear Regression

3.3.1 Formulation

Let us extend the notion of linear regression from the bivariate case to a multivariate case, when several features can be used as predictors for a target feature.

The problem of multivariate linear regression can be formulated as a particular case of the correlation learning problem with just one quantitative target variable u and linear admissible rules so that

$$u = w_1 x_1 + w_2 x_2 + \ldots + w_p x_p + w_0$$

where $w_0, w_1, \ldots, w_p$ are unknown weights, parameters of the model.

For any entity $i = 1, 2, \ldots, N$, the rule-computed value of u

$$\hat{u}_i = w_1 x_{i1} + w_2 x_{i2} + \ldots + w_p x_{ip} + w_0$$

differs from the observed one by $d_i = |\hat{u}_i - u_i|$, which may be zero—when the prediction is exact. To find $w_1, w_2, \ldots, w_p, w_0$, one can minimize the summary square error

$$D^2 = \Sigma_i d_i^2 = \Sigma_i \left(u_i - w_1 * x_{i1} - w_2 * x_{i2} - \ldots - w_p * x_{ip} - w_0 \right)^2 \qquad (3.12)$$

over all possible parameter vectors $w = (w_0, w_1, \ldots, w_p)$.

To make the problem treatable in terms of linear operations, a fictitious feature x_0 is introduced such that all its values are 1: $x_{i0} = 1$ for all $i = 1, 2, \ldots, N$. Then criterion D^2 can be expressed as $D^2 = \Sigma_i (u_i - \langle w_i, x_i \rangle)^2$ using the inner products $\langle w, x_i \rangle$ where $w = (w_0, w_1, \ldots, w_p)$ and $x_i = (x_{i0}, x_{i1}, \ldots, x_{ip})$ are $(p + 1)$-dimensional vectors of which all x_i are known whereas w is not. From now on, the unity feature x_0 is assumed to be part of data matrix X in all correlation learning problems.

The criterion D^2 in (3.12) is but the squared Euclidean distance between the N-dimensional target feature column $u = (u_i)$ and vector $\hat{u} = Xw$ whose components are $\hat{u}_i = \langle w, x_i \rangle$. Here X is $N \times (p + 1)$ matrix whose rows are x_i (augmented with the component $x_{i0} = 1$, thus being $(p + 1)$-dimensional) so that Xw is the matrix product of X and w. Vectors defined as Xw for all possible w's form $(p + 1)$-dimensional vector space, referred to as X-span.

Thus, the problem of minimization of (3.12) can be reformulated as follows: given a target vector u, find its projection $\hat{u}$ in the X-span space. The global solution to this problem is well-known: it is provided by a matrix P_X applied to u:

$$\hat{u} = P_X u \qquad (3.13)$$

where P_X is the so-called orthogonal projection operator, an $N \times N$ matrix, defined as:

$$P_X = X \left(X^T X \right)^{-1} X^T \qquad (3.14)$$

so that

$$\hat{u} = X(X^TX)^{-1}X^Tu \text{ and } w = (X^TX)^{-1}X^Tu. \tag{3.15}$$

Matrix P_X projects every N-dimensional vector u to its nearest match in the $(p+1)$-dimensional X-span space. The inverse $(X^TX)^{-1}$ does not exist if the rank of X, as it may happen, is less than the number of columns in X, $p+1$, that is, if matrix X^TX is singular or, equivalently, the dimension of X-span is less than $p+1$. In this case, the so-called pseudo-inverse matrix $(X^TX)^+$ can be used as well. This is not a big deal computationally: for example, in MatLab one just puts pinv(X^TX) instead of inv(X^TX).

The quality of approximation is evaluated by the minimum value D^2 in (3.12) averaged over the number of entities and related to the variance of the target variable. Its complement to 1, the determinacy coefficient, is defined by the equation

$$\rho^2 = 1 - D^2/(N\sigma^2(u)) \tag{3.16}$$

The determinacy coefficient shows the proportion of the variance of u explained by the linear regression. Its square root, ρ, is referred to as the coefficient of multiple correlation between u and $X = \{x_0, x_1, x_2, ..., x_p\}$.

3.3.2 Case Studies

Case-Study 3.3. Linear Regression for Market Town Data
Consider feature Post expressing the number of post offices in Market towns (Table 1.2 on p. 15) and try to relate it to other features in the table. It obviously relates to the population. For example, towns with population of 15,000 and greater are those and only those where the number of post offices is 5 or greater. This correlation, however, is not as good as to give us more guidance in predicting Post from the Population. For example, at the seven towns whose population is from 8000 to 10,000 any number of post offices from 1 to 4 may occur, according to the table. This could be attributed to effects of services such as a bank or hospital present at the towns. Let us specify a set of features in Table 1.2 that can be thought of as affecting the feature Post, to include in addition to Population some other features—PS—Primary schools, Do—General Practitioners, Hos—Hospitals, Ba—Banks, Sst—Superstores, and Pet—Petrol Stations; seven features altogether, taken as the set of input variables (predictors).
What we want is to establish a linear relation between this set and target feature Post. A linear relation is an equation representing Post as a weighted sum of input features plus a constant intercept; the weights can be any reals, not necessarily positive. If the relation is supported by the data, it can be used for various purposes such as analysis, prediction and planning.

Table 3.5 Weight coefficients of input features at Post Office as target variable for Market towns data

Feature	Weight
Pop_Res	0.0002
PSchools	0.1982
Doctors	0.2623
Hospitals	−0.2659
Banks	0.077
Superstores	0.0028
Petrol	−0.3894
Intercept	0.5784

In the example of seven Market town features used for linearly relating them to Post Office feature, the least-squares optimal weight coefficients are presented in Table 3.5. Each weight coefficient shows how much the target variable would change on average if the corresponding feature is increased by a unity, while the others do not change. One can see that increasing population by a thousand would give a similar effect as adding a primary school, about 0.2, which may seem absurd in the example as Post Office variable can have only integer values. Moreover, the linear function format should not trick the decision maker into thinking that increasing different input features can be done independently: the features are obviously not independent so that increase of, say, the population will lead to respectively adding new schools for the additional children. Still, the weights show relative effects of the features—according to Table 3.5, adding a doctor's surgery in a town would lead to maximally possible increase in post offices. The maximum value is assigned to the intercept in this case. What this may mean? Is it the number of post offices in an empty town with no population, hospitals or petrol stations? Certainly not. The intercept expresses that part of the target variable which is relatively independent of the features taken into account. It should be also pointed out that the weight values are relative not to just feature concepts but specific scales in which features measured. Change of a feature scale, say 10-fold, would result in a corresponding, inverse, change of its weight (due to the linearity of the regression equation). This is why in statistics the relative weights are considered for the scales expressed in units of the standard deviation. To find them, one should multiply the weight for the current scale by the feature's standard deviation (see Table 3.6).

The standardized weights are well justified when input features are mutually uncorrelated—indeed, they show the pair-wise correlation with the target feature. Yet in a situation of correlated features, like this, they seem to have much less definite interpretation, except for showing the changes of the target in units of the standard deviations, although some claim that they also reflect feature's correlation with the target or even importance for predicting the target. An argument against their usage as a correlation measure is that, in fact, a regression coefficient multiplied by the standard deviation loses its "purity" as a measure of correlation to the target at constant levels of the other features because the standard deviation does not pertain to constant features. An argument against their usage as measures of

Table 3.6 Standardized weight coefficients of input features at Post Office as target variable for Market towns

Feature	Weights in natural scales, w	Standard deviations, s	Weights in standardized scales, w *s
Pop_Res	0.0002	6193.2	1.3889
PSchools	0.1982	3.7344	0.5419
Doctors	0.2623	1.3019	0.3414
Hospitals	−0.2659	0.58	−0.1542
Banks	0.077	3.384	0.3376
Superstores	0.0028	1.7242	0.0048
Petrol	−0.3894	1.637	−0.6375

importance for prediction is that the standardized coefficient has nothing to do with the change of the coefficient of determinacy when the corresponding feature is removed from the equation of regression.

Bring (1994) proposes to kill two birds with one stone: to clean up the standard deviations from the non-constancy of the other features, which are claimed to reflect the changes in the coefficients of determinacy. Specifically, take the variance of a feature and take off the proportion of it unexplained by the linear regression of it through the other features. The square root of the result represents the partial standard deviation, which is proportional to the so-called "t-value", and, in the original squared form, to the change of the coefficient of determinacy inflicted by the removal of the feature from the list of the explanatory variables (Bring 1994). Unfortunately, this is not that simple, as the next Case-Study 3.4 shows.

Case-Study 3.4. Using Feature Weights Standardized

Table 3.7 presents the feature weights standardized with both the original and partial standard deviations as well as the absolute reductions of the original coefficient of determinacy 0.8295 after removal of the corresponding variables. There is a general agreement between the absolute values of the first column and those in the third column, but the second column has little in common with either of them.

Table 3.7 Different indexes to express the idea of importance of a feature in the Post regression problem

Feature	Weights expressed in standard deviations	Weights with partial standard deviations	Determinacy coefficient reduction
Pop_Res	1.3889	1603	0.0247
PSchools	0.5419	1.02	0.0077
Doctors	0.3414	0.64	0.0055
Hospitals	−0.1542	0.41	0.0023
Banks	0.3376	3.27	0.0059
Superstores	0.0048	1.07	0
Petrol	-0.6375	0.96	0.0251

Table 3.8 Weight coefficients for reduced set of features at Post Office as target variable for Market towns data

Feature	Weight
POP_RES	0.0003
PSchools	0.1823
Hospitals	−0.3167
Banks	0.0818
Petrol	−0.4072
Intercept	0.5898

A general analysis of a simpler problem of relation between the regression coefficients and correlation coefficients between the target and input features can be found in Waller and Jones (2010).

Amazingly, the convenient standardization involves negative weights, specifically at features Petrols and Hospitals. This can be an artifact of the method, related to the effect of "replication" of features. One can think of Hospitals being a double for Doctors, and Petrol, for Superstores. Thus, before jumping to conclusions, one should check whether the minus disappears if the "replicas" are removed from the set of features. As Table 3.8 shows, not in this case: the negative weights remain, though they slightly change, as well as other weights. This illustrates that the interpretation of linear regression coefficients as weights should be cautious and restrained.

In our example, coefficient of determinacy $\rho^2 = 0.83$, that is, the seven features explain 83% of the variance of Post Office feature, and the multiple correlation is $\rho = 0.91$. Curiously, the reduced set of five features (see Table 3.8) contributes almost the same, 83.4% of the variance of the target variable. This may make one wonder whether just one Population feature could suffice for doing the regression. This can be tested with the 2D method described in Sect. 3.1 or with the nD method of this section.

According to the formulation of the multivariate linear regression method, the estimated parameters must be feature weight coefficients—no room for an intercept in the formula. To accommodate the intercept, a fictitious feature whose all values are unities is introduced. That is, an input data matrix X with two columns is to be used: one for the Population feature, the other for the fictitious variable of all ones. According to (3.10), this leads to the slope 0.0003 and intercept 0.4015, though with somewhat reduced coefficient of determinacy, which is $\rho^2 = 0.78$ in this case. From the prediction point of view this may be all right, but the reduced feature set looses on interpretation.

3.4 Linear Discrimination and SVM

3.4.1 Linear Discrimination

Discrimination is an approach to address the problem of drawing a rule to distinguish between two classes of entity points in the feature space, a "yes" class and "no" class, such as for instance a set of banking customers in which a, typically very small, subset of fraudsters constitutes the "yes" class and that of the others the "no" class. On Fig. 3.16, entities of "yes" class are presented by circles and of "no" class by squares.

The problem is to find a function $u = f(x)$ that would separate the two classes in such a way that $f(x)$ is positive for all entities in the "yes" class and negative for all the entities in the "no" class. When the discriminant function $f(x)$ is assumed to be linear, the problem is of linear discrimination. It differs from that of the linear regression in that aspect that the target values here are binary, either "yes" or "no", so that this is a classification rather than regression, problem.

The classes on Fig. 3.16 can be discriminated by a straight—dashed—line indeed. The dotted vector w, orthogonal to the "dashed line" hyperplane, represents a set of coefficients at the linear classifier represented by the dashed line. Vector w also shows the direction at which function $f(x) = \langle w, x \rangle - b$ grows. Specifically, $f(x)$ is 0 on the separating hyperplane, and it is positive above, and negative beneath, that. With no loss of generality, w can be assumed to have its length equal to unity. Then, for any x, the inner product $\langle w, x \rangle$ would express the length of the projection of vector x along the direction of w.

To find an appropriate w, even in the case when "yes" and "no" classes are linearly separable, various criteria can be utilized. A most straightforward classifier is defined as follows: put 1 for "yes" and -1 for "no" and apply the least-squares criterion of linear regression. This produces a theoretically sound solution approximating the best possible—Bayesian—solution in a conventional statistics model. Yet, in spite of its good theoretical properties, least-squares solution may be

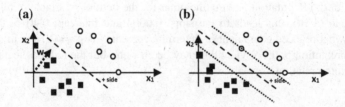

Fig. 3.16 A geometric illustration of a separating hyper-plane between classes of circles and squares. The dotted vector w on **a** is orthogonal to the hyper-plane: its elements are hyper-plane coefficients, so that it is represented by equation $\langle w, x \rangle - b = 0$. Vector w points at the direction: at all points above the dashed line, the circles included, function $f(x) = \langle w, x \rangle - b$ is positive. The dotted lines on, **b** show the margin, and the squares and circle on them are support vectors

not necessarily the best at some data configurations. In fact, it may even fail to separate the positives from negatives when they are linearly separable. Consider the following example.

Worked Example 3.4. A Failure of Fisher Discrimination Criterion
Let there be 14 2D points that are presented in Table 3.9 (first line) and displayed in Fig. 3.17a. Points 1,2,3,4,6 belong to the positive class (dots on Fig. 3.17a), and the others to the negative class (stars on Fig. 3.17a). Another set, obtained by adding to each of the components a random number, according to the normal distribution with zero mean and 0.2 the standard deviation, is presented in the bottom line of Table 3.9 and Fig. 3.17b. The class assignment for the disturbed points remains the same.

The optimal vectors w according to formula (3.15) are presented in Table 3.10 as well as that for the separating, dotted, line in Fig. 3.17d.

Note that the least-squares solution depends on the values assigned to classes, leading potentially to an infinite number of possible solutions under different numerical codes for "yes" and "no". A popular discriminant criterion of minimizing the ratio of a "within-class error" over "out-of-class error", proposed by R. Fisher in his founding work of 1936, in fact, can be expressed with the least-squares criterion as well. Just change the target as follows: assign N/N_1, rather than +1, to "yes" class

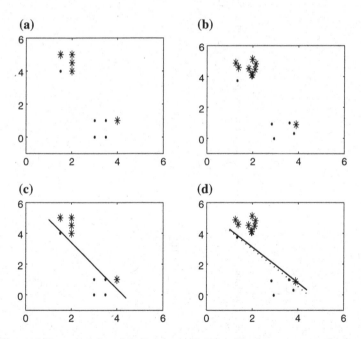

Fig. 3.17 **a** and **b** represent the original and perturbed data. The least squares optimal separating line is added in **c** and **d** shown by solid. Entity 5 is wrongly assigned to the "dot" class according to the solid line in (**d**); a separating line is shown dotted there

Table 3.9 x-y coordinates of 14 points as given originally and perturbed with a white noise of standard deviation 0.2, that is, generated from the Gaussian distribution N(0,0.2)

Entity #		1	2	3	4	5	6	7	8	9	10	11	12	13	14
Original data	x	3.00	3.00	3.50	3.50	3.00	1.50	3.00	3.00	3.00	1.50	3.00	3.00	3.00	1.50
	y	0.00	1.00	1.00	0.00	1.00	3.00	3.00	5.00	3.50	5.00	3.00	5.00	3.50	5.00
Perturbed data	x	3.93	3.83	3.60	3.80	3.89	1.33	1.95	3.13	1.83	1.26	1.98	1.99	3.10	1.38
	y	−0.03	0.91	0.98	0.31	0.88	3.73	3.09	3.82	3.51	3.87	3.11	5.11	3.46	3.59

Table 3.10 Coefficients of straight lines on Fig. 3.17

	Coefficients at		
	x	y	Intercept
LSE at original data	−1.2422	-0.8270	5.2857
LSE at perturbed data	−0.8124	−0.7020	3.8023
Dotted at perturbed data	−0.8497	−0.7020	3.7846

and $-N/N_2$ to "no" class, rather than −1 (see Duda et al. 2001, pp. 242–243). This means that Fisher's criterion may also lead to a failure in a linearly separable situation (see Fig. 3.17d).

By far the most popular set of techniques, Support Vector Machine (SVM), utilize a different criterion—that of maximum margin. The margin of a point x, with respect to a hyperplane, is the distance from x to the hyperplane along its perpendicular vector w (Fig. 3.17a), which is measured by the absolute value of inner product $\langle w, x \rangle$. The margin of a class is defined by the minimum value of the margins of its members. Thus the criterion requires, like L_∞, finding such a hyperplane that maximizes the minimum of class margins, that is, crosses the middle of line between the nearest entities of two classes. Those entities that fall on the margins, shown by dotted lines on Fig. 3.17b, are referred to as support vectors; this explains the method's title.

It should be noted that the classes are not necessarily linearly separable; moreover, in most cases they are not. Therefore, the SVM technique is accompanied with a non-linear transformation of the data into a high-dimensional space which is more likely to make the classes linear-separable. Such a non-linear transformation is provided by the so-called kernel function. The kernel function imitates the inner product in the high-dimensional space and is represented by a between-entity similarity function such as that defined by formula (3.20) on p. 202.

The intuition behind the SVM approach is this: if the population data—those not present in the training sample—concentrate around training data, then having a wide margin would keep classes separated even after other data points are added (see Fig. 3.18). One more consideration comes from the Minimum Description Length principle: the wider the margin, the more robust the separating hyperplane

Fig. 3.18 Illustrative example of 2D entities belonging to two classes, circles and squares. The separating line in the space of Gaussian kernel is shown by the dashed oval. The support entities are shown by black

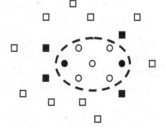

is and the less information of it needs to be stored. A criticism of the SVM approach is that the support vector machine hyperplane is based on the borderline objects—support vectors—only, whereas the least-squares hyperplanes take into account all the entities so that the further away an entity is the more it may affect the solution, because of the quadratic nature of the least-squares criterion. Some may argue that both borderline and far away entities can be rather randomly represented in the sample under investigation so that neither should be taken into account in distinguishing between classes: it is some "core" entities of the patterns that should be separated—however, there has been no such an approach taken in the literature so far.

3.4.2 Support Vector Machine (SVM) Criterion

Another criterion would put the separating hyperplane just in the middle of an interval drawn through closest points of the different patterns. This criterion produces what is referred to as the support vector machine since it heavily relies on the points involved in the drawing of the separating hyperplane (as shown on the right of Fig. 3.16). These points are referred to as support vectors. A natural formulation would be like this: find a hyperplane $H : \langle w, x \rangle = b$ with a normed w to maximize the minimum of absolute values of distances $|\langle w, x_i \rangle - b|$ to H from points x_i belonging to each of the classes. This, however, is rather difficult to associate with a conventional formulation of an optimization problem because of the following irregularities:

(i)   an absolute value to maximize,
(ii) the minimum over points from each of the classes, and
(iii) w being of the length 1, that is, normed.

However, these all can be successfully tackled. The issue (i) is easy to handle, because there are only two classes, on the different sides of H. Specifically, the distance is $\langle w, x_i \rangle - b$ for "yes" class and $-\langle w, x_i \rangle + b$ for "no" class—this removes the absolute values. The issue (ii) can be taken care of by uniformly using inequality constraints

$\langle w, x_i \rangle - b \geq \lambda$ for x_i in "yes" class and
$-\langle w, x_i \rangle + b \geq \lambda$ for x_i in "no" class

and maximizing the margin λ with respect to these constraints. The issue (iii) can be addressed by dividing the constraints by λ so that the norm of the weight vector becomes $1/\lambda$, thus inversely proportional to the margin λ. Moreover, one can change the criterion now because the norm of the ratio w/λ is minimized when λ is maximized. Denote the "yes" class by $u_i = 1$ and "no" class by $u_i = -1$. Then the problem of deriving a hyperplane with a maximum margin can be reformulated, without the irregularities, as follows: find b and w such that the norm of w or its square, $\langle w, w \rangle$, is minimum with respect to constraints

$$u_i(\langle w, x_i \rangle - b) \geq 1 \quad (i = 1, 2, \ldots, N) \tag{3.17}$$

This is a problem of quadratic programming with linear constraints, which is easier to analyze in the format of its dual optimization problem. The dual problem can be formulated by using the so-called Lagrangian, a common concept in optimization, that is, the original criterion penalized by the constraints weighted by the so-called Lagrange multipliers that are but penalty rates. Denote the penalty rate for the violation of i-th constraint by α_i. Then the Lagrangian can be expressed as

$$L(w, b, \alpha) = \langle w, w \rangle / 2 + \Sigma_i \alpha_i (u_i(\langle w, x_i \rangle - b) - 1), \tag{3.18}$$

where $\langle w, w \rangle$ has been divided by 2 with no loss of generality, just for the sake of convenience. The optimum solution minimizes L over w and b, and maximizes L over non-negative α. The first order optimality conditions require that all partial derivatives of L are zero at the optimum, which leads to equations $\Sigma_i \alpha_i u_i = 0$ and $w = \Sigma_i \alpha_i u_i x_i$. Multiplying the latter expression by itself leads to equation $\langle w, w \rangle = \Sigma_{ij} \alpha_i \alpha_j u_i u_j \langle x_i, x_j \rangle$. The second item in Lagrangian L becomes equal to $\Sigma_i \alpha_i u_i \langle w, x_i \rangle - \Sigma_i \alpha_i u_i b - \Sigma_i \alpha_i = \langle w, w \rangle - 0 - \Sigma_i \alpha_i$. This leads us to the following, dual, problem of optimization regarding the Lagrangian multipliers, which is equivalent to the original problem: Maximize criterion

$$\Sigma_i \alpha_i - \Sigma_{ij} \alpha_i \alpha_j u_i u_j \langle x_i, x_j \rangle / 2 \tag{3.19}$$

subject to $\Sigma_i \alpha_i u_i = 0$ and $\alpha_i \geq 0$.

Support vectors are defined as those x_i for which penalty rates are positive, $\alpha_i > 0$, in the optimal solution—only they contribute to the optimal vector $w = \Sigma_i \alpha_i u_i x_i$; the others have zero coefficients and disappear.

It should be noted that the margin constraints can be violated, which is not difficult to take into account—by using non-negative values η_i expressing the extent of violations:

$$u_i(\langle w, x_i \rangle - b) \geq 1 - \eta_i \quad (i = 1, 2, \ldots, N)$$

in such a way that they are minimized in a combined criterion $\langle w, w \rangle / 2 + C \Sigma_i \eta_i$ where C is a large "reconciling" coefficient that is a user-defined parameter. The dual problem for the combined criterion remains almost the same as above, in spite of the fact that an additional set of dual variables, β_i, needs to be introduced as corresponding to the constraints $\eta_i \geq 0$. Indeed, the Lagrangian for the new problem can be expressed as

$$L(w, b, \alpha, \beta) = \langle w, w \rangle / 2 - \Sigma_i \alpha_i (u_i (\langle w, x_i \rangle - b) - 1) - \Sigma_i \eta_i (\alpha_i + \beta_i - C),$$

which differs from the previous expression by just the right-side item. This implies that the same first-order optimality equations hold, $\Sigma_i \alpha_i u_i = 0$ and $w = \Sigma_i \alpha_i u_i x_i$, plus additionally $\alpha_i + \beta_i = C$. These latter equations imply that $C \geq \alpha_i \geq 0$ because β_i are non-negative.

Since the additional dual variables are expressed through the original ones, $\beta_i = C - \alpha_i$, the dual problem can be shown to remain unchanged and it can be solved by using quadratic programming algorithms (see Vapnik 2006; Schölkopf and Smola 2005). Recently, approaches have appeared for solving the original problem as well (see Groenen et al. 2008).

3.4.3 Kernels

Situations at which classes are linearly separable are very rare; in real data, classes are typically well intermingled with each other. To attack these typical situations with linear approaches, the following trick can be applied. The data are nonlinearly transformed into a much higher dimensional space in which, because of both nonlinearity and higher dimension, the classes may be linearly separable. The transformation can be performed only virtually because of specifics of the dual problem: dual criterion (3.19) depends not on individual entities but rather just inner products between them. This property obviously translates to the transformed space, that is, to the transformed entities. The inner products in the transformed space can be computed with the so-called kernel functions $K(x,y)$ so that in criterion (3.19) inner products $\langle x_i, x_j \rangle$ are substituted by the kernel values $K(x_i, x_j)$. Moreover, by substituting the expression $w = \Sigma_i \alpha_i u_i x_i$ into the original discrimination function $f(x) = \langle w, x \rangle - b$ we obtain its different expression $f(x) = \Sigma_i \alpha_i u_i \langle x, x_i \rangle - b$, also involving inner products only, which can be used as a kernel-based decision rule in the transformed space: x belongs to "yes" class if $\Sigma_i \alpha_i u_i K(x, x_i) - b > 0$.

It is convenient to define a kernel function over vectors $x = (x_v)$ and $y = (y_v)$ through the squared Euclidean distance $d^2(x,y) = (x_1 - y_1)^2 + ... + (x_V - y_V)^2$ because matrix $(K(x_i, x_j))$ in this case is positive definite—a defining property of matrices of inner products. Arguably, the most popular is the Gaussian kernel defined by:

$$K(x, y) = \exp\left(-d^2(x, y)\right) \tag{3.20}$$

Q.3.8. Consider a full set B_n of 2^n binary 1/0 vectors of length n like those presented by columns below for $n = 3$:

$$
\begin{array}{llllllllll}
1 & 0 & 0 & 0 & 0 & 1 & 1 & 1 & 1 \\
2 & 0 & 0 & 1 & 1 & 0 & 0 & 1 & 1 \\
3 & 0 & 1 & 0 & 1 & 0 & 1 & 0 & 1
\end{array}
$$

These columns can be considered as integers coded in the binary number system; moreover, they are ordered from 0 to 7. Prove that this set shutters any subset of n (or less) points.

A. Indeed, let S be a set of elements $i_1, i_2, ..., i_n$ in B_n that are one-to-one labeled by numbers from 1 to n. Consider any partition of S in two classes, S_1 and S_3. Assign 0 to each element of S_1 and 1 to each element of S_3. The partition follows that vector of B_n that corresponds to the assignment.

Q.3.9. Consider set B_n defined above. Prove that its rank is n, that is, there are n columns in matrix B_n that form a base of the space of n-dimensional vectors.

A. Take, for example, n columns e_p that contain unity at p-th position whereas other n-1 elements are zero ($p = 1, 2, ...n$). These obviously are mutually orthogonal and any vector $x = (x_1,...,x_n)$ can be expressed as a linear combination $x = \Sigma_p x_p e_p$, which proves that vectors e_p form a base of the n-dimensional space.

Q.3.10. What is VC-dimension of the linear discrimination problem at an arbitrary dimension $p \geq 2$?

A. The VC = dimension in this case is $p + 1$, because each subset of p points can be separated from the others by a hyperplane, but there can be such $(p + 1)$-point configurations that cannot be shattered using linear separators.

Worked Example 3.5. SVM for the Iris Dataset

Consider Iris dataset standardized by subtracting, from each feature column, its midrange and dividing the result by the half-range.

Take the Gaussian kernel in (3.20) to find a support vector machine surface separating Iris class 3 from the rest. The resulting solution embraces 21 supporting entities (see Table 3.11), along with their "alpha" prices reaching into hundreds and even, on two occasions, to the maximum boundary 500 embedded in the algorithm.

There is only one error with this solution, entity 78 wrongly recognized as belonging to taxon 3. The errors increase when we apply a cross-validation techniques, though. For example, "leave-all-one-out" cross-validation leads to nine errors: entities 63, 71, 78, 82 and 83 wrongly classified as belonging to taxon 3 (false positives), while entities 127, 133, 135 and 139 are classified as being out of taxon 3 (false negatives).

Table 3.11 List of support entities in the problem of separation of taxon 3 (entities 101 to 150) in Iris data set from the rest (thanks to V. Sulimova for the computation)

N	Entity	Alpha	N	Entity	Alpha
1	18	0.203	12	105	3.492
2	28	0.178	13	106	15.185
3	37	0.202	14	115	53.096
4	39	0.672	15	118	15.724
5	58	13.63	16	119	449.201
6	63	209.614	17	127	163.651
7	71	7.137	18	133	500
8	78	500	19	135	5.221
9	81	18.192	20	139	16.111
10	82	296.039	21	150	26.498
11	83	200.312			

Q.3.11. Why only 10, not 14, points are drawn on Fig. 3.17b?

A. Because each of the points 11–14 doubles a point 7–10.

Q.3.12. What would change if the last four points are removed so that only points 1–10 remain?

A. The least-squares solution will be separating again

3.5 Learning Correlation with Neural Networks

3.5.1 Artificial Neuron and Neural Network: Presentation

Neural network is one of the most popular structures used for predictions of target features. It is a network of artificial neurons modeling the neuron cell in a living organism. A neuron cell fires an output when its summary input becomes higher than a threshold. Dendrites bring signal in, axons pass it out, and the firing occurs via a synapse, a gap between neurons, that makes the threshold (see Fig. 3.19).

This is modeled in the concept of artificial neuron as follows (see Fig. 3.20). A neuron model is drawn as a set of input elements connected to an output. The connections are assigned with wiring weights.

The input signals are data features or other neurons' outputs. The output element receives a combined signal, the sum of feature values weighted by the wiring weights. The output compares this with a firing threshold, otherwise referred to as a bias, and fires an output depending on the result. Ideally, the output is 1 if the combined signal is greater than the threshold, and −1 if it is smaller than that. This is, in fact, what is called the sign function of the difference, $sign(x)$, which is 1, 0 or −1 if x is positive, zero or negative, respectively. This activation function is overly straightforward sometimes. Instead, the so-called sigmoid and symmetric sigmoid

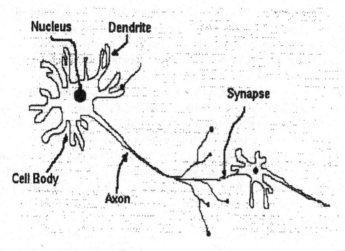

Fig. 3.19 Scheme of a neural cell

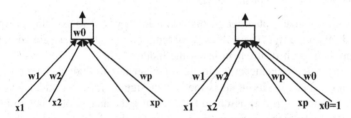

Fig. 3.20 A scheme of an artificial neuron, on the left. The same neuron with the firing threshold shown as a wiring weight on the fictitious input always equal to 1 is on the right

Fig. 3.21 Graphs of sign (**a**), sigmoid (**b**) and symmetric sigmoid (**c**) functions

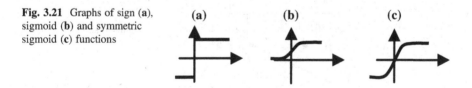

functions are considered as smooth exponent-based counterparts to sign(x). Their graphs are shown alongside with that for *sign(x)* on Fig. 3.21. Sometimes the output element is assumed as doing no transformation at all, just passing the combined signal as the neuron's output, which is referred to as a linear activation function.

The firing threshold, or bias, hidden in the box in neuron on the left on Fig. 3.20, can be made explicit if one more, fictitious, input is added to the neuron.

This input is always equal to 1 so that its wiring weight is always added to the combined input to the neuron. It is assumed to be equal to minus the bias so that the total sum is the difference between the combined signal and the bias. In the remainder, we assume that the bias, with the minus sign, is always explicitly present among the wiring weights in this way (see Fig. 3.20 on the right).

Artificial neurons can be variously combined in neural networks. There have been defined many specific types of neural network structures, referred to as architectures, of which the most generic is a three-layer structure with no feedback connections, referred to as a feedforward neural network. Such a network is presented on Fig. 3.22. There are two outbound layers, the input and output ones, and one intermediate layer which is referred to as a hidden layer. This is why such a structure is referred to as a one hidden-layer neural network (NN).

Network on Fig. 3.22 is designed as a one-hidden-layer NN for predicting petal sizes of Iris features from their sepal sizes. Recall that in Iris data set, each of 150 specimens is presented with four features which are the length and width of petals (features w3 and w4) and sepals (features w1 and w2). It is likely that the sepal sizes and petal sizes are related.

In fact, the further material can be used for building an NN for modeling correlation between any inputs and outputs—the only possible difference, in numbers of input and/or output units, plays no role in the organization of computations.

This neural network consists of the following layers:

(a) Input layer that accepts three inputs: a bias input $x_0 = 1$ as explained above (see Fig. 3.20 on the right) as well as sepal length and width; these are combined to be inputs to each of the neurons at the hidden layer.
(b) Output layer producing an estimate for petal length and width with a linear activation function. Its input is the output signals from the hidden layer. No fictitious input $x_0 = 1$ is assumed here because the activation function here just passes the combined signal through without a threshold.

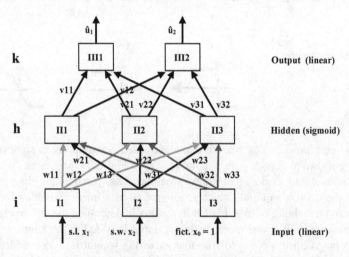

Fig. 3.22 A feed-forward network with two input and two output features (no feedback loops). Layers: Input (I, indexed by i), Output (III, indexed by k) and Hidden (II, indexed by h)

(c) Hidden layer consisting of three neurons. Each of them takes a combined input from the first layer and applies to it its sigmoid activation function. The output signals of these three neurons constitute inputs to the output layer. The architecture allows for any number of hidden neurons with no changes in the computations.

The one-hidden-layer structure is generic in NN theory. It has been proven, for instance, that such a structure can exactly learn any subset of the set of entities. Moreover, any pre-specified mapping of inputs to outputs can be approximated up to a pre-specified precision with such a one-hidden-layer network, if the number of hidden neurons is large enough (Haykin 1999). This property, for many years, served as a justification for the specialists to consider one-layer neural networks only. The past decade saw an explosion of research on deep neural networks, that is, networks with a dozen or more layers (see, for example, LeCun et al. 2015). It appears deep neural networks are capable to capture non-linear hidden features and, in this way, dramatically improve precision of the neural network decision rules. Deep learning is an area of intense research efforts which should bring novel breakthroughs in machine learning. There is a caveat though. The hidden features are hidden indeed. There is no way to explicitly describe relations between given features and the hidden one. That brings forward the I. Asimov's controversy pointed to in his "I, Robot" series: unlike in data analysis, machine learning may generate unexpected and unpredictable consequences.

3.5.2 Activation Functions and Network Function

Two popular activation functions, besides the *sign* function $\mathring{u}_i = sign(\hat{u}_i)$, are the *linear* activation function, $\mathring{u}_i = \hat{u}_i$ and *sigmoid* activation function $\mathring{u}_i = s(\hat{u}_i)$ where

$$s(x) = (1 + e^{-x})^{-1} \tag{3.21}$$

is a smooth analogue to the sign function, except for the fact that its output is between 0 and 1 rather than -1 and 1 (see Fig. 3.21b). To imitate the perceptron with its $sign(x)$ output, between -1 and 1, we first double the output interval and then subtract 1 to obtain what is referred to as a symmetric sigmoid or hyperbolic tangent:

$$th(x) = 2s(x) - 1 = 2(1 + e^{-x})^{-1} - 1 \tag{3.21'}$$

This function, illustrated on Fig. 3.21c, in contrast to sigmoid $s(x)$, is symmetric: $th(-x) = -th(x)$, like $sign(x)$, which can be useful in some contexts.

The sigmoid activation functions have nice mathematical properties; they are not only smooth, but their derivatives can be expressed through the functions themselves, see Q.3.13 and (3.28) further on.

Let us express now the function of the one-hidden-layer neural network presented on Fig. 3.22. Its wiring weights between the input and hidden layer form a matrix $W = (w_{ih})$, where i denotes an input, and h a hidden neuron, $h = 1, 2,...,$ H where H is the number of hidden neurons. The wiring weights between the hidden and output layers form matrix $V = (v_{hk})$, where h denotes a hidden neuron and k an output.

Layers I and III are assumed to be linear giving no transformation to their inputs; all of the hidden layer neurons will be assumed to have a symmetric sigmoid as their activation function.

To find out an analytic expression for the network's output, let us work it out layer by layer. Neuron h in the hidden layer receives, as its input, a combined signal

$$z_h = w_{1h}x_1 + w_{2h}x_2 + w_{3h}x_0$$

which is h-th component of vector $z = \Sigma_i x_i * w_{ih} = x * W$ where x is a 1×3 input vector. Then its output will be $th(zh)$. These constitute an output vector $th(z) = th(x * W)$ that is input to the output layer. Its k-th node receives a combined signal $\Sigma_j v_{jk} * th(z_j)$ which is k-th component of the matrix product $th(z) * V$, that is passed as the NN output $\hat{u}$. Therefore, the NN on Fig. 3.22 transforms input x into output $\hat{u}$ according to the following formula

$$\hat{u} = th(x * W) * V \tag{3.22}$$

which combines linear operations of matrix multiplication with a nonlinear symmetric sigmoid transformation. If matrices W, V are known, (3.22) computes the function $u = F(x)$ in terms of th, W, and V. The problem is to fit this model with training data provided, at this instance, by the Iris data set.

3.5.3 Learning a Multi-layer Network

Given all the wiring weights W, between the input and hidden layers, and wiring weights V, between the hidden and output layers, as well as pre-specified hidden layer activation functions, the NN on Fig. 3.22 takes an input of the sepal length and width and transforms it into estimates of the corresponding petal length and width.

The quality of the estimates can be measured by the average squared error. The better adapted weights W and V are, the smaller the error. Where the weights come from? They are learnt from the training data.

Thus, the problem is to estimate weight matrices W and V at the training data in such a way that the average squared error is minimized.

The machine learning paradigm is based on the assumption that a learning device adapts itself incrementally by facing entities one by one. This means that the full sample is assumed to be never known to the device so that global solutions, such as the orthogonal projection used in linear discrimination, are not applicable. In such a situation an optimization algorithm that processes entities one by one should be applied. Such is the gradient method, also referred to as the steepest descent.

This method relies on the so-called gradient of the function to be optimized (see Sect. A.2 in Appendix). The gradient is a vector that can be derived or estimated at any admissible solution, that is, matrices W and V. This vector shows the direction of the steepest ascent over the optimized function considered as a surface. Its elements are the so-called partial derivatives of the optimized function that can be derived according to rules of calculus. The gradient is useful for maximizing a criterion, but how one can do minimization with the steepest ascent? Easily, by moving in the opposite direction, that is, taking minus gradient.

Assume, we have some estimates of matrices W and V as well as their gradients, that is, matrices gW and gV, whose components express the steepest ascent direction of changes in W and V. Then, according to the method of steepest descent, the matrices V and W should be moved in the direction of $-gW$ and $-gV$ with the control of the length of the step by a factor referred to as the learning rate. The equations expressing the move from the old state to the new one are as follows:

$$V(new) = V(old) - \mu * gV, W(new) = W(old) - \mu * gW \qquad (3.23)$$

where μ is the learning rate (step size). The importance of properly choosing the step size is illustrated on Fig. 3.23.

The gradient of the criterion of squared error is defined by: (a) the matrices W and V, (b) the error value itself, and (c) the input feature values. This is why it is convenient to apply this approach when entities come in a sequence so that each individual entity gives an estimate of the gradient and, accordingly, the move to a new state of matrices W and V according to Eq. (3.23). The sequence of entities is natural when the learning is done on the fly by processing entities in the order of their arrival. In the situations when all the entities have been already collected in a data set, as the Iris data set, the sequence is organized artificially in a random order. Moreover, as the number of entities is typically rather small (as it is in the case of just 150 Iris specimens) and the gradient process is rather slow, it is usually not enough to process all the entities just once. The processing of all the entities in a

Fig. 3.23 The importance of properly choosing the step in the steepest descent process: if the leap is too big, the new state may be worse than the old one

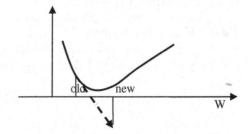

Table 3.12 Relative error values in the predicted petal dimensions with full Iris data after 5000 epochs

Number of hidden neurons	Relative error, %	
	Petal length	Petal width
3	5.36	8.84
6	3.99	8.40
10	3.98	8.15

random order constitutes *an epoch*. A number of epochs need to be executed until the matrices *V* and *W* are more or less stabilized.

Worked Example 3.6. Learning Iris Petal Sizes

Consider, at any Iris specimen, its two sepal sizes as the input and its two petal sizes as the output. We are going to find a decision rule relating them in the format of a one-hidden-layer NN.

At the Iris data, the architecture presented on Fig. 3.22 and program nnn.m implementing the error back-propagation algorithm leads to the average errors at each of the output variables presented in Table 3.12 at different numbers of hidden neurons *h*. Note that the errors are given relative to feature ranges.

The number of elements in matrices V and W here are five-fold of the number of hidden neurons, thus ranging from 15 at the current setting of three hidden neurons to 50 when this grows to 10. One can see that the increase in the numbers of hidden neurons does bring some improvement, but not that great—probably not worth doing.

Here are a few suggestions for further work on this example:

1. Find values of *E* for the errors reported in Table above.
2. Take a look at what happens if the data are not normalized.
3. Take a look at what happens if the learning rate is increased, or decreased, ten times.
4. Extend the table above for different numbers of hidden neurons.
5. Try petal sizes as input with sepal sizes as output.
6. Try predicting only one size over all input variables.

Worked Example 3.7. Predicting Marks at Student Dataset

Let us embark on an ambitious task of predicting student marks at the Students data—we partially dealt with this in Sect. 3.3. The nnn.m program leads to the average errors in predicting student marks over three subjects, as presented in Table 3.13 at different numbers of hidden neurons *h*. Surprisingly, the prediction works rather well: the errors are on the level of 3 points only, more or less independently of the number of hidden neurons utilized.

Table 3.13 Average absolute error values in the predicted student marks over all three subjects, with full Student data after 5,000 epochs

H	|e1|	|e2|	|e3|	# param.
3	3.65	3.16	3.17	27
6	3.29	3.03	3.75	54
10	3.17	3.00	3.64	90

3.5.4 Steepest Descent for the Square Error Criterion with Linear Rules

In machine learning, the assumption is that the decision rule is learnt incrementally by using entities one by one. That is, the global solutions involving the entire sample are not applicable. In such a situation an optimization algorithm that processes entities one by one should be applied. The most popular is the gradient method, also referred to as the steepest descent.

This method relies on the gradient of the function to be optimized. If we are to minimize function $f(x)$ over x spanning a subspace D of the n-dimensional vector space R^n, we can utilize its gradient gf for this purpose. The gradient gf at $x \in D$ is a vector consisting of the f's partial derivatives over all components of x, under the assumption that a full derivative, geometrically corresponding to the tangential hyperplane, does exist. This vector shows the direction of the steepest ascent of $f(x)$, so that its opposite vector $-gf$ shows the opposite direction which is considered as that of the steepest descent of $f(x)$. The method of steepest descent produces a sequence of points $x(0)$, $x(1)$, $x(2)$, … starting from an arbitrary $x(0)$ by using recursive equation

$$x(t+1) = x(t) - \mu_t * gf(x(t)) \qquad (3.23')$$

where parameter μ_t denotes the length of the step to go from $x(t)$ in the direction of the steepest descent, referred to as the learning rate in machine learning. The sequence $x(t)$ is guaranteed to converge to the minimum point at a constant $\mu_t = \mu$ if $f(x)$ is strictly convex, so that there is a sphere of a finite radius such that $f(x)$ is always greater than its lower part, as shown on the right of Fig. 3.24 (see Polyak 1987).

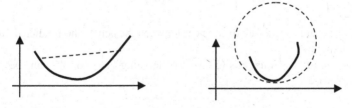

Fig. 3.24 A convex function, on the left and strictly convex function, on the right

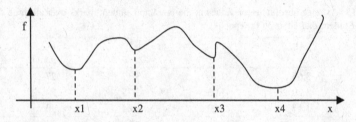

Fig. 3.25 Points x1–x4 are points of local minimum for the function whose graph is drawn with the line. The global minimum is only one of them, x4

The process converges if $f(x)$ is a convex function and μ_t tends to 0 when t grows to infinity, but not too fast so that the sum of the series $\Sigma_t \mu_t$ is infinity. This guarantees that the moves from $x(t)$ to $x(t + 1)$ are small enough to not over-jump the point of minimum but not that small to stop the sequence short of reaching the optimum by themselves.

If $f(x)$ is not convex however, the sequence reaches just one of the local optima depending on the starting point $x(0)$ (see Fig. 3.25). Luckily, the square error in the problem of linear discriminant analysis is strictly convex so that the steepest descent sequence converges to the optimum from any initial point. This gives rise to the algorithm described in the following section.

3.5.5 *Learning Wiring Weights with Error Back-Propagation*

The problem of learning a neural network is to find weight matrices W and V minimizing the squared difference between u observed and $\hat{u}$ computed:

$$E = d(u, \hat{u}) = \langle u - th(x * W) * V, u - th(x * W) * V \rangle / 2, \qquad (3.24)$$

over the training entity set. The division by 2 is made to avoid factor 2 in the derivatives of E.

Specifically, with just two outputs in Fig. 3.22, the error function is

$$E = \left[(u_1 - \hat{u}_1)^2 + (u_2 - \hat{u}_2)^2 \right] / 2 \qquad (3.24')$$

where $u_1 - \hat{u}_1$ and $u_2 - \hat{u}_2$ are differences between the actual and predicted values of the two outputs.

Steepest descent equations (3.23) for learning V and W can be written component-wise:

$$v_{hk}(t+1) = v_{hk}(t) - \mu * \partial E / \partial v_{hk},$$
$$w_{ih}(t+1) = w_{ih}(t) - \mu * \partial E / \partial w_{ih}(i \in I, h \in II, k \in III) \qquad (3.25)$$

To make these computable, let us express the derivatives explicitly; first those at the output, over v_{hk}:

$$\partial E / \partial v_{hk} = -(u_k - \hat{u}_k) * \partial \hat{u}_k / \partial v_{hk}.$$

To advance, notice that $\partial \hat{u}_k / \partial v_{hk} = th(zh)$, since $\hat{u}_k = \Sigma_j th(zh) * v_{hk}$. Putting this into equation above makes

$$\partial E / \partial v_{hk} = -(u_k - \hat{u}_k) * th(zh). \qquad (3.26)$$

Regarding the second layer, of W, let us find the derivative $\partial E / \partial w_{ih}$ which requires more chain-based derivations. Specifically,

$$\partial E / \partial w_{ij} = \Sigma_k \left[-(u_k - \hat{u}_k) * \partial \hat{u}_k / \partial w_{ij} \right].$$

Since $\hat{u}_k = \Sigma_j th\left(\Sigma_i x_i * w_{ij} \right) * v_{jk}$, this can be expressed as

$$\partial \hat{u}_k / \partial w_{ij} = v_{jk} * th'\left(\Sigma_i x_i * w_{ij} \right) * x_i.$$

The derivative $th'(z)$ can be expressed through $th(z)$ as explained in Q.3.13 later, which leads to the following final expression for the partial derivatives:

$$\partial E / \partial w_{ij} = -\Sigma_k \left[-(u_k - \hat{u}_k) * v_{jk} \right] * \left(1 + th(z_j) \right) \left(1 - th(z_j) \right) * x_i / 2 \qquad (3.27)$$

Equations (3.23), (3.26) and (3.27) lead to the following rule for processing an entity, or instance, in the error back-propagation algorithm as applied to neural network in Fig. 3.22.

1 *Forward computation (of the output $\hat{u}$ and error)*. Given matrices V and W, upon receiving an instance (x,u), the estimate $\hat{u}$ of vector u is computed according to the neural network as formalized in Eq. (3.22), and the error $e = u - \hat{u}$ is calculated.
2 *Error back-propagation (for estimation of the gradient elements)*. Each neuron receives the relevant error estimate, which is

   - $e_k = -(u_k - \hat{u}_k)$, for (3.26) at output neurons k ($k = III1, III2$) or
   - $\Sigma_k [(u_k - \hat{u}_k) * v_{hk}]$, for (3.27) at hidden neurons h ($j = II1, II2, II3$) [the latter can be seen as the sum of errors arriving from the output neurons according to the corresponding synapse weights].

These are used to adjust the derivatives (3.26) and (3.27) by multiplying them with local data depending on the input signal, which is $th(zh)$, for neuron k's source h in (3.26), and $th'(zh)x_i$ for neuron h's source i in (3.27).

3 *Weights update.* Matrices V and W are updated according to formula (3.23).

What is nice in this procedure is that the computation can be done locally, so that every neuron processes only the data that are available to this neuron, first from the input layer, then backwards, from the output layer. In particular, the algorithm does not change if the number of hidden neurons is changed from $h = 3$ on Fig. 3.22, to any other integer $h = 1, 2, ...$; nor does it change if the number of inputs and/or outputs changed.

3.5.6   *Error Back-Propagation: Computation*

For a data set available as a whole, "offline", due to the specifics of the binary target variables and activation functions, such as $th(x)$ and $sign(x)$, which have -1 and 1 as their boundaries, the data in the NN context are frequently pre-processed to make every feature's range to lie between -1 and 1 and the midrange to be 0. This can be done by using the conventional shifting and rescaling formula for each feature v, $yi_v = (xi_v - a_v)/b_v$, at which b_v is equal to the half-range, $b_v = (M_v - m_v)/2$, and shift coefficient a_v, to the mid-range, $a_v = (M_v + m_v)/2$. Here M_v denotes the maximum and m_v the minimum of feature v.

The practice of digital computation, with a limited number of digits used for representation of reals, shows that it is a good idea to further expand the ranges into a $[-10,10]$ interval by multiplying afterwards all the entries by 10: in this range, digital numbers stored in computer arguably lead to smaller computation errors than in the range $[-1,1]$, if they are closer to 0.

The implementation of the method of gradient descent for learning neural networks cannot be straightforward because the minimized squared error depends both on the wiring weight matrices V and W and input/output pairs (x,u), yet there is no way to freely change the latter—the process is bound by the set of observations. This is why the observed pairs (x_i,u_i), the instances, are used as triggers to the steepest descent changes in matrices V and W. Specifically, given V and W, the instances are put one by one, in a random order, to see what are the discrepancies between the observed u and computed $\hat{u}$. When all of the instances have been entered, their order is randomly changed and they are ready to be put all over again—this is referred to as a new "epoch". The matrices V and W are changed either at each (x_i,u_i) instance, using the errors $\hat{u}-u$ locally, or after an epoch, using the accumulated errors.

The error back-propagation algorithm, with the local changes of matrices V and W, can be formulated as follows.

A. Initialize weight matrices $W = (w_{ih})$ and $V = (v_{hk})$ by using random normal distribution $N(0,1)$ with the mean at 0 and the variance 1.
B. Standardize data to $[-10,10]$ ranges and 0 averages as described above.
C. Formulate halting criterion as explained below and run a loop over epochs.
D. Randomize the order of entities within an epoch and run a loop of the error back-propagation *instance processing procedure*, below, in that order.
E. If Halt-criterion is met, end the computation and output results: W, V, $\hat{u}$, e, and E. Otherwise, execute D again.

The best halting criterion, according to the nature of the steepest descent process should be at

(i) Matrices V and W stabilized. Unfortunately, in real world computations this criterion requires by far too many iterations, so that in practice the matrices fail to converge. Thus, other stopping criteria are used.
(ii) The difference between the average values (over iterations within an epoch) of the error function becomes smaller than a pre-specified threshold, such as 0.0001.
(iii) The number of epochs performed reaches a pre-specified threshold such as 5000.

Instance Processing Procedure

Specifics of the NN structure and function provide for simple and effective rules for processing individual entities in the procedure of the steepest descent. Before updating the wiring weights according to Eqs. (3.23), two following steps are executed:

1. *Forward computation of the estimated output and its error.* Upon receiving a training instance input feature values, they are processed by the neural network to produce an estimate of the output, after which the error is computed as the difference between real and estimated output values.
2. *Error back-propagation for estimation of the gradient.* The computed error of the output is back-propagated through the network. Each neuron of the output layer corresponds to a specific output feature and, thus, receives the error in this feature. Each neuron of the hidden layer receives a combined error signal from all output neurons weighted by the corresponding wiring weights. These are used to adjust the gradient elements by using the hidden neuron activation function.

In the Appendix A.4, a Matlab code nnn.m is presented for learning NN weights with the error back-propagation algorithm according to the NN of Fig. 3.22. Two parameters of the algorithm, the number of neurons in the hidden layer and the learning rate, are its input parameters. The output is the minimum level of error achieved and the corresponding weight matrices V and W.

The code includes the following steps:

1. *Loading data.* It is assumed that all data are in subfolder Data. According to the task, this can be either iris.dat or stud.dat or any other dataset.
2. *Normalizing data.* This is done by shifting each column to its midrange with the follow-up dividing it by the half-range, after which all data set is multiplied by 10, to have them in [−10,10] scale as described above.
3. *Preparing input and output* training sub-matrices. This is done after the decision has been made of what features fall in the former and what features fall in the latter categories. In the case of Iris data, for example, the target is petal data (features w3 and w4) and input is sepal measurements (features w1 and w2) as described. In the case of Students data, the target can be students' marks on all three subjects (CI, SP and OOP), whereas the other variables (occupation categories, age and number of children), input.
4. *Initializing the network.* This is done by: (a) specifying the number of hidden neurons H, (b) filling in matrices W and V with random (0,1) normally distributed values, and (c) setting a loop over epochs with the counter initialized at zero.
5. *Organizing a loop over the entities.* For setting a random order of entities to be processed, the Matlab command randperm(n) for making a random permutation of integers 1, 2,..., n can be used.
6. *Forward pass.* Given an entity, the output is calculated, as well as the error, using the current V, W and activation functions. The program uses the symmetric sigmoid (3.21) as the activation function of hidden neurons.
7. *Error back-propagation.* Gradient matrices for V and W according to formulas (3.26) and (3.27) are computed.
8. *Weights V and W update.* Having the gradients computed and learning rate accepted as the input, updated W and V are computed according to (3.23).
9. *Halt-condition.* This includes both the level of precision, say 0.01, and a threshold to the number of epochs, say, 5,000. If either is reached, the program halts.

Q.3.13. Prove that the derivatives of sigmoid (3.21) or hyperbolic tangent (3.21) functions appear to be simple polynomials of selves. Specifically,

$$s'(x) = \left((1 + e^{-x})^{-1}\right)' = (-1)(1 + e^{-x})^{-2}(-1)e^{-x} = s(x)(1 - s(x)), \quad (3.28)$$

$$\begin{aligned}th'(x) &= [2 * s(x) - 1]' = 2 * s(x)' = 2 * s(x) * (1 - s(x)) \\ &= (1 + th(x)) * (1 - th(x))/2\end{aligned} \quad (3.28')$$

Q.3.14. Find a way to improve the convergence of the process, for instance, with adaptive changes in the step size values.

Q.3.15. Use k-fold cross validation to provide estimates of variation of the results regarding the data change.

Q.3.16. Develop a scoring function for learning a category by using the contribution of the partition to be built to the category.

3.6 Association Between Nominal Features: Elementary and Linear Modeling

3.6.1 Elementary Analysis: Quetelet Index and Chi-Squared

3.6.1.1 Conceptual Relations from Statistics

To analyze interrelations between two nominal features, they are cross-classified in the so-called contingency table. A contingency table has its rows corresponding to categories of one feature and columns to categories of the other feature, while the entries are counts of entities falling in the overlap of the corresponding row and column categories. The contingency table's structure may well serve for the analysis of association between the nominal features summarized in the table.

Worked Example 3.8. Contingency Table on Market Towns Data

To cross-classify features Banks and Farmer's Market on Market towns data, we first need to categorize the quantitative feature Banks. Consider, for example, the four-category partition of the range of Banks feature at Market towns set presented in Table 3.14.

These categories are cross-classified with FM "yes" and "no" categories in Table 3.15. Besides the cross-classification counts, the table also contains summary within category counts, the totals, on the margins of the table, the last row and last

Table 3.14 Definition of Ba categories on the Market towns dataset

Category	Definition	Notation
1	Ba ≥ 10	10+
2	10>Ba≥ 4	4+
3	4>Ba≥ 2	2+
4	Ba = 0 or 1	1−

Table 3.15 Cross classification of the Ba categories with FM categories

FarmMarket	Bank/building society categories				Total
	10+	4+	2+	1−	
Yes	2	5	1	1	9
No	4	7	13	12	36
Total	6	12	14	13	45

Table 3.16 BA/FM cross-classification relative frequencies, percent

FM | Ba	10+	4+	2+	1−	Total
Yes	3.44	11.11	3.22	3.22	20
No	8.89	15.56	28.89	26.67	80
Total	13.33	26.67	31.11	28.89	100

Table 3.17 Protocol/attack contingency table for Intrusion data

Category	Apache	Saint	Smurf	Norm	Total
Tcp	23	11	0	30	64
Udp	0	0	0	26	26
Icmp	0	0	10	0	10
Total	23	11	10	56	100

column—this is why they are referred to as marginal frequencies. The total count balances the sheet in the bottom-right corner.

The same contingency data converted to relative frequencies by relating them to the total number of entities are presented in Table 3.16.

Q.3.17. Build a contingency table for features "Protocol-type" and "Attack type" in Intrusion data. **A.** See Table 3.17

A contingency table can be used for assessment of correlation between two sets of categories. The highest level of correlation is that of a conceptual association. A conceptual association may exist if a row, k, has all its entries, not marginal of course, except just one, say l, equal to 0, which would mean that all of the extent of category k belongs to the column category l. The data, thus, indicate that the category k may logically imply the category l.

Worked Example 3.9. Equivalence and Implication from a Contingency Table

Such are rows "Udp" and "Icmp" in Table 3.17. There is a perfect match in this table: a row category k = "Icmp" and a column category l = "Smurf", that contains the only non-zero count. No other combination (k, l′) or (k′, l) is possible according to the table. In such a situation, one may claim that, subject to the sampling error, category l may occur if and only if k does, that is, k and l are equivalent.

A somewhat weaker, but still very much valuable is the case of "Udp" row in Table 3.17. It appears, Udp protocol implies "Norm" column category—a no-attack situation, though there is no equivalence here because the "Norm" column contains another positive count, in row "Tcp".

Case-Study 3.5. Trimming Contingency Data: A Bad Option

Unfortunately, there are no zeros in Table 3.15, and thus, no conceptual relation between the number of Banks and the presence of a Farmer's market. But some of the entries are really close to 0, which may make us tempted to trim the data a bit. Imagine, for example, that in row "Yes" of Table 3.15, two last entries are 0, not 1s.

Table 3.18 A trimmed BA/ FM cross classification "cleaned" of 13 towns, to sharpen the view

| FMarket | Number of banks/build, societies | | | | |
	10+	4+	2+	1−	Total
Yes	2	5	0	0	7
No	0	0	13	12	25
Total	2	5	13	12	32

This would imply that a Farmers Market may occur only in a town with 4 or more Banks. A logical implication, that is, a production rule, "If BA is 4 or more, then a Farmer's market must be present", could be derived then from thus modified table. One may try taking this path and cleaning the data of smaller entries, by removing corresponding entities from the table, of course, to not obscure our "vision" of the pattern of correlation. Thus trimmed Table 3.18 is obtained from Table 3.15 by removing just 13 entities from "less popular" entries. This latter table expresses, with no exception, a very simple conceptual statement "A town has a Farmer's market if and only if the number of Banks in it is 4 or greater". However nice the rule may sound, let us not forget the cost of the trimming which is the 13 towns, almost 30% of the sample, that have been removed as those not fitting the stated perspective. Such a data doctoring borders with forgery—one of the reasons for a famous quip usually attributed to B. Disraeli, a celebrated British politician of XIX century: "There are three gradations of lies: lies, damned lies and statistics." The issue of sample adjustment so far has received no reasonable solution, even with respect to outliers—values falling way beyond the feature range one would expect normally. Anyway, the conclusion of the trimming exercise is that one should try finding ways of expressing conceptual relations without doctoring the sample.

3.6.1.2   Capturing Relationships with Quetelet Indexes

Quetelet index provides for a strategy for visualization of correlation patterns in contingency tables without removal of "not-fitting" entities. In 1832, A. Quetelet (1796–1874), a founding father of statistics, proposed to measure the extent of association between row and column categories in a contingency table by comparing the local count with an average one.

Let us consider correlation between the presence of a Farmer's Market and the category "10 or more Banks" according to Table 3.15. We can see that their joint probability/ frequency is the entry in the corresponding row and column: $P(Ba = 10 + \& FM = Yes) = 2/45 = 4.44\%$ (joint probability/frequency rate). Of the 20% entities that fall in the row "Yes", this makes the proportion of "Ba = 10+" under condition "FM = Yes" equal to $P(Ba = 10+ /FM = Yes) = P(Ba = 10+ \& FM = Yes)/P(FM = Yes) = 0.0444/0.20 = 0.222 = 22.2\%$. Such a ratio expresses the conditional probability/rate.

Is this high or low? Hard to tell without comparing this with the unconditional rate, that is, with the frequency of category "Ba = 10+" in the whole dataset, which is P(Ba = 10+) = 13.33%. Let us compute the (relative) difference between the two, which is referred to as Quetelet index q:

$$q(\text{Ba} = 10+/\text{FM} = \text{Yes}) = [P(\text{Ba} = 10+/\text{FM} = \text{Yes})$$
$$-P(\text{Ba} = 10+)]/P(\text{Ba} = 10+)$$
$$= [0.2222 - 0.1333]/0.1333 = 0.6667 = 66.7\%.$$

That means that condition "FM = Yes" raises the frequency of the Bank category by 66.7%.

This logic concurs with our everyday intuition. Consider, for example, the risk of getting a serious illness, say tuberculosis, which may be, say, about 0.1%, one in a thousand, in a given region. Take a condition such as "Bad housing" and count the rate of tuberculosis under this condition, amounting to, say, 0.5%—which is very small by itself, yet a five-fold increase over the average tuberculosis rate. This is exactly what Quetelet index measures: $q(l/k) = (0.5 - 0.1)/0.1 = 400\%$ to show that the change of the average rate is 4 times.

Worked Example 3.10. Quetelet Index in a Contingency Table

Let us apply the general Quetelet index formula (3.29) to entries in Table 3.15. This leads to Quetelet index values presented in Table 3.19. By highlighting positive values in the table, we obtain the same pattern as on the "purified" data as in Case-Study 3.5, but this time in a somewhat more realistic manner, keeping the sample intact. Specifically, one can see that "Yes" FM category provides for a strong increase in the probabilities, whereas "No" category leads to much weaker changes.

Table 3.19 BA/FM Cross classification Quetelet coefficients, % (positive entries highlighted using bold font)

FMarket	10+	4+	2+	1−
Yes	**66.67**	**108.33**	−63.29	−61.54
No	−16.67	−27.08	**16.07**	**15.38**

Q.3.18. Compute Quetelet coefficients for Table 3.17.
A. See Table 3.20 in which positive entries are highlighted in bold

Table 3.20 Quetelet indices for the protocol/attack contingency Table 3.17, %; positive values are highlighted using bold font

Category	Apache	Saint	Surf	Norm
Tcp	**56.25**	**56.25**	−100.00	−16.29
Udp	−100.00	−100.00	−100.00	**78.57**
Icmp	−100.00	−100.00	**900.00**	−100.00

Case-Study 3.6. Has There Been Any Bias in S'n'S' Policy?

Take on the case of Stop-and-Search policy in England and Wales 2005 represented according to race (B—black, A—Asian and W—white), by numbers in Table 1.12 in Sect. 1.3.2—these are overwhelmingly in category W. The criticism of this policy came out of comparison of this distribution with the distribution of the entire population. Such a distribution, according to the latest pre-2005 census 2001, can be easily found on web. By subtracting from that the numbers of Stop-and-Search occurrences, under the assumption that nobody has been subjected to this more than once, Table 3.21 has been drawn. Its last column gives the numbers that were used for the claim of a racial bias: indeed category B members have been subjects of the policy six times more frequently than category W members. A similar picture emerges when Quetelet coefficients are used (see Table 3.22). Category B is subject to Stop-and-Search policy 400% more frequently than on average, whereas category W, 15% less.

Yet some would consider drawing a table like Table 3.21, and of course the derived Table 3.22, as something nonsensical, because it is based on an implicit assumption that the Stop-and-Search policy applies to the population randomly. They would argue that police apply the policy only when they deem it necessary, so that the comparison should involve not all of the total population but only that criminal. Indeed, the distribution of subjects to Stop-and-Search policy by race has been almost identical to that of the imprisoned population of the same year. Therefore, the claim of a racial bias by police should be declared incorrect, provided that the system of justice is not biased overall.

Table 3.21 Distribution of stop-and-search policy cross-classified with race

	S'n'S	Not S'n'S	Total	S'n'S-to-Total
Black	131,723	1,377,493	1,509,216	0.0873
Asian	70,252	2,948,179	3,018,431	0.0233
White	676,178	46,838,091	47,514,269	0.0142
Total	878,153	51,163,763	52,041,916	0.0169

Table 3.22 Relative Quetelet coefficients for cross-classification in Table 3.21, %

	S'n'S	Not S'n'S
Black	417.2	−7.2
Asian	37.9	−0.6
White	−15.7	0.3

3.6.1.3 Conditional Probabilities, Quetelet Indexes and Pearson's Chi-Squared

Consider two sets of disjoint categories on an entity set I, K and L. Using $k = 1,\ldots,$ K (for example, occupation of individuals constituting I) and $l = 1,\ldots, L$ (say, family or housing type) as category indices should not bring any confusion here. Both, K and L, make a partition of the entity set I; these partitions are crossed to see if there is any correlation between them. For a a pair of categories $(k,l) \in K \times L$, count the number of entities that fall in both. The (k,l) co-occurrence count is denoted by N_{kl}. Obviously, these counts sum to N because the categories are not overlapping and cover the entire dataset. A table housing these counts, N_{kl}, or their relative values, relative frequencies $p_{kl} = N_{kl}/N$, is referred to as a contingency table or just cross-classification. The totals, that is, within-row sums $N_{k+} = \Sigma_l N_{kl}$ and within-column sums $N_{+l} = \Sigma_k N_{kl}$ (as well as their relative frequency counterparts) are referred to as marginals (because they are located on the margin of the contingency table).

The (empirical) probability that category l occurs under condition k can be expressed as $P(l/k) = p_{kl}/p_{k+} = N_{kl}/N_{k+}$. The probability $P(l)$ of the category l with no condition is just $p_{+l} = N_{+l}/N$. Similar notation is used when l and k are swapped. The relative difference between the conditional and unconditional probabilities is referred to as the (relative) Quetelet index (Mirkin 2001):

$$q(l/k) = \frac{P(l/k) - P(l)}{P(l)} \qquad (3.29)$$

where $P(l) = N_{+l}/N$, $P(k) = N_{k+}/N$, $P(l/k) = N_{kl}/N_{k+}$. That is, Quetelet index expresses correlation between categories k and l as the relative change in the probability of l when k is taken into account.

With little algebra, one can derive a simpler—and symmetric—expression

$$\begin{aligned} q(l/k) &= [N_{kl}/N_{k+} - N_{+l}/N]/(N_{+l}/N) \\ &= N_{kl}N/(N_{k+}N_{+l})-1 = \frac{p_{kl}}{p_{k+}p_{+l}} - 1 \end{aligned} \qquad (3.29')$$

Highlighting high positive and negative values in a Quetelet index table, such as Tables 3.19 and 3.22, visualizes the pattern of association between the two sets of categories.

This visualization can be extended to a theoretically sound presentation. Let us define the average Quetelet association index Q as the sum of pair-wise Quetelet indexes weighted by their frequencies/probabilities:

$$Q = \sum_{k=1}^{K} \sum_{l=1}^{L} p_{kl} q(l,k) = \sum_{k=1}^{K} \sum_{l=1}^{L} p_{kl} \left(\frac{p_{kl}}{p_{k+}p_{+l}} - 1 \right) = \sum_{k=1}^{K} \sum_{l=1}^{L} \frac{p_{kl}^2}{p_{k+}p_{+l}} - 1$$

$$(3.30)$$

The right-hand expression for Q in (3.30) is popular in the statistical analysis of contingency data. In fact, this Q is equal to the chi-squared correlation coefficient proposed by Pearson (1900) in a very different context—as a measure of deviation of the contingency table entries from the statistical independence.

To explain this in more detail, let us first introduce the concept of statistical independence. The sets of k and l categories are said to be statistically independent if $p_{kl} = p_{k+} p_{+l}$ for all k and l. Obviously, such a condition is hard to fulfill in reality K. Pearson suggested using relative squared errors to measure the deviations of observed frequencies from the statistical independence. Specifically, he introduced the following coefficient usually referred to as Pearson's chi-squared association coefficient:

$$X^2 = N \sum_{k=1}^{K} \sum_{l=1}^{L} \frac{(p_{kl} - p_{k+}p_{+l})^2}{p_{k+}p_{+l}} = N \left(\sum_{k=1}^{K} \sum_{l=1}^{L} \frac{p_{kl}^2}{p_{k+}p_{+l}} - 1 \right) \qquad (3.31)$$

The equation on the right can be proven with little algebra. Consider, for example, this part of the expression on the left in (3.31):

$$\sum_{l=1}^{L} \frac{(p_{kl} - p_{k+}p_{+l})^2}{p_{k+}p_{+l}} = \sum_{l=1}^{L} \frac{p_{kl}^2 - 2p_{kl}p_{k+}p_{+l} + (p_{k+}p_{+l})^2}{p_{k+}p_{+l}}$$

$$= \sum_{l=1}^{L} \frac{p_{kl}^2}{p_{k+}p_{+l}} - 2 \sum_{l=1}^{L} p_{kl} + \sum_{l=1}^{L} p_{k+}p_{+l} = \sum_{l=1}^{L} \frac{p_{kl}^2}{p_{k+}p_{+l}} - p_{k+}$$

The expression on the right in the above is derived by using equations $\Sigma_l p_{kl} = p_{k+}$ and $\Sigma_l p_{+l} = 1$. Summing all these equations over k will produce (3.31). On the other hand, the expression on the right in (3.31) is obviously equal to $\Sigma_l p_{kl} q(l/k)$ so that

$$\sum_{l=1}^{L} \frac{(p_{kl} - p_{k+}p_{+l})^2}{p_{k+}p_{+l}} = \sum_{l=1}^{L} p_{kl} q(l/k) \qquad (3.32)$$

By comparing the right-hand parts of (3.30) and (3.31), it is easy to see that $X^2 = NQ$. The same follows from summing all the equations (3.32) over k.

The popularity of X^2 index in statistics and related fields rests on the following theorem proven by K. Pearson: if the contingency table is based on a sample of entities independently drawn from a population in which the statistical independence holds (so that all deviations are due to just randomness in the sampling), then the probabilistic distribution of X^2 converges to the chi-squared distribution (when N tends to infinity) introduced by Pearson earlier for similar analyses. The probabilistic chi-squared distribution (with p degrees of freedom) is defined as the distribution of the sum of squares of p random variables, each distributed according to the standard Gaussian $N(0,1)$ distribution.

This theorem is not always of interest to a computational data analyst, because they analyze data that are not necessarily random or not necessarily independently sampled. However, Pearson's chi-squared coefficient is frequently used just for scoring correlation in contingency tables. The equation $X^2 = NQ$ gives a credible support to this practice. According to this equation, the value of X^2 is not only a measure of deviation from the statistical independence. It also has a different meaning as a measure of association between categories: that of the averaged Quetelet coefficient. If, for example $X^2/N = 0.25$, one may credibly claim that the knowledge of L-category at an object improves the chance of its corresponding K-category by 25% on average.

To get more intuition on the underlying correlation concept, let us take a look at the extreme values that X^2 can take and situations at which the extreme values are reached (Mirkin 2001). It appears that at $K \le L$, that is, if the number of columns is not smaller than that of rows, X^2 ranges between 0 and $K - 1$. It reaches 0 if there is a statistical independence at all (k,l) entries so that all $q_{kl} = 0$, and it reaches $K - 1$ if each column l contains only one non-zero entry $p_{k(l)l}$, which is thus equal to p_{+l}. Such a structure of the contingency table can be interpreted as an empirical evidence that the logical implication $l(k) \rightarrow k$ has place for all $k = 1,2,\ldots, K$.

Representation of the chi-squared through Quetelet coefficients,

$$X^2 = \sum_{k=1}^{K} \sum_{l=1}^{L} Np_{kl}q(l/k) \tag{3.33}$$

amounts to decomposition of X^2 into the sum of $N_{kl}\, q(l/k)$ items and allows for visualization of the items within the contingency table format, such as that presented in Table 3.23.

In fact, not only the total sum of these items coincide with that of the original chi-squared items $N(p_{kl} - p_{k+}p_{+l})^2/p_{k+}p_{+l}$, but also the within-column and within-row sums coincide too, as the derivation of (3.32) above clearly demonstrates for the latter case.

However all the original chi-squared items in (3.31) are positive and cannot show whether the correlation expressed by an individual entry is positive or negative. To overcome this shortcoming, another visualization of X^2 is in use. That visualization involves the square roots of the chi-squared items

$$r(k, l) = \frac{p_{kl} - p_{k+}p_{+l}}{\sqrt{p_{k+}p_{+l}}} \tag{3.34}$$

Table 3.23 Four-fold contingency table between binary features			Feature Y		Total
			Yes	No	
Feature X	Yes		a	b	$a + b$
	No		c	d	$c + d$
Total			$a + c$	$b + d$	$N = a + b + c + d$

that are convenient to refer to as Pearson indexes. Obviously, $X^2 = N\Sigma_{k,l}r(k,l)^2$. Pearson indexes indeed have the same signs as $q(l/k)$, and in fact are closely related: $q(l/k) = r(k,l)[(p_{k+}p_{+l})]^{1/3}$. It is less clear what interpretation of its own the values $r(k,l)$ may have, although they are useful in the Correspondence analysis of contingency tables (Sect. 3.6.3), see also normalized Laplacian in Sect. 5.2.

Q.3.19. Take two binary features presented as 1/0 variables and build their contingency table, sometimes referred to as a four-fold table (Table 3.23) in which symbols a, b, c, and d are used to denote the co-occurrence counts.

Prove that Quetelet coefficient $q(Yes/Yes)$ expressing the relative difference between $a/(a + c)$ and $(a + b)/N$ is equal to

$$q(Yes/Yes) = \frac{ad - bc}{(a+c)(a+b)},$$

and the summary Quetelet coefficient Q, or Pearson's X^2/N, is equal to

$$Q = \frac{(ad - bc)^2}{(a+c)(b+d)(a+b)(c+d)}.$$

Q.3.20. Prove that the correlation coefficient between two 1/0 binary features can be expressed in terms of the four-fold table as $\rho = \sqrt{Q}$, that is,

$$\rho = \frac{ad - bc}{\sqrt{(a+c)(b+d)(a+b)(c+d)}}.$$

Q.3.21. Given a $K \times L$ contingency table P and a pair of categories, $k \in K$ and $l \in L$, consider an absolute Quetelet index $a(l/k) = P(l/k) - P(l)$ – the change from the frequency of $l \in L$ on the whole entity set I to the frequency of l on entities falling in category $k \in K$. In terms of P, $P(l) = p_{+l}$ and $P(l/k) = p_{kl}/p_{+l}$. Prove that the summary Quetelet index $A = \Sigma_{k,l}p_{kl}a(l/k) = \Sigma_{k,l}p_{kl}^2/p_{k+} - \Sigma_l p_{+l}^2$ is equal to the following expression, an asymmetric analogue to Pearson chi-squared:

$$A = \sum_{k=1}^{K} \sum_{l=1}^{L} \frac{(p_{kl} - p_{k+}p_{+l})^2}{p_{k+}} \tag{3.35}$$

which also is the numerator of the so called Goodman-Kruskal "tau-b" association index (Kendall and Stewart 1967).

A. Indeed, by taking the square of the numerator, expression in (3.35) becomes equal to $\Sigma_{k,l}(p_{kl}^2 - 2p_{kl}p_{k+}p_{+l}+p_{k+}^2p_{+l}^2)/p_{k+}$, which is $\Sigma_{k,l}p_{kl}^2/p_{k+} - 2\Sigma_{k,l}p_{kl}p_{+l} + \Sigma_{k,l}p_{k+}p_{+l}^2 = \Sigma_{k,l}p_{kl}^2/p_{k+} - 2\Sigma_{k,l}p_{+l}^2 + \Sigma_l p_{+l}^2$ because $\Sigma_k p_{kl} = p_{+l}$ and $\Sigma_k p_{k+} = 1$. This is obviously $\Sigma_{k,l}p_{kl}^2/p_{k+} - \Sigma_l p_{+l}^2 = \Sigma_{k,l}p_{kl}a(l/k) = A$, which proves the statement.

Worked Example 3.11. Visualization of Contingency Table Using Weighted Quetelet Coefficients

Let us multiply Quetelet coefficients in Table 3.19 by the frequencies of the corresponding entries in Table 3.15. Quetelet coefficients in Table 3.19 are taken relative to unity, not per cent. This leads us to Table 3.24 whose entries sum to the value of Pearson's chi-square coefficient for Table 3.15, 6.86. Note that entries in Table 3.24 can be both positive and negative; those with absolute value greater than $6.86/4 = 1.72$ are highlighted in bold—they show the entries of an extraordinary deviation from the average. Of them, column 4+ supplies the highest positive impact and the highest negative impact.

Worked Example 3.12. A Conventional Decomposition of Chi-square Coefficient

Let us consider a conventional way of visualization of contingency tables, by putting Pearson indexes, the square roots $r(k,l)$ of the chi-square coefficient items in (3.34) as the table's elements. These are in Table 3.25. The table does show a similar pattern of positive and negative associations. However, it is not the entries of the table that sum to the chi-square coefficient but rather the squares of the entries. The fact that the summary values on the margins in Tables 3.24 and 3.25 are the same is not by chance: it exemplifies a mathematical property (see Eq. (3.32)).

Table 3.24 BA/FM chi-squared (NQ = 6.86) and its decomposition according to (3.33); extreme values are highlighted using bold font

FMarket	10+	4+	2+	1−	Total
Yes	1.33	**5.41**	−0.64	−0.62	5.48
No	−0.67	**−1.90**	**3.09**	**1.85**	1.37
Total	0.67	3.51	1.45	1.23	**6.86**

Table 3.25 Square roots of the items in Pearson chi-squared (X^2 = 6.86); the items themselves are in parentheses; those positive are highlighted using bold font

FMarket	10+	4+	2+	1-	Total
Yes	**0.73**(0.53)	**1.68(3.82)**	−1.08 (1.16)	−0.99 (0.98)	(5.49)
No	−0.36 (0.13)	−0.84 (0.70)	**0.54 (0.29)**	**0.50 (0.25)**	(1.37)
Total	(0.67)	(3.52)	(1.45)	(1.23)	(6.86)

Q.3.22. In Table 3.24, all marginal values, the sums of rows and columns, are positive, in spite of the fact that many within-table entries are negative. Is this just due to specifics of the distribution in Table 3.15 or a general property?
A. A general property: the within-row or within-column sums of the elements, $N_{lk}\ q(l/k)$, must be positive, see (3.32).
Q.3.23. Find a similar decomposition of chi-squared for OOPmarks/Occupation in Student data. Hint: First, categorize quantitative feature OOPmarks somehow: you may use equal bins, or conventional boundary age points such as 35, 65 and 75, or any other considerations.
Q.3.24. Can any logical production rules come from the columns of Table 3.17?
A. Yes, both Apache and Saint attacks may occur at the tcp protocol only.
Q.3.25. Of 100 Christmas shoppers in Q.2.25, 50 spent £60 each, 20 spent £100 each, and 30 spent £150 each. Those who spent £60 each are males only and those who spent £100 each are females only, whereas among the rest 30 individuals half are men and half are women. Build a contingency table for the two features, gender and spending. Find and interpret the value of Quetelet coefficient for females who spent £100 each.
A. The contingency table (of co-occurrence counts):

	Spending, £			
Gender	60	100	150	Total
Female	0	20	15	35
Male	50 .	0	15	65
Total	50	20	30	100

This table of absolute co-occurrence counts coincides with that of proportions expressed per cent because the number of shoppers is 100.
  Quetelet coefficient for (Female/£100) entry is

$$Q = 100 * 20/(20 * 35) - 1 = 3.86 - 1 = 1.86$$

This means that being female in this category of spending is more likely than the average, by 186%.

3.6.1.4 Different Association Measures

Given two nominal features represented by partitions $S = \{S_1, S_2,..., S_K\}$ and $T = \{T_1, T_2,...., T_L\}$ of the entity set I, summarized in a $K \times L$ contingency table $P = (p_{kl})$ where p_{kl} is the proportion of entities in $S_k \cap T_l$. Let us describe a few approaches to scoring the association between S and T that are used in popular data analysis programs.
  One idea for assessing the extent of association is to use a correlation measure over the contingency table entries, such as averaged Quetelet coefficients, Q and A, or chi-squared X^2, as discussed above in Sect. 3.6.1.

Seemingly another idea is to score the extent of reduction of uncertainty over T obtained when S becomes available. This idea works like this: take a measure of uncertainty of a feature, in this case partition, $v(T)$, and evaluate it within each of S-classes, $v(T(S_k))$, $k = 1,\ldots, K$. Then the average uncertainty on these classes will be $\sum_{k=1}^{K} p_{k+} v(T(S_k))$, where p_{k+} are proportions of categories k, that is, classes S_k, so that the total reduction of uncertainty is equal to

$$v(T/S) = v(T) - \sum_{k=1}^{K} p_{k+} v(T(S_k)) \qquad (3.36)$$

Of course, a function like (3.36) can be considered a measure of association over the contingency table P as well, but a nice feature of this approach is that it can be extended from nominal features to quantitative ones—by using an uncertainty index over quantitative T-features.

Two very popular measures defined according to (3.36) are the so-called impurity function (Breiman et al. 1984) and information gain (Quinlan 1993).

The impurity function builds on Gini coefficient as a measure of variance (see Sect. 2.3.1). Let us recall that Gini index for partition T is $G(T) = 1 - \sum_{l=1}^{L} p_l^2$ where p_l is the proportion of entities in T_l. If I is partitioned in clusters S_k, $k = 1,\ldots,$ K, partitions T and S form a contingency table of relative frequencies $P = (p_{kl})$. Then the reduction (3.36) of the value of Gini coefficient due to partition S is equal to

$$\Delta(T, S) = G(T) - \sum_{k=1}^{K} p_{k+} G(T(S_k)). \qquad (3.37)$$

This index $\Delta(T, S)$ is referred to as impurity of S over partition T. The greater the impurity, the better the split S.

It is not difficult to prove that $\Delta(T, S)$ relates to Quetelet indexes from Sect. 3.6.1. Indeed, $\Delta(T, S) = A(T, S)$ where $A(T, S)$ is the average absolute Quetelet index defined by Eq. (3.35) in Q.3.21. This implies indeed that $\Delta(T, S) = \sum_l p_{kl}^2/p_{k+} - \sum_l p_{+l}^2$, which proves the following statement: The impurity function is equal to the average absolute Quetelet index.

The information gain function builds on entropy as a measure of uncertainty (see Sect. 2.3.1). Let us recall that entropy of partition T is $H(T) = -\sum_{l=1}^{L} p_l \log(p_l)$ where p_l is the proportion of category l, that is, part T_l. If the entity set is partitioned in clusters S_k, $k = 1,\ldots, K$, partitions T and S form a contingency table of relative frequencies $P = (p_{kl})$. Then the reduction (3.36) of the value of entropy due to partition S is equal to $I(T, S) = H(T) - \sum_{k=1}^{K} p_{k+} H(T(S_k)$. This index $I(T, S)$ is referred to as the information gain due to S. In fact, it is equal to a popular characteristic of the cross-classification of T and S, the mutual information defined as $I(T, S) = H(T) + H(S) - H(ST)$ where $H(ST)$ is entropy of the bivariate distribution represented by the contingency table P. Please note that the mutual information is symmetric with regard to S and T, in contrast to the impurity function.

To prove the statement let us just put forward the definition of the information gain and use the property of logarithm that $\log(a/b) = \log(a) - \log(b)$:

$$I(T,S) = H(T) - \sum_k p_k H(T(S_k)) = H(T) + \sum_k p_{k+} + \sum_l p_{kl} \log(p_{kl}/p_{k+})$$
$$= H(T) - \sum_k p_{k+} \log(p_{k+}) + \sum_{k,l} p_{kl} \log(p_{kl}) = H(T) + H(S) - H(ST),$$

which completes the proof.

The reduction of uncertainty measures are absolute differences that much depend on the measurement scale and, also, on values of $\upsilon(T)$ and $\upsilon(S)$. This is why it can be of advantage to use relative versions of the reduction of uncertainty measures normalized by $\upsilon(T)$ or $\upsilon(S)$ or both. For example, the popular program C4.5 (Quinlan 1993) uses the information gain normalized by $H(S)$ and referred to as the information gain ratio.

3.6.2   Least-Squares Analysis of Association Between Dummy Matrices

3.6.2.1   Linear Regression of One Dummy Matrix Over the Other One

Consider a nominal feature over an entity set I of cardinality N represented by partition $T = \{T_l\}$, and another nominal feature represented by partition $S = \{S_k\}$. This can be a cluster partition derived from available features to approximate T. Rather than directly concentrating on their contingency table $P = (p_{kl})$, let us take a look at the association between T and S from a different perspective.

Let us define an $N \times L$ dummy matrix X corresponding to partition T by assigning each category T_l with a binary variable x_l, a dummy, which is just a 1/0 N-dimensional vector whose elements $x_{il} = 1$ if $i \in T_l$ and $x_{il} = 0$, otherwise ($l = 1,..., L$). Similarly define an $N \times K$ dummy matrix Y whose columns y_k are 0/1-vectors corresponding to categories k of S.

Let us consider linear regression of y_k over set of all X-categories which is achieved by using the orthogonal projector $P_X = X(X^TX)^{-1}X^T$, so that $\hat{y}_k = P_X y_k$. Let us take a look at the components of the computed vector $\hat{y}_k$. Let us recall that the projector's matrix P_X consists of diagonal blocks with (i,j)-th elements $1/N_l$ for $i,j \in T_l$. whereas all the other elements are zero. Then the inner product of its i-th row and vector is the number of unities in it multiplied by $1/N_l$ for that l for which $i \in T_l$. That is exactly N_{lk}/N_l where N_{lk} is the cardinality of intersection $S_k \cap T_l$. In other words, i-th element of $\hat{y}_k$ is the conditional probability $N_{lk}/N_l = p(k/l)$ where l is the T-category of object i, $i = 1, 2,..., N$.

This gives the structure of regression of the dummy matrix Y over dummy matrix X, $\hat{Y} = P_X Y$ in terms of the conditional frequency contingency table.

Let us now standardize each feature y_k by a scale shift a_k and rescaling factor $1/b_k$, according to the conventional formula $y'_k = (y_k - a_k)/b_k$. This will correspondingly change the $\hat{Y}$-values, so that the conditional probabilities will change to $c_{kl} = (p(k/l) - a_k)/b_k$. In mathematical statistics, the issue of standardization is just a routine transforming the probabilistic density function to a standardized format. Things are different in data analysis, since no density function is assigned to data usually. The scale shift is considered as positioning the data against a backdrop of the "norm", whereas the act of rescaling is to balance feature "weights" (see Sect. 2.2 for discussion). Therefore, choosing the feature means as the 'norm" should be reasonable. The mean of feature y_k is obviously the proportion of unities in it, which is p_{k+} in notations related to the contingency table P. In fact, the remainder of this section can be considered as another reason for using $a_k = p_{k+}$. The choice of rescaling factors is somewhat less certain, though using all $b_k = 1$ should seem reasonable too because all the dummies are just 1/0 variables measured in the same scale. Incidentally, 1 is the range of any dummy. Some other values related to y_k's dispersion could be used as well. Especially suitable *is* $b_k = (p_{k+})^{1/2}$ which is the standard deviation of the Poisson distribution of a 1/0 variable (see Sect. 2.3.3), as will be seen in the end of this section. With the scale shift value specified, the standardized conditional probabilities in $\hat{Y}$ can be expressed as

$$p'(k/l) = \frac{p_{lk} - p_{k+}p_{+l}}{p_{+l}b_k} \tag{3.38}$$

Let us compute the sum of squares of all the $\hat{Y}$-elements (3.38). Within the k-th column, there are N_{+l} values p(k/l) (3.38), which leads us to the value

$$B_{lk} = N_{+l}p'(k/l)^2 = N\frac{(p_{lk} - p_{k+}p_{+l})^2}{b_k^2 p_{+l}} \tag{3.39}$$

as the contribution of (k,l)-pair to the sum of squares, $\langle \hat{Y}', \hat{Y}' \rangle$, where symbol "'" refers to the fact that the data matrix Y has been pre-standardized.

Accordingly, the total contribution of partition S to the total scatter of the set of standardized dummies representing partition T is equal to

$$B(S,T) = \left\langle \hat{Y}', \hat{Y}' \right\rangle = \sum_{l=1}^{L}\sum_{k=1}^{K}B_{lk}^2 = N\sum_{l=1}^{L}\sum_{k=1}^{K}\frac{(p_{lk} - p_{k+}p_{+l})^2}{b_k^2 p_{+l}} \tag{3.40}$$

The term "contribution" comes from the regression model under consideration:

$$Y' = P_X Y + E = \hat{Y}' + E,$$

which satisfies the Pythagorean decomposition property because of the orthogonality of the projector P_X:

$$\langle Y', Y' \rangle = \langle \widehat{Y}', \widehat{Y}' \rangle + \langle E, E \rangle,$$

so that $\langle \widehat{Y}', \widehat{Y}' \rangle$ is the contribution indeed.

It should be noted that the value $\langle E, E \rangle$ can be considered a measure of distance, Euclidean squared, as usual, between partitions T and S corresponding to the nominal features and their dummy matrices X and Y under consideration

$$\delta(X, Y) = \langle E, E \rangle = \|Y' - P_X Y'\|^2 \qquad (3.41)$$

This can be referred to as the dummy matrix regression distance.

The total contribution (3.40) reminds us of both the averaged relative Quetelet coefficient Q in (3.30), as well as the impurity function $\Delta(T, S)$ in (3.37) and the averaged absolute Quetelet coefficient in (3.35). The latter two indexes, up to the constant N of course, emerge at the rescaling factors being unity, $b_k = 1$. The former emerges when rescaling factors $b_k = \sqrt{p_{k+}}$. The square root of the frequency has an appropriate meaning—this is a good estimate of the standard deviation in Poisson model of the variable: according to this model, N_{k+} unities are thrown randomly into the fragment of memory assigned for the storage of vector y_k. In fact, at this scaling system, $B(T/S) = X^2$, the Pearson chi-squared!

Let us summarize the proven facts.

Statement 3.6.2.1 The impurity function in (3.37) can be equivalently expressed as

(a) The reduction of Gini uncertainty index of partition T when partition S is taken into account;
(b) The averaged absolute Quetelet index $a(l/k) = p_{kl}/p_{k+} - p_{+l}$ of the same effect;
(c) The total contribution of partition S to the summary data scatter of the set of dummy 1/0 features corresponding to classes of T and standardized by subtracting the mean with no rescaling.

Statement 3.6.2.2 The Pearson chi-squared association index can be equivalently expressed as

(a) A measure of statistical independence between partitions T and S;
(b) The averaged relative Quetelet index $q(l/k) = (p_{kl}/p_{k+} - p_{+l})/p_{+l}$ between partitions T and S;
(c) The total contribution of partition S to the summary data scatter of the set of dummy 1/0 features x_l corresponding to classes T_l and standardized by subtracting the mean and dividing the result by $b_l = \sqrt{p_l}$.

The claims of equivalence in Statements 3.6.2.1 and 3.6.2.2, although having been described by this author earlier (see, for example Mirkin 1996, 2012) remain virtually unknown, in spite of the fact that they point to useful relations between different association measures and, as well, between statistical association measures and preferred data normalization options.

3.6.2.2   Canonical Analysis of Dummy Matrices: Dual Scaling, "L'analyse Des Correspondences" and the Chi-Squared

Consider two nominal features over an entity set I of cardinality N represented by partitions $T = \{T_l\}$ and $S = \{S_k\}$. Define $N \times L$ dummy matrix X and $N \times K$ dummy matrix Y corresponding to partitions T and S, respectively.

Consider the linear subspaces $L(X)$ and $L(Y)$ spanning matrices X and Y. The problem of canonical correlation is to find in $L(X)$ and $L(Y)$ normed vectors x and y maximally oriented towards each other so that $\langle x, y \rangle$ is maximum with respect to $x \in L(X)$ and $y \in L(Y)$ such that $\langle x, x \rangle = a^T X^T X a = 1$ and $\langle y, y \rangle = b^T Y^T Y b = 1$. In fact the problem is of finding vectors a and b maximizing $a^T X^T Y b$ such that $a^T X^T X a = 1$ and $b^T Y^T Y b = 1$. Since matrices $X^T X$ and $Y^T Y$ are diagonal with diagonal elements equal to N_{+l} and N_{k+}, respectively, the normalizing constraints can be reformulated as

$$\sum_{l=1}^{L} N_{+l} b_l^2 = 1, \quad \sum_{k=1}^{K} N_{k+} a_k^2 = 1,$$

A mathematical solution to this problem is described in Sect. A.3.3 of Appendix. It is related to spectral analysis of the product $P_X P_Y$ where P_X and P_Y are $N \times N$ orthogonal projector matrices, defined as $P_X = X(X^T X)^{-1} X^T$, $P_Y = Y(Y^T Y)^{-1} Y^T$. The general (i,j)-th element of $P_X P_Y$ is the inner product of i-th row of P_X and j-th row/column of P_Y. The former consists of $1/N_{+l}$ for all the objects belonging to the same l-th category of T as i, and zeros at other objects. The latter consists of $1/N_{k+}$ for all the objects belonging to the same k-th category of S as j, and zeros at other objects. Non-zero values meet at objects belonging to both T_{+l} and S_{k+}, that is, $S_k \cap T_l$. Therefore the inner product is equal to $N_{kl}/(N_{k+} N_{+l})$.

The spectrum of matrix $P_X P_Y$ is the same as of matrix AB where $A = (X^T X)^{-1} X^T Y$, $B = (Y^T Y)^{-1} Y^T X$ as defined in Sect. A.3.3. The matrices A and B are of sizes $L \times K$ and $K \times L$, respectively, and their entries are conditional probabilities. It is easy to find out that (l,k)-th element of A is equal to N_{lk}/N_{+l}, and (k,l)-th element of B is equal to N_{kl}/N_{k+}. Here $N_{lk} = N_{kl}$ is the number of objects in the intersection $S_k \cap T_l$.

A version of the power method for finding the maximum eigenvalue of $P_X P_Y$ or of AB, which is the same, may be defined by using matrices A and B separately. The method begins with any K-dimensional $a = a(0)$ and finds L-dimensional $b' = Aa(0)$, which is then normalized by computing $\beta = (b'^T Y^T Y b')^{1/2}$ and taking $b(1) = b'/\beta$. Now next a-vector is computed as $a' = Bb(1)$ and then normalized similarly as $a(1) = a'/(a'^T X^T X a')^{1/2}$. Next iterations run similarly. The convergence is warranted if the maximal eigenvalue is greater than the other eigenvalues indeed. The maximum eigenvalue is computed as the product of the norms of a' and b'. The corresponding eigenvectors a and b can be considered as the best numerical codes for the categories; they are mutually oriented towards each other.

Moreover, the total of the canonical eigenvalues is related to the Pearson chi-squared coefficient between the two nominal variables. Since both spaces, $L(X)$ and $L(Y)$, contain the bisector subspace of all N-dimensional vectors with equal elements, this generates a trivial eigenvalue 1, which should not be taken into account when considering the total of canonical eigenvalues. As is well known, the sum of all eigenvalues of a square matrix is invariant under linear transformations of the space; it is equal to the sum of the diagonal elements. As was shown above the (i,i)-th element of matrix $P_X P_Y$ is equal to $N_{kl}/(N_{k+}N_{+l})$, where k and l are those indices for which $i \in T_l$ and $i \in S_k$. The number of objects i such that $i \in T_l$ and $i \in S_k$, is equal to N_{kl}. Therefore, the total of diagonal elements in matrix $P_X P_Y$ is equal to

$$\sum_{k=1}^{K}\sum_{l=1}^{L} N_{kl} \frac{N_{kl}}{N_{k+}N_{+l}} = \sum_{k=1}^{K}\sum_{l=1}^{L} \frac{N_{kl}^2}{N_{k+}N_{+l}} = \sum_{k=1}^{K}\sum_{l=1}^{L} \frac{p_{kl}}{p_{k+}p_{+l}} = X^2/N + 1,$$

according to Eq. (3.31). By subtracting the trivial eigenvalue 1, we conclude that the total of non-trivial canonical eigenvalues is equal to X^2/N.

The phenomenon of canonical correlation between nominal features has attracted considerable attention of researchers. In particular, there are two techniques of numerical analysis of nominal features based on the canonical correlation. One is referred to as dual scaling (see Nishisato 2014); that utilizes a version of the iterations according to the power method, as described above. Another, referred to as the correspondence analysis, builds on the simultaneous consideration of both spaces, $L(X)$ and $L(Y)$, and processes related to their interrelation (Benzecri 1992; Greenacre 2017; Lebart et al. 1995). Specifically, an attention is given to equations relating the canonical vectors, $a = \mu B b$ and $b = \mu A a$, the so-called transition formulas. Since elements of A and B are conditional probabilities $p(k/l) = p_{lk}/p_{+l}$, and $p(l/k) = p_{kl}/p_{k+}$, respectively, a and b appear to be averaged versions of each other (up to the singular value μ), which leads to a joint display of both S-categories and T-categories.

This author developed a symmetric version of the correspondence analysis which involves no dual spaces; this is described in the next Sect. 3.6.3.

3.6.3 Correspondence Analysis

3.6.3.1 Correspondence Analysis: Presentation

Correspondence Analysis is an extension of PCA to contingency tables taking into account the specifics of co-occurrence data: they are not only comparable across the table but also can be meaningfully summed together across the table. This leads to a unique way of standardization of such data—by using the Quetelet coefficients rather than the original frequencies, which is an advantage over the common situations in which the data standardization is rather arbitrary.

Correspondence Analysis (CA) is a method for visually displaying both row and column categories of a contingency table $P = (p_{ij})$, $i \in I$, $j \in J$, in such a way that distances between the presenting points reflect the patterns of co-occurrences in P. This method is usually introduced as a set of dual heuristics applied simultaneously to rows and columns of the contingency table (see, for example, Lebart et al. 1995; Greenacre 2017). Yet there is a way for introducing CA as an encoder-decoder based data recovery technique similar to that used for introducing PCA above. According to this perspective (Lebart and Mirkin 1993; Mirkin 1996), CA is a version of PCA differing from PCA due to the specifics of contingency data, in the following aspects:

(i) The CA method obtains hidden factors representing the relative Quetelet indexes rather than the original frequency data;

(ii) Both rows and columns are not equivalently contributing; each is assigned with a weight reflecting its frequency; the greater the frequency, the greater the weight. These weights are used in the approximation problem throughout; both in the least-squares criterion and the mutual orthogonality conditions;

(iii) Both rows and columns are visualized on the same display, thus referred to as a biplot, in such a way that the geometric distances between the representing points reflect the so-called chi-square distances between row (or column) conditional frequency profiles;

(iv) The data scatter is defined as the sum of squared entries weighted by the products of the row and column weights, that is equal to the Pearson chi-squared association coefficient.

Worked Example 3.13. Correspondence Analysis of a Theft/Age Contingency Table

Consider Table 3.26, cross-classifying cases of attempted theft from shops and supermarkets in the Netherlands 1979 from the book by Israëls 1987; see also Lombardo et al. (2016).

Two classification bases are: categories of stolen goods and age groups of perpetrators. There are 13 goods categories, from closing to household items to perfumes; see the list in the left column of Table 3.26. There are 9 age groups (less than 12 years old, 12–14 years old, 15–17, 18–20, 21–29, 30–39, 40–49, 50–64, 65 and over years old) coded accordingly in the first row of Table 3.26.

To apply the method to data in Table 3.26, we first transform that into Quetelet coefficients format (see Table 3.27).

This standardization does make the data structure somewhat. It suffices to mention pairs (tobacco, AGold) and (toys, AGmin) whose q-values exceed 200% increase from the average. But the transformation $p \Rightarrow q$ is not alone in CA. It is coupled with the weighting of each row and column by its corresponding marginal

Table 3.26 Cross-classification of theft cases over stolen goods and age groups

Category	AGmin	AG13	AG16	AG19	AG25	AG35	AG45	AG57	AGold	Total
Clothing	81	138	304	384	942	359	178	137	45	2568
Accessories	66	204	193	149	297	109	53	68	28	1167
Tobacco	150	340	229	151	313	136	121	171	145	1756
Stationery	667	1409	527	84	92	36	36	37	17	2905
Books	67	259	258	146	251	96	48	56	41	1222
Records	24	272	368	141	167	67	29	27	7	1102
Household	47	117	98	61	193	75	50	55	29	725
Candy	430	637	246	40	30	11	5	17	28	1444
Toys	743	684	116	13	16	16	6	3	8	1605
Jewelry	132	408	298	71	130	31	14	11	10	1105
Perfumes	32	57	61	52	111	54	41	50	28	486
Hobby	197	547	402	138	280	200	152	211	111	2238
Other	209	550	454	252	624	195	88	90	34	2496
Total	2845	5622	3554	1682	3446	1385	821	933	531	20,819

Table 3.27 Quetelet indexes for the goods/age contingency Table 3.26

Category	AGmin	AG13	AG16	AG19	AG25	AG35	AG45	AG57	AGold	Total
Clothing	-0.77	-0.80	-0.31	0.85	1.22	1.10	0.76	0.19	-0.31	0.1233
Accessories	-0.59	-0.35	-0.03	0.58	0.54	0.40	0.15	0.30	-0.06	0.0561
Tobacco	-0.37	-0.28	-0.24	0.06	0.08	0.16	0.75	1.17	2.24	0.0843
Stationery	0.68	0.80	0.06	-0.64	-0.81	-0.81	-0.69	-0.72	-0.77	0.1395
Books	-0.60	-0.22	0.24	0.48	0.24	0.18	0.00	0.02	0.32	0.0587
Records	-0.84	-0.09	0.96	0.58	-0.08	-0.09	-0.33	-0.45	-0.75	0.0529
Household	-0.53	-0.40	-0.21	0.04	0.61	0.56	0.75	0.69	0.57	0.0348
Candy	1.18	0.63	0.00	-0.66	-0.87	-0.89	-0.91	-0.74	-0.24	0.0694
Toys	2.39	0.58	-0.58	-0.90	-0.94	-0.85	-0.91	-0.96	-0.80	0.0771
Jewelry	-0.13	0.37	0.58	-0.20	-0.29	-0.58	-0.68	-0.78	-0.65	0.0531
Perfumes	-0.52	-0.57	-0.26	0.32	0.38	0.67	1.14	1.30	1.26	0.0233
Hobby	-0.36	-0.09	0.05	-0.24	-0.24	0.34	0.72	1.10	0.94	0.1075
Other	-0.39	-0.18	0.07	0.25	0.51	0.17	-0.11	-0.20	-0.47	0.1199
Total	0.1367	0.27	0.1707	0.0808	0.1655	0.0665	0.0394	0.0448	0.0255	1.0000

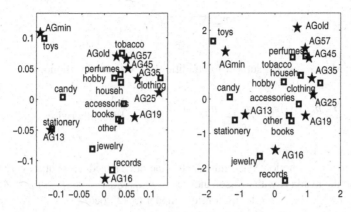

Fig. 3.26 Visualization of thefts/age contingency table in Table 3.26 using Correspondence Analysis, on the left, and PCA, on the right. Squares stand for good types and stars for thief categories

probability so that the squared errors in the criterion are weighted by products of the marginal probabilities. Moreover, the vector norm is weighted by them too.

Figure 3.26 represents a CA visualization of Table 3.26 derived as described above, on the left, and PCA, on the right. The visualizations do not differ that much, although the CA display gives a more clear picture. They clearly show which good types an age group tends to steal. AGmin is close to toys, whereas AG16 is near jewelry and records.

3.6.3.2   Correspondence Analysis: Formulation

Correspondence Analysis (CA) is a method for visually displaying both row and column categories of a contingency table $P = (p_{ij})$, $i \in I$, $j \in J$, in such a way that distances between the presenting points reflect the pattern of co-occurrences in P. To be specific, let us take on the issue of visualization of P on a 2D plane so that we are looking for just two approximating factors, $u_1 = (v_1, w_1)$ where $v_1 = (v_1(i))$ and $w_1 = (w_1(j))$ and $u_2 = (v_2, w_2)$ where $v_2 = (v_2(i))$ and $w_2 = (w_2(j))$, with $I \cup J$ as their domain, such that each row $i \in I$ is displayed as point $u(i) = (v_1(i), v_2(i))$ and each column $j \in J$ as point $u(j) = (w_1(j), w_2(j))$ on the plane as shown in Fig. 3.26.

The $|I|$-dimensional vectors v_t and $|J|$-dimensional vectors w_t constituting u_t ($t = 1, 2$) are calculated to approximate the relative Quetelet coefficients $q_{ij} = p_{ij}/(p_{i+}p_{+j}) - 1$ rather than the co-occurrences p_{ij} themselves, according to equations:

$$q_{ij} = \mu_1 v_1(i)w_1(j) + \mu_2 v_2(i)w_2(j) + e_{ij} \qquad (3.42)$$

where μ_1 and μ_2 are positive reals, by minimizing the weighted least-squares criterion

$$E^2 = \sum_{i \in I} \sum_{j \in J} p_{i+} p_{+j} e_{ij}^2 \tag{3.43}$$

with regard to μ_t, v_t, w_t, subject to conditions of weighted orthonormality:

$$\sum_{i \in I} p_{i+} v_t(i) v_{t'}(i) = \sum_{j \in J} p_{+j} w_t(j) w_{t'}(j) = \begin{cases} 1 & \text{if } t = t' \\ 0 & \text{otherwise} \end{cases} \tag{3.44}$$

where $t, t' = 1, 2$.

The weighted criterion E^2 is equivalent to the unweighted least-squares criterion L^2 applied to the matrix R with Pearson indexes $r_{ij} = q_{ij}(p_{i+}p_{+j})^{1/2} = (p_{ij} - p_{i+}p_{+j})/(p_{i+}p_{+j})^{1/2}$ as its entries. To be exact, let us consider model (3.42′) below:

$$r_{ij} = \alpha_1 f_1(i) g_1(j) + \alpha_2 f_2(i) g_2(j) + \varepsilon_{ij} \tag{3.42′}$$

and try minimize the sum of squared residuals $L^2 = \sum_{i,j} \varepsilon_{ij}^2$. According to the SVD theory, the solution to this problem is constituted by the two maximal singular values α_1 and α_2 of matrix $R = (r_{ij})$ and corresponding pairs of the normed singular vectors (f_1, g_1) and (f_2, g_2), respectively. Let us put $r_{ij} = q_{ij}(p_{i+}p_{+j})^{1/2}$ in (3.42′) and divide the equation by $(p_{i+}p_{+j})^{1/2}$, which leaves it invariant:

$$q_{ij} = \alpha_1 f_1(i) g_1(j)/(p_{i+}p_{+j})^{1/2} + \alpha_2 f_2(i) g_2(j)/(p_{i+}p_{+j})^{1/2} + \varepsilon_{ij}/(p_{i+}p_{+j})^{1/2}$$

This can be equivalently rewritten as:

$$q_{ij} = \alpha_1 v_1(i) w_1(j) + \alpha_2 v_2(i) w_2(j) + e_{ij} \tag{3.42″}$$

where $v_t(i) = f_{it}/(p_{i+}^{1/2})$, $w_t(j) = g_{jt}/(p_{+j}^{1/2})$, and $e_{ij} = \varepsilon_{ij}/(p_{i+}p_{+j})^{1/2}$. Equation (3.42″) is similar to Eq. (3.42). What is about criterion? One can see that $\varepsilon_{ij} = e_{ij}(p_{i+}p_{+j})^{1/2}$. Therefore, the $L^2 = E^2$. That means that the α_1 and α_2 are the maximal singular values of R, and the above definitions provide for the solutions of the problem in (3.42), (3.43), and (3.44).

Therefore, the factors v and w are determined by the singular-value decomposition of matrix $R = (r_{ij})$. More explicitly, the two maximal singular values μ_t and corresponding singular vectors $f_t = (f_{it})$ and $g_t = (g_{jt})$ of matrix R, defined by equations $R g_t = \mu_t f_t$, $R^T f_t = \mu_t g_t$ ($t = 1, 2$) determine the optimal values μ_t and optimal solutions to the problem of minimization of (3.43)–(3.44), according to equations $v_t(i) = f_{it}/(p_{i+}^{1/2})$ and $w_t(j) = g_{jt}/(p_{+j}^{1/2})$.

The singular triplet equations can be rewritten in terms of v_t and w_t, as follows:

$$\sum_{j \in J} \frac{p_{ij}}{p_{i+}} w_t(j) = \mu_t v_t(i), \qquad \sum_{i \in I} \frac{p_{ij}}{p_{+j}} v_t(i) = \mu_t w_t(j) \tag{3.45}$$

To prove the left-hand equation, take equation $Rg_t = \mu_t f_t$ in its component-wise form, $\sum_{j \in J} r_{ij} g_j = \mu f_i$ (index t omitted for the sake of convenience) and substitute by vectors v and w defined above: $\sum_{j \in J} r_{ij} \left(\frac{p_{+j}}{p_{i+}}\right)^{1/2} w(j) = \mu v(i)$. This is equivalent to $\sum_{j \in J} \left(\frac{p_{ij}}{p_{i+}} - p_{+j}\right) w(j) = \mu v(i)$. To complete the proof, equation $\sum_j p_{+j} w(j) = 0$ is to be proven. To do that, let us first prove that vector g_0 whose components are $p_{+j}^{1/2}$ is a singular vector of R corresponding to singular value 0 (the other component of the singular triplet is equal to $f_0 = (p_{i+}^{1/2})$). Indeed,

$$\sum_{j \in J} r_{ij} p_{+j}^{1/2} = (1/p_{i+}^{1/2}) \sum_{j \in J} (p_{ij} - p_{i+} p_{+j}) = (1/p_{i+}^{1/2})(p_{i+} - p_{i+}) = 0.$$

Then the equation $\sum_j p_{+j} w(j) = 0$ follows from the fact that all the singular vectors are mutually orthogonal so that singular vector g corresponding to w is orthogonal to g_0, which proves the statement. The right-hand equation can be proven in a similar way, from equation $R^T f_t = \mu_t g_t$.

Equations (3.45) are referred to as transition equations and considered to justify the joint display of rows and columns because the row-points $v_t(i)$ appear to be averaged column-points $w_t(j)$ and, vice versa, the column-points appear to be averaged versions of the row-points, up to the singular value of μ_t course.

The mutual location of the row-points is considered as justified by the fact that between-row-point squared Euclidean distances $d^2(u(i), u(i'))$ approximate the chi-square distances between corresponding rows of the contingency table. Specifically, chi-square distance is defined as a weighted squared Euclidean distance:

$$\chi^2(i, i') = \sum_{j \in J} p_{+j} (q_{ij} - q_{i'j})^2 = \sum_{j \in J} (p_{ij}/p_{i+} - p_{i'j}/p_{i'+})^2 / p_{+j}. \quad (3.46)$$

Here $u(i) = (v_1(i), v_2(i))$ for v_1 and v_2 rescaled in such a way that their norms are equal to μ_1 and μ_2, respectively. A similar property holds for columns j, j'. In fact, it is the right-hand item in (3.46) which is used to define the chi-squared distance between either columns or rows of a contingency table (Lebart et al. 1995), but the definition in terms of Quetelet coefficients in the middle of (3.46) (Mirkin 1996) looks more natural. The distance is dubbed chi-square distance because of its links to the chi-square coefficient for the table P. First of all, if we take the weighted chi-square summary distance to 0, $\sum_{i \in I} p_{i+} c^2(i, 0)$ where 0 is put instead of $q_{i'j}$ in (3.46), it is easy to see that this is the Pearson chi-squared coefficient, without the factor N of course, which is simultaneously the expression for the data scatter according to criterion E^2 in (3.43):

$$\sum_{i \in I} p_{i+} c^2(i, 0) = \sum_{i \in I} \sum_{j \in J} p_{i+} p_{+j} e_{ij}^2 = X^2/N \quad (3.47)$$

The weighted data scatter is equal to the scatter of R, the sum of its squared entries $T(R)$, which can be easily proven from the definition of R. Indeed, $T(R) = \sum_{i \in I} \sum_{j \in J} (p_{ij} - p_{i+}p_{+j})^2/(p_{i+}p_{+j}) = X^2/N$. This implies that

$$X^2/N = \mu_1^2 + \mu_2^2 + E^2 \tag{3.48}$$

which can be seen as a decomposition of the contingency data scatter, expressed by X^2, into contributions of the individual factors, μ_1^2 and μ_2^2, and unexplained residuals, E^2. (Only two factors are considered here, but the number of factors to be found can be raised up to the rank of matrix R with no other changes).

In a common situation, the first two singular values account for a major part of X^2, thus justifying the use of the plane of the first two factors for visualization of the interrelations between I and J.

3.6.3.3   Correspondence Analysis: Computation

Given a contingency table P, the computation of correspondence analysis factors can go in three steps: (a) computing Pearson index matrix R, (b) finding the singular decomposition of R and the two first correspondence analysis factors, and (c) visualization of the joint display of rows and columns of P. Here are MatLab commands for these.

(a) Computing Pearson index matrix R

```
≫Pc=sum(P); Pr=sum(P'); total=sum(Pc);
≫P=P/total; %relative frequencies
≫Pc=Pc/total; %column relative frequencies
≫Pr=Pr/total; %row relative frequencies
≫Prod=Pr'*Pc; % matrix of products
≫ rProd=Prod.^(0.5); % square roots of products
≫ r=(P-Prod)./rProd; % Pearson index matrix
```

(b) Finding the correspondence analysis factors:

```
≫[a,mu,b]=svd(r);
≫% finding first factor
≫x1=a(:,1)./sqrt(Pr');
≫y1 = b(:,1)./sqrt(Pc');
≫% finding second factor
≫x2= a(:,2)./sqrt(Pr');
≫y2= b(:,2)./sqrt(Pc');
```

As a bonus, one can estimate the proportion of data scatter, the chi-squared, taken into account by the factors, and display it on the screen:

```
≫yy=r.*r; chi=sum(sum(yy))% data scatter
```

```
≫ccn=(mu(1,1)^2+mu(2,2)^2)*100/chi;
%contribution of the first two
≫disp('Contribution of the solution:'); ccn
```

(c) Visualization of the joint display of rows and columns of P. The plot is easy to do with command

```
≫plot(x1,x2,'ks', y1,y2,'kp');
```

Yet to make the points annotated with row and column names, which are to be available in a string cell termed say 'names', the joint set of rows and columns should get their x-coordinate and y-coordinate vectors, z1 and z2 below:

```
≫z1=[x1' y1']; z2=[x2' y2']; text(z1,z2,names);
≫v=axis; axis(1.5*v);
```

The last line is to make the plot to look tighter by extending its boundaries.

3.6.4 Correlation Between Projection Matrices

Consider two nominal features over an entity set I of cardinality N represented by partitions $T = \{T_l\}$ and $S = \{S_k\}$. Define the $N \times L$ dummy matrix X and $N \times K$ dummy matrix Y corresponding to partitions T and S, respectively.

Consider the linear subspaces $L(X)$ and $L(Y)$ spanning matrices X and Y, as well as corresponding orthogonal projection matrices $P_X = X(X^T X)^{-1} X^T$ and $P_X = Y (Y^T Y)^{-1} Y^T$. Such a matrix expresses the orientation of the space $L(X)$ or $L(Y)$ in a concise way, like the normal vector a for a hyperplane defined by the equation $<a, x> = 0$. Recall that the spaces $L(X)$ or $L(Y)$ overlap over the unidimensional bisector line. To take this parasitic subspace out, one should subtract the parasitic subspace from the matrices—by simply subtracting $1/N$ from all the elements of the matrices, which transforms them to $\bar{P}_X$ and $\bar{P}_Y$.

The similarity between the corrected subspaces $L(X)^-$ and $L(Y)^-$ can be measured by the inner product, or even a correlation index, between $\bar{P}_X$ and $\bar{P}_Y$.

The inner product is equal to

$$
\langle \bar{P}_X, \bar{P}_Y \rangle = \sum_{k=1}^{K} \sum_{l=1}^{L} \sum_{i \in S_k} \sum_{j \in T_l} \left(\frac{1}{N_{k+}} - \frac{1}{N} \right) \left(\frac{1}{N_{+l}} - \frac{1}{N} \right)
$$
$$
= \sum_{k=1}^{K} \sum_{l=1}^{L} \sum_{i \in S_k} \sum_{j \in T_l} \left(\frac{1}{N_{k+}N_{+l}} - \frac{1}{NN_{+l}} - \frac{1}{NN_{k+}} + \frac{1}{N^2} \right).
$$

The latter expression is the result of multiplication of the expressions in the parentheses in the former one. Let us figure out what are the summation results for each of the four items. The item $1/(N_{k+}N_{+l})$ appears in those (i,j) at which $i \in S_k$ and $j \in S_l$. that is, at those (i,j) at which both $i,j \in S_k \cap S_1$. That implies that the sum of these items is equal to $\sum_{k=1}^{K} \sum_{l=1}^{L} \frac{N_{kl}^2}{N_{k+}N_{+l}}$. The second item does not depend on i, so that there are N of them to sum. Within any category T_l, there are N_l items leading to the summary value $-1/N$ within each $l = 1,2...,\ N$. Summing these over l leads to -1 as the total. Similarly, the total of summation of the third item is -1 as well. The item number four, $1/(N^2)$, summed N^2 times, over all i and all j, will produce 1 as the total. Altogether, the inner product of the projection matrices is equal to

$$\langle P_X^-, P_Y^- \rangle = \sum_{k=1}^{K} \sum_{l=1}^{L} \frac{N_{kl}^2}{N_{k+}N_{+l}} - 1 = Q. \tag{3.49}$$

That is, rather unexpectedly, the inner product of orthogonal projector's matrices is the average Quetelet index, or Pearson chi-squared related to N.

Q.3.26. Prove that the sum of elements of matrix P_X^- considered as an $N{\times}N$ vector is 0; so is the mean of P_X^-.

Q.3.27. Prove that the sum of squares of elements of matrix P_X^- considered as an $N{\times}N$ vector is $K - 1$ where K is the number of X-categories. **Hint**: Follow the way of the previous paragraph. Consider the sum of all the elements of P_X^- to be multiplied by themselves, in real, and see what are the results of summation of them over all $i,j = 1,...,\ N$.

Now we are ready to see what is the correlation coefficient between matrices P_X^- and P_Y^-:

$$\rho(P_X^-, P_Y^-) = \frac{Q}{\sqrt{(K-1)(L-1)}} = \frac{\sum_{k=1}^{K} \sum_{l=1}^{L} \frac{(N_{kl} - N_{k+}N_{+l})^2}{N_{k+}N_{+l}}}{\sqrt{(K-1)(L-1)}} \tag{3.50}$$

That is an interesting index. According to Kendall and Stewart (1967, Eq. (33.64)), that is the squared value of the so-called Tschuprow association coefficient. A. Tschuprow (1874–1926) proposed it in early 20th century based on the fact that the product $(K - 1)(L - 1)$ is the mean value for Pearson's chi-squared under the hypothesis that the features are statistically independent.

Q.3.28. Give an analytic expression for the quadratic distance between P_X^- and P_Y^-. **A.** That is $d(P_X^-, P_Y^-) = \langle P_X^-, P_X^- \rangle + \langle P_Y^-, P_Y^- \rangle - 2\langle P_X^-, P_Y^- \rangle = K - 1 + L - 1 - 2Q = 2((K + L)/2 - 1 - Q)$ where Q is the average Quetelet coefficient. In the case when $K = L$, this works fine. However, the greater the difference between L and K, the greater the minimum value of d differs from 0, which perhaps correctly reflects the mismatch between the two features.

3.7 Distance Between Relations Corresponding to Tied Rankings and Partitions

This material follows that by Mirkin and Fenner (2019). The topic of comparing rankings was initiated by Charles Spearman (1863−1945) (a junior collaborator of the founding fathers of multivariate statistics, Francis Galton and Karl Pearson), who was hired to further pursue the golden dream of Galton, a proof that human talent is inherited, mainly from one's parents and, partly, from even more distant ancestors. Although ranking is non-quantitative, Spearman proposed using ranks as numerical values, so that the Pearson correlation coefficient could be employed. This is straightforward when the observations being compared are linearly ordered. However, different observations can sometimes be assigned the same numerical rank value, which then led to the introduction of the term "tied observations", subsequently replaced by "tied rankings". Formally, a *tied ranking* can be represented as an ordered partition $R = (R_1, R_2, ..., R_p)$, that is, a partition whose parts are linearly ordered by their indices 1, 2, ..., p. We say that an element i *precedes* an element j in the ranking R if the part containing i precedes the part containing j. A clear-cut case of an ordered partition is given by the rank features in social surveys. A ranked feature asks respondents to classify alternatives using an ordered set of categories, such as "strongly agree", "agree", "neutral", "disagree", "strongly disagree". The term ranking is used here as a synonym of the ordered partition; when considering a series of objects with no ties, that is referred to as a strict ranking.

In 1938, a British statistician Maurice Kendall (1907–1983) introduced a different representation for rankings by using the relation of precedence between ranks rather than the ranks themselves. Given a tied ranking R, we define a square observation-to-observation matrix (the *Kendall matrix*) in which the (i, j) entry is +1 if i precedes j in R, 0 if i and j have the same rank, or −1 if j precedes i. The Kendall rank correlation coefficient between two tied rankings is the Pearson correlation coefficient between the corresponding Kendall matrices, considered as vectors in an $N \times N$-dimensional space.

In the 1950s, John Kemeny (1926–1992) approached the issue of comparing rankings from a social consensus perspective. Given a set of ordered partitions, a consensus ordered partition should represent the major tendency in the set. A conventional approach, the majority rule, may fail when determined by voting on pairs of alternatives.

Specifically, the so-called Condorcet paradox holds: if there are three parties at a meeting, each supporting one of three cyclically related linear orderings of three alternatives, say, (a) [i,j,k], (b) [j,k,i], and (c) [k,i,j], respectively, then the majority rule would lead to a cycle in the precedence relation: i would precede j because this is so for the majority, (a) and (c); similarly, j would precede k, and k would precede i. This contradicts the requirement that the precedence relation corresponding to the majority consensus ranking should be transitive. This paradox is a basis of the celebrated "social choice impossibility theorem" by an American economist Kenneth Arrow (1921–2017), a Nobel Prize winner.

John Kemeny proposed a different definition for consensus ranking using a distance measure between rankings. Rather than defining any specific distance measure ab initio, he formulated four axioms that should hold for any admissible distance measure. These axioms led Kemeny to derive the unique distance measure satisfying them. The *Kemeny distance* turned out to be the L_1-*distance* between the Kendall matrices [see (Kemeny and Snell 1962) for a convincing exposition].

Here, we are going to describe a joint geometric space of ordered and unordered partitions using the corresponding weak order and equivalence relations on the set of observations (as it is done in (Mirkin 1979; Mirkin and Fenner 2019).

a.   Weak orders and equivalence relations

Given a finite set I of N elements, a collection of its subsets $R = \{R_1, R_2, ..., R_p\}$ is referred to as a *partition* if the subsets R_s are all non-empty, non-overlapping, and cover the entire set I, so that each $i \in I$ belongs to a unique subset R_s, $1 \le s \le p$. The subsets are called the parts of the partition R. A partition is said to be *ordered* if there is a linear order relation of precedence between its parts, $R_s < R_t$, that is transitive, anti-reflexive and complete. If the order coincides with the natural order between indices 1, 2, ..., p, we use parentheses to denote this, viz. $R = (R_1, R_2, ..., R_p)$. In Decision Theory, an ordered partition is referred to as a *ranking*.

Each ordered partition $R = (R_1, R_2, ..., R_p)$ generates a *binary preference relation*

$$\rho = \{(i,j) : i \in R_s, j \in R_t, \text{and } s \le t\}. \tag{3.51}$$

Usually, two non-overlapping binary relations are defined with respect to a ranking $R = (R_1, R_2, ..., R_p)$: the *strict preference* relation $P = \{(i, j): i \in R_s, j \in R_t, \text{and } s < t\}$ and the *indifference* relation $E = \{(i, j): i, j \in R_s \text{ for some } s\}$. The indifference relation E here is transitive, reflexive and symmetric, thus E is the equivalence relation corresponding to the unordered partition $\check{R}$ having the same parts as R. Obviously, $\rho = P \cup E$, that is, ρ in (3.51) is a non-strict preference relation in which the strict preference and indifference relations are merged together. Usually, researchers try to avoid such a "mix"; but we will see later that there is no problem with this merger. The next part of this section is a brief reminder of some conventional concepts and facts about preference relations [see, for example, (Steele and Stefánsson 2015)].

If ρ is a binary relation, its inverse ρ^{-1} is defined as $\rho^{-1} = \{(i, j): (j, i) \in \rho\}$. If ρ is the preference relation corresponding to a tied ranking $R = (R_1, R_2, ..., R_p)$, then its inverse ρ^{-1} corresponds to the reverse tied ranking $R^{-1} = (R_p, ..., R_2, R_1)$. It is easy to see that the indifference relation E corresponding to any tied ranking R satisfies $E = \rho \cap \rho^{-1}$. Thus, the strict preference relation is the difference $P = \rho - E = \rho - \rho^{-1}$.

It is clear that ρ in (3.51) is

–   Reflexive, that is, $(i, i) \in \rho$ for any $i \in I$,

- Transitive, that is, if $(i, j) \in \rho$ and $(j, k) \in \rho$, then $(i, k) \in \rho$ for any i, j, k $\in A$, and
- Complete, that is, $(i, j) \in \rho$ or $(j, i) \in \rho$, or both, for any i, j $\in A$.

Of course, reflexivity can be considered as a special case of completeness for which i = j. A binary relation satisfying these properties is usually referred to as a *weak order*. In fact, a converse statement also holds: A preference relation ρ corresponds to an ordered partition **R** if and only if it is a weak order.

To prove that, assume ρ to be a binary relation on the set *I* that is reflexive, transitive and complete. Consider any i $\in I$ and define the subset $\rho(i) = \{j \in I: (i, j) \in \rho\}$. Then, for any pair i, k $\in I$, if $(i, k) \in \rho$ then $\rho(k) \subseteq \rho(i)$. This holds because whenever j $\in \rho(k)$, i.e. $(k, j) \in \rho$, then $(i, j) \in \rho$ also, because ρ is transitive. Therefore, since ρ is complete, for any pair i, k $\in I$, either $\rho(k) \subseteq \rho(i)$ or $\rho(i) \subseteq \rho(k)$, or both. It follows that the collection of sets $\rho(i)$ is linearly ordered by set-theoretic inclusion, so they can be ordered as a sequence of sets S_t for t = 1, 2, ..., p, where $S_1 \supset S_2 \supset ... \supset S_p$. Then the subsets $R_t = S_t - S_{t+1}$, t = 1, 2, ..., p-1, and $R_p = S_p$, form a ranking **R** = $(R_1, R_2, ..., R_p)$. It is quite easy to check that its corresponding preference relation (3.51) coincides with the given relation ρ. The reverse implication, that the relation (3.51) corresponding to an ordered partition is reflexive, transitive and complete, has already been established above. This completes the proof.

The subsets $R_t = S_t - S_{t+1}$ in the proof each satisfy $R_t = \rho(i) \cap \rho^{-1}(i)$ for some i $\in I$. This establishes that a binary relation ρ is a weak order if and only if its strict part *P* is anti-reflexive and transitive, its indifference part *E* is an equivalence relation, and P, P^{-1}, E form a partition of the Cartesian product $I \times I$.

b. Refinement and betweenness

A ranking **R'** is a *refinement* of a ranking **R** if it is obtained from the latter by subdividing some of its parts into smaller ones, and some ordering is defined between the smaller parts of each subdivided part of **R**. The corresponding preference relations, ρ' and ρ, are related by set-theoretic inclusion: A tied ranking **R'** is a refinement of a tied ranking **R** if and only if $\rho' \subset \rho$.

Indeed, if **R'** is a refinement of a tied ranking **R** then, for some pairs i, j of elements of A such that both $(i,j) \in \rho$ and $(j,i) \in \rho$, only one of these holds for ρ'. Conversely, suppose that ρ and ρ' correspond to tied rankings **R** and **R'**, respectively, and that $\rho' \subset \rho$. Then $\rho'(i) \subseteq \rho(i)$ for any i $\in I$, and, moreover, the inclusion is proper for some i $\in I$. Consider any such i. Let {i1, i2, ..., ik} be a maximal subset of *I* such that $\rho(i) \supset \rho'(i1) \supset \rho'(i2) \supset ... \supset \rho'(ik)$. Then, by the previous analysis, every equivalence class $R'_{tu} = \rho'(iu) \cap \rho'^{-1}(iu)$ will be part of the equivalence class $R_t = \rho(i) \cap \rho^{-1}(i)$, which completes the proof.

We say that ρ is *coarser* than ρ', if ρ' is a refinement of ρ.

A binary relation τ on *I* is said to be *between* binary relations ρ and ρ' if and only if $\rho \cap \rho' \subseteq \tau \subseteq \rho \cup \rho'$. A ranking **T** is said to be between tied rankings **R** and **R'** if, for any i, j $\in I$, the ordering between them in **T** is compatible with their ordering in both R and R': that is, (i) if i precedes j in both R and R' then i precedes j in T; (ii) if i precedes j in one of R and R', and i and j are indifferent in the other, then i either precedes j or is indifferent to j in T; (iii) if i and j are indifferent in both R and R', then i and j are indifferent in T; lastly, (iv) if i precedes j in R but j precedes i in R', then anything can be true of the ordering between i and j in T: i may precede j, or j may precede i, or i and j may be indifferent in T (Kemeny and Snell 1962). It is easy to prove that T is between R and R' if and only if the same is true for their weak orders.

In the general case of two arbitrary tied rankings R and R', the relation $\rho \cap \rho'$ is a partial preference relation because there can be i, j $\in I$ such that i strictly precedes j in R, whereas j strictly precedes i in R', so that neither (i, j) nor (j, i) belongs to $\rho \cap \rho'$.

Such a case, which is not uncommon, is exemplified by a proverbial question: "What is better: being poor but healthy or being rich but ill?" (with a proverbial answer that to be both rich and healthy is better indeed.)

What is appealing about $\rho \cap \rho'$ is that its indifference relation is always an equivalence relation, thus corresponding to the partition that is just the intersection of the unordered partitions $\check{R}$ and $\check{R}'$ that correspond to the ordered partitions R and R', respectively. The intersection $\check{R} \cap \check{R}'$ is the partition of I in which the parts are the intersections $R_s \cap R_t'$ of some part R_s of R and some part R_t' of R' for which R_s and R_t' are not disjoint.

Both ordered and unordered intersections can be visualized as a block matrix in which the blocks are formed by the subsets of rows and columns corresponding to the parts of the ordered partitions R' and R, respectively (see Fig. 3.27). Of course, the blocks of the intersections are only partially ordered so that, for example, blocks $R_2' \cap R_3$ and $R_3' \cap R_2$ are not comparable. However, a linear order can be imposed naturally by ordering the blocks first by rows and then by columns, so that any block of the first row precedes the blocks in all other rows. This is the so-called lexicographic product $R' * R$ introduced in (Mirkin 1979). Similarly, an alternative

	R_1	R_2	R_3	R_4	R_5
R_1'	$R_1' \cap R_1$	$R_1' \cap R_2$	$R_1' \cap R_3$	$R_1' \cap R_4$	$R_1' \cap R_5$
R_2'	$R_2' \cap R_1$	$R_2' \cap R_2$	$R_2' \cap R_3$	$R_2' \cap R_4$	$R_2' \cap R_5$
R_3'	$R_3' \cap R_1$	$R_3' \cap R_2$	$R_3' \cap R_3$	$R_3' \cap R_4$	$R_3' \cap R_5$
R_4'	$R_4' \cap R_1$	$R_4' \cap R_2$	$R_4' \cap R_3$	$R_4' \cap R_4$	$R_4' \cap R_5$

Fig. 3.27 A visual representation of the intersection of two rankings $R' \cap R$, where R' relates to rows and R to columns. It is assumed that the rows and columns are permuted according to the rankings R' and R, respectively

lexicographic product $R * R'$ is defined by ordering blocks first by columns and then by rows. Curiously, in the ordered series R', $R' * R$, $R * R'$ and R, the middle term of each triplet is between the other two (Mirkin 1979). A similar statement holds for the corresponding relations ρ', $\rho' * \rho$, $\rho * \rho'$ and ρ.

Q.3.29. Consider two rankings on an 8-element set, $R' = (1\text{–}2\text{–}3\text{–}4, 5\text{–}6\text{–}7\text{–}8)$ and $R = (1\text{–}4\text{–}5, 2\text{–}3\text{–}7, 6\text{–}8)$. Here symbol – joins elements of the same part. Prove that $E' \cap E = \{1\text{–}4, 2\text{–}3, 5, 7, 6\text{–}8\}$ and give examples of rankings between R' and R. **A.** For example, any ordering of $E' \cap E$ which is compatible with both R and R' as well as its further aggregations, say, $S = (1\text{-}4, 2\text{-}3, 5, 6\text{-}8,7)$ and $T = (1\text{-}4\text{-}2\text{-}3,5\text{-}6\text{-}8,7)$.

c. Correlation by Spearman and Kendall

Consider the Spearman rank correlation, that is, the Pearson correlation coefficient between ranks taken as numerical values. To deal with the case of tied rankings, each element of an equivalence class of the indifference relation is assigned with the average within-class rank. The average rank of the elements in part R_s of the tied ranking $R = (R_1, R_2, \ldots, R_p)$ is $L + (|R_s| + 1)/2$, where L is the cardinality of $R_1 \cup R_2 \cup \ldots \cup R_{s-1}$, and $|\cdot|$ denotes the number of elements in a set. The *Kendall rank correlation* is based on the representation of tied rankings on I by N×N matrices. Given a tied ranking R and the corresponding preference relation $\rho = P \cup E$, we now define a skew-symmetric matrix $\mathbf{K} = (k_{ij})$, for i, j $\in I$, such that $k_{ij} = 1$ if (i, j) $\in P$, $k_{ij} = 0$ if (i, j) $\in E$, and $k_{ij} = -1$, if (j, i) $\in P$. The Kendall rank correlation coefficient between R and R' is the correlation coefficient between their Kendall matrices, $\mathbf{K}$ and $\mathbf{K}'$, considered as vectors in an N^2-dimensional space. This is compatible with the non-quantitative nature of tied rankings, especially since the mean of a skew-symmetric matrix is always 0.

It should be noted that, soon after the Kendall matrix was defined, a somewhat similar skew-symmetric representation for quantitative features was proposed by Daniels (1944), who proved that, given a quantitative feature x on I, the matrix $\mathbf{X} = (x_{ij})$, where $x_{ij} = x_i - x_j$, can be used to represent the feature in statistical computations. For example, the inner product of the matrices $\mathbf{X}$ and $\mathbf{X}'$ corresponding to features $\mathbf{x}$ and $\mathbf{x}'$ is proportional to the inner product of $\mathbf{x}$ and $\mathbf{x}'$ after they have been centered by subtracting their means, viz. $\langle \mathbf{X}, \mathbf{X}' \rangle = 2N \langle \mathbf{x} - m(\mathbf{x}), \mathbf{x}' - m(\mathbf{x}') \rangle$, where m(x) is the mean of $\mathbf{x}$. This implies that the correlation coefficient between $\mathbf{X}$ and $\mathbf{X}'$ is equal to the correlation coefficient between $\mathbf{x}$ and $\mathbf{x}'$. Therefore, the Spearman rank correlation coefficient can also be defined as the Pearson correlation coefficient between the corresponding matrices of rank differences (r_{ij}), where $r_{ij} = r_i - r_j$. The Kendall matrix is then just the matrix of signs in the Daniels matrix $\mathbf{X} = (x_{ij})$, where $x_{ij} = x_i - x_j$.

d. **Kemeny distance**

Rather than defining an ad hoc distance measure, Kemeny formulated four axioms that should hold for any acceptable distance measure $d(R, R')$ between rankings R and R'. These axioms require that the acceptable distance measures should:

A1. Be mathematical metrics, that is, have the following properties:

   (a) Symmetry: $d(R, R') = d(R', R)$;
   (b) Non-negativity and definiteness: $d(R, R') \geq 0$ and $d(R, R') = 0$ if and only if $R = R'$;
   (c) Strict triangle inequality: for any rankings R, R' and R'', $d(R, R'') \leq d(R, R') + (R', R'')$; moreover, equality holds if and only if R' is between R and R''.

A2. If R' is obtained from R by a permutation of the set I and S' from S by the same permutation, then $d(R', S') = d(R, S)$.

To formulate the next axiom, let us say that a subset $B \subset I$ is a *segment* of a tied ranking R if its complement $I - B \neq \varnothing$ and each element i $\in I - B$ either precedes all the elements of B or is situated after all the elements of B. The tied ranking R restricted to a segment B will be denoted by R_B.

A3. If R and R' coincide on $I - B$ and B is a segment of both R and R', then $d(R, R') = d(R_B, R_B')$.

A4. Unit of scale: The minimum positive distance is equal to 1.

Kemeny proved that the only distance satisfying all four axioms is the L_1-metric between the corresponding skew-symmetric Kendall matrices divided by 2 (Kemeny 1959), namely:

$$\mathrm{kd}(R, R') = \frac{1}{2} \sum_{i,j \in I} \left| k_{ij} - k'_{ij} \right| \tag{3.52}$$

We see from (3.52) that the pairs of elements (i, j) in I can be divided into three subsets:

(a) those contributing 1 to $\mathrm{kd}(R, R')$: pairs (i, j) such that i precedes j in either R or R' while j precedes i in the other;
(b) those contributing ½ to $\mathrm{kd}(R, R')$: pairs (i, j) such that i and j are indifferent in either R or R' whilst one precedes the other in the other ranking;
(c) those contributing 0 to $\mathrm{kd}(R, R')$: pairs (i, j), that are similarly related in both rankings—either i precedes j, or j precedes i, or i and j are indifferent.

e. The mismatch distance between binary relations and the corresponding binary matrices

Binary relations considered as subsets of the Cartesian product $I \times I$ may be compared using any of the many measures of dissimilarity between subsets that have been introduced over the years (Morlini and Zani 2012). One particularly simple measure is the number of pairs for which they differ, the so-called mismatch distance, i.e., the number of pairs in their symmetric difference:

$$d(\rho, \rho') = |(\rho - \rho') \cup (\rho' - \rho)| \qquad (3.53)$$

The mismatch distance between unordered partitions was described in earlier publications by B. Mirkin in Russian from 1969 onwards [see, for example, (Mirkin and Cherny 1972)]; it is sometimes referred to as Mirkin's distance (Meilă 2007).

We note that $d(\rho, \rho')$ is a metric on the space of all binary relations on I and satisfies Axiom A1, including the strict triangle inequality, even for binary relations that do not correspond to tied rankings.

The mismatch distance can easily be translated into a distance between $N \times N$ matrices. Given a binary relation $\rho \in I \times I$, we define its binary matrix $\mathbf{r} = (r_{ij})$ by:

$$r_{ij} = \begin{cases} 1 & \text{if } (i,j) \in \rho \\ 0 & \text{if } (i,j) \notin \rho \end{cases}$$

Then the mismatch distance between R and R' is the mismatch (Hamming) distance between the corresponding binary relations ρ and ρ', and is thus given by:

$$d(R, R') = d(\rho, \rho') = |(\rho - \rho') \cup (\rho' - \rho)|$$
$$= \sum_{i,j \in I} \left| r_{ij} - r'_{ij} \right| = \sum_{i,j \in I} \left(r_{ij} - r'_{ij} \right)^2 \qquad (3.54)$$

The right-hand equality allows the original L_1-distance to be transformed into the square of the more conventional, Euclidean or L_2-distance because the absolute differences are either 1 or 0.

Obviously, the mismatch distance (3.54) is much simpler than the Kemeny distance (3.52) because the only possible non-zero contribution to $d(R, R')$ by an ordered pair (i, j) is 1, and this only occurs when j precedes i in one of the tied rankings but not in the other ranking. This happens when $r_{ij} = 0$ and $r_{ii}' = 1$ or, vice versa, $r_{ij}' = 0$ and $r_{ij} = 1$. It may therefore be somewhat of a surprise that these two distance measures are, in fact, equal, that is, the Kemeny distance (3.52) is equal to the mismatch distance (3.54).

To prove that, let us first analyse the contributions of pairs of elements $i, j \in I$ to the Kemeny distance between R and R' depending on their relative positions in the rankings R and R'; the various cases are shown in Table 3.28. We note that the contribution of the pair (j, i) is exactly the same as that of the pair (i, j).

Table 3.28 The contribution of a pair $(i, j) \in I \times I$ to the Kemeny distance (3.52) between R and R'

	Cases	R		
		i precedes j	i and j are indifferent	j precedes i
R'	i precedes j	0	½	1
	i and j are indifferent	½	0	½
	j precedes i	1	½	0

Now we need to take into account a subtle difference between the concepts of ranking and preference relation. The Kemeny distance is between two rankings—it records disagreements in the relative positions between a pair of elements in the two rankings; the symmetry between i and j accounts for the factor ½ in the expression (3.52) for the Kemeny distance.

In contrast, the mismatch distance is between binary relations and counts the disagreements between the relations in respect of ordered pairs of elements. We, therefore, must distinguish between the ordered pair (i, j) and the inverse pair (j, i), relative to the corresponding relation, ρ or ρ'. The various cases of the contributions to the mismatch distance are shown in Tables 3.29 and 3.30, respectively.

Returning to the analysis of the interrelation between two elements i, j $\in I$, we need to combine Tables 3.29 and 3.30 by summing them, which produces Table 3.31.

Table 3.29 The contribution of the ordered pair $(i, j) \in I \times I$ to the mismatch distance (3.54) between R and R'

	Cases	R		
		i precedes j	i, j are indifferent	j precedes i
R'	i precedes j	0	0	1
	i and j are indifferent	0	0	1
	j precedes i	1	1	0

Table 3.30 The contribution of the ordered pair $(j, i) \in I \times I$ to the mismatch distance (3.54) between R and R'

	Cases	R		
		i precedes j	i, j are indifferent	j precedes i
R'	i precedes j	0	1	1
	i and j are indifferent	1	0	0
	j precedes i	1	0	0

Table 3.31 Summary contribution of the ordered pairs (i, j) and (j, i) to the mismatch distance (3.54)

		R		
	Cases	i precedes j	i, j are indifferent	j precedes i
R'	i precedes j	0	1	2
	i and j are indifferent	1	0	1
	j precedes i	2	1	0

If we double the values in Table 3.28 to account for both ordered pairs (i, j) and (j, i), we observe that the resulting entries are identical to those in Table 3.31, which completes the proof.

Consider a simple example where I consists of three elements, 1, 2, and 3 that are linearly ordered in R and all tied in R', so that $R = (\{1\}, \{2\}, \{3\})$ and $R' = (\{1, 2, 3\})$. Their respective Kendall matrices are

$$\mathbf{k} = \begin{matrix} 0 & 1 & 1 \\ -1 & 0 & 1 \\ -1 & -1 & 0 \end{matrix} \quad \text{and} \quad \mathbf{k'} = \begin{matrix} 0 & 0 & 0 \\ 0 & 0 & 0 \\ 0 & 0 & 0 \end{matrix},$$

so that the Kemeny distance $kd(R, R') = 6/2 = 3$.

On the other hand, their respective weak order matrices are

$$\mathbf{r} = \begin{matrix} 1 & 1 & 1 \\ 0 & 1 & 1 \\ 0 & 0 & 1 \end{matrix} \quad \text{and} \quad \mathbf{r'} = \begin{matrix} 1 & 1 & 1 \\ 1 & 1 & 1 \\ 1 & 1 & 1 \end{matrix},$$

so that the mismatch distance $d(R, R') = 3$ as well.

Q.3.30. Consider three rankings on a 7-element set $I = \{1,2,3,4,5,6,7\}$: R1 = (1–2–3, 4–5, 6–7), R2 = (2–3–6, 4, 1–5–7), and R3 = (1–2, 3–4, 5–6–7). Build their relation and Kendall matrices, and compute mismatch and Kemeny distances between all the three

f. The mismatch distance expressed in terms of the contingency table

Although the following results can be established directly, we now rely on Axiom A1(c), which states that $d(R, R') = d(R, R'') + d(R'', R')$ if and only if R'' is between R and R', that is $\rho \cap \rho' \subseteq \rho'' \subseteq \rho \cup \rho'$ for the corresponding preference relations. By virtue of (3.5.4), we may use $d(R, R')$ and $d(\rho, \rho')$ interchangeably.

Let us first consider a tied ranking R and its reverse R^{-1}. Obviously, $\rho \cap \rho^{-1} = E$, where E is the indifference relation of R, which is an equivalence relation, stripped of all ranking information. Similarly, $\rho \cup \rho^{-1} = U$, the universal relation $U = I \times I$, which contains all possible ordered pairs of elements of I. Both E and U are, therefore, between ρ and ρ^{-1} for any weak order ρ.

Let N_s be the number of elements in part R_s of the tied ranking $R = (R_1, R_2, ..., R_p)$. Then the mismatch distance between U and E is easily seen to be

$$d(E, U) = N^2 - \Sigma_s N_s^2, \qquad (3.55)$$

since the first term on the right is the number of ones in the binary matrix of U and the second term is the number of ones in the binary matrix of E.

Curiously, the mismatch distance between U and R itself is exactly half the distance in (3.55). That is, the mismatch distance between a tied ranking R and the universal relation U is given by

$$d(R, U) = 1/2 \left(N^2 - \Sigma_s N_s^2 \right). \qquad (3.56)$$

To prove that, we first notice that $d(R, U) = d(R^{-1}, U)$ and $d(R, E) = d(R^{-1}, E)$. Indeed, neither U nor E depend on the ranking information in R, and, moreover, the number of pairs in ρ and ρ^{-1}, which is the number of ones in their respective matrices r and r^{-1}, is the same. Since both U and E are between R and R^{-1}, we have: $d(R, R^{-1}) = d(R, U) + d(U, R^{-1}) = 2d(R, U)$ and $d(R, R^{-1}) = d(R, E) + d(E, R^{-1}) = 2d(R, E)$.

This implies that $d(R, U) = d(R, E)$. So, since R is between E and U, $d(E, U) = d(E, R) + d(R, U) = 2d(R, U)$. Equation (3.56) now follows from this and (3.55), which completes the proof.

We also have proved that the distance $d(R, R^{-1})$ is equal to $d(E, U)$ given by (3.55), whereas the distance $d(R, E)$ is equal to $d(R, U)$, given by (3.56).

Now we are in a position to prove a formula for the mismatch distance between a ranking R and its arbitrary refinement R'. Like the previous results in this sub-section, this does not depend on the ranking information.

Specifically, the mismatch distance between a ranking $R = (R_1, R_2, ..., R_p)$ and its arbitrary refinement $R' = (R_1', R_2', ..., R_q')$, where q > p, is given by

$$d(R, R') = \left(\Sigma_s N_s^2 - \Sigma_t N_t'^2 \right), \qquad (3.57)$$

where N_s and N_t' are the numbers of elements in the parts R_s of R and R_t' of R', respectively.

Indeed, since R is between R' and U, we have $d(R', U) = d(R', R) + d(R, U)$, so $d(R', R) = d(R', U) - d(R, U)$. Both distances $d(R', U)$ and $d(R, U)$ are determined by Eq. (3.56), adjusted for the corresponding parts of R' and R, respectively. This immediately yields (3.57), completing the proof.

Consider now two ordered partitions, R and R', and their lexicographic products $R * R'$ and $R'* R$. We shall show that the entire ranking component contributing to the distance between R and R' is accounted for by the distance between $R * R'$ and $R' * R$. First, consider the intersection $R \cap R'$, as presented in Fig. 3.27.

Letting $N_{st} = |R_s \cap R_t'|$, for s = 1, 2, ..., p and t = 1, 2, ..., q, denote the numbers of elements in the parts of the intersection, we can present these cardinalities as a contingency table, or cross-classification, between R and R'.

The distance between $R * R'$ and $R' * R$ is equal to half of the total of the products of the cardinalities of those parts in the intersection $R \cap R'$ for which the orderings in R and R' are contradictory:

$$d(R * R', R' * R) = \frac{1}{2}\Sigma_{s > s'}\Sigma_{t > t'}N_{st}N_{s't'} \tag{3.58}$$

Considering the rankings R and R' as unordered partitions, denoted above by $\check{R}$ and $\check{R}'$, respectively, the mismatch distance between the corresponding equivalence relations, E and E', can be expressed as

$$d(E, E') = \Sigma_s N_s^2 + \Sigma_t N_t'^2 - \Sigma_{s,t}N_{st}^2 \tag{3.59}$$

where N_s, N_t', and N_{st} are, as above, the numbers of elements in parts R_s of R, R_t' of R' and $R_s \cap R_t'$ of $R \cap R'$, respectively. The mismatch distance between tied rankings R and R' can be decomposed into ranking and equivalence parts as follows:

$$d(R, R') = \frac{1}{2}d(E, E') + d(R * R', R' * R). \tag{3.60}$$

To prove that, consider the corresponding binary relations ρ, ρ', and $\rho \cap \rho'$. Since the intersection $\rho \cap \rho'$ is between ρ and ρ', $d(\rho, \rho') = d(\rho, \rho \cap \rho') + d(\rho \cap \rho', \rho')$. On the other hand, $\rho * \rho'$ is between $\rho \cap \rho'$ and ρ, and $\rho' * \rho$ is between $\rho \cap \rho'$ and ρ', so $d(\rho, \rho \cap \rho') = d(\rho, \rho * \rho') + d(\rho * \rho', \rho \cap \rho')$ and $d(\rho \cap \rho', \rho') = d(\rho \cap \rho', \rho' * \rho) + d(\rho' * \rho, \rho')$. But $\rho \cap \rho'$ is between $\rho * \rho'$ and $\rho' * \rho$, so $d(\rho * \rho', \rho' * \rho) = d(\rho * \rho', \rho \cap \rho') + d(\rho \cap \rho', \rho' * \rho)$. Substituting these in the equation $d(\rho, \rho') = d(\rho, \rho \cap \rho') + d(\rho \cap \rho', \rho')$, we obtain

$$d(\rho, \rho') = d(\rho, \rho * \rho') + d(\rho * \rho', \rho' * \rho) + d(\rho' * \rho, \rho').$$

Since $\rho * \rho'$ is a refinement of ρ, and $\rho' * \rho$ is a refinement of ρ', $d(\rho, \rho * \rho') = 1/2(\Sigma_s N_s^2 - \Sigma_{s,t}N_{st}^2)$ and $d(\rho' * \rho, \rho') = 1/2(\Sigma_t N_t'^2 - \Sigma_{s,t}N_{st}^2)$. This implies, by (3.59), that $d(\rho, \rho * \rho') + d(\rho' * \rho, \rho') = \frac{1}{2}d(E, E')$. Together with the equation above, this completes the proof.

This section can be looked at as an attempt to find some structure in the -1 entries in the Kendall matrices occurring in the formula for the Kemeny distance between tied rankings. These entries appear whenever a pair of elements, i and j, are inversely related. First, we showed that the Kemeny distance can be expressed in terms of the mismatch distance between the preference relations (weak orders) corresponding to the rankings, in which no negative entries appear. The mismatch distance can be properly defined in terms of the non-negative 0–1 matrices of weak orders, rather than the Kendall matrices of the rankings, containing entries 1, 0 and -1.

Q.3.31. What is the meaning of the mismatch distance between partitions?
A. This is the probability of two random objects to belong to a same part in one partition and to different parts in the other partition.
Q.3.32. Frequently, when dealing with partitions, pairs of objects are considered as subsets rather than elements of the cartesian square of the set of objects. Then the number of pairs in set S is not $|S|^2$, but rather what is called the binomial coefficient $|S|(|S|-1)/2$, equal to the number of two-element subsets of S. Reformulate the expression for the mismatch distance for this case.

3.8   Decision Trees

3.8.1   General

Decision tree is a structure used for learning and predicting quantitative or nominal target features. In the former case it is referred to as a regression tree, in the latter, classification tree. This structure can be considered a multivariate extension of contingency tables in such a way that only meaningful combinations of feature categories are involved.

As illustrated on Fig. 3.28, a decision tree recursively partitions the entity set into smaller clusters by splitting a parental cluster over a single feature. The root of a decision tree corresponds to the entire entity set. Each node corresponds to a subset of entities, cluster, and its children are the cluster's parts defined by values of a single predictor feature x. Note that the trees on Fig. 3.28 are binary: each interior node is split in two parts. This is a most convenient format, currently used in most popular programs. Only binary trees are considered in this section.

Decision trees are built from top to bottom in such a way that every split is made to maximize the homogeneity of the resulting subsets with respect to a desired target feature. The splitting stops either when the homogeneity is enough for a reliable prediction of the target feature values or when the set of entities is too small to consider its splits reliable. A function scoring the extent of homogeneity to decide of the stopping is, basically, a measure of correlation between the partition of the entity set being built and the target feature.

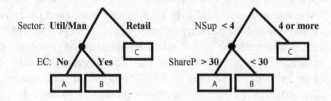

Fig. 3.28 Decision trees for three product based classes of Companies, A, B, and C, made using categorical features, on the left, and quantitative features, on the right

When the process of building a tree is completed, each terminal node is assigned with a value of the target that is determined to be characteristic for that node, and thus should be predicted at the conditions leading to the node. For example, both trees on Fig. 3.28 are precise—each terminal class corresponds to one and only one product, which is the target feature, so that each of the trees give a precise conceptual description of all products by conjunctions of the corresponding branch values. For example, product A can be described as that which is not in Utility sector, nor E commerce utilized in the production process (left-side tree) or as that in which less than 4 suppliers are involved and the share price is greater than 30. Both descriptions are fitting here since both give no errors at all.

Decision trees are very popular because they are simple to understand, use, and interpret. However, one should use them properly, because the decision rules produced with them can be overly simplistic and frequently imprecise. Their effectiveness much depends on the features and samples selected for the analysis. As always in learning correlation, a simpler tree is preferred to a complex one because of the over-fitting problem: a complex tree is more likely to reflect noise in the data rather than the true tendencies.

In the Sect. 3.6 we have described popular association indexes between partitions. Many of them are used as homogeneity scoring functions (to be) utilized in the process of classification tree building. This will be described next.

To build a binary decision tree, one needs the following information:

(a) a dataset of input features X,
(b) an output feature u over the set of objects,
(c) a scoring function $W(S,u)$ that scores admissible partitions S against the output feature,
(d) a rule for splitting a subset of objects, corresponding to the terminal node under consideration, in more homogeneous clusters,
(e) split-stopping criterion
(f) rule for pruning long or unreliable branches, and
(g) rule for the assignment of u-values to terminal nodes.

Let us comment on each of these items:

(a) The input features are, typically, quantitative or nominal. Quantitative features are handled rather easily by testing all possible splits of their ranges. More problematic are categorical features, especially those with many categories because the number of possible binary splits can be very large. However, this issue does not emerge at all if categorical features are preprocessed into the quantitative format of binary dummy variables corresponding to individual categories (which is advocated in this text too, see more detail in Sect. 2.3.2). Indeed, each of the dummy variables admits only one split—that separating the corresponding category from the rest, which reduces the number of possible splits to the number of categories—an approach advocated by Loh and Shih (1997), among many others. A number of such splits can be done in sequence to warrant that any combination of categories is admissible in this approach too.

Since this approach involves one feature at a time only, missing values are not of an issue here, because all the relevant information such as means and frequencies can be reasonably well estimated from those values that are available —this is a stark contrast with the other multivariate techniques.

(b) In principle, the decision tree format does not prevent from using multiple target features—just single-target criteria should be summed when there are several targets; this approach was successfully applied to sociology survey mixed scale data by P. Rostovtsev and B. Mirkin back in seventies (Mirkin 1985). However, all current internationally available programs involve only single target feature. Depending on the scale of the target feature, the learning task may differ, as well as the terminology. Specifically, if the target feature is quantitative, a decision tree is referred to as a regression tree, and if the target feature is categorical, a decision tree is referred to as a classification tree. Yet classification trees may differ with regards to the learning task: (a) learning a whole partition, if the target is nominal, or (b) learning just a category. This section focuses only on the task of learning a classification tree with a partitional target.

(c) Given a decision tree, its terminal nodes (leaves) form a partition S, which is considered then against the target feature u with a scoring function measuring the overall correlation $W(S,u)$. This suggests a context of the analysis of correlation between two features. If the target u is quantitative, then a tabular regression of u over S should be analyzed and scored. This approach involving the concept of the correlation ratio, as a natural scoring function, is described in the next section.

Unfortunately, in the data mining literature, this natural approach is not appreciated; thus, the correlation ratio is not popular. In contrast, at a categorical target, two most popular scoring functions, Gini index and Pearson chi-squared, fit perfectly in the framework of contingency tables and Quetelet indexes as described in Sect. 3.6.2. Moreover, it is mathematically proven in that section that these two can be considered as implementations of the same approach of maximizing the contribution to the data scatter of the target categories—the only difference being the way the dummy variables representing the categories are normalized: (i) no normalization to make it Gini index or (ii) normalization by Poissonian standard deviations so that less frequent categories get more important, to make it Pearson chi-squared. This sheds a fresh light on the criteria and suggests the user a way for choosing between the two indexes depending on user's preferences over the importance of being rare.

(d) Admissible partitions conventionally are obtained by splitting the entity subset corresponding to one of the current terminal nodes over one of the features. To make it less arbitrary, most modern programs do only binary splits. That means that any node may be split only in two parts: (i) that corresponding to a category and the rest, for a categorical feature or (ii) given an a, those "less than *a*" and those "greater than *a*", for a quantitative feature. This text attends to this approach as well. All possible splits are tested and that split which leads to the largest value of the criterion is actually made, after which the process is reiterated.

(e) Stopping rule typically assumes a degree of homogeneity of sets of entities, that is, clusters, corresponding to terminal nodes and, of course, their sizes: too small clusters are not stable and should be excluded.

(f) Pruning: In some programs, the size of a cluster is unconstrained so that in the process of splitting nodes over features, some split parts may become very small and, thus, unreliable as terminal nodes. This makes it useful to prune the tree after it is computed, usually by merging the small subset nodes into greater agglomerations. This is typically done not according to the splitting criterion *W* (*S*,*u*) but according to more local considerations such as testing whether proportions of the target categories in a cluster are similar to those used at the assignment of *u* values to terminal nodes or by removing nodes with small chi-squared values (see, for a review, Esposito et al. 1997).

(g) Assigning a terminal node with a *u* category conventionally is done by just averaging its values over the node entities if *u* is quantitative or according to the maximum probability of an *u* category. Then the quality of quantitative prediction is accessed, as usual, by computing the differences between observed and predicted values of *u*, and their variance of course. In the nominal target case, this leads to an obvious estimate of the probability of the error: unity minus the maximum probability; these then are averaged over the terminal nodes of the decision tree. To make the error's estimate more robust, cross-validation techniques are used. Consider, say, a tenfold cross validation. The entity set is randomly divided into ten equal-sized subsets. Each of them is used as a testing ground for a decision tree built over the rest: these errors are averaged and given as the error's estimate to the tree built over the entire entity set. These techniques are beyond the scope of the current text.

It should be mentioned that the assignment of a category to a terminal cluster in the tree can be of an issue in some situations: (i) if no obvious winning category occurs in the cluster, (ii) if the category of interest is quite rare, that is, when *u*'s distribution is highly skewed. In this latter case using Quetelet coefficients relating the node proportions with those in the entire set may help by revealing some great improvements in the proportions, thus leading to interesting tendencies discovered.

3.8.2 Three Approaches to Scoring Correlation for Decision Trees

The process of building a classification tree is, basically, a process of splitting clusters into smaller parts driven by a measure of correlation between the partition

S being built and the target feature *u*. Since our focus here is the case of nominal *u*'s only, the target feature is represented by a partition *T* which is known to us on the training set.

How to define a function $w(S,T)$ to score correlation between the target partition *T* and partition *S* being built? Three possible approaches are:

1. A popular idea is to use a measure of uncertainty, or impurity, of a partition and score the goodness of split S by the reduction of uncertainty achieved when the split is made. If it is Gini index, or nominal variance, which is taken as the measure of uncertainty, the reduction of uncertainty is the popular impurity function utilized in a popular decision tree building program CART (Breiman et al. 1984). If it is entropy, which is taken as the measure of uncertainty, the reduction of uncertainty is the popular Information gain function utilized in another popular decision tree building program C3.5 (Quinlan 1993).

2. Another idea would be to use a popular correlation measure defined over the contingency table between partitions S and T such as Pearson chi-squared. Indeed Pearson chi-squared is used for building decision trees in one more popular program, SPSS (Green and Salkind 2003), as a criterion of statistical independence criterion, though, rather than a measure of association. Yet because Pearson chi-squared is equal to the summary relative Quetelet index (see Sect. 3.6.2), it is a measure association, and it is in this capacity that Pearson chi-squared is used in this text. Moreover, both the impurity function and Information gain mentioned above also are correlation measures defined over the contingency table as shown in the formulation part of this section. Indeed, the Information gain is just the mutual information between *S* and *T*, a symmetric function, and the impurity function, the summary absolute Quetelet index.

3. One more idea comes from the discipline of analysis of variance in statistics: the correlation can be measured by the proportion of the target feature variance taken into account by the partition *S*. How come? The variance is a property of a quantitative feature, and we are talking of a target partition here. The trick is that each class of the target partition is represented by the corresponding dummy feature, which is equal to 1 at entities belonging to the class and 0 at the rest. Each of them can be treated as quantitative, as explained in Sect. 3.6, so that the summary explained proportion would make a measure of correlation between *S* and *T*. What is nice in this approach, that it is uniform across different types of feature scales: both categorical and quantitative features can be treated the same, which is not the case with other approaches. Although this approach has been advocated by the author for a couple of decades already (see, for example, Mirkin 1996, 2012), no computational program has come out of that so far. There is a good news though: both the impurity function and Pearson chi-squared can be expressed as the summary explained proportion of the target variance, under different normalizations of the dummy variables course (see Sect. 3.6). To get the impurity function (Gini index), no normalization is needed at all, and Pearson chi-squared emerges if each of the dummies is normalized by

the square root of its frequency. That means that Pearson chi-squared is underlied by the idea that more frequent classes are less contributing. This might suggest the user to choose Pearson chi-squared if they attend to this idea, or, in contrast, the impurity function if they think that the frequencies of target categories are irrelevant to their case.

There have been developed a number of myths about classification tree building programs and correlation scoring functions involved in them. The following comments are purported to shed light on some of them.

Comment 1. Difference between CART and CHAID.

There is an opinion lurking in some comments on the web that of two popular programs, CART (Breiman et al. 1984) and CHAID (Green and Salkind 2003), the former is more oriented at prediction whereas the latter, at description. The reason for this perhaps can be traced to the fact that CART involves the impurity function that is defined as the reduction in uncertainty whereas CHAID involves Pearson chi-squared as a measure of the deviation from statistical independence. Yet this opinion is completely undermined by the fact that the measures have very similar predictive powers shaped as the summary Quetelet indexes, the only difference being that one of them involves the relative Quetelet indexes, and the other absolute ones (see Statements 3.5.3.1(b) and 3.5.3.2(b)).

Comment 2. Difference between Pearson chi-squared index and impurity function.

The difference between impurity function and Pearson chi-squared amounts to just different scaling options for the dummy variables representing classes of the target partition T (see items (c) in Statements 3.5.3.1 and 3.5.3.2). The smaller T classes get rescaled to larger values, thus contributing more, when using Pearson chi-squared.

Comment 3. Zeros in contingency tables.

Pearson chi-squared introduced to measure the deviation of a bivariate distribution from the statistical independence, appears also to signify a purely geometric concept, the contribution to the data scatter (see (a) and (c) in Statement 3.6.2.2 on p. 231). This leads to a different advice regarding the zeros in a contingency table. According to classical statistics, the presence of zeros in a contingency table contradicts the hypothesis of statistical independence so that the data are to be trimmed to avoid zeros. However, in the context of data scatter decompositions, the chi-squared is just a contribution with no statistical independence involved so that the presence of zeros is of no issue in this context: thus, no data trimming is needed.

Consider an entity set I with a pre-specified partition $T = \{T_l\}$—which can be set according to categories l of a nominal feature—that is to be learnt by producing a classification tree. At each step of the tree building process, a subset $J \subseteq I$ is to be split into a partition $S = \{S_k\}$ in such a way that S is as close as possible to $T(J)$ which is the overlap of T and J. The question is: how the similarity between S and $T(J)$ is to be measured? When $S = T(J)$, there is no confusion between the two. Otherwise, it is the contingency table between S and $T(J)$, $P = (p_{kl})$ where p_{kl} is the proportion of J- entities in $S_k \cap T_l$, that expresses the confusion, which is why it is frequently referred to as a confusion table in this context.

One idea for assessing the extent of similarity is to use a correlation measure over the contingency table such as the averaged Quetelet coefficients, Q and A, or chi-squared X^2, as discussed in Sect. 3.6.2.

Seemingly another idea is to score the extent of reduction of uncertainty over $T(J)$ obtained when S becomes available. This idea works like this: take a measure of uncertainty of a feature, in this case partition $T(J)$, $v(T(J))$, and evaluate it at each of S-classes, $v(T(S_k))$, $k = 1,\ldots, K$. Then the average uncertainty on these classes will be $\sum_{k=1}^{K} p_{k+} v(T(S_k))$, where p_{k+} are proportions of entities in classes S_k, so that the reduction of uncertainty is equal to

$$v(T(J)/S) = v(T(J)) - \sum_{k=1}^{K} p_{k+} v(T(S_k)) \qquad (3.61)$$

Of course, a function like (3.61) can be considered a measure of correlation over the contingency table P as well. One more nice feature of this approach is that it can be extended from nominal features to quantitative ones—just with an uncertainty index over quantitative T-features,

Two very popular measures defined according to (3.61) are the so-called impurity function and information gain. The impurity function builds on Gini coefficient as a measure of variance (see Sect. 3.6). Let us recall that Gini index for partition T is $G(T) = 1 - \sum_{l=1}^{L} p_l^2$ where p_l is the proportion of entities in T_l. If J is partitioned in clusters S_k, $k = 1,\ldots, K$, partitions T and S form a contingency table of relative frequencies $P = (p_{kl})$. Then the reduction (3.61) of the value of Gini coefficient due to partition S is equal to $\Delta(T(J), S) = G(T(J)) - \sum_k p_k G(T(S_k))$. This index $\Delta(T(J),S)$ is referred to as impurity of S over partition T. The greater the impurity the better the split S. As established in Statement 3.6.2.1 in Sect. 3.6.2.1, $\Delta(T, S) = A(T, S)$ where $A(T, S)$ is the summary absolute Quetelet index defined by Eq. (3.35) in Q.3.21, p. 225.

Three functions discussed above, Gini index, Pearson chi-squared, and Information gain can be coded as presented in columns of the box below where input p is a contingency table. Due to a holistic nature of MatLab computation, it is possible to organize the computation without looping through the matrix elements. The subroutines gini, chi and ing in the box can be considered pseudocodes of the functions for coding in any other language as well.

```
function a=gini(p)      function a=chi(p)      function a=ing(p)
  tot=sum(sum(p));        tot=sum(sum(p));       p=p+1;
% total                 % total                % to avoid zeros
  pr=p/tot;               pr=p/tot;              tot=sum(sum(p));
  rp=sum(pr');            rp=sum(pr');           pr=p/tot;
% row sums                cp=sum(pr);            rp=sum(pr');
  cp=sum(pr);             ir=find(rp>0);         cp=sum(pr);
%column sums            % nonzero rows           pl=log2(pr);
  ps=pr.*pr;             ic=find(cp>0);          pp=pr.*pl;
  rps=sum(ps');         %nonzero columns         rpp=sum(pp');
  ir=find(rp>0);         ps=pr.*pr;              a1=sum(rpp.*rp);
  tr=rps(ir)./rp(ir)     ip=rp'*cp;              tp=cp.*log2(cp);
  a1=sum(tr);            psi=ps(ir,ic);          a2=sum(tp);
  a2=sum(cp.*cp);        ipi=ip(ir,ic);          a=a1-a2;
  a=a1-a2;               tp= psi./ipi;           return
  return                 a1=sum(sum(tp));
                         a=a1-1;
                         return
```

Q.3.33. What is the formula of summary contribution B of partition S to the set of dummy features representing partition T when they have been normalized by dividing by their Bernoullian standard deviations $b_l = \sqrt{p_{+l}(1 - p_{+l})}$?

Q.3.34. Consider a partition $S = \{S_k\}$ ($k = 1, 2, ..., K$) on J and a set of categorical features $v \in V$, each with a set of categories $L(v)$. The category utility function (Fisher 1987) scores partition S against the feature set according to formula:

$$u(S) = \frac{1}{K} \sum_{k=1}^{K} p_k \left[\sum_{v \in V} \sum_{l \in L(v)} p(v = l/S_k)^2 - \sum_{v \in V} \sum_{l \in L(v)} p(v = l)^2 \right] \qquad (3.62)$$

The term in the square brackets is the increase in the expected number of attribute values that can be predicted given a class, S_k, over the expected number of attribute values that could be predicted without using the class. The assumed prediction strategy follows a probability-matching approach. According to this approach, entities arrive one-by-one in a random order, and the category l is predicted for them with the frequency reflecting its probability, $P(l/k)$ if the class S_k is known, or $p_k = N_k/N$ if information of the class S_k is not provided. Factors p_k weigh classes S_k according to their sizes, and the division by K takes into account the differences in the numbers of clusters: the smaller the better. Prove that the category utility function $u(S)$ is the sum of impurity functions $\Delta(l,S)$ over all features $l \in L$ related to the number of clusters, that is, $u(S) = \sum_{l \in L} \Delta(l, S)/K$.

3.8.3 Tabular Regression for Regression Trees and the Correlation Ratio

This section concerns yet another invention by Karl Pearson, the concept of piecewise constant regression and its version, tabular regression.

Consider x a categorical feature on the same entity set as a quantitative feature y, such as Occupation and Age at Students data set, or when building a regression tree, x being a partition to be built, and y, the target feature. The within-category distributions of y can be used to investigate the correlation between x and y.

The correlation between x and y is higher when the within-category spreads are tighter because the tighter the spread within an x-category, the more precise is prediction of the value(s) of y at it. Figure 3.29 illustrates an ideal case of a perfect correlation—all within-category y-values are the same leading to an exact prediction of Age when Occupation is known.

Figure 3.30 presents another extreme, when knowledge of an Occupation category does not lead to a better prediction of Age than when the Occupation is unknown.

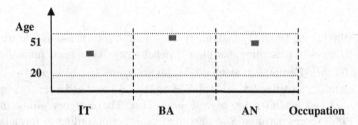

Fig. 3.29 In a situation of ideal correlation, with zero within-category variances, knowledge of the Occupation category would provide an exact prediction of the age within it

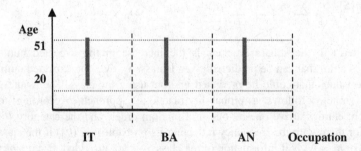

Fig. 3.30 Wide within-category distributions: the case of full variance within categories in which the knowledge of Occupation would give no information of age

A simple statistical model extending that for the mean will be referred to as tabular regression. The tabular regression of quantitative y over categorical x is a table comprising three columns corresponding to:

(1) Category of x
(2) Within category mean of y
(3) Within category standard deviation of y.

The number of rows in the tabular regression thus corresponds to the number of x-categories; there should be a marginal row as well, with the mean and standard deviation of y on the entire entity set.

Worked Example 3.14. Tabular Regression of Age (Quantitative Target) Over Occupation (Categorical Predictor) in Students Data

Let us draw a tabular regression of Age over Occupation in Table 3.32. The table suggests that if we know the Occupation category, say IT, then we can safely predict the Age as being 28.2 within the margin of plus/minus 5.6 years. With no knowledge of the Occupation category, we could only say that the Age is on average 33.7 plus/minus 8.5, a somewhat less precise estimate.

The table can be visualized in a manner similar to those in box-plots (see Fig. 3.31).

Table 3.32 Tabular regression of age over occupation in Students data

Occupation	Age mean	Age Std
IT	28.2	5.6
BA	39.3	7.3
AN	33.7	8.7
Total	33.7	8.5

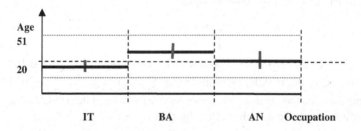

Fig. 3.31 Tabular regression visualized with the within-category averages and standard deviations represented by the position of solid horizontal lines and vertical line sizes, respectively. The dashed line's position represents the overall average (grand mean)

There is an integral characteristic of the tabular regression, the correlation ratio, which is akin to the determinacy coefficient at linear regression. This coefficient scores the extent at which the within group variance is smaller on average than the variance of the feature on the set before the split—a determinacy coefficient for the tabular regression.

Given a quantitative feature y, with no further information, its average, $\bar{y} = \sum_{i \in I} y_i/|I|$, would represent a proper summarization of the data. If, however, a set of categories of another variable, x, is additionally present, a more detailed summarization can be provided: the within category averages. Let S_k denote the set of entities falling in k category of x, then the within-category averages are $\bar{y}_k = \sum_{i \in S_k} y_i/|S_k|$.

This can be considered the least-squares solution to the model of tabular regression which extends the data recovery model for the average in Sect. 2.2.2 as follows. Find a set of c_k values such that the summary square error $L = \sum_{i \in I} e_i^2$ is minimized, where $e_i = y_i - c_k$ according to equations

$$y_i = c_k + e_i \text{ for all } i \in S_k \qquad (3.63)$$

The equations underlie the tabular regression and are referred to sometimes as the piece-wise constant regression. It is not difficult to prove that the optimal c_k in (3.61) is the within category average $\bar{y}_k$, which implies that the minimum value of L is equal to $L_m = \sum_{k=1}^{K} \sum_{i \in S_k} (y_i - \bar{y}_k)^2$. By dividing and multiplying the interior sum by the number of elements in S_k, $|S_k|$, we can see that in fact $L_m = N\sigma_w^2$ where σ_w^2 is the average within category variance defined as

$$\sigma_w^2 = \sum_K p_k s_k^2 \qquad (3.64)$$

where $p_k = |S_k|/N$ is the proportion of category k and σ_k^2 the variance of y within S_k.

To further analyze this, consider equation

$$(y_i - \bar{y}_k)^2 = y_i^2 + \bar{y}_k^2 - 2y_i\bar{y}_k$$

and sum it over all $i \in S_k$. This would lead to the summary right-hand item being similar to that in the middle, thus producing $\sum_{i \in S_k} (y_i - \bar{y}_k)^2 = \sum_{i \in S_k} y_i^2 - |S_k|\bar{y}_k^2$. Summing these equations over k and moving the right-hand item to the other side of the equation, would lead to the following decomposition:

$$\sum_{i \in I} y_i^2 = \sum_{k=1}^{K} |S_k|\bar{y}_k^2 + \sum_{k=1}^{K} \sum_{i \in S_k} (y_i - \bar{y}_k)^2 \qquad (3.65)$$

Note that the right-hand item in (3.65) is the summary least-squares criterion of model in (3.63) L_m. This allows us to interpret the Eq. (3.65) as a decomposition of the scatter of variable y, the item on the left, in two parts on the right: the explained part, in the middle, and the unexplained part L_m.

The explained part sums contributions of individual categories k, $|S_k|\bar{y}_k^2$. The value of the contribution is proportional to both the category frequency and its squared value—the greater the better.

Another expression of decomposition (3.65) can be obtained under the assumption that variable y is centered, so that its mean is 0, by relating it to N:

$$\sigma^2 = \sum_{k=1}^{K} p_k \bar{y}_k^2 + \sum_{k=1}^{K} p_k \sigma_k^2 \tag{3.66}$$

where σ^2 is the variance of y, the item on the right the minimum value L_m/N from (3.64), and the item in the middle, the weighted summary squared distance between the grand mean $\bar{y} = 0$ and within-category means $\bar{y}_k$.

Equation (3.66) is very popular in statistics as the decomposition of the variance into the within-group variance, the item on the right, and the between-group variance, the item in the middle, as the base of a popular method for comparison of within-category means which is referred to as ANOVA (ANalysis Of VAriance). In the context of the tabular regression model (3.65) viewed as a data recovery model, the original decomposition (3.65) of the quantitative feature scatter into part explained by the nominal feature and part remaining unexplained is more appropriate. Viewed in this light, decomposition (3.66) shows that the category k contribution to the total variance of y is proportional to its frequency multiplied by the squared difference between within-category mean $\bar{y}_k$ and grand mean $\bar{y} = 0$.

The correlation ratio shows the relative drop in the variance of y when y is predicted according to model (3.63) or, in other words, the relative proportion of the explained part of the variance. The correlation ratio is usually denoted by η^2 and defined by the following formula:

$$\eta^2 = 1 - \sigma_w^2 / \sigma^2 \tag{3.67}$$

The notation, η^2, is well accepted in the literature, although some authors prefer using the term, correlation ratio, for the squared root of η^2, but most authors leave the term as is.

The definition implies the following properties:

- The range of η^2 is between 0 and 1.
- Correlation ratio $\eta^2 = 1$ when all within-category variances σ_k^2 are zero (that is, when y is constant within each group S_k).
- Correlation ratio η^2 is about 0 when all σ_k^2 are of the order of σ_k^2.

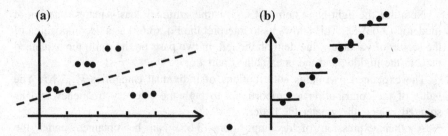

Fig. 3.32 Different patterns of linear and piecewise constant association between features corresponding to x-axis ad y-axis: almost perfect piecewise constant match against a highly non-linear pattern in (**a**), and almost linear arrangement against a highly non-constant pattern in (**b**)

Case-Study 3.7. Is There Any Relation Between Correlation Coefficient and Correlation Ratio?
Consider two quantitative features x and y. Divide the range of x in four equal-sized bins to produce a categorical variable xc. Is there any relation between the correlation coefficient between x and y and the correlation ratio coefficient between xc and y?

Some claim that η^2 should be always greater than ρ because the correlation ratio captures any type of functional relation whereas the correlation coefficient relates to linear functions only.

In general, no relation between η^2 and ρ can be claimed. The former can be greater than the latter in some cases, and smaller in some others, as presented in Fig. 3.32a at which $\eta^2 \gg \rho$ and 3.32b at which $\eta^2 \ll \rho$.

Q.3.35. Consider the variance to be an uncertainty measure for a quantitative feature y. Define the uncertainty reduction measure according to formula (3.16), with T changed for y of course, and prove that it is equal to the numerator of the correlation measure—the part of variance of y explained by its tabular regression over S.

A. The summary contribution of S to the data scatter is equal to $B = \sum_{k=1}^{K} c_k^2 |S_k| = \sum_{i\in I} y_i^2 - \sum_{k=1}^{K} \sigma_k^2 |S_k|$ where σ_k^2 is the within-cluster variance of y (see (3.13) in Sect. 3.2). Then $B = N(\sigma^2 - \sum_{k=1}^{K} p_k \sigma_k^2)$ where σ^2 is the variance of the standardized feature y (note that the mean of y is 0!) and p_k the proportion of entities in cluster S_k. The last equation clearly shows that the explained part of v is $B = N\sigma^2\eta^2$. If y has been z-score standardized so that $\sigma^2 = 1$, B equals the correlation ratio.

3.8.4 Building Classification Trees

Building of a classification tree is a recursive process: starting from the entire data set, partition a cluster into a number of parts according to one of the features. To make the partitions less arbitrary, only binary splits are involved in most of the update programs. That means that any node may be split only in two parts: (i) that corresponding to a category and the rest, for a categorical feature, or (ii) given a threshold a, those "less than or equal to a" and those "greater than a", for a quantitative feature. This approach naturally comes when the data are preprocessed by "enveloping" categories into the corresponding "quantitative" dummy features, that assign a unity to every object falling into the category, and a zero to all the rest. Indeed, at $a = 0$, such a dummy feature would split the set in two parts—that for the corresponding category and the rest. Given a cluster, the choice of feature and threshold a for doing the split is driven by a correlation scoring function, be it Information gain, Pearson chi-squared, Gini index or anything else.

A cluster is not to be split anymore if it is smaller than a user defined threshold TS (TS = 10 is set further on) or is homogeneous enough. We use two different homogeneity tests: (a) large enough proportion of a target category in the cluster, say, above 80%, and (b) small enough value of the scoring function which is set to be 0.03 for Gini index, 0.08 for Pearson chi-squared, and 0.15 for Information gain. These levels of magnitude reflect the functions' ranges: Gini index is very close to 0 hardly reaching 0.5 at all, Pearson chi-squared, related to N, changes between 0 and 1 because it cannot be greater than the number of split parts minus 1, and Information gain can have larger values when the number of target categories is 3 or more. This sets the stopping conditions.

Worked Example 3.15 Classification Tree for the Iris Dataset
At Iris dataset with its three taxa, Iris setosa and Iris versicolor and Iris virginica, taken as target categories, all the three scoring functions—Impurity (Gini) function, Pearson chi-squared and Information gain—lead to the same classification tree, presented on Fig. 3.33.

The tree of Fig. 3.33 was found with program clatree.m. It comprises three leaf clusters: A, consisting of all 50 Iris setosa specimens; B, containing 54 entities of which 49 are of Iris versicolor and 5 of Iris virginica; C, containing 46 entities of which 45 are of Iris virginica and 1 of Iris versicolor. Altogether, this misplaces 6 entities leading to the accuracy of 96%. Of course, the accuracy would somewhat diminish if a cross-classification scheme is applied (see Loh and Shih 1997, who draw a slightly different tree for Iris dataset).

Let us take a look at the action of each variable at each of the two splits in Table 3.33. Each time features w3 and w4 appear to be most contributing, so that at

Fig. 3.33 Classification tree for the three-taxon partition at Iris dataset found by using each of the Gini, Pearson chi-squared and Information gain scoring functions

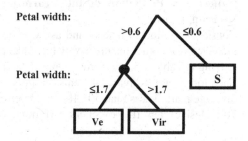

Table 3.33 Values of Gini index at the best split of each feature on the Iris dataset clusters in Fig. 3.33

Feature	First split		Second split	
	Value	Gini	Value	Gini
w1	5.4	0.228	6.1	0.107
w2	3.3	0.127	3.4	0.036
w3	1.9	0.333	3.7	0.374
w4	0.6	0.333	1.7	0.390

Table 3.34 Confusion tables between a split and target partition on the Iris dataset

Target partition classes	Iris setosa	Iris versicolor	Iris virginica	Total
Full set				
w4 ≤ 1.7	50	49	5	104
w4>1.7	0	1	45	46
Total	50	50	50	150
First cluster removed				
w4 ≤ 1.7	0	49	5	54
w4 > 1.7	0	1	45	46
Total	0	50	50	100

the first split, at which w3 and w4 give the same impurity value, w4 made it through just because it is the last maximum, which is remembered by the program.

The tree involves just one feature, w4: Petal width, used for splitting twice, first at w4 = 0.6 and then at w4 = 1.7. The Pearson chi-squared value (related to N of course) is 1 at the first split and 0.78 at the second. The Impurity function grows by 0.33 at the first split and 0.39 at the second. The fact that the second value is greater than the first one may seem to be somewhat controversial. Indeed, the first split is supposed to be the best, so that it is the first value that ought to be maximum. Nevertheless, this opinion is wrong: if the first split was at w4 = 1.7 that would generate just 0.28 of impurity value, less than the optimal 0.33 at w4 = 0.6. Why? Because the first taxon has not been extracted yet and grossly contributes to a higher confusion (see the top part in Table 3.34).

Project 3.3. Prediction of the Learning Outcome at Student Data Using Decision Trees

Consider the Student dataset and ask whether students' learning successes can be predicted from other features available (Occupation, Age, Number of children)? By looking at Table 1.5, one hardly can expect that marks can be predicted in this way. Therefore, let us divide students in three groups: I—not so good performers (average mark is less than 50), II—good performers (average mark between 50 and 70 inclusive), and III—excellent performers (average mark higher than 70). To do this, we compute the average mark over the three subjects (SE, OOP, and CI) and create a partition of students T as described; the distribution of T appears to be I-25, II-58, III-17.

We have a 100×5 matrix X to explore the correlation between X and T, the three columns, 1,2,3, being dummy variables for Occupation categories (IT, BA, AN), column 4 for Age, and column 5 for Number of children. The two conventional stopping criteria, the cluster's size and prevalence of a target class, are not sufficient at this data, because after one or two splits, the program just chips away small fragments of clusters without much improving them. This corresponds to the situations at which the scoring function does not show much improvements either. Therefore, we utilize one more criterion—the minimum value of the scoring function, a threshold below which there is no splitting. Since the three scoring functions we use have different ranges, the thresholds must be different too. At this study, the threshold is set at 0.03 for the Gini index, 0.08 for the Pearson chi-squared, and 0.15 for the Information gain. The minimum cluster size is taken at 10, and the prevalence of a target class is set at 80%.

The classification tree found with Gini index is presented on Fig. 3.34. The distributions of target categories in clusters in Fig. 3.34 are presented in Table 3.35. Bold font highlights four terminal clusters as well as high or low proportions of target classes in clusters. High proportions here are those greater than 70% and low proportions are those smaller than 5%.

Tree on Fig. 3.34 is driven by two features: AN Occupation, that structures the set rather well—one split part, those of AN occupation, get more than 70% of category I, and none of category III, and the other of category II. All further divisions are over feature Age; the 12 students in cluster 8 are rather specific—these are of AN occupation aged between 22 and 28, so that 75% of them are in category I, an improvement over parental cluster 6. Cluster 4 of younger not-AN students seems an attempt at drawing a cluster to predict category III—it has a highest jump in its proportion, to 36.4% from 17% in the entire set (cluster 1). The 25 older people among not-AN students are overwhelmingly, 92%, in category II. More splits would have followed if we had decreased the minimum acceptable value of Gini index, say from 0.03 to 0.01.

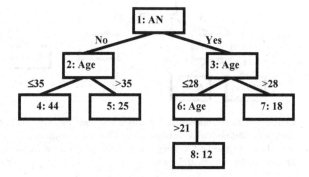

Fig. 3.34 Classification tree on Students data, targeting partition T of students in three categories, found using Gini index. The legend *Number: A* presents, at a split cluster labeled by Number, A as the split variable or, at an unsplit cluster, A as the size (the number of students in it)

Table 3.35 Distributions of target classes in clusters of tree on Fig. 3.34, %

Target categories	Clusters in tree on Figure 3.34							
	1	2	3	4	5	6	7	8
I	25.0	3.9	73.2	3.3	3.0	69.2	77.8	75.0
II	58.0	73.5	25.8	61.4	93.0	30.8	23.2	25.0
III	17.0	23.6	0	36.4	3.0	0	0	0
Gini index at split	0.168	0.046	0.035			0.048		
Cluster size	100	69	31	44	25	13	18	12

How well this tree would fare at prediction? To address this question properly, one should either conduct a cross-validation test or set aside a random testing set before using the rest for building a tree, after which see the levels of errors on the testing set.

Yet for the illustrative purposes, let us calculate the prediction error by using tree on Fig. 3.34. This is done by using the terminal clusters 4, 5, 7, 8 comprising 44, 25, 18, 12 elements, respectively. They total to 99, not 100, because of chipping off an element from cluster 6 to make it into cluster 7. That means: for students in AN category aged 21 or less, no prediction of their learning success level will be made; the classifier takes what is referred to as reject option (comprising approximately 1% of future cases if our sample is representative). According to the data in Table 3.35, the optimal prediction rule would predict then category II at Cluster 4 (with error $100 - 61.4 = 38.6\%$), category II again, at cluster 5 (with error $100 - 92 = 8\%$), and category I at clusters 7 and 8 (with errors 23.2 and 25.0%, respectively). The average error is the sum of the individual cluster errors weighted by their relative sizes, $(38.6 * 44 + 8 * 25 + 23.2 * 18 + 25 * 12)/99 = 26.2\%$.

What happens, if we use the parental cluster 6 instead of the chipped cluster 8? First thing—no reject option is involved then. Second, the error somewhat increases as should be expected: $(38.6 * 44 + 8 * 25 + 23.2 * 18 + 30.8 * 13)/100 = 27.0\%$.

Figure 3.35 presents trees found by using the Pearson chi-squared (a) and Information gain (b). In contrast to Gini index, decreasing the increment threshold

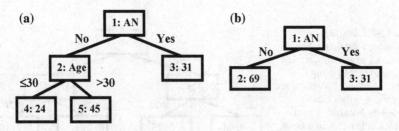

Fig. 3.35 Classification trees on Student dataset targeting partition T of students in the three categories, defined above, found using Pearson chi-squared (**a**) and Information gain (**b**). The legends are of format "Number: A" where "A", at a split cluster, is the split variable or, at an unsplit cluster, the cluster's size

does not much help at Information gain: chipping here and there rather than splits will be added. The change of splitting Age value to 30 at cluster 2 on tree (a) does lead to some improvements: the 45 older students are 83.2% in category II. Yet among the 24 younger students, 45.8% belong to category III (leaving 53.2% in category II and 0 in category I).

With this example, one can see that the 90–100% precision is not that easy to achieve. That is, a terminal node may have rather modest proportions of target categories, like cluster 5 on Fig. 3.35a: about 54% of II category and 46% of III category. Conventional thinking would label the node as an II category predictor because the share of II is greater than half.

However, one should note that, in fact, the proportion 54% is smaller than that, 58%, in the entire set, which means that in fact these conditions, Not_AN and younger age, less than 31, wash out some of II category. It is a case when the style of Quetelet's thinking may produce a better description. This thinking goes beyond proportions in the terminal node and requires comparing the category shares at the node with that in the whole sample. In contrast to a reduction of II category, this cluster boasts a dramatic increase of III category—from 17% in the entire set to 46% in the cluster, 29%. This difference would be picked up by using the absolute Quetelet coefficient which is equal to Gini index. Even more dramatic would be the relative increase, $(45.8 - 17)/17 = 170\%$. It is this increase that has been picked up by Pearson chi-squared scoring function, because it is driven by the relative Quetelet coefficient.

Q.3.36. Drawing a lift chart in marketing research. Consider a marketing campaign advertising a product. There is a 1000 strong sample from the set of targeted customers whose purchasing behavior is known because of prior campaigns. The sample is composed of clusters of a classification tree with different response (that is, purchasing) rates (see Table 3.36). To plan an effective campaign, marketing researchers use what is called a lift chart—a visual representation of the response rate.

The x-axis of a lift chart shows the percentiles of the sample, say, from 10 to 100%. On y-axis, the so-called lifts are presented. Given a group of customers, the lift is defined as the ratio of the group's response rate to the baseline response rate, which is the response rate for the entire sample. On the lift chart, the percentiles of the sample are taken in the descending order of the lift. Both baseline and percentile lifts are presented on the chart. Build a lift chart for the sample. **A.** First, we calculate the baseline rate which is the average of the response rates in Table 3.36 weighted by the cluster proportions: $r = 0.1 * 30 + 0.4 * 10 + 0.25 * 4 + 0.25 * 0 = 8\%$.

Table 3.36 Proportions of four clusters in a sample of 1000 customers and their purchasing behavior (response rate)

Cluster share, %	10	40	25	25
Response rate, %	30	10	4	0

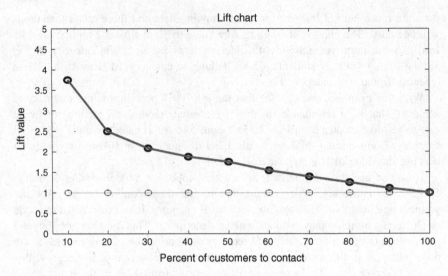

Fig. 3.36 Lift chart for data in Table 3.36

Now we take the most responsive 10% of the customers and calculate their lift value: 30/8 = 3.75. Next, we take the most responsive 20% of the sample, that is the first cluster plus a hundred customers from the second cluster and see their response rate—there should be 30 customers from the first cluster plus 10 from the second who have purchased the product, which gives 40/200 = 20% response rate leading to the lift value of 20/8 = 3.5. Next percentile, 30% of the sample is composed of the first cluster plus 200 customers from the second cluster leading to 50/300 = 17.7% response rate and lift 3.3. In this way, the chart presented on Fig. 3.36 is computed.

3.8.5   Building Classification Trees: Computation

Consider an entity set I with a nominal target feature represented by partition T of I as well as a set of quantitative input features X (some or all of X-features may be binary dummy variables corresponding to categories). At each step of the process of building a classification tree a cluster $J \subseteq I$ is to be split according to a feature x_v from X in two clusters, S_1 and S_2 so that $S_1 = \{i|i \in J \text{ and } x_{iv} \leq a\}$ and $S_2 = \{i|i \in J \text{ and } x_{iv} > a\}$ where a is a value of x_v. The choice of x_v and a is guided by a scoring function $W(S,T)$ defined over the contingency table P cross-classifying T by S.

That implies that a cluster, as an element of the hierarchical structure being built, should maintain at least the following data:

- (i)   its entity set,
- (ii) its parental cluster,
- (iii) feature x_v over which it has been split,
- (iv) splitting value a,
- (v)   the inequality, $\leq$ or $>$, in the cluster defining predicate.

The process starts at the universal cluster consisting of the entire set I. The process stops if either of two conditions holds:

- (a) $|J| < n$, where n is a pre-specified threshold on the minimum number of entities in a cluster, and
- (b) if the frequency of a T-cluster is greater than a pre-specified threshold α. To make testing of (b) easier, each cluster should bear one more feature

- (vi) the distribution of T in it. One more useful piece of data supplied with a cluster would be
- (vii) a signal of whether it may or may not be split again.

The recursive nature of the process, as well as the presence of a set of data to accompany each cluster, would make it a fitting subject of an object oriented code. Yet since the object oriented part of MatLab is not quite native in it, a procedural construction will be described in this section. This construction involves two parts, provided that computing scoring function $W(T,S)$ over contingency table P, has been implemented: (A) finding the best split over a feature, and (B) building a hierarchy of the best splits.

A pseudocode, or MatLab, function, msplit.m, takes in a column-feature x, partition of the set of its indices, t, as a cell array of t-classes, and a string with the name of a scoring method. It produces partition s, the feature splitting value y, and the value of scoring function ma. The stages of computation are annotated within the code.

```
function [g,ma,y]=msplit(x,t,method)

n=length(x);
%-------preparing the set of split value candidates
xv=union(x,x);%set of x values sorted
ll=length(xv);
rl=length(t);
if ll==1 %feature x is constant
    g{1}=[1:n];
    ma=0;
    y=max(x);
else
    for k=1:(ll-1) %loop over splitting values
        f{1}=find(x<=xv(k)); %first split set
        f{2}=setdiff([1:n],f{1}); % the rest
        for ik=1:2; for il=1:rl
                p(ik,il)=length(intersect(f{ik},t{il}));
            end
        end %   contingency table p
        switch method
            case 'gini'
                res=gini(p);
            case 'chi'
                res=chi(p);

            case 'ing'
                res=ing(p);
            otherwise
                disp('The method is wrong ');
                pause(10);
        end
    %-----------looking for the best split
        if res>ma
            ma=res;
            g=f;
            y=xv(k);
        end
    end
end
```

The computation is organized in code clatree.m printed in the appendix. Here are just a few comments on its structure. Consider a set of ss clusters stored in a cell structure indexed from 1 to ss; in the beginning, the structure stores just the universal cluster I and its features at ss = 1. Of these clusters, those in the end, starting from index tt $\geq$ ss are eligible for splitting. The newly split clusters are indexed by index bb starting from bb = ss + 1. (Note that with this system of indexing, there is no need to assign clusters with a label informing that they should not be split anymore: the clusters to split can only be fresh ones!) After split parts are put in the structure, the indices are updated.

There can be a number of stopping criteria that are to be set in the very beginning of the program: it stops when no clusters eligible for splitting remain. In the current version of program clatree.m, three types of stopping criteria are employed. First is the number of entities, TS: a cluster with a smaller number of entities cannot make it into the tree and of course cannot be split further. Second, the dominant proportion of the target classes, ee: a cluster is not split anymore if this has been reached. And the third stopping criterion is tin, a threshold on the scoring function value: if it is less then tin at a split, the cluster is not split.

3.8.6 Random Forest Classifier

Although the decision tree structure is very good for interpretation, it is not as good, in general, for prediction; I guess, because of its instability. This is why one of the most prominent authors in the field proposed to diversify the tree structure by using a large number of decision trees generated by the bootstrap (Breiman 2001). To be more exact, let us specify a number P of bootstrap trials and generate, randomly with resampling, P series of the length N of indices from 1 to N, where N is the number of objects. Then we specify a number $m < V$ where V is the number of features and, for each of the P bootstrap series, generate a data table with objects corresponding to the indices in the series and k features randomly taken from the original V features. Then a decision tree is drawn based on each of the P data tables. In a refined version, the k-element feature subset is randomly drawn at each consecutive splitting step. This set of decision trees is what is called a random forest. Given an entity at which all the input variables are defined, one can decide over its class, in a classification problem, as follows. Use every tree in the forest, identify the location of the entity in the tree, and predict the target class accordingly. Count the number of votes, that is, the trees, for each of the target classes and predict that one with the largest number of votes. If the task at hand is not of classification but rather of prediction of the quantitative feature values, that is, of regression, the predicted outcome is the average of the values predicted by each of the regression trees in the forest.

This procedure works quite well, so that the random forest voting has become one of the most popular tools used by the practitioners.

However, the random forest concept loses its interpretability while gaining in the accuracy. This is why one needs a measure of feature importance according to a random forest, to compensate the loss albeit partly.

Consider a most popular feature importance measure, the valence (Breiman 2001; Louppe et al. 2013), which can be defined for any measure of impurity of a subset S with regard to a target feature u, $\gamma(S)$. (we skip the output partition in this notation). Limiting ourselves to the classification problem, that can be any measure considered in Sects. 3.6.1 and 3.6.2, as for example, Gini index. Then the impurity loss at partitioning a node s of the tree being built in its left part, s_L, and its right part, s_R, will be

$$\Delta\gamma(s) = \gamma(s) - p_L\gamma(s_L) - p_R\gamma(s_R) \tag{3.68}$$

where p_L and p_R are the proportions of S-entities in the left and right split parts, s_L and s_R, respectfully.

The importance of a variable w for predicting u is scored by summing the weighted impurity losses $p(s)\Delta\gamma(s)$, where $p(s)$ is the proportion of objects at the node s, for all nodes s at which w is used as the splitting criterion, averaged over all the trees in the forest:

$$\nabla_v(w) = \frac{1}{P}\sum_{t=1}^{P}\sum_{s_t \leftarrow w} p(s_t)\Delta\gamma(s_t) \tag{3.69}$$

Here the symbol $s_t \leftarrow w$ denotes the fact that the t-th tree node s_t involves splitting of the variable w, so that the second summation involves all the nodes s_t formed by splitting the variable w ($t = 1,2,\ldots, P$).

Assume that all the features, both the target u and input V, are categorical and the decision trees in the ensemble are "fully developed and balanced" (see Louppe et al. 2013) so that all the divisions possible have been made—for the binary features that would mean that all the branches involve all the V features so that the tree has 2^V leaves. Then the importance weights can be estimated according to the following formulas. Given a subset W of k features, define

$$\gamma(u|W) = \sum_{w0} P(W = w0)\gamma(u|W = w0)$$

where P is the probability, and summation runs over all possible combinations $w0$ of the features in W. Given a feature v and subset of features W such that $v \notin W$, define $G(u|W + v) = \gamma(u|W) - \gamma(u|W + v)$. Then the following formula holds (Louppe et al. 2013):

$$\nabla_\gamma(w) = \sum_{k=0}^{V-1} \frac{1}{(V - k)C_V^k} \sum_{W \in P_k^{-v}} G(u|W + v) \tag{3.70}$$

Table 3.37 The table of digit-to-edge features e1–e7 according to the Fig. 1.2 in Sect. 1.2.3

Numeral	e1	e2	e3	e4	e5	e6	e7
0	1	1	1	0	1	1	1
1	0	0	1	0	0	1	0
2	1	0	1	1	1	0	1
3	1	0	1	1	0	1	1
4	0	1	1	1	0	1	0
5	1	1	0	1	0	1	1
6	1	1	0	1	1	1	1
7	1	0	1	0	0	1	0
8	1	1	1	1	1	1	1
9	1	1	1	1	0	1	1
(3.70)	0.412	0.581	0.531	0.542	0.656	0.225	0.372
K = 1	0.414	0.583	0.532	0.543	0.658	0.221	0.368
K = 4	0.309	0.757	0.489	0.445	0.810	0.122	0.387

The bottom rows contain the feature importance weights computed according to f-la (3.70), as well as those estimated over a sample of 10,000 fully randomized trees, generated at k = 1 and k = 4, against the target feature Numeral, see on the left, by Louppe et al 2013

where P_k^{-v} is the set of all k-element subsets of features such that v is not among them, and C_V^k is the binomial coefficient.

Case-Study 3.8. Importance of Digit Features
Consider Table 3.37 of the ten numeral digits characterized by the seven binary features corresponding to the presence or absence of an edge in the numeral drawing over the seven-line rectangle in Fig. 1.2 in Sect. 1.2.3. The numerals themselves occupy the left-most column in the table.

The bottom rows contain the feature importance weights computed according to f-la (3.70), as well as those estimated over a sample of 10,000 fully randomized trees, generated at k = 1 and k = 4 according to the random forest algorithm, against the target feature Numeral. As one can see, the sampling estimates at k = 1 follow the theoretical estimates rather closely.

3.9 Naïve Bayes Approach

3.9.1 Bayes Decision Rule

Consider a situation in which there is only one target, a binary feature labeling two states of the world corresponding to "positive" and "negative" classes of entities. According to Thomas Bayes (c. 1701–1761), all relevant knowledge of the world should be shaped by the decision maker in the form of probability distributions. Then, whatever new data may be observed, they may lead to changing the probabilities—hence the difference between prior probabilities and posterior, data-updated, probabilities. Specifically, assume that, $P(1) = p_1$ and $P(2) = p_2$ are prior probabilities of the two states so that p_1 and p_2 are positive and sum to unity. Assume furthermore that there are two probability density functions, $f_1(x_1, x_2, ..., x_p)$ and $f_2(x_1, x_2, ..., x_p)$, defining the generation of observed entity points $x = (x_1, x_2, ..., x_p)$ for each of the classes. That gives us, for any point $x = (x_1, x_2, ..., x_p)$ to occur, two probabilities, $P(1\&x) = p_1 f_1(x)$ and $P(2\&x) = p_2 f_2(x)$, of x being generated at either class. Therefore, the total probability of x to occur is $f(x) = p_1 f_1(x) + p_2 f_2(x)$. If an $x = (x_1, x_2, ..., x_p)$ is actually observed, it leads to a change in probabilities of the classes, from the prior probabilities $P(1) = p_1$ and $P(2) = p_2$ to posterior probabilities $P(1/x)$ and $P(2/x)$, respectively. These can be computed as conditional probabilities

$$P(1|x) = p_1 f_1(x)/f(x) \text{ and } P(2|x) = p_2 f_2(x)/f(x). \qquad (3.71)$$

The decision of which class the entity x belongs to depends on what value, $P(1/x)$ or $P(2/x)$ is greater. The class is considered to be the positive if $P(1/x) > P(2/x)$ or, equivalently,

$$f_1(x)/f_2(x) > p_2/p_1 \qquad (3.72)$$

or, the negative, if the reverse inequality holds. This rule is referred to as Bayes decision rule. Another expression of the Bayes rule can be drawn by using the difference $B(x) = P(1/x)-P(2/x)$ rather than the ratio: x is taken to belong to the positive class if $B(x) > 0$, and the negative class if $B(x) < 0$. Equation $B(x) = 0$ defines the so-called separating surface between the two classes.

The proportion of errors admitted by the Bayes rule is $1 - P(1/x)$ when 1 is predicted and $1 - P(2/x)$ when 2 is predicted. These are the minimum error rates achievable when both within-class distributions $f_1(x)$ and $f_2(x)$ and priors p_1 and p_2 are known.

Unfortunately, the distributions $f_1(x)$ and $f_2(x)$ are typically not known. Then some simplifying assumptions are to be made so that the distributions could be estimated from the observed data. Among most popular assumptions are:

(i) Gaussian probability and
(ii) Local independence.

Let us consider them in turn:

(i) Gaussian probability

The class probability distributions $f_1(x)$ and $f_2(x)$ are assumed to be Gaussian, so that each can be expressed as

$$f_k(x) = \left[\exp -(x - \mu_k)^T \Sigma_k^{-1}(x - \mu_k)/2\right]/[(2\pi)^p |\Sigma_k|]^{1/2} \qquad (3.73)$$

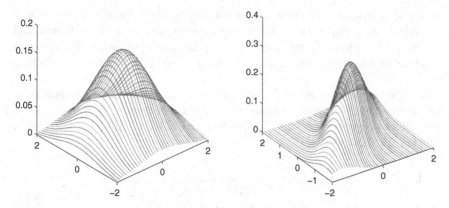

Fig. 3.37 Gaussian bivariate density functions over the origin as the expectation point—with zero correlation on the left and 0.8 correlation on the right

where μ_k is the central point, Σ_k the $p \times p$ covariance matrix and $|\Sigma_k|$ its determinant ($k = 1, 2$).

The Gaussian distribution is tremendously popular. There are at least two reasons for that. First, it is treatable theoretically and, in fact, may frequently lead to the least squares criterion within the probabilistic approach. Second, some real-world stochastic processes, especially in physics, can be thought of as having the Gaussian distribution. This is justified theoretically with the so-called probability limit theorems. These theorems state that the sum of a multitude of independent probabilistic distributions converges to a Gaussian distribution. Typical shapes of a 2D Gaussian density function are illustrated on Fig. 3.37: that with zero correlation on the left and 0.8 correlation on the right.

In the case at which the within-class covariance matrices are equal to each other, the Bayes decision function $B(x)$ is linear so that the separating surface $B(x) = 0$ is a hyperplane.

(ii) Local independence (Naïve Bayes)

The assumption of local independence states that all variables are independent within each class so that the within-cluster distribution is a product of one-dimensional distributions:

$$f_k(x_1, x_2, \ldots, x_p) = f_{k1}(x_1) f_{k2}(x_2) \ldots f_{kp}(x_p) \qquad (3.74)$$

This postulate much simplifies the matters because usually it is not difficult to produce rather reliable estimates of the one-dimensional density functions $f_{kv}(x_v)$ from the training data. Especially simple such a task is when features $x_1, x_2, \ldots, x_p$ are binary themselves. In this case the Bayes rule is referred to as a naïve Bayes rule because in most cases the assumption of independence (3.74) is obviously wrong in practical situations. Take, for example, the cases of text categorization or genomic analyses—constituents of a text or a protein, serving as the features, are necessarily interrelated according to the syntactic and semantic structures, in the former, and biochemical reactions, in the latter. Yet the decision rules based on the wrong assumptions and distributions appear surprisingly good (see discussion in Manning et al. 2008).

Combining the assumptions of local independence and Gaussian distributions in the case of binary variables, one can arrive at equations expressing the conditional probabilities through exponents of linear functions of the variables (as described in Mitchell 2010) so that:

$$
\begin{aligned}
P(1/x) &= \frac{1}{1 + \exp(c_0 + c_1 x_1 + \ldots + c_p x_p)}, \\
P(2/x) &= \frac{\exp(c_0 + c_1 x_1 + \ldots + c_p x_p)}{1 + \exp(c_0 + c_1 x_1 + \ldots + c_p x_p)}
\end{aligned}
\tag{3.75}
$$

Equations (3.75) express what is referred to as the logistic regression. Logistic regression is a popular decision rule that can be applied to any data on its own right as a model for the conditional probability, and not necessarily derived from the restrictive independence and normality assumptions.

3.9.2 Naïve Bayes Classifier

Consider a learning problem related to data in Table 3.38 further on: there is a set of entities, which are newspaper articles, divided into a number of categories—there are three categories in Table 3.38 according to the three subjects: Feminism, Entertainment and Household. Each article is characterized by its set of keywords presented in the corresponding line. The entries are either 0—no occurrence of the keyword, or 1—one occurrence, or 2—two or more occurrences of the keyword.

The problem is to form a rule according to which any article, including those outside of the collection in Table 3.38, can be assigned to one of these categories using its profile—the data on occurrences of the keywords in the corresponding line of Table 3.38.

Consider the Naïve Bayes decision rule. It assigns each category k with its conditional probability $P(k/x)$ depending on the profile x of an article in question according to equations in (3.71):

$$
P(k/x) = p_k f_k(x)/f(x)
$$

where $f(x) = \sum_l p_l f_l(x)$. According to the Bayes rule, the category k, at which $P(k/x)$ is maximum, is selected. Obviously, the denominator does not depend on k and can be removed: that category k is selected, at which $p_k f_k(x)$ is maximum.

Table 3.38 An illustrative database of 12 newspaper articles along with 10 keywords

Article	Keyword									
	Drink	Equal	Fuel	Play	Popular	Price	Relief	Talent	Tax	Woman
F1	1	2	0	1	2	0	0	0	0	2
F2	0	0	0	1	0	1	0	2	0	2
F3	0	2	0	0	0	0	0	1	0	2
F4	2	1	0	0	0	2	0	2	0	1
E1	2	0	1	2	2	0	0	1	0	0
E2	0	1	0	3	2	1	2	0	0	0
E3	1	0	2	0	1	1	0	3	1	1
E4	0	1	0	1	1	0	1	1	0	0
H1	0	0	2	0	1	2	0	0	2	0
H2	1	0	2	2	0	2	2	0	0	0
H3	0	0	1	1	2	1	1	0	2	0
H4	0	0	1	0	0	2	2	0	2	0

The articles are labeled according to their main subjects—F for feminism, E for entertainment, and H for household

According to the Naïve Bayes approach, $f_k(x)$ is assumed to be the product of the probabilities of occurrences of the keywords in category k. How can one estimate such a probability? This is not that simple as it sounds.

For example, what is the probability of term "drink" in H category according to Table 3.38? Probably, it can be taken as ¼—since the term is present in only one of four members of H. But what's about term "play" in H—it occurs thrice but in two documents only; thus its probability cannot be taken ¾; yet ²/₄ does not seem right either. A popular convention accepts the "bag-of-words" model for the categories. According to this model, all occurrences of all terms in a category are summed, to produce 31 for category H in Table 3.38. Then each term's probability in category k would be its summary occurrence in k divided by the bag's total. This would lead to a fairly small probability of the "drink" in H, just 1/31. This bias is not that important, however, because what matters indeed in the Naïve Bayes rule is the feature relative contributions, not the absolute ones.

And the relative contributions are all right with "drink", "fuel" and "play" contributing 1/31, 6/31 and 3/31, respectively, to H. Moreover, taking the total account of all keyword occurrences in a category serves well for balancing the differences between categories according to their sizes.

There is one more issue to take care of: zero entries in the training data. Term "equal" does not appear at all in H, leading thus to its zero probability in the category. This means that any article with an occurrence of "equal" would not be classed into H category, however heavy evidence from other keywords may be. One could make a point of course that term "equal" has not been observed in H just because the sample of four articles in Table 3.38 is too small, which is a strong argument indeed. To make up for these, another, a "uniform prior" assumption is widely accepted. According to this assumption each term is present once at any category before the count is started. For the case of Table 3.38, this adds 1 to each numerator and 10 to each denominator, which means that the probability of "drink", "equal", "fuel" and "play" in category H will be $(1 + 1)/(31 + 10) = 2/41$, $(0 + 1)/(31 + 10) = 1/41$, $(6 + 1)/(31 + 10) = 7/41$ and $(3 + 1)/(31 + 10) = 4/41$, respectively.

To summarize, the "bag-of-words" model represents a category as a bag containing all occurrences of all keywords in the documents of the category plus one occurrence of each keyword, to be added to every count in the data table.

Table 3.39 contains the prior probabilities of categories, that are taken to be just proportions of categories in the collection, 4 of each in the collection of 12, as well as within-category probabilities of terms (the presence of binary features) computed as described above. Logarithms of these are given too.

Now we can apply Naïve Bayes classifier to any entity presented in the format of Table 3.38 including those in Table 3.38 itself (the training set). Because the probabilities in Table 3.39 are expressed in thousands, we may use sums of their logarithms rather than the probability products; this seems an intuitively appealing operation. Indeed, after such a transformation the score of a category is just the inner product of the row representing the tested entity and the feature scores corresponding to the category. Table 3.40 presents the logarithm scores of article E1 for each of the categories.

Q.3.37. Apply Naïve Bayes classifier in Table 3.39 to article X = (2 2 0 0 0 0 2 2 0 0) which involves items "drink", "equal", "relief" and "talent" frequently.
A. The category scores are: $s(F/X) = 35.2$, $s(E/X) = 35.6$, and $S(H/X) = 29.4$ pointing to Entertainment or, somewhat less likely, Feminism.
Q.3.38. Compute Naïve Bayes category scores for all entities in Table 3.38 and prove that the classifier correctly attributes them to their categories.
A. See Table 3.41

It should be mentioned that the Naïve Bayes computations here, as applied to the text categorization problem, follow the so-called multinomial model in which only terms present in the entities are considered—as many times as they occur. Another popular model is the so-called Bernoulli model, in which terms are assumed to be generated independently as binomial variables. The Bernoulli model based computations differ from these on two counts: first, the features are binary indeed so that only binary information, yes or no, of term occurrence is taken, and, second, for each term the event of its absence, along with its probability, is counted too (for more detail, see Manning et al. 2008; Mitchell 2010).

Table 3.39 Prior probabilities for Naïve Bayes rule for the data in Table 3.39 according to the bag-of-word conventions

Category	Prior probability / its logarithm	Total count	Term counts / Term probabilities (in thousands) / Logarithms of the probabilities									
F	333	37	4	6	1	3	3	4	1	6	1	8
	5.809		108	162	27	81	81	108	27	162	27	216
			4.7	5.1	3.3	4.4	4.4	4.7	3.3	5.1	3.3	5.4
E	333	42	4	3	4	7	7	3	4	6	2	2
	5.809		95	71	95	167	167	71	95	143	48	48
			4.6	4.3	4.6	5.1	5.1	4.3	4.6	5.0	3.9	3.9
H	333	41	2	1	7	4	4	8	6	1	7	1
	5.809		49	24	171	98	98	195	146	24	171	24
			3.9	3.2	5.1	4.6	4.6	5.3	5.0	3.2	5.1	3.2

There are three lines for each of the categories representing, from top to bottom, the term counts from Table 3.38, their probabilities multiplied by 1000 and rounded to an integer, and the natural logarithms of the probabilities

Table 3.40 Computation of category scores for entity E1 (first line) from Table 3.38 according to the logarithms of within-class feature probabilities; the maximum score is highlighted by bold font

Entity E1			2	0	1	2	2	0	0	0	1	0	0	
Category	Log(p_k)	Feature weights (probability logarithms)												Score
		Inner product												
F	5.809		3.6	5.1	3.3	3.4	3.4	3.7	3.3	3.3	5.1	3.3	5.4	35.2
			2 * 3.6	0	1 * 3.3	2 * 3.4	2 * 3.4	0	0	0	1 * 5.1	0	0	
E	5.809		3.6	3.3	3.6	5.1	5.1	3.3	3.6	3.6	5.0	3.9	3.9	**39.2**
			2 * 3.6	0	1 * 3.6	2 * 5.1	2 * 5.1	0	0	0	1 * 5.0	0	0	
H	5.809		3.9	3.2	5.1	3.6	3.6	5.3	5.0	5.1	3.2	5.1	3.2	33.5
			2 * 3.9	0	1 * 5.1	2 * 3.6	2 * 3.6	0	0	0	1 * 3.2	0	0	

There are two lines for each of the categories: that on top replicates the logarithms from Table 3.39 and that on the bottom computes the inner product

Table 3.41 Naïve Bayes category scores for the items in Table 3.38 with maxima highlighted using bold font

Articles	Category scores		
	F	E	H
F1	**37.7006**	35.0696	29.3069
F2	**28.9097**	25.9362	21.5322
F3	**23.9197**	20.1271	13.8723
F4	**38.276**	33.6072	30
E1	33.2349	**37.9964**	33.3322
E2	37.244	**43.1315**	40.2435
E3	43.1957	**43.5672**	40.8398
E4	21.1663	**23.9203**	19.4367
H1	25.8505	29.394	**33.5895**
H2	33.929	40.4527	**43.749**
H3	29.9582	35.3573	**38.3227**
H4	23.7518	28.8344	**33.8408**

3.10 Metrics of Accuracy

Consider a generic problem of learning a binary target feature, so that all entities belong to either class 1 or class 2. A decision rule, applied to an entity, generates a "prediction" which of these two classes the entity belongs to. The classifier may return some decisions correct and some erroneous. Let us pick one of the classes as that of our interest, say 1, then there can be two types of errors: false positives (FP)—the classifier says that an entity belongs to class 1 while it does not, and false negatives (FN)—the classifier says that an entity does not belong to class 1 while it does.

Let it be, for example, a lung screening device for testing against a lung cancer. Whilst established in a hospital cancer ward, on a selected sample of 200 patients sent by local surgeries for investigation, it may produce results that are presented in Table 3.42. Its rows correspond to the diagnosis by the screening device and the columns to the results of further, more elaborate and definitive, tests. This is a cross-classification contingency table, and it is frequently referred to as a confusion table.

There are 94 true positives TP and 98 true negatives TN in the table so that the total accuracy of the device can be rated as $(94 + 98)/200 = 0.96 = 96\%$.

Table 3.42 Confusion table of patients' lung screening test results

		True lung cancer		Total
		Yes	No	
Device's diagnosis	Yes	94	7	101
	Not	1	98	99
Total		95	105	200

Table 3.43 Contingency
table of volunteers' lung
screening test results

		True lung cancer		Total
		Yes	No	
Device's diagnosis	Yes	2	2	4
	Not	1	195	196
Total		3	197	200

Respectively, the numbers of false positives FP = 7, and false negatives FN = 1 sum up to 8 leading to 4% error rate. Yet there are significant differences between these two showing that the device is in fact better than the totals show. Indeed, the 7 FP are not that important, because patients with the suspected cancer will be investigated further in depth anyway so that their No-status will be restored, with the cost of further testing. In contrast, 1 FN may go out of the medical system and get their cancer untreated with the potential loss of life because of the error. This is an example of different costs associated with FP and FN errors. The device made just one serious error out of 95 true cancer cases. The TP rate, the proportion of correctly identified true cases, frequently referred to as recall or sensitivity, 94/95 = 98.9%, is impressive indeed. On the other hand, the precision, that is, the proportion of the 94 TP cases related to all cancer predicted cases, 101, is somewhat smaller, just 93% to reflect that FP rate is 7%. The difference between precision and sensitivity is somewhat averaged in the value of accuracy rate, 96% in this case, so that the accuracy rate works reasonably well here as a single characteristic of the quality of the testing device.

Yet in a situation in which there is a great disparity in the sizes of Yes and No classes, the accuracy rate fails to reflect the results properly. Consider, for example, results of the same device at a random sample of 200 individuals who have not been sent for the screening by doctors but rather volunteered to be screened from public at large (Table 3.43).

The accuracy rate at Table 3.43 is even greater than that at Table 3.42, (2 + 195)/200 = 98.5%. Yet both sensitivity, 2/3 = 66.7%, and precision, 2/4 = 50%, are quite mediocre. The high accuracy rate is caused by the very high specificity, the proportion of correctly identified No cases, 195/197 = 98.9%, and by the fact that there are very few Yes cases.

As to a single measure adequately reflecting sensitivity and precision, the one most popular is their harmonic mean, the F-measure, which is equal to $F = 2/(1/(2/3) + 1/(2/4)) = 2/(3/2 + 4/2) = 4/7 = 57.1\%$.

Case-Study 3.9. Prevalence and Quetelet Coefficients

If one looks at the record of the screening device according to Table 3.43, out of 4 cancer cases diagnosed, 2 are correct, and compares that with the prevalence of the cancer at the sample, 3 cases of 200—the difference is impressive indeed. This difference is exactly what is caught up in the concept of Quetelet coefficient $q(l/k)$ (see Sect. 3.6.1) at row $k = 1$ and column $l = 1$. This takes the relative difference between the conditional probability $P(1/1) = 2/4$ and the average probability

$P(l = 1) = 3/200$ which is referred to sometimes as the prevalence: $q(1/1) = (2/4-3/200)/(3/200) = 2 * 200/(3 * 4) - 1 = 33.33 = 3333\%$, quite a change. This high value probably explains the difference in sensitivity and specificity between Tables 3.43 and 3.42.

Indeed, a similar Quetelet coefficient at Table 3.42 is $q(1/1) = 94 * 200/(101 * 95) - 1 = 0.96 = 96\%$, a less than a 100% increase, which may convey the idea that Table 3.42 is much more balanced than Table 3.43. The accuracy measure works well at balanced tables and it does not at those that are not.

In general, the situation can be described by a confusion, or contingency, table between two sets of categories related to the class being predicted (1 or not) and the true class (1 or not), see Table 3.44 further on. Of course, if one changes the class of interest, the errors will remain errors, but their labels will change: false positives regarding class 1 are false negatives when the focus is on class 2, and vice versa.

Among popular indexes scoring the error or accuracy rates are the following:

FP rate = FP/(FP + TN)—the proportion of false positives among those not in 1; 1-FP rate is referred to sometimes as specificity—it shows the proportion of correct predictions among other, not class 1, entities.
TP rate = TP/(TP + FN)—the proportion of true positives in class 1; in information retrieval, this frequently is referred to as recall or sensitivity.
Precision = TP/(TP + FP)—the proportion of true positives in the predicted class 1.

These reflect each of the possible errors separately. There are indexes that try to combine all the errors, too. Among them the most popular are:

Accuracy = (TP + TN)/N—the total proportion of accurate predictions. Obviously, 1—Accuracy is the total proportion of errors.
F-measure = 2/(1/Precision + 1/Recall)—the harmonic average of Recall and Precision.

The latter measure is getting more popularity than the former because the Accuracy counts both types of errors equally, which may be at odds with the common sense in those frequent situations at which errors of one type are "more expensive" than the others. Consider, for example, the case of medical diagnostics in Tables 3.42, 3.43 and 3.44: a tumor wrongly diagnosed as malignant would cost much less than the other way around when a deadly tumor is diagnosed as benign. F-measure, to some extent, is more conservative because it, first, combines rates rather than counts, and, second, utilizes the harmonic mean which tends to be close to the minimum of the two, as can be seen from the statements in the next questions, Q.3.39 and Q.3.40.

Q.3.39. Consider two positive reals, a and b, and assume, say that $a < b$. Prove that the harmonic mean, $h = 2/(1/a + 1/b)$ stays within the interval between a and $2a$ however large the difference $b - a$ is.
A. Take b be $b = ka$ at some $k > 1$. Then $h = 2/(1/a + 1/(ka)) = 2ka/(1+ k)$. The coefficient at a, $2k/(1+ k)$, is less than 2, which proves the statement.

Table 3.44 A statistical representation of the match between the true class and predicted class. The entries are counts of the numbers of co-occurrences

		True class		
		1	Not	Total
Predicted Class	1	True Positives	False Positives	TP + FP
	Not	False Negatives	True Negatives	FN + TN
Total		TP + FN	FP + TN	N

Q.3.40. Consider two positive real values, a and b, and prove that their mean, $m = (a + b)/2$, and harmonic mean, $h = 2/(1/a + 1/b)$, satisfy equation $mh = ab$. **A.** Take the product $mh = [(a + b)/2][2/(1/a + 1/b)]$ and perform elementary algebraic operations.

More elaborate representation of errors of the two types can be achieved with the so-called receiver operating characteristics (ROC) graph analysis (see, for example, Fawcett 2006). ROC graphs are especially suitable in the cases of classifiers that have a continuous output such as Bayes classifiers. ROC graph is a 2D Cartesian plane plotting TP rate against FP rate so that the latter is shown on x-axis and the former, on y-axis (see Fig. 3.38).

To be specific, let us take a Bayes classifier's rule in (3.66) and change the ratio p_2/p_1 for an arbitrary threshold $d > 0$. Take now $d = d_1$ for a specific d_1, so that the rule now predicts class 1 if $f1(x)/f2(x) > d_1$. Count the proportions of true and false positives, $tp1$ and $fp1$, at this threshold and put the point $(fp1, tp1)$ onto a ROC graph. Then change d to d_2 and count the rates, $tp2$ and $fp2$, at this threshold. If, say, $d_2 > d_1$, then the TP rate can only decrease, because the number of positive

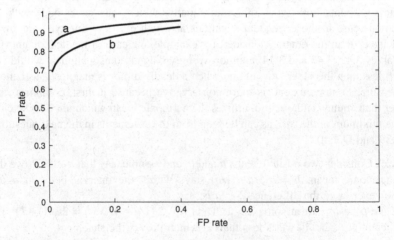

Fig. 3.38 ROC curves for two classifiers; that of a is superior to that of b

predictions can only decrease. The FP rate, in a regular case, should increase at $d_2 > d_1$ so that point $(fp2, ft2)$ would go to the right and above the former point on the ROC plot. In this way, by step-by-step changing the threshold d, one can obtain a ROC curve such as curves "a" and "b" on the plot of Fig. 3.38. Such a curve can be utilized as a characteristic of the classifier under consideration that can be used, for instance, for selection of suitable levels of TP and FP rates. In the case shown on Fig. 3.38, one can safely claim that classifier "a" is superior to that of "b", because at each FP rate level, TP rate of "a" is greater than that of "b".

There is a quantitative characteristic of the quality of a ROC, called the area under the curve (AUC), which is indeed the area between the ROC and the FP-rate-axis: the greater the AUC, the better the classifier. The AUC expresses the probability that for a random pair of objects in which one is a "yes" and the second a "no" object, the classifier correctly predicts their belongingness to the "yes" and "no" classes. Experiments show that the AUC of 0.9 or greater can be considered an excellent one.

3.11 Summary

The goal of this chapter is to present a significant variety of techniques for learning correlation from data. Most popular concepts—regression, correlation, chi-squared, discrimination, Bayes classifiers, decision trees, neural networks, support vector machine, and correspondence analysis—are presented. Some of these are accompanied with concepts that are interesting on their own such as the bag-of-words model or kernel. The description, though, is rather fragmentary, except perhaps the classification trees for which a number of theoretical results is invoked to show their firm relations to bivariate analysis: first, summary Quetelet indexes in contingency tables and, second, normalization options for dummy variables representing target categories.

Overall, the chapter contents reflect basics of the art of learning correlations from data. Perhaps the subject is by far too complex and major advances are a matter of the future rather than the past. One such advance, though, should be mentioned here—deep learning based on neural networks with many layers (see, for example, Schmidhuber 2015).

The Chapter outlines several important characteristics of summarization and correlation between two features, and displays some of the properties of those. They are:

- linear regression and correlation coefficient for two quantitative variables;
- tabular regression and correlation ratio for the mixed scale case; and
- contingency table, Quetelet index, statistical independence, and Pearson's chi-squared for two nominal variables.

They all are applicable in the case of multidimensional data as well.

Some of the characteristics described here are rather unconventional. For example, the concepts of tabular regression and correlation ratio are not terribly popular in data mining. The Quetelet indexes are recognized by neither community, the more so the idea that Pearson chi-squared is a summary correlation measure, not necessarily a criterion of statistical independence.

Some examples of non-linear regression and nature-inspired approaches for fitting that are outlined. Computational bootstrap based validation is considered.

References

J.-P. Benzecri, *Correspondence Analysis Handbook* (CRC Press, 1992). ISBN 10 0824784375

M. Berthold, D. Hand, *Intelligent Data Analysis* (Springer, Berlin, 2003)

L. Breiman, J.H. Friedman, R.A. Olshen, C.J. Stone, *Classification and Regression Trees* (Wadswarth, Belmont, Ca, 1984)

A.C. Davison, D.V. Hinkley, *Bootstrap Methods and Their Application*, 7th edn. (Cambridge University Press, Cambridge, 2005)

H.B. Demuth, M.H. Beale, O. De Jess, M.T. Hagan, *Neural network design* (Martin Hagan, 2014)

R.O. Duda, P.E. Hart, D.G. Stork, *Pattern Classification* (Wiley-Interscience, 2001). ISBN 0-471-05669-3

S.B. Green, N.J. Salkind, *Using SPSS for the Windows and Macintosh: Analyzing and Understanding Data* (Prentice Hall, 2003)

M. Greenacre, *Correspondence Analysis in Practice* (CRC Press, 2017)

P.D. Grünwald, *The Minimum Description Length Principle* (MIT Press, 2007)

J.F. Hair, W.C. Black, B.J. Babin, R.E. Anderson, *Multivariate Data Analysis*, 7th edn. (Prentice Hall, 2010). ISBN-10: 0-13-813263-1

J. Han, M. Kamber, J. Pei, *Data Mining: Concepts and Techniques*, 3rd edn. (Elsevier, 2011). ISBN: 978-9380931913

S.S. Haykin, *Neural Networks*, 2nd edn. (Prentice Hall, 1999). ISBN: 0132733501

A.Z. Israëls, *Eigenvalue Techniques for Qualitative Data* (Leiden, DSWO Press, 1987)

J. Kemeny, L. Snell, *Mathematical Models in Social Sciences* (New-York, Blaisdell, 1962)

M.G. Kendall, A. Stewart, *Advanced Statistics: Inference and Relationship*, 2nd edn. (Griffin, London, 1967)

L. Lebart, A. Morineau, M. Piron, *Statistique Exploratoire Multidimensionelle* (Dunod, Paris, 1995). ISBN 2-10-002886-3

C.D. Manning, P. Raghavan, H. Schütze, *Introduction to Information Retrieval* (Cambridge University Press, Cambridge, 2008)

B. Mirkin, *Group Choice* (Halsted Press, Washington, DC, 1979)

B. Mirkin, *Grouping in Socio-Economic Research* (Finansy i Statistika Publishers, Moscow, Russia, 1985)

B. Mirkin, *Mathematical Classification and Clustering* (Kluwer, AP, Dordrecht, 1996)

B. Mirkin, *Clustering: A Data Recovery Approach* (Chapman & Hall/CRC, 2012). ISBN 978-1-4398-3841-9

F. Murtagh, *Correspondence Analysis and Data Coding with Java and R* (Chapman & Hall/CRC, Boca Raton, FL, 2005)

T.M. Mitchell *Machine Learning* (McGraw Hill, 2010)

S. Nishisato, *Elements of Dual Scaling: An introduction to practical data analysis* (Psychology Press, 2014)

B. Polyak, *Introduction to Optimization* (Optimization Software, Los Angeles, 1987). ISBN 0911575146

J.R. Quinlan, *C4. 5: Programs for Machine Learning* (Morgan Kaufmann, San Mateo, 1993)

B. Schölkopf, A.J. Smola, *Learning with Kernels* (The MIT Press, 2005)

V. Vapnik, *Estimation of Dependences Based on Empirical Data*, 2nd edn. (Springer Science + Business Media Inc., 2006)

A. Webb, *Statistical Pattern Recognition* (Wiley, 2002). ISBN-0-470-84514-7

Articles

L. Breiman, Random forests. Mach. Learn. **45**(1), 5–32 (2001)

J. Bring, How to standardize regression coefficients. Am. Stat. **48**(3), 209–213 (1994)

J. Carpenter, J. Bithell, Bootstrap confidence intervals: when, which, what? A practical guide for medical statisticians. Stat. Med. **19**, 1141–1163 (2000)

H.E. Daniels, The relation between measures of correlation in the universe of sample permutations. Biometrika, **33**(2), 129–135 (1944)

F. Esposito, D. Malerba, G. Semeraro, A comparative analysis of methods for pruning decision trees. IEEE Trans. Pattern Anal. Mach. Intell. **19**(5), 476–491 (1997)

T. Fawcett, An introduction to ROC analysis. Pattern Recogn. Lett. **27**, 861–873 (2006)

D.H. Fisher, Knowledge acquisition via incremental conceptual clustering. Mach. Learn. **2**, 139–173 (1987)

P.J.F. Groenen, G. Nalbantov, J.C. Bioch, SVM-Maj: a majorization approach to linear support vector machines with different hinge errors. Adv. Data Anal. Classif. **2**(1), 17–43 (2008)

J.G. Kemeny, Mathematics without numbers. Daedalus **88**(4), 577–591 (1959)

L. Lebart, B.G. Mirkin, Correspondence analysis and classification, in *Multivariate Analysis: Future Directions*, vol. 2 ed. by C. Cuadras, C.R. Rao (North Holland, 1993), pp. 341–357

Y. LeCun, Y. Bengio, G. Hinton, Deep learning. Nature **521**(7553), 436 (2015)

W.Y. Loh, Y.S. Shih, Split selection methods for classification trees. Stat. Sin. 815–840 (1997)

E. Lombardo, F. Beh, P. Kroonenberg, Modelling trends in ordered correspondence analysis using orthogonal polynomials. Psychometrika **81**(325–349), 2016 (2016)

G. Louppe, L. Wehenkel, A. Sutera, P. Geurts, Understanding variable importances in forests of randomized trees. in *Advances in Neural Information Processing Systems* (*NIPS*), (2013), pp. 431–439

M. Meilă, Comparing clusterings—an information based distance, J. Multivar. Anal. **98**(5), 873–895 (2007)

B. Mirkin, L. Cherny, Some properties of the partition space, in K. Bagrinovsky, E. Berland (Eds.), Math. Anal. Econ. Models III, Institute of Economics of the Siberian Branch of the USSR's Academy of the Sciences, Novosibirsk, 126–147 (1972)

B. Mirkin, Eleven ways to look at the chi-squared coefficient for contingency tables. Am. Stat. **55** (2), 111–120 (2001)

B. Mirkin, T.I. Fenner Tied rankings, ordered partitions, and weak orders: Distance and consensus. J. Classif. **36**(2), (2019)

J.N. Morgan, J.A. Sonquist, Problems in the analysis of survey data, and a proposal. J. Am. Stat. Assoc. **58**, 415–435 (1963)

I. Morlini, S. Zani, A new class of weighted similarity indices using polytomous variables. J. Classif. **29**(2), 199–226 (2012)

K. Pearson, On a criterion that a given system of deviations from the probable in the case of a correlated system of variables is such that it can be reasonably supposed to have arisen in random sampling. Phil. Mag. **50**, 157–175 (1900)

J. Schmidhuber, Deep learning in neural networks: an overview. Neural Netw **61**, 85–117 (2015)

K. Steele, H.O. Stefánsson, Decision Theory, The Stanford Encyclopedia of Philosophy (Winter 2015 edn.), Edward N. Zalta (Ed.), http://plato.stanford.edu/archives/win2015/entries/decision-theory/

N.G. Waller, J.A. Jones, Correlation weights in multiple regression. Psychometrika **75**(1), 58–69 (2010)

Chapter 4
Core Partitioning: K-means
and Similarity Clustering

Abstract K-means is arguably the most popular cluster-analysis method. The method's output is twofold: (1) a partition of the entity set into clusters, and (2) centers representing the clusters. The method is rather intuitive and usually requires just a few pages to get presented. In contrast, this text includes a number of less popular subjects that are much important when using K-means for real-world data analysis:

- Data standardization, especially, at nominal or mixed scales
- Innate and other tools for interpretation of clusters
- Analysis of examples of K-means working and its failures
- Initialization—the choice of the number of clusters and location of centers.

Versions of K-means such as incremental K-means, nature inspired K-means, and entity-center "medoid" methods are presented. Three modifications of K-means onto different cluster structures are given: Fuzzy K-means for finding fuzzy clusters, Expectation-Maximization (EM) for finding probabilistic clusters, and Kohonen's self-organizing maps (SOM) that tie up the sought clusters to a visually convenient two-dimensional grid. An equivalent reformulation of K-means criterion is described to yield what we call the complementary criterion. This criterion allows to reinterpret the method as that for finding big anomalous clusters. In this formulation, K-means is shown to extend the Principal component analysis criterion to the case at which the scoring factors are supposed to be binary. This allows to address a haunting issue at K-means, finding the "right" number of clusters K, by one-by-one building Anomalous clusters. Section 4.6 is devoted to partitioning over similarity data. First of all, the complementary K-means criterion is equivalently reformulated as the so-called semi-average similarity criterion. This criterion is maximized with a consecutive merger process referred to as SA-Agglomeration clustering to produce provably tight, on average, clusters. This method stops merging clusters when the criterion does not increase anymore if the data has been pre-processed by zeroing the similarities of the objects to themselves. A similar process is considered for another natural criterion, the summary within-cluster similarity, for which two pre-processing options are considered. These are: a popular "modularity" clustering option, based on subtraction of random interactions,

B. Mirkin, *Core Data Analysis: Summarization, Correlation,*
and Visualization, Undergraduate Topics in Computer Science,
https://doi.org/10.1007/978-3-030-00271-8_4

and "uniform" partitioning, based on a scale shift, a.k.a. soft thresholding. Using either pre-processing option, the summary clustering also leads to an automated determination of the number of clusters. The chapter concludes with Sect. 4.7 on consensus clustering, a more recent concept. In the context of central partition for a given ensemble of partitions, two distance-between-partitions measures apply, both involving the so-called consensus matrix. The consensus similarity is defined, for any two objects, by the number of clusters in the ensemble to which both objects belong. This brings the issue of consensus into the context of similarity clustering, in the form of either the semi-average criterion or uniform partitioning criterion.

4.1 General: Clustering as Categorical Summarization

Clustering is a set of methods for finding and describing cohesive groups in data, typically, as "compact" clusters of entities in the feature space.

Consider data patterns in Fig. 4.1: a clear-cut cluster structure in part (a), a blob in (b), and an ambiguous "cloud" in (d).

Some argue that term "clustering" applies only to structures of the type presented on Fig. 4.1a and c, moreover, depending on the level of resolution, one may distinguish 3 or 7 clusters in (c). There are no "natural" clusters in the other two cases, Fig. 4.1b and d, however. Indeed, initially, the term was used to express a clear-cut clustering. But currently clustering has become synonymous to building a classification over empirical data, and as such it embraces all the situations in which data can be structured into cohesive chunks.

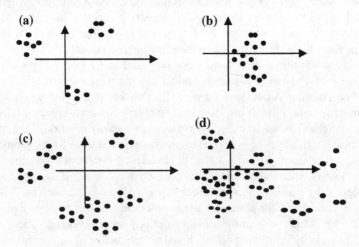

Fig. 4.1 Clear-cut cluster structures at (**a**) and (**c**); data clouds with no clear structure at (**b**) and (**d**)

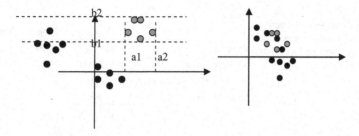

Fig. 4.2 Cluster of blank circles on the left is well described by the predicate $a1 < x < a2$ and $b1 < y < b2$. A similar cluster on the right cannot be accurately described by interval predicates without false positive and false negative errors

To serve as data summaries, clusters are to be not only found, but conceptually described as well.

A "class" always expresses a concept embedded into a fragment of knowledge—this is what is referred to, in logics, as the class' "intension", in contrast to empirical instances of the class constituting what is referred to as the class' "extension", e.g., the concept of "tree" versus real plants growing here and there. Therefore, two dual intelligent activities—cluster finding and cluster describing—should be exercised both when clustering.

As Fig. 4.2 illustrates, a cluster is rather easy to describe by combining corresponding feature intervals when it is clear-cut geometrically. This knowledge-driven data analysis perspective can be reflected in dividing all cluster finding techniques in the following categories:

(a) clusters are to be found directly in terms of features—this is frequently referred to as conceptual clustering;
(b) clusters are to be found simultaneously with a transformation of the feature space making them clear-cut—this direction started very recently and is not well shaped yet;
(c) clusters are to be found as subsets of entities first, so that the description comes as a follow-up stage—this is the genuine clustering which covers most of the clustering activities so far.

There have been developed several different clustering approaches of which K-means is by far most popular. Table 4.1 presents the numbers of pages claimed by popular search engines as returned in response to querying several popular clustering techniques—they all will be explained in this text further on (the query experiment took place June 6, 2018, in Moscow Russia). Only "single linkage" may challenge the overall superiority of K-means, having mysteriously received support by the Bing search engine; almost 2.8 million pages.

Table 4.1 The numbers of pages, in thousands, claimed by popular search engines as returned in response to querying cluster-analysis approaches

Query	Search engine		
	Google	Bing	Yandex
K-means	915,000	779	64,000
Single linkage	43,900	2770	23,000
Agglomerative clustering	418	88	8000
Spectral clustering	8320	271	12,000

4.2   K-means Clustering

4.2.1   Batch K-means Partitioning

K-means is a major clustering technique, of type (c), that is present, in various forms, in all major statistical packages such as SPSS and SAS, as well as data mining packages such as Clementine, iDA tool and DBMiner. It is very popular in many application areas such as image analysis, marketing research, bioinformatics, and medical informatics.

In general, the cluster finding process according to K-means is run in a Euclidean space. It starts from K tentative centers and repeatedly applies two steps:

(a) collecting clusters of points around centers,
(b) updating centers as within cluster means,

  – until convergence.

This makes much sense—whichever centers are suggested first, as hypothetical cluster tendencies, they are checked then against the real data and moved to the areas of higher density.

In its generic, so-called Batch mode, K-means can be formulated as comprising the following steps 0–3 illustrated on Fig. 4.3 for $K = 3$ and entity set I:

Batch K-means

0. *Initialization*: The user chooses the number K of clusters and puts K hypothetic cluster centers among the entity points, see Fig. 4.3a.
1. *Cluster update*: Given K centers c_k ($k = 1, 2, ..., K$), each of the entities $i \in I$ is assigned to one of the centers according to Minimum distance rule: distances between i and each c_k are calculated, and i is assigned to the nearest c_k, see Fig. 4.3b. For each center c_k, the entities assigned to it form cluster S_k ($k = 1, 2, ..., K$), see Fig. 4.3c.
2. *Center update*: At each of the given K clusters S_k, its gravity center is computed and set as the new center c'_k ($k = 1, 2, ..., K$), see Fig. 4.3d.
3. *Halting test*: New centers c'_k are compared with those from the previous iteration. If $c'_k = c_k$ for all $k = 1, 2, ..., K$, stop and output both c'_k and S_k for all $k = 1, 2, ..., K$. Otherwise, set c'_k as c_k and go to "1. Cluster update step".

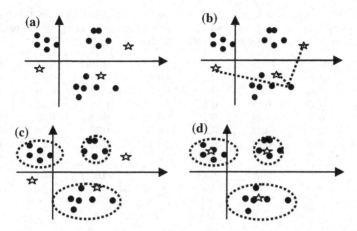

Fig. 4.3 Main steps of Batch K-means: **a** initialization of centers, **b** cluster update using Minimum distance rule (the pointed lines show distances from an entity to all centers), **c** cluster update completed, **d** center update completed

The algorithm is appealing in several aspects. Conceptually, it may be considered a model for the human process of typology making, with types represented by clusters S_k and centers c_k. Also, it has nice mathematical properties. This method is computationally easy, fast and memory-efficient. However, researchers and practitioners point to some less desirable properties of K-means. Specifically, they refer to lack of advice with respect to

(a) the initial setting, i.e. the number of clusters K and initial positioning of centers,
(b) instability of clustering results with respect to the initial setting and data standardization, and
(c) insufficient interpretation aids.

These issues can be alleviated, to an extent, by looking more attentively into the properties of the mathematical criterion driving it, as will be explained later in this section.

A decoder-based summarization model underlying the method is that the entities are assigned to clusters in such a way that each cluster is represented by its center, sometimes referred to as the cluster's standard point or prototype. This point expresses, intensionally, the typical tendencies of the cluster.

Worked Example 4.1. K-means Clustering of Company Data
Consider the standardized Company data in Table 2.18 copied here as Table 4.2.

This data set can be visualized with two principal components as presented in Project 2.1 (Sect. 2.3).

Let entities An, Br and Ci be initial centers of three clusters. One can computationally compare each of the entities with each of the centers to decide which center better represents an entity. To compare two points, Euclidean squared distance is a natural choice (see Table 4.3).

Table 4.2 The Company data standardized by: (i) shifting to the feature averages, (ii) dividing by the feature ranges, and (iii) further dividing the category based three columns by $\sqrt{3}$. Contributions of the features to the data scatter are presented in the bottom

Av	−0.20	0.23	−0.33	−0.63	0.36	−0.22	−0.14
An	0.40	0.05	0	−0.63	0.36	−0.22	−0.14
As	0.08	0.09	0	−0.63	−0.22	0.36	−0.14
Ba	−0.23	−0.15	−0.33	0.38	0.36	−0.22	−0.14
Br	0.19	−0.29	0	0.38	−0.22	0.36	−0.14
Bu	−0.60	−0.42	−0.33	0.38	−0.22	0.36	−0.14
Ci	0.08	−0.10	0.33	0.38	−0.22	−0.22	0.43
Cy	0.27	0.58	0.67	0.38	−0.22	−0.22	0.43
Cnt	0.74	0.69	0.89	1.88	0.62	0.62	0.50
Cnt, %	12.42	11.66	14.95	31.54	10.51	10.51	8.41

Table 4.3 Computation of squared Euclidean distance between rows Av and An in Table 4.2 as the sum of squared differences between corresponding components

Points	Coordinates							d(An, Av)
An	0.40	0.05	0.00	−0.63	0.36	−0.22	−0.14	
Av	−0.20	0.23	−0.33	−0.63	0.36	−0.22	−0.14	
An-Av	0.60	−0.18	0.33	0	0	0	0	
Squares	0.36	0. 03	0.11	0	0	0	0	0.50

Table 4.4 Distances between three company entities chosen as centers and all the companies; each company column shows three distances between the company and centers—the highlighted minima present best matches between the centers and companies

Point	Av	An	As	Ba	Br	Bu	Ci	Cy
An	**0.50**	**0.00**	**0.77**	1.55	1.82	2.99	1.90	2.41
Br	2.20	1.82	1.16	**0.97**	**0.00**	**0.75**	0.83	1.87
Ci	2.30	1.90	1.81	1.22	0.83	1.68	**0.00**	**0.61**

According to the Minimum distance rule, an entity is assigned to its nearest center (see Table 4.4 in which all distances between the entities and centers are presented; those chosen according to the Minimum distance rule are highlighted in bold.) entities assigned to the same center form a tentative cluster around it. Clusters found at Table 4.4 are S1 = {Av, An, As}, S2 = {Ba, Br, Bu}, and S3 = {Ci, Cy} as illustrated in Fig. 4.4. These are the product classes already, but this is irrelevant to the computation. K-means procedure has its own logic that needs to ensure that the new tentative clusters lead to the same centers.

One needs to proceed further on and update centers by using the information of the assigned clusters. New centers are defined as centers of the tentative clusters, whose components are the averages of the corresponding components within the clusters; these are presented in Table 4.5.

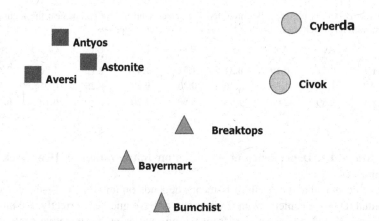

Fig. 4.4 Table 4.2 rows on the plane of two first principal components: it should not be difficult to discern clusters formed by products: distances within A, B and C groups are smaller than between them

Table 4.5 Tentative clusters from Table 4.2 and their centers

Av	−0.20	0.23	−0.33	−0.63	0.36	−0.22	−0.14
An	0.40	0.05	0	−0.63	0.36	−0.22	−0.14
As	0.08	0.09	0	−0.63	−0.22	0.36	−0.14
Center 1	0.10	0.12	−0.11	−0.63	0.17	−0.02	−0.14
Ba	−0.23	−0.15	−0.33	0.38	0.36	−0.22	−0.14
Br	0.19	−0.29	0	0.38	−0.22	0.36	−0.14
Bu	−0.60	−0.42	−0.33	0.38	−0.22	0.36	−0.14
Center 2	−0.21	−0.29	−0.22	0.38	−0.02	0.17	−0.14
Ci	0.08	−0.10	0.33	0.38	−0.22	−0.22	0.43
Cy	0.27	0.58	0.67	0.38	−0.22	−0.22	0.43
Center 3	0.18	0.24	0.50	0.38	−0.22	−0.22	0.43

The updated centers differ from the previous ones. Thus, we must update their cluster lists by using the distances between updated centers and entities; the distances are presented in Table 4.6. As it is easy to see, the Minimum distance rule assigns centers again with the same entity lists.

Therefore, the process has stabilized—if we repeat it all over again, nothing new would ever come—the same centers and the same assignments. The process stops at this point, and the found clusters along with their centers are returned (they are, in the standardized format, in Table 4.5).

The result obviously depends on the standardization of the data, performed beforehand, as the method heavily relies on the squared Euclidean distance and, thus, on the relative weighting of the features, just like PCA.

Table 4.6 Distances between the three updated centers and all the companies; the highlighted column minima present best matches between centers and companies

Point	Av	An	As	Ba	Br	Bu	Ci	Cy
Center 1	**0.22**	**0.19**	**0.31**	1.31	1.49	2.12	1.76	2.36
Center 2	1.58	1.84	1.36	**0.33**	**0.29**	**0.25**	0.95	2.30
Center 3	2.50	2.00	1.95	1.69	1.20	2.40	**0.15**	**0.15**

Case-Study 4.1. Dependence of K-means on Initialization: A Drawback and Advantage

The bad news is that the result of K-means depends on initialization—the choice of the initial tentative centers, even if we know, or have guessed correctly, the number of clusters K. Indeed, if we start from wrong entities as the tentative centers, the result can be rather disappointing.

In some packages, such as SPSS (Green and Salkind 2003), K first entities are taken as the initial centers. Why not start from rows for Av, An and As then? Taking these three as the initial tentative centers will stabilize the process at wrong clusters S1 = {Av, Ba, Br}, S2 = {An}, and S3 = {As, Bu, Ci, Cy} (see Q.4.4). But what else can be expected if all centers are taken from the same cluster?

However, even if the initial centers are taken from right clusters, this would not necessarily warrant good results either. Start, for example, from Av, Ba and Ci (note, these relate to different products!); the final result will be rather odd— S1 = {Av, An, As}, S2 = {Ba, Bu}, and S3 = {Br, Ci, Cy} because Br should be in B-cluster S2.

Figure 4.5 illustrates the fact that such instability is not because of a specially designed example but rather an ordinary phenomenon. There are two clear-cut clusters on Fig. 4.5, that can be thought of as uniformly distributed sets of points,

Fig. 4.5 Case of two clear-cut clusters and two different initializations: (a) and (b). Case (a) results in a correct separation of the clusters, case (b) not

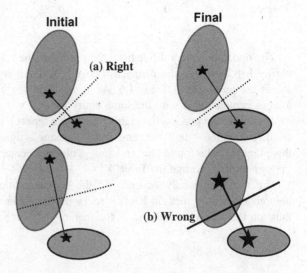

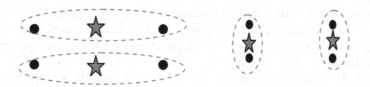

Fig. 4.6 An example of K-means failure: two clusterings at a four-point-set with K-means—that intuitive on the right and that counter-intuitive on the left, with stars as centers

and two different initializations, symmetric one on the (a) part and not symmetric one on (b) part. The Minimum distance rule at $K = 2$ amounts to drawing a hyperplane that orthogonally cuts through the middle of the line between the two centers; the hyperplane is shown on Fig. 4.5 as the line separating the centers. In Fig. 4.5, the case (a) presents initial centers that are more or less symmetric so that the line through the middle separates the clusters indeed. In the case (b), initial centers are highly asymmetric so that the separating line cuts through one of the clusters, thus distorting the position of the further centers; the final separation still cuts through one of the clusters and, therefore, is utterly wrong.

There is one more property of K-means clusters illustrated by Fig. 4.5: they are convex. Indeed, the Minimum Distance rule assigns each center with the intersection of half-spaces formed by the orthogonal cutting hyperplanes.

Another example of non-optimality of K-means is presented in Fig. 4.6 which involves four points only. There are two settings there, that on the left and that on the right. In both settings, centers represented by stars, are final in the algorithm, but that on the right leads to a much smaller value for the criterion (4.3) further on.

However, the non-optimality of K-means can be of an advantage, too—in those cases when K-means criterion leads to solutions that are counterintuitive such as those in which the fact that K-means favors equal cluster sizes brings good results, as described in Case-Study 4.2 below.

Case-Study 4.2. Uniform Clusters Can Be Too Costly

Here is an example when the square-error clustering criterion leads to a solution which is at odds with intuition, however that cannot be reached with batch K-means algorithm because of its local nature.

Consider the case of Fig. 4.7 that presents three sets of points, two consisting of big clumps of say 100 entities each, around points A and B, and a small one around point C, consisting say of just one entity located at that point. Assume that the distance between A and B is 2, and between B and C, 10. There can be only two 2-cluster partitions possible: (I) 200 of A and B entities together in one cluster while

Fig. 4.7 Three clumps of points to be clustered using the K-means clustering at $K = 2$: which two will join together, A and B or B and C?

the second cluster consists of just one entity in C; (II) 100 A entities for one cluster while 101 entities in B and C for the other. The third partition, consisting of cluster B and cluster A + C, cannot be optimal because cluster A + C is more outstretched than a similar cluster B + C in (II).

Let us compare the values of K-means criterion using the squared Euclidean distance between entities and their centers.

In the case (I), center of cluster A + B will be located in the middle of the interval between A and B, thus on the distance 1 from each, leading to the total squared Euclidean distance 200*1 = 200. Since cluster C contains just one entity, it adds 0 to the value of K-means criterion, which is 200 in this case.

In the case (II), cluster B + C has center, which is the gravity center, between B and C distanced from B by d = 10/101. Thus, the total value of K-means criterion here is $100*d^2 + (10 - d)^2$ which is less than $100*(1/10)^2 + 10^2 = 101$ because d < 1/10 and 10 - d < 10. Cluster A contributes 0 because all 100 entities are located in A which is, therefore, the center of this cluster.

Case (II) wins by a great margin: K-means criterion, in this case, favors more equal distribution of entities between clusters in spite of the fact that case (I) is intuitive and case (II) is not: A and B are much closer to each other than B and C.

Yet Batch K-means algorithm leads to non-optimal, but intuitive, case (I) rather than optimal, but odd, case (II), if started from the most distant points A and C as initial centers—an intuitively appealing option. Indeed, according to Minimal distance rule, all entities in B will join A, thus resulting in (I) clustering.

Case-Study 4.3. Robustness of K-means Criterion with Data Normalization

Let us generate two 2D clusters, of 100 and 200 elements, respectively. First cluster— a Gaussian spherical distribution with the mean in point (1, 1) and the standard deviation 0.5. The second cluster, of 200 elements, is uniformly randomly distributed in the rectangle of the length 40 and width 1, put over axis x either at $y = 5$ (Fig. 4.8, on the left) or at $y = 3$ (Fig. 4.8, on the right). K-means criterion of course cannot separate these two if applied in the original space; its criterion value will be minimized

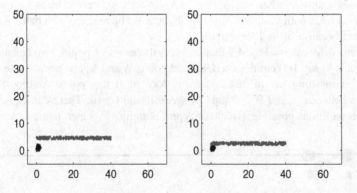

Fig. 4.8 A set of two differently shaped clusters, a circle and rectangle; the y-coordinates of their centers are 1 (circle) and 4.5 (rectangle) on the left, and 1 and 3.5, on the right

by dividing the set somewhere closer to one fourth of the strip of the rectangular cluster so that the split parts will have approximately 150 points each.

Yet after the data standardization, with the grand means subtracted and range-normalized, the clusters on the left part of Fig. 4.8 are perfectly separable with K-means criterion, that does attain its minimum value, at $K = 2$, on the cluster-based partition of the set. This holds with z-scoring, too.

This tendency changes, though, at a less structured case on the right of Fig. 4.8. The best split indeed holds at about $x = 10$ in this case. At a random sample, 32 points of the rectangular cluster join circular cluster in the optimal split. Curiously, the z-scoring standardization, in this case, works towards a better recovery of the structure so that the optimal 2-cluster partition, at the same data sample, merges only 5 rectangular cluster elements into the circular cluster, thus splitting the rectangular cluster over a mark $x = 5$.

These results do not much change when we go to Gaussian similarities (affinity data) defined by formula $G(x, y) = e^{-d(x,y)/2\sigma^2}$ where $d(x, y)$ is the squared Euclidean distance between x and y if $x \neq y$, and σ^2 is equal to 10 at the original data and 0.5 at the standardized data, in the manner of spectral clustering (see Sect. 5.2) and apply algorithm AddRem from Sect. 5.4.

4.2.2 Batch K-means and Its Criterion

4.2.2.1 Batch K-means as Alternating Minimization

The cluster structure in K-means is specified by a partition S of the entity set in K non-overlapping clusters, $S = \{S_1, S_2, \ldots, S_K\}$ represented by lists of entities S_k, and cluster centers $c_k = (c_{k1}, c_{k2}, \ldots, c_{kV})$, $k = 1, 2, \ldots, K$.

There is a model that can be thought of as that driving K-means algorithm. According to this model, each entity, represented by the corresponding row of Y matrix as $y_i = (y_{i1}, y_{i2}, \ldots, y_{iV})$, belongs to a cluster, say S_k, and is equal, up to small residuals, or errors, to the cluster's center:

$$y_{iv} = c_{kv} + e_{iv} \text{ for all } i \in S_k \text{ and all } v = 1, 2, \ldots, V \qquad (4.1)$$

Equations (4.1) define as simple a decoder as possible: whatever entity belongs to cluster S_k, its data point in the feature space is decoded as center c_k.

The problem is to find such a partition $S = \{S_1, S_2, \ldots, S_K\}$ and cluster centers $c_k = (c_{k1}, c_{k2}, \ldots, c_{kV})$, $k = 1, 2, \ldots, K$, that minimize the square error of decoding

$$L^2 = \sum_{i \in I} \sum_{v \in V} e_{iv}^2 = \sum_{k=1}^{K} \sum_{i \in S_k} \sum_{v \in V} (y_{iv} - c_{kv})^2 \qquad (4.2)$$

Fig. 4.9 The distances—
lines connecting centers with
entity points—in criterion W
(S, c)

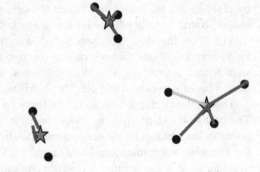

Criterion (4.2) can be equivalently reformulated in terms of the squared Euclidean distances as the summary distance between entities and their cluster centers (see (4.3) and Fig. 4.9). Please note that the number of distances in the sum is N and does not depend on the number of clusters.

$$L^2 = W(S, c) = \sum_{k=1}^{K} \sum_{i \in S_k} d(y_i, c_k) \qquad (4.3)$$

This is because the distance referred to as squared Euclidean distance is defined, for any V-dimensional $x = (x_v)$ and $y = (y_v)$ as $d(x, y) = (x_1 - y_1)^2 + (x_2 - y_2)^2 + \ldots + (x_V - y_V)^2$ so that the rightmost summation symbol in (4.2) leads to $d(y_i, c_k)$ indeed.

This criterion depends on two groups of variables, S and c, and thus can be minimized by the alternating minimization method which proceeds by repetitively applying the same minimization step: Given one group of variables, optimize criterion over the other group, and so forth, until convergence.

Specifically, given centers $c = (c_1, c_2, \ldots, c_K)$, find a partition S minimizing the summary distance (4.3). Obviously, to choose a partition S, one should choose, for each entity $i \in I$, one of K distances $d(y_i, c_1), d(y_i, c_2), \ldots, d(y_i, c_K)$. The choice to minimize (4.3) is according to the Minimum distance rule: for each $i \in I$, choose the minimum of $d(y_i, c_k), k = 1, \ldots, V$, that is, assign any entity to its nearest center.

When there are several nearest centers, the assignment is taken among them arbitrarily. In general, some centers may be assigned no entity at all with this rule.

The other step in the alternating minimization would be minimizing (4.3) over c at a given S. The solution to this problem comes from the additive format of criterion (4.2) that provides for the independence of cluster center components from each other. As was indicated in Sect. 2.2.2, it is the mean that minimizes the square error, and thus the within-cluster mean vectors minimize (4.3) over c at given S.

Thus, starting from an initial set of centers, $c = (c_1, c_2, \ldots, c_K)$, the alternating minimization method for criterion (4.3) will consist of a series of repeated applications of two steps: (a) clusters update—find clusters S according to the Minimum distance rule, (b) centers update—make centers equal to within cluster mean vectors. The computation stops when new clusters coincide with those on the previous step. This is exactly the K-means in its Batch mode.

The convergence of the method follows from two facts: (i) at each step, criterion (4.3) can only decrease, and (ii) the number of different partitions S is finite.

Q.4.1. How many distances are summed in $W(S, c)$? (**A**: This is equal to the number of entities N.) Does this number depend on the number of clusters K? (**A**: No.) Does the latter imply: the greater the K, the less the $W(S, c)$? (**A**: Yes.) Why?

Q.4.2. Why is convergence guaranteed for K-means?

A. Because K-means is alternating minimization process at which criterion $W(S, c)$ may only decrease at each step. Convergence follows from the fact that there are only a finite number of different partitions on I.

Q.4.3. Assume that $d(y_i, c_k)$ in $W(S, c)$ is city-block distance rather than Euclidean squared. Could K-means be adjusted to make it alternating minimization algorithm for the modified $W(S, c)$?

A. (Yes, just use the city-block distance through, as well as within cluster median points rather than gravity centers.) Would this make any difference? (Yes, it will; especially at skewed distributions of the variables.)

Q.4.4. Demonstrate that, at Companies data, value $W(S, c)$ at product-based partition $\{1-2-3, 4-5-6, 7-8\}$ is lower than at partition $\{1-4-6, 2, 3-5-7-8\}$ found at seeds 1, 2 and 3.

A. Indeed the sums of within-cluster distances to cluster centers in the product based clusters are 0.7193, 0.8701, 0.3070, respectively, totaling to 1.8964, whereas the sums of the second partition are 1.4411, 0, 2.1789 and sum up to 3.62.

Q.4.5. Demonstrate that, at Companies data, value $W(S, c)$ at product-based partition $\{1-2-3, 4-5-6, 7-8\}$ is lower than at partition $\{1-2-3, 4-6, 5-7-8\}$ found at seeds 1, 4 and 7.

A. Indeed, the sums of within-cluster distances to cluster centers in the product-based clusters are 0.7193, 0.8701, 0.3070, respectively, totaling to 1.8964, whereas the sums of the second partition are 0.7193, 0.4413, 1.1020 that total to 2.2624.

Q.4.6. Can example of Fig. 4.6 or its modification lead to a similar effect for the case of least-modules criterion related to the city-block distance and median rather than average centers? Can it be further extended to PAM method which uses city-block distance and "medoid" entities rather than coordinates, as described in Sect. 4.2.5?

4.2.2.2 A Pseudo-code for Batch K-means: Computation

To summarize, an application of K-means clustering involves the following steps:

0. Select a data set.
1. Standardize the data.
2. Choose number of clusters K.
3. Define K hypothetical centers (seeds).

4. Clusters update: Assign entities to the centers according to Minimum distance rule.

5. Centers update: define centers as the gravity centers of thus obtained clusters.

6. Iterate 4. and 5. until convergence.

MatLab codes for the items 4 and 5 can be put as follows.

4. Clusters update: Assign points to the centers according to Minimum distance rule:

Given data matrix X and a $K \times V$ array of centers cent, produce an N-dimensional array of cluster labels for the entities and the summary within cluster distance to centers, wc:

```
function [labelc,wc]=clusterupdate(X,cent)
    [K,m]=size(cent);
    [N,m]=size(X);
    for k=1:K
            cc=cent(k,:); %center of cluster k
            Ck=repmat(cc,N,1);
            dif=X-Ck;
            ddif=dif.*dif; %Nxm matrix of squares
            dist(k,:)=sum(ddif');
    %distances from entities to cluster center
    end
    [aa,bb]=min(dist); %Minimum distance rule
    wc=sum(aa);
    labelc=bb;
    return
```

4. Centers update: Put centers in gravity centres of clusters defined by the array of cluster labels *labelc* according to data in matrix X, to produce $K \times V$ array centres of the centers:

```
function      centres=ceupdate(X,labelc)
            K=max(labelc);
            for k=1:K
        clk=find(labelc==k);
        elemk=X(clk,:);
        centres(k,:)=mean(elemk);
    end
    return
```

Batch K-means with MatLab, therefore, is to embrace steps 3–6 above and output a clustering in cell array termed, say, Clusters, along with the proportion of unexplained data scatter found by using preliminarily standardized matrix X and set of initial centers, cent, as input. This can be put like this:

```
function [Clusters,uds]=k_means(X,cent)
  [N,m]=size(X);
  [K,m1]=size(cent);
  flag=0; %-- stop-condition
  membership=zeros(N,1);
  dd=sum(sum(X.*X)); %-- data scatter
  %--- clusters and centers updates
  while flag==0
        [labelc,wc]=clusterupdate(Y,cent);
        if isequal(labelc,membership)
        %--stop-condition's working
                flag=1;
                centre=cent;
                w=wc;
        else
                cent=ceupdate(Y,labelc);
                membership=labelc;
        end
  end
  %-----preparing the output --------------
  uds=w*100/dd;
  Clusters{1}=membership;
  Clusters{2}=centre;
  return
```

Q.4.7. Check the values of criterion (4.3) at each initial setting considered for Company data above. Find out which is the best among them.

Q.4.8. Prove that the square-error criterion (4.2) can be reformulated as the sum of within cluster variances $\sigma_{kv}^2 = \Sigma_{i \in Sk} (y_{iv} - c_{kv})^2 / N_k$ weighted by the cluster cardinalities N_k:

$$W(S, c) = \sum_{k=1}^{K} N_k \sum_{v \in V} \sigma_{kv}^2 \tag{4.4}$$

Q.4.9. Prove that the following criterion (4.5) is equivalent to criterion (4.3) in terms of entity-to-entity distances only—without any reference to centers at all:

Table 4.7 Cross classification of the original Iris taxa and 3-cluster clustering found starting from entities 1, 51 and 101 as initial seeds. The clustering does separate Iris Setosa but misplaces 14 + 3=17 specimens between two other taxa

Cluster	Setosa	Versicolor	Virginica	Total
S1	50	0	0	50
S2	0	47	14	61
S3	0	3	36	39
Total	50	50	50	150

$$D(S) = \sum_{k=1}^{K} \sum_{i,j \in S_k} d(y_i, y_j)/N_k \qquad (4.5)$$

where $d(y_i, y_j)$ is the squared Euclidean distance between i and j's rows.

A. This can be proven by substituting the definition of center c_{kv}, $c_{kv} = \sum_{i \in Sk} y_{kv}/N_k$, through feature values in formula (4.2).

Q.4.10. Prove that if Batch K-means is applied to Iris data, preprocessed by subtracting the feature means only, with K = 3 and specimens 1, 51, and 101 taken as the initial centers, the resulting clustering cross-classified with the prior three 50-strong classes forms contingency table presented in Table 4.7.

Q.4.11. Prove that if Batch K-means is applied to Iris data, mean-range normalized with K = 3 and specimens 1, 51, and 101 taken as the initial centers, the resulting clustering cross-classified with the prior three classes forms contingency table presented in Table 4.7.

In the following two sections we describe two approaches at reaching deeper minima of K-means criterion (4.3): (a) an incremental version and (b) nature inspired versions.

4.2.3   Incremental K-means

An incremental version of K-means uses the Minimum distance rule not for all of the entities but for one of them only. There can be two different reasons for doing so:

(**Ri**) The user is not able to operate over the entire data set and takes in the entities one by one, because of the data protocol, so that entities are to be clustered on the fly as, for instance, in an on-line application process.

(**Rii**) The user operates with the entire data set, but wants to smooth the action of the algorithm so that no drastic changes in the cluster contents may occur. To do this, the user may specify an order of the entities and run entities one-by-one in this order for a number of epochs like it is done in a neural network learning process.

The result of such a one-by-one entity processing may differ from that of Batch K-means because each version finds a locally optimal solution on a different structure of locality neighborhoods.

Q.4.12. What is the difference in neighborhoods between Batch and incremental versions of K-means?

When an entity y_i joins cluster S_k whose cardinality is N_k, center c_k changes to c'_k, according to the following formula:

$$c'_k = N_k c_k/(N_k + 1) + y_i/(N_k + 1)$$

When y_i moves out of cluster S_k, the formula remains valid if all the pluses are changed for minuses. To extend the formula so that it holds for both cases, let us introduce variable z_i which is equal to $+1$ when y_i joins the cluster and -1 when it moves out of it. Then the extended formula is:

$$c'_k = N_k c_k/(N_k + z_i) + y_i z_i/(N_k + z_i)$$

Accordingly, the distances from other entities change to $d(y_j, c'_k)$.

Because of the incremental setting, the stopping rule of the straight version (reaching a stationary state) may be not necessarily applicable here. In **Ri** case, the natural stopping rule is to end when there are no new entities observed. In **Rii** case, the process of running through the entities one-by-one stops when all entities remain in their clusters. The process may be stopped as well when a pre-specified number of runs through the entity set, that is, epochs, is reached.

Q.4.13. Consider a run of incremental K-means at situation **Rii** on the Companies data, at which the order of entities follows the order of their distances to nearest centers. Let K = 3 and entities Av, Ba and Ci initial centers.
A. Sequential steps of the incremental computation are presented in Table 4.8. In this table, cluster updates are provided as well as their centers after each single update. The column on the right presents squared Euclidean distances between centers and entities yet unclustered, with the minima highlighted in bold. The minimum distance determines, in this version, which of the entities joins the clustering next. One can see that on iteration 2 company Br switches to center Ba after center Ci of the third cluster had been updated to the mean of Ci and Cy— because its distance to the new center increased from the minimum 0.83–1.20. This leads to correct, product-based, clusters.
Q.4.14. Prove that the same initialization leads to wrong, that is, non-product based, clusters with Batch K-means.

Table 4.8 Iterations of incremental K-means on standardized Company data starting with centers Av, Ba and Ci; minima are highlighted in bold to signify candidates for joining

Incremental one-by-one entity clustering

Iteration	Cumulative clusters	Centers							Distances				
									An	As	Br	Bu	Cy
0	Av	-0.20	0.23	-0.33	-0.62	0.36	-0.22	-0.14	0.51	0.88	2.20	2.25	3.01
	Ba	-0.23	-0.15	-0.33	0.38	0.36	-0.22	-0.14	1.55	1.94	0.97	0.87	2.46
	Ci	0.08	-0.10	0.33	0.38	-0.22	-0.22	0.43	1.90	1.81	0.83	1.68	0.61
1	Av, An	0.10	0.14	-0.17	-0.62	0.36	-0.22	-0.14	0.70	1.88	2.50	2.59	
	Ba	-0.23	-0.15	-0.33	0.38	0.36	-0.22	-0.14	1.94	0.97	0.87	2.46	
	Ci	0.08	-0.10	0.33	0.38	-0.22	-0.22	0.43	1.81	0.83	1.68	**0.61**	
2	Av, An	0.10	0.14	-0.17	-0.62	0.36	-0.22	-0.14		**0.70**	1.88	2.50	
	Ba	-0.23	-0.15	-0.33	0.38	0.36	-0.22	-0.14		1.94	0.97	0.87	
	Ci, Cy	0.18	0.24	0.50	0.38	-0.22	-0.22	0.43		1.95	1.20	2.40	
3	Av, An, As	0.10	0.12	-0.11	-0.62	0.17	-0.02	-0.14		1.49	2.12		
	Ba	-0.23	-0.15	-0.33	0.38	0.36	-0.22	-0.14		0.97	**0.87**		
	Ci, Cy	0.18	0.24	0.50	0.38	-0.22	-0.22	0.43		1.20	2.40		
4	Av, An, As	0.10	0.12	-0.11	-0.62	0.17	-0.02	-0.14			1.49		
	Ba, Bu	-0.42	-0.29	-0.33	0.38	0.07	0.07	-0.14			**0.64**		
	Ci, Cy	0.18	0.24	0.50	0.38	-0.22	-0.22	0.43			1.20		
5	Av, An, As	0.10	0.12	-0.11	-0.62	0.17	-0.02	-0.14					
	Ba, Bu, Br	-0.21	-0.29	-0.22	0.38	-0.02	0.17	-0.14					
	Ci, Cy	0.18	0.24	0.50	0.38	-0.22	-0.22	0.43					

4.2.4   *Nature Inspired Algorithms for K-means*

4.2.4.1   Nature Inspired Algorithms

In real-world applications, K-means typically does not move far away from the initial setting of centers. Considered in the perspective of minimization of criterion (4.3), this leads to the strategy of multiple runs of K-means starting from randomly generated sets of centers to reach as deep a minimum of (4.3) as possible. This strategy works well on illustrative small data sets but it may fail when the data set is large because in this case random settings cannot cover the space of solutions in a reasonable time. Nature inspired approach provides a well-defined framework for using random centers in parallel, rather than in sequence, to channel them to deeper minima as an evolving population of admissible solutions. The main difference of the nature inspired optimization from the classical optimization is that the latter reaches for a single solution, provably optimal, whereas the former runs a population of solutions and does not much care for the provability.

A nature inspired algorithm mimics some natural process to set rules for the population behavior and/or evolution. Among the nature inspired approaches, the following are especially popular:

A. Genetic
B. Evolutionary
C. Particle swarm optimization

A K-means method according to each of these will be described in this section.

A nature inspired algorithm proceeds as a sequence of steps of evolution for a population of possible solutions, that is, clusterings represented by specific data structures. A K-means clustering comprises two items: a partition S of the entity set I in K clusters and a set of clusters' K centers $c = \{c_1, c_2, ..., c_K\}$. Typically, only one of them is carried out in a nature-inspired algorithm. The other is easily recovered according to K-means rules. Given a partition S, centers c_k are found as vectors of within cluster means. Given a set of centers, each cluster S_k is determined as the set of points nearest to its center c_k, according to the Minimum distance rule $(k = 1, 2, ..., K)$. Respectively, the following two representations are most popular in nature inspired algorithms:

 (i) Partition as a string,
(ii) Centers as a string.

Consider them in turn.

(i) **Partition as a string**

Having pre-specified an order of entities, a partition S can be represented as a string of cluster labels $k = 1, 2, ..., K$ of the entities thus ordered. If, for instance, there are eight entities ordered as e1, e2, e3, e4, e5, e6, e7, e8, then the string 12333112 represents partition S with three parts according to the labels,

S1 = {e1, e6, e7}, S2 = {e2, e8}, and S3 = {e3, e4, e5}, which can be easily seen
from the diagram relating the entities and labels:

$$
\begin{array}{cccccccc}
e1 & e2 & e3 & e4 & e5 & e6 & e7 & e8 \\
1 & 2 & 3 & 3 & 3 & 1 & 1 & 2
\end{array}
$$

A string of N integers from 1 to K is considered not admissible, if some integer
between 1 and K is absent from it (so that the corresponding cluster is empty). Such
a not admissible string for the entity set above would be 11333111, because it lacks
label 2 and, therefore, makes class S2 empty.

(ii) **Centers as a string**

Consider the same partition as in (i) on the set of eight objects, the companies in
Company data Table 4.2 in their order. Clusters S1 = {e1, e6, e7}, S2 = {e2, e8},
and S3 = {e3, e4, e5}, as well as their centers, are presented in Table 4.9. The three
centers form a sequence of $7 \times 3 = 21$ numbers c = (−0.24, −0.09, −0.11, 0.04,
−0.02, −0.02, 0.05, 0.34, 0.31, 0.33, −0.12, 0.07, −0.22, 0.14, 0.01, −0.12, −0.11,
0.04, −0.02, 0.17, −0.14), which suffices for representing the clustering: the
sequence can be easily converted back in three 7-dimensional center vectors to
recover then clusters with the Minimum distance rule. It should be pointed out that
the original clusters may be somewhat weird and not recoverable in this way. For
example, entity e4, which is Ba, appears to be nearer to center 1 rather than to
center 3 so that the Minimum distance rule would produce clusters S1 = {e1, e4,
e6, e7}, S2 = {e2, e8}, and S3 = {e3, e5} rather than those original ones, but such
a loss makes no difference, because K-means clusters necessarily satisfy the
Minimum distance rule so that all the entities are nearest to their cluster's centers.

What is important is that any 21-dimensional sequence of real values can be
treated as the clustering code for its centers.

Table 4.9 Centers of clusters S1 = {e1, e6, e7}, S2 = {e2, e8}, and S3 = {e3, e4, e5} according
to data in Table 4.2

Av	−0.20	0.23	−0.33	−0.63	0.36	−0.22	−0.14
Bu	−0.60	−0.42	−0.33	0.38	−0.22	0.36	−0.14
Ci	0.08	−0.10	0.33	0.38	−0.22	−0.22	0.43
Center 1	−0.24	−0.09	−0.11	0.04	−0.02	−0.02	0.05
An	0.40	0.05	0	−0.62	0.36	−0.22	−0.14
Cy	0.27	0.58	0.67	0.38	−0.22	−0.22	0.43
Center 2	0.34	0.31	0.33	−0.12	0.07	−0.22	0.14
As	0.08	0.09	0.00	−0.62	−0.22	0.36	−0.14
Ba	−0.23	−0.15	−0.33	0.38	0.36	−0.22	−0.14
Br	0.10	−0.29	0.00	0.38	−0.22	0.36	−0.14
Center 3	0.01	−0.12	−0.11	0.04	−0.02	0.17	−0.14

4.2.4.2   GA for K-means Clustering

Genetic algorithms work over a population of strings, each representing an admissible solution and referred to as a chromosome. The optimized function is referred to as the fitness function. Let us use a string representation for partitions $S = \{S_1, ..., S_K\}$ of the entity set. The minimized fitness function is the summary within-cluster distance to centers, the function $W(S, c)$ in (4.3):

0. *Initial setting.* Specify an even integer P for the population size (no rules exist for this), and randomly generate P chromosomes, that is, strings $s_1, ..., s_P$ of K integers 1, ..., K in such a way that all K integers 1, 2, ..., K are present within each chromosome. For each of the strings, define corresponding clusters, cal-culate their centers as gravity centres and the value of criterion, $W(s_1), ..., W(s_P)$, according to formula (4.3).

1. *Mating selection.* Choose P/2 pairs of strings; each pair to mate and produce two "children" strings. The mating pairs usually are selected randomly (with replacement, so that the same string may appear in several pairs and, moreover, can form both parents in a pair). To mimic Darwin's "survival of the fittest" law, the probability of selection of string s_t (t = 1, ..., P) should reflect its fitness value $W(s_t)$. Since the fitness is greater for the smaller W value, some make the probability inversely proportional to $W(s_t)$ (see Murthy and Chowdhury 1996) and some to the difference between a rather large number and $W(s_t)$ (see Lu et al. 2004). This latter approach can be taken further with the probability proportional to the explained part of the data scatter—in this case the "rather large number" taken to be the data scatter rather than any arbitrary real.

2. *Cross-over.* For each of the mating pairs, generate a random number r between 0 and 1. If r is smaller than a pre-specified probability p (typically, p is taken about 0.7–0.8), then perform a cross-over; otherwise the mates themselves are considered the result. A (single-point) cross-over of string chromosomes $a = a_1a_2...a_N$ and $b = b_1b_2...b_N$ is performed as follows. A random number n between 1 and N − 1 is selected and the strings are crossed over to produce children $a_1a_2...a_nb_{n+1}...b_N$ and $b_1b_2...b_na_{n+1}...a_N$. If a child is not admissible (like, for instance, strings $a = 11133222$ and $b = 32123311$ crossed over at $n = 4$ would produce $a' = 11133311$ and $b' = 32123222$ so that a' is inad-missible because of absent 2), then various policies can be applied. Some authors suggest the cross-over operation to be repeated until an admissible pair is produced. Some say inadmissible chromosomes are ok, just they must be assigned with a smaller probability of selection.

3. *Mutation.* Mutation is a random alteration of a character in a chromosome. This provides a mechanism for jumping to different "ravines" of the minimized fitness function. Every character in every string is subject to the mutation pro-cess, with a low probability q which can be constant or inversely proportional to the distance between the corresponding entity and corresponding center.

4. *Elitist survival.* This strategy suggests keeping the best fitting chromosome(s) stored separately. After the cross-over and mutations have been completed, find

fitness values for the new generation of chromosomes. Check whether the worst of them is better than the record or not. If not, put the record chromosome instead of the worst one into the population. Then find the record for thus obtained population. If this record is better than the stored elit value, change the stored value for the current one.

5. *Halt condition.* Check the stop condition (typically, a limit on the number of iterations). If this doesn't hold, go to 1; otherwise, halt.

Lu et al. (2004) note that such a GA works much faster if after step 3. Mutation the labels are changed according to the Minimum distance rule. They apply this instead of the elitist survival.

Thus, a GA algorithm operates with a population of chromosomes representing admissible solutions. To update the population, mates are selected, undergone a cross-over process generating offspring which then is subjected to mutation process. Elite maintenance completes the update. In the end, the elite is output as the best solution.

A computational shortcoming of the GA algorithm is that the length of the chromosomes is the size of the entity set N, which may run in millions in contemporary applications. Can this be overcome? Sure, by using center not partition strings to represent a clustering. Center string sizes depend on the number of features and number of clusters, not the number of entities. Another advantage of center strings is in the mutation process. Rather than an abrupt swap between literals, they can be changed softly, in a quantitative manner by adding or subtracting a small change. This is utilized in evolutionary and particle swarm algorithms.

4.2.4.3   Evolutionary K-means

Here, the chromosome is represented by a set of K centers $c = (c_1, c_2, \ldots c_K)$ which can be considered a string of KV real ("float") numbers. In contrast to the partition-as-string representation, the length of the string here does not depend on the number of entities that can be of advantage when the number of entities is massive. Each center in the string is analogous to a gene in a chromosome.

The cross-over of two center strings c and c', each of the length KV, is performed at a randomly selected place n, $1 \leq n < KV$, exactly as it is in the genetic algorithm above. Chromosomes c and c' exchange the portions lying to the right of n-th component to produce two offsprings. This means that, a number of centers in c is substituted by corresponding centers in c'. Moreover, if n cuts across a center, its components change in each of the offspring chromosomes.

The process of mutation, according to Bandyopadhyay and Maulik (2002), can be organized as follows. Given the fitness W values of all the chromosomes, let $minW$ and $maxW$ denote their minimum and maximum respectively. For each chromosome, its radius R is defined as a proportion of $maxW$ reached at it: $R = (W - minW)/(maxW - minW)$. When the denominator is 0, that is, if $minW = maxW$, define $R = 1$ in all chromosomes. Here, W is the fitness value of the chromosome under consideration. Then the mutation intensity δ is generated randomly in the interval between $-R$ and $+R$.

Let *minv* and *maxv* denote the minimum and maximum values in the data set along feature v ($v = 1, ..., V$). Then every v-th component xv of each center c_k in the chromosome changes to

$$xv + \delta * (maxv - xv) \text{ if } \delta \geq 0 \text{ (increase), or}$$

$$xv + \delta * (xv - minv), \text{ otherwise (decrease)}.$$

The perturbation leaves chromosomes within the hyper-rectangle defined by boundaries *minv* and *maxv*. Please note that the best chromosome, at which $W = minW$, does not change in this process because its $R = 0$.

Elitism is maintained in the process as well.

The algorithm follows the scheme outlined for the genetic algorithm.

Based on little experimentation, this algorithm is said to outperform the previous one, GA, many times in terms of the speed of convergence.

The evolutionary approach can be further modified such as, for example, the so-called Differential evolution (see Paterlini and Krink 2006 who claim that this method outperforms the others in K-means). In Differential evolution, the cross-over, mutation and elite maintenance are merged together by removing the mating stage and changing those for the following. An offspring chromosome is created for every chromosome t in the population ($t = 1, ..., P$) as follows. Three other chromosomes, k, l and m, are taken randomly from the population. Then, for every component (gene) $x.t$ of the chromosome t, a uniformly random value r between 0 and 1 is drawn. This value is compared to the pre-specified probability p (somewhat between 0.5 and 0.8). If $r > p$ then the component goes to the offspring unchanged. Otherwise, this component is substituted by the linear combination of the same component in the three other chromosomes: $x.m + \alpha*(x.k - x.l)$ where α is a small scaling parameter. After the offspring's fitness is evaluated, it substitutes chromosome t if it is better; otherwise, t remains as is and the process applies to the next chromosome.

4.2.4.4 Particle Swarm Optimization for K-means

Particle swarm mimics a drift of a bee population so that the population members here are not crossbred, nor they mutate. They just move randomly by drifting in random directions having an eye on the best places visited so far, individually and socially. This can be done because they are vectors of real numbers. Because of the change, the genetic metaphor is abandoned here, and the elements are referred to as particles rather than chromosomes, and the set of them as a swarm rather than a population.

Each particle comprises:

- a position vector x that is an admissible solution to the problem in question (such as the KV center vector in K-means),
- the evaluation of its fitness $f(x)$ (such as the summary distance W in (4.3)),
- a velocity vector z of the same dimension as x, and
- a record of the best position b reached by the particle so far.

The swarm best position bg is determined as the best among all the individual best positions b.

At iteration t ($t = 0, 1, \ldots$) the next iteration's position is defined as the current position shifted by the velocity vector:

$$x(t+1) = x(t) + z(t+1)$$

where $z(t+1)$ is computed as a change in the direction of personal and population's best positions:

$$z(t+1) = z(t) + \alpha(b - x(t)) + \beta(bg - x(t))$$

where

- α and β are uniformly distributed random numbers (typically, within the interval between 0 and 2, so that they are around unity),
- item $\alpha(b - x(t))$ is referred to as the cognitive component and
- item $\beta(bg - x(t))$ as the social component of the process.

Initial values $x(0)$ and $z(0)$ are generated randomly within the manifold of admissible values.

In some implementations, the group best position bg is changed for that of local best position bl that is defined by the particle's neighbors only so that some pre-defined neighborhood topology makes its effect. There is a report that the local best position works especially well, in terms of the depth of the minimum reached, when it is based on just two Euclidean neighbors.

Q.4.15. Formulate a particle swarm optimization algorithm for K-means clustering.

4.2.5   Partition Around Medoids PAM

K-means centers are average points rather than individual entities, which may be considered artificial in contexts in which the user may wish to consider but only genuinely occurring real world entities rather the "synthetic" averages. Estates or art objects or countries are examples of entities for which this restriction seems adequate. To implement the idea, let us change the concept of cluster prototype from center to *medoid* (Kaufman and Rousseeuw 1990). An entity in a cluster S, $i* \in S$, is referred to as its medoid if it is the nearest in S to all other elements of S, that is, if i^* minimizes the sum of distances $D(i) = \Sigma_{j \in S} d(i, j)$ over all $i \in S$. The symbol $d(i, j)$ is used here to denote any dissimilarity function, which may or may not be squared Euclidean distance, between observed entities $i, j \in I$.

The method of partitioning around medoids PAM (Kaufman and Rousseeuw 1990) works exactly as Batch K-means with the only difference that medoids, not centers, are used as cluster prototypes. It starts, as usual, with choosing the number of clusters K and initial medoids $c = (c_1, c_2, ..., c_K)$ that are not just M-dimensional points but individual entities. Given medoids c, clusters S_k are collected according to the Minimum distance rule—as sets of entities that are nearest to entity c_k for all $k = 1, 2, ..., K$. Given clusters S_k, medoids are updated according to the definition. This process reiterates again and again, and halts when no change of the clustering occurs. It obviously will never leave a cluster S_k empty. If the size of the data set is not large, all computations can be done over the entity-to-entity distance matrix without ever changing it.

Worked Example 4.2. PAM Applied to Company Data
Let us apply PAM to the Company data displayed in Table 4.2 with K = 3 and entities Av, Br and Cy as initial medoids. We can operate over the distance matrix, presented in Table 4.10.

With seeds, Av, Br, Cy, the Minimum distance rule would obviously produce the product-based clusters A, B, and C. At the next iteration, clusters' medoids are computed: they are obviously An in A cluster, Bu in B cluster and either of the two entities in C cluster—leave it thus at the less controversial Cy. With the set of medoids changed to An, Bu and Cy, we apply the Minimum distance rule again, leading us to the product-based clusters again. This halts the process.

Note that PAM can lead to instability in results because the assignment depends on distances to just a single entity.

Q.4.16. Why Cy is less controversial than Ci in Table 4.10?
A. Because Cy unequivocally relates to Ci only, whereas Ci is close to Br as well.

Table 4.10 Distances between standardized company entities

Entities	Av	An	Ast	Ba	Br	Bu	Ci	Cy
Av	**0.00**	**0.51**	**0.88**	1.15	2.20	2.25	2.30	3.01
An	**0.51**	**0.00**	**0.77**	1.55	1.82	2.99	1.90	2.41
As	**0.88**	**0.77**	**0.00**	1.94	1.16	1.84	1.81	2.38
Ba	1.15	1.55	1.94	**0.00**	**0.97**	**0.87**	1.22	2.46
Br	2.20	1.82	1.16	**0.97**	**0.00**	**0.75**	**0.83**	1.87
Bu	2.25	2.99	1.84	**0.87**	**0.75**	**0.00**	1.68	3.43
Ci	2.30	1.90	1.81	1.22	**0.83**	1.68	**0.00**	**0.61**
Cy	3.01	2.41	2.38	2.46	1.87	3.43	**0.61**	**0.00**

For the sake of convenience, those smaller than 1, are highlighted in bold

Q.4.17. Assume that the distance d(Bre, Bum) in Table 4.10 is 0.85 rather than 0.75. Show that then if one chooses Civ to be medoid of C cluster, then the Minimum distance rule would assign to Civ not only Cyb but also Bre, because its distance to Civ, 0.83, would be less than its distance to Bum, 0.85. Show that this cluster, {Civ, Cyb, Bre} will remain stable over successive iterations.

4.2.6   Initialization of K-means: Conventional Approaches

To initialize K-means, one needs to specify:

  (i)   the number of clusters, K, and
 (ii)   initial centers, $c = (c_1, c_2, ..., c_K)$.

Each of these can be of an issue in practical computations. Both depend on the user's expectations related to the level of granularity and typological attitudes, which remain beyond the scope of the theory of K-means. This is why some suggest relying on the user's view of the substantive domain to specify the number and positions of initial centers as hypothetical prototypes. There have been however a number of approaches for specifying the number and location of the initial centers by exploring the structure of the data, of which we describe the following three:

(a)   multiple runs of K-means;
(b)   distant representatives;
(c)   algorithm "Build"

Further on we are going to present yet another approach based on the complementary criterion derived in Sect. 4.2.

(a)   Multiple runs of K-means

According to this approach, at a given K, a number of K-means' runs R is pre-specified; each run starting with K randomly selected entities as the initial seeds (randomly generated points within the feature ranges have proven to give inferior results in experiments reported by several authors). Then the best result in terms of the square-error criterion $W(S, c)$ (4.3) is output. This can be further extended to choosing the "right" number of clusters K. Let us denote by W_K the minimum value of $W(S, c)$ found after R runs of K-means over random initializations. Then the series W_K found at different K, from a pre-specified range say between 2 and 20, is usually taken to see which K would lead to the best W_K over the range. Unfortunately, the best W_K is not necessarily minimum W_K, because the minimum value of the square-error criterion cannot increase when K grows, which should be reflected in the empirically found W_K's. In the literature, a number of stop criteria utilizing W_K have been suggested based on some simplified data models and intuition such "Gap" and "Jump" statistics. Unfortunately, they all may fail even in the relatively simple situations of controlled computation experiments (see Chiang and Mirkin 2010 for a review).

A relatively simple heuristic rule is based on the intuition that when there are K^* well separated clusters, then for $K < K^*$ a $(K + 1)$-cluster partition should be the K-cluster partition with one of its clusters split in two, which would drastically decrease W_{K+1} from W_K. On the other hand, at $K > K^*$, both K- and $(K + 1)$-cluster partitions are to be the "right" K^*-cluster partition with some of the "right" clusters split randomly, so that W_K and W_{K+1} are not that different. Therefore, as "a crude rule of thumb", Hartigan (1975, p. 91) proposed calculating index

$$H_K = (W_K/W_{K+1} - 1)(N - K - 1),$$

where N is the number of entities, while increasing K, so that the very first K at which H_K becomes smaller than 10 is to be taken as the estimate of K^*. It should be noted that, in the experiments by Chiang and Mirkin (2010), this rule came as the best of a set of nine different criteria and, moreover, the threshold 10 in the rule appeared to be not very sensitive to 10–20% changes.

Case-Study 4.4. Hartigan's Index for Choosing the Number of Clusters

Consider values of H_K for Iris and Town datasets computed after the results of 100 runs of Batch K-means using the mean/range standardization starting from random K entities taken as seeds (Table 4.11). Each of the computations has been repeated twice (see 1st and 2nd sets in Table 4.11) to illustrate typical variations of H_K values due to the fact that empirical values of W_K may be not optimal. In particular, at the 2nd set of K-means over Town data we can see a break of the rule that H_K is positive because of the monotonic relation between K and the optimal W_K that are to decrease when K grows. The monotonic relation here is broken because the values of W_K after 100 runs are not necessarily minimal indeed.

The "natural" number of clusters in Iris data, according to Hartigan's criterion is not 3 as claimed because of substantive considerations but much greater, 11! In Town data set, the criterion would indicate 4 naturally occurring clusters. However, one should argue that the exact value of 10 in Hartigan's rule does not bear much credibility—it should be accompanied by a significant drop in H_K value. We can see such a drop at K = 5, which should be taken, thus, as the "natural" number of

Table 4.11 Values of Hartigan's H_K index for two data sets at K ranging from 2 to 11 as based on two different sets of 100 clusterings from random K entities as initial centers

Dataset		$K = 2$	3	4	5	6	7	8	9	10	11
Iris	1st set	108.3	38.8	29.6	24.1	18.6	15.0	14.1	15.4	15.4	**9.4**
	2nd set	108.3	38.8	29.6	24.1	18.7	15.4	15.6	15.7	14.0	**7.2**
Town	1st set	13.2	10.5	**9.3**	5.0	4.7	3.1	3.0	3.2	3.2	1.6
	2nd set	13.2	10.5	**9.3**	5.8	4.1	2.5	3.0	7.2	−0.2	1.8

clusters in Town data. Similarly, a substantial drop of H_K on Iris data occurs at K = 3, which is the number of natural clusters, taxa, in this set.

Altogether, making multiple runs of K-means seems a sensible strategy, especially when the number of entities is not that high. With the number of entities growing into thousands, the number of tries needed to reach a representative value of W_K may become prohibitively large. Deeper minima can be sought by using the evolutionary schemes described above. On the other hand, the criterion $W(S, c)$ has some intrinsic flaws and should be used only along some domain-knowledge or data-structure based strategy.

(b) Mutually distant points

An attractive idea would be to take a number of entities one by one on the principle that each is as far away from the others as possible. These faraway points then are assigned to be the initial centers. For example, one may wish to start from two entities such that the distance between them is the greatest, then take an entity whose average distance to these is maximum. Et cetera. Unfortunately, this method does not work in practical computations (see, for example, Steinley and Brusco 2007). Probably, too often the distant points do not represent a cluster: it is not enough to have a distant point; to make it into a cluster center, it must be surrounded by some other data points. This is taken into account in the next approach.

(c) "Build" algorithm for a pre-specified K (Kaufman and Rousseeuw 1990)

This process involves only actual entities. It starts with choosing the medoid of set I, that is, the entity whose summary distance to the others is minimum, and takes it as the first medoid c_1. Assume that a subset of m initial seeds have been selected already $(K > m \geq 1)$ and proceed to selecting c_{m+1}. Denote the set of already selected seeds by c and consider all remaining entities $i \in I-c$. Define distance $d(i, c)$ as the minimum of the distances $d(i, c_k)$ $(k = 1, ..., m)$ and form an auxiliary cluster A_i consisting of such j that are closer to i than to c so that $E_{ij} = d(j, c)-d_{ij} > 0$. The summary value $E_i = \sum_{j \in A_i} E_{ij}$ reflects both the number of points in A_i and their remoteness from c. That $i \in I-c$ for which E_i is maximum is taken as the next seed c_{m+1}.

Worked Example 4.3. Selection of Initial Medoids in Company Data

Let us apply Build algorithm to the matrix of entity-to-entity distances for Company data displayed in Table 4.10, at K = 3. First, we calculate the summary distances from all the entities to the others, see Table 4.12, and notice that Bre is the medoid of the entire set I, because its total distance to the others, 9.60, is the minimum of total distances in Table 4.12. Thus, we set Bre as the first initial seed.

Table 4.12 Summary distances for entities according to Table 4.10

Entity	Av	An	As	Ba	Br	Bu	Ci	Cy
Distance to others	12.30	11.95	10.78	10.16	9.60	13.81	10.35	14.17

Now we build auxiliary clusters A_i around all other entities. To form A_{Av}, we take the distance between Av and Br, 2.20, and see, in Table 4.11, that distances from Av to entities An, As, and Ba are smaller than that, which makes them Av's auxiliary cluster with $E_{Av} = 4.04$. Similarly, A_{An} is set to consist of the same entities, but it is less remote than Ave because $E_{An} = 2.98$ is less than E_{Av}. Auxiliary cluster A_{As} consists of Av and An with even smaller $E_{As} = 0.67$. Auxiliary clusters for Ba, Ci and Cy consist of one entity each (Bu, Cy and Ci, respectively) and have much smaller the levels of remoteness; cluster A_{Bu} is empty because Br is its nearest. This makes the most remote entity Ave the next selected seed. Now, we can start building auxiliary clusters on the remaining six entities again. Of them, clusters A_{An} and A_{Bu} are empty and the others are singletons, of which A_{Cy} consisting of Ci is the remotest, with $E_{Cy} = 1.87-0.61 = 1.24$. This completes the set of initial seeds: Br, Av, and Cy. Note, these are companies producing different products. It is this set that was used to illustrate PAM in Sect. 4.2.5.

4.2.7   Notes on Software for K-means

This is based on this author's chapter in Hennig et al. (2015). Although the K-means method deserves to be coded as a standalone application supplied with powerful tools for getting a meaningful cluster structure by using a wide spectrum of computation and interpretation devices, the method rather appears in a generic version in most popular computational platforms and packages. In this section, a brief description of implementations of the method will be given in computational platforms:

- Matlab (see http://www.mathworks.com/products/matlab/),
- Weka (see http://www.cs.waikato.ac.nz/ml/weka/),
- R (see http://www.r-project.org/),
- SPSS (see http://www-01.ibm.com/software/analytics/spss/).

K-means clustering in Matlab takes data matrix X and the number of clusters K as its inputs and performs a run of batch K-means clustering with the squared Euclidean distances by default. It allows using some different distances, first of all the city block and cosine, as well. Other user selected parameters include:

- An initialization method: selecting K random rows from X, or a matrix of user-specified K points in the feature space, or results of preliminary clustering of a 10% random subsample of the dataset, or the so-called K-means++ initialization.
- Number of runs with different initializations.
- The way of reacting to the fact that some cluster happens to get empty: this can be an "error" message or removal of empty clusters or making a singleton instead.

- The maximum number of iterations so that the computation halts upon reaching that number (100, by default).
- A possibility of further minimization of the criterion by moving individual entities between clusters.

In Weka, K-means is implemented as SimpleKMeans command. This command automatically handles a mixture of categorical and numerical attributes. To this end, the program converts all nominal attributes into binary numeric attributes and normalizes scales of all the numeric features in the dataset to lie within the interval [0, 1]. When prompted, the program replaces all missing values for nominal and numeric attributes with the modes and means of features in the training data. Initialization is made randomly from among the observations. Instance weighting can be included if needed. The Weka SimpleKMeans algorithm uses Euclidean distance measure to compute distances between instances and centers. It can visualize the results with figures drawn for each cluster so that features are represented by lines radiating from the center.

In R, Project for Statistical Computing, K-means clustering is available in either batch version or incremental one. It works either with random K observations taken as initial centers or with a user-specified set of centers. The maximum number of iterations can be specified, as well as the number of runs from random initializations. In the Weka version, $K = 1$ is made possible. If an empty cluster emerges, an "error" message appears.

The statistics package SPSS allows to use its rich system for both data pre-processing and interpretation of results.

Its implementation of K-means is similar to that in Matlab—all the parameters available in Matlab are available in SPSS, except for the way for handling missing values. There is a different system for handling missing data in SPSS. Additional features include the possibility for storing the clusters set as a nominal variable, ANOVA table for clusters to allow finding most contributing variables, and the matrix of distances between found cluster centers.

Currently, some efforts are devoted to parallelization of K-means computations to apply it to big data, that is, data with millions or more entities. The method is well suited for distributed computations. Say, a big-data method is based on the idea that the dataset is partitioned into portions processed at different processors. An iteration is then performed as follows: a central processor supplies a set of centers so that each processor computes: (a) distances from its portion entities to the centers and (b) within cluster sums, after which it supplies the central processor with the sums and number of entities. Then the central processor sums all the received input within each cluster and computes the cluster centers.

Further development of cloud computations as a commodity will show what versions of the algorithm will be required in the future.

4.3 Complementary Criterion for K-means: Spectral and Anomalous Clusters

4.3.1 Complementary K-means Criterion

Let us consider any cluster S_k and define c_k as the vector of within-cluster means so that $c_{kv} = \Sigma_{i \in Sk} y_{kv}/N_k$ where N_k is the number of entities in S_k ($k = 1, 2, ..., K$). Take the square error criterion $L^2 = W(S, c)$ (4.2) and apply little algebraic manipulation to it:

$$W(S,c) = \sum_{k=1}^{K} \sum_{i \in S_k} \sum_{v=1}^{V} (y_{iv} - c_{kv})^2$$

$$= \sum_{k=1}^{K} \sum_{i \in S_k} \sum_{v=1}^{V} (y_{iv}^2 - 2y_{iv}c_{kv} + c_{kv}^2) = \sum_{i=1}^{N} \sum_{v=1}^{V} y_{iv}^2 - \sum_{k=1}^{K} N_k <c_k, c_k > .$$

The expression

$$B(S,c) = \sum_{k=1}^{K} N_k <c_k, c_k > \tag{4.6}$$

on the right has appeared after the sum $\Sigma_{i \in Sk} y_{kv}$ has been substituted by its equal, $N_k c_{kv}$.

Let us denote

$$T(Y) = \sum_{i=1}^{N} \sum_{v=1}^{V} y_{iv}^2, \tag{4.7}$$

the data scatter. Then the equation above can be rewritten as

$$T(Y) = B(S, c) + W(S, c) \tag{4.8}$$

Equation (4.8) decomposes data scatter $T(Y)$ in two parts: that one explained by the cluster structure (S, c), which is $B(S, c)$, and the unexplained part which is $W(S, c)$. The larger the explained part, the smaller the unexplained part, and the better the match between clustering (S, c) and data. This equation is fully analogous to the similar decomposition (2.20) and (2.24) of the data scatter over the Principal Component Analysis model, which is not surprising at all because the models are much similar. The comment on the Pythagorean meaning of the decomposition remains valid as well. The only difference is that here items $B(S, c)$ and $W(S, c)$ standing for the two sides of a right angle triangle, stand for the model and its error. Another associated fact is that Eq. (4.8) is well known in the analysis of variance in statistics; items $B(S, c)$ and $W(S, c)$ are referred to in that other context as between-group and within-group variance.

What is most important here, though, is that the clustering's contribution to the data scatter $B(S, c)$ in (4.6) is a complementary criterion to $W(S, c)$. Since the data scatter is constant regarding the issue of minimizing $W(S, c)$, then one may equivalently maximize $B(S, c)$ instead, to find an optimal clustering. Can we find out, what is the meaning of this complementary criterion? Easily. Here is a purely geometric view, as there are other meanings as well.

Since the sum of c_{kv}^2 over v is but the squared Euclidean distance between 0 and c_k,

$$B(S, c) = \sum_{k=1}^{K} \sum_{v \in V} c_{kv}^2 N_k = \sum_{k=1}^{K} N_k d(0, c_k) \qquad (4.9)$$

The criterion on the right in formula (4.9) was first mentioned, under the name of "criterion of distant centers", by Mirkin (1996, p. 292). To maximize the criterion on the right in formula (4.9), two conditions must be met. First, the clusters should be big, consist of many entities (maximizing N_k). Second, the clusters should be as far away from 0 as possible (maximizing $d(0, c_k) = \langle c_k, c_k \rangle$). There can be two approaches to (sub)maximization of (4.9) considered. One is to build clusters one by one, which is implemented in the method of Anomalous clusters (see Sect. 4.3.3). The other would be to build clusters in parallel—an option glossed over in this text.

4.3.2   The Complementary Criterion and Spectral Approach to Clustering

Another expression of the cluster-explained part of the data scatter is

$$B(S, c) = \sum_{k=1}^{K} \sum_{i \in S_k} \langle y_i, c_k \rangle \qquad (4.10)$$

which can be derived from (4.9) by substituting expression $\sum_{i \in S_k} y_{iv}/N_k$ in (4.9) instead of one of the c_{kv}'s. Formula (4.10) shows that the complementary criterion is to maximize within-cluster inner products of entities and centers. Note that the distance based criterion $W(S, c)$ makes sense at any set of centers, whereas the inner product based criterion $B(S, c)$ makes sense only when centers are within-cluster averages. As is well known, the distance does not depend on the location of the space origin whereas the inner product heavily depends on that—only special arrangements are suitable for the latter.

By further substituting c_k in (4.10) by its equal $\sum_{i \in S_k} y_{iv}/N_k$, one obtains yet one more equivalent expression:

$$B(S,c) = \sum_{k=1}^{K} \frac{1}{N_k} \sum_{i,j \in S_k} <y_i, y_j>, \qquad (4.11)$$

which appeals to entity-to-entity similarity values $<y_i, y_j>$ where $i, j = 1, 2, \ldots,$ N. Let us denote by A the $N \times N$ matrix of similarity values $a_{ij} = <y_i, y_j>$. Obviously,

$$A = YY^T. \qquad (4.12)$$

Given a clustering $S = \{S_1, S_2, \ldots, S_K\}$, let us introduce N-dimensional cluster membership vectors $z_k = (z_{ik})$ such that $z_{ik} = 1$ if $i \in S_k$ and $z_{ik} = 0$, otherwise $(k = 1, \ldots, K)$. Obviously, $<z_k, z_k> = z_k^T z_k = N_k$. Therefore, using this notation, one can easily see that Eq. (4.11) can be further rewritten as

$$B(S,c) = \sum_{k=1}^{K} \frac{1}{z_k^T z_k} \sum_{i,j=1}^{N} <y_i, y_j> z_{ik} z_{jk}.$$

This can be reformulated, finally, in matrix terms, as

$$B(S,c) = \sum_{k=1}^{K} \frac{z_k^T A z_k}{z_k^T z_k}. \qquad (4.13)$$

the sum of Rayleigh quotients $z_k^T A z_k / z_k^T z_k$, each corresponding to a cluster $k = 1, 2, \ldots, K$. Let us remind that, to achieve the global maximum of (4.13) with respect to arbitrary real-valued vectors $z_1, z_2, \ldots, z_K$, one needs to take just the K eigenvectors of matrix A corresponding to the first K eigenvalues of A sorted in the descending order, The eigenvalues are equal to the values of the corresponding Rayleigh quotients themselves.

This shows that the complementary K-means criterion $B(S, c)$ admits what is referred to as the spectral approach to clustering.

The spectral approach to clustering involves:

(1) Relaxing the combinatorial problem from, in this case, 0/1-values of components of z_k, to any real values and finding a spectral solution to thus relaxed problem;
(2) Adjusting the found solution back to the required format, 1/0 components, in our case. This latter step can be done differently. A simplest decision for the adjustment would be by putting ones for largest components, and zeros for the others. This, however, may lead to rather odd clusters sometimes. Therefore, some propose performing a clustering process but not in the raw space but rather in the space of found K eigenvectors.

Since clusters are not overlapping, model in (4.1) can be rewritten differently in such a way that no explicit references are made over individual clusters. To do that, let us again use the N-dimensional membership vectors $z_k = (z_{ik})$ such that $z_{ik} = 1$ if $i \in S_k$ and $z_{ik} = 0$, otherwise. Then the clustering model can be reformulated as follows. For any data entry, the following equation holds:

$$y_{iv} = \sum_{k=1}^{K} c_{kv} z_{ik} + e_{iv},\qquad(4.14)$$

Indeed, since any entity $i \in I$ belongs to one and only one cluster S_k, only one of $z_{i1}, z_{i2}, \ldots, z_{iK}$ can be non-zero, that is, equal to 1, at any given i, which makes (4.14) equivalent to (4.1).

Yet (4.14) makes the clustering model similar to that of PCA in (2.18) except that z_{ik} in PCA are arbitrary values to score hidden factors, whereas in (4.14) z_{ik} are to be 1/0 binary values: it is clusters, not factors, that are of concern here. That is, clusters in model (4.14) correspond to factors in model (2.18).

The decomposition (4.8), (4.6) of the data scatter into explained and unexplained parts is similar to that in (2.24) making the contributions of individual clusters $\Sigma_{v \in V} c_{kv}^2 N_k$ akin to contributions μ_k^2 of individual principal components. More precisely, μ_k^2 in (2.24) are eigen-values of YY^T, that can be expressed thus with an analogous formula,

$$\mu_k^2 = z_k^T YY^T z_k / z_k^T z_k = \sum_v c_{kv}^2 |S_k|\qquad(4.15)$$

in which the latter equation is due to the fact that vector z_k here consists of binary 1/0 entries.

4.3.3   Anomalous Pattern and Intelligent K-means

This method involves remote clusters, as Build does, too, but it does not discard them after finding, which allows for obtaining the number of clusters K as well. Besides, it is less computationally intensive. The method employs the concept of reference point. A reference point is chosen to exemplify an "average" or "normal" entity, not necessarily among the dataset. For example, when analyzing student marks over different subjects, one might choose a "normal student" point which would indicate levels of marks in tests and work in projects that are considered normal for the contingent of students under consideration, and then see what patterns of observed behavior deviate from this. Or, a bank manager may set as his reference point, a customer having specific assets and backgrounds, to see what patterns of customers deviate from this. In engineering, a moving robotic device should be able to segment the environment into homogeneous chunks according to the robot's location as its reference point, with objects that are nearer to it having finer resolution than objects that are farther away. In many cases the gravity center of the entire entity set, its "grand mean", can be taken as a reference point of choice.

Using the chosen reference point allows for the comparison of entities with it, not with each other, which drastically reduces computations: instead of mulling over all the pair-wise distances, one may focus on entity-to-reference-point distances only—a reduction to the order of N from the order of N^2.

An anomalous pattern is found by building a cluster which is most distant from the reference point. To do this, the cluster's seed is defined as the entity farthest away from the reference point. Now a version of K-means at $K = 2$ is applied with two seeds: the reference point which is never changed in the process and the cluster's seed, which is updated according to the standard procedure. In fact, only the anomalous cluster is of interest here. Given a center, the cluster is defined as the set of entities that are closer to it than to the reference point. Given a cluster, its center is found as the gravity center, by averaging all the cluster entities. The procedure is reiterated until convergence (see Fig. 4.10).

Obviously, the Anomalous pattern method is a version of K-means in which:

(i)   the number of clusters K is 2;
(ii) center of one of the clusters is forcibly kept at 0, through all the iterations;
(iii) the initial center of the anomalous cluster is taken as an entity furthest away from 0.

Property (iii) mitigates the issue of determining appropriate initial seeds. This provides for using Anomalous pattern algorithm iteratively to obtain an initial setting for K-means.

Worked Example 4.4. Anomalous Pattern in Market Towns

Let us apply the Anomalous pattern method to Town data assuming the grand mean as the reference point and scaling by range. That means that after mean-range standardization the reference point is 0.

The point farthest from 0 to be taken as the initial "anomalous" center, appears to be entity 35 (St Austell) whose distance from zero (note—after standardization!) is

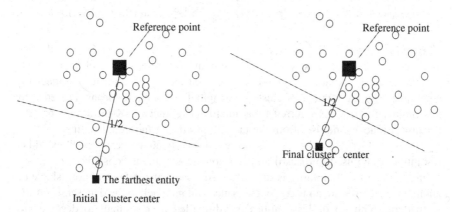

Fig. 4.10 Extracting an anomalous pattern cluster with the reference point in the gravity center: the first iteration is on the left and the final one on the right

4.33, the maximum. There are only three entities, 26, 29 and 44 (Newton Abbot, Penzance and Truro) that are closer to the seed than to 0, thus forming the cluster along with the original seed, at this stage. After one more iteration, the anomalous pattern cluster stabilizes with 8 entities 4, 9, 25, 26, 29, 35, 41, 44. Its center is displayed in Table 4.13.

As follows from the fact that all the standardized centers are positive and mostly fall within the range of 0.3–0.5, the anomalous cluster, according to Table 4.13, consists of better off towns—all the center values are larger than the grand mean by 30–50% of the feature ranges. This probably relates to the fact that they comprise eight out of the eleven towns that have a resident population greater than 10,000. The other three largest towns have not made it into the cluster because of their deficiencies in services such as Hospitals and Farmers' Markets. The fact that the scale of population is by far the largest in the original table doesn't much affect the computation here as it runs with the range standardized scales at which the total contribution of this feature is not high, just about 8.5% only. It is rather the concerted action of all the features associated with a greater population which makes the cluster.

This process is illustrated in Fig. 4.11. The stars show the origin and the anomalous seed at the beginning of the iteration. Curiously, a picture may not fit well into the concept of anomalous pattern cluster, as illustrated in Fig. 4.11—the anomalous pattern is dispersed here across the plane, which is at odds with the property that the entities in it must be closer to the seed than to the origin. The cause is not an error, but the fact that this plane represents all 12 original variables and presents them rather selectively. It is not that the plane makes too little of the data scatter—on the contrary, it makes a decent 76% of the data scatter. The issue here is the second axis in which the last feature FM expressing whether there is a Farmers market or not takes a lion share—thus stratifying the entire image over y axis.

Figure 4.11 is produced with commands:

```
> > subplot(1, 2, 1); plot(x1, x2,'k.', 0, 0,'kp', x1(35), x2(35),'kp'); text(x1(fir), x2(fir), ftm);
> > subplot(1, 2, 2); plot(x1, x2,'k.', 0, 0,'kp', x1a, x2a,'kp'); text(x1(sec), x2(sec), fsm);
```

Here fir and sec are lists of indices of towns belonging to the pattern after the first and second iterations, respectively, while ftm and fsm refer to lists of their names.

The Anomalous pattern method can be used as a procedure to automatically determine both the number of clusters and initial seeds for K-means. Preceded by this option, K-means is referred to as intelligent K-means, iK-means for brevity, because it relieves the user from the task of specifying the initial setting.

In iK-means method, the user is required to specify an integer, t, the threshold of resolution, to be used to discard all the Anomalous patterns consisting of t or less entities. When $t = 0$, nothing is discarded. At $t = 1$—the default option, singleton anomalous patterns are considered a nuisance and put back to the data set. If $t = 10$, all patterns with 10 or less entities are discarded as too small to deserve any

Table 4.13 Center of the anomalous pattern cluster of Town data in real and standardized scales

Center	P	PS	Do	Ho	Ba	Sm	Pe	DIY	Sp	Po	CAB	FM
Real	18,484	7.6	3.6	1.1	11.6	4.6	4.1	1.0	1.4	4.4	1.2	4.0
Std'zed	0.51	0.38	0.56	0.36	0.38	0.38	0.30	0.26	0.44	0.47	0.30	0.18

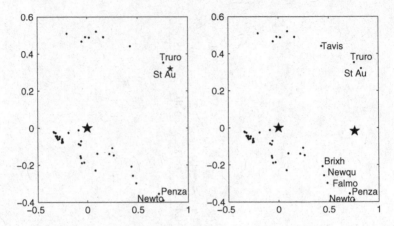

Fig. 4.11 The first and second iterations of anomalous pattern cluster on the principal component plane; the visual separation of the pattern over y axis is due to a very high loading of the presence (top) or absence (bottom) of a Farmer's market

attention at all—the level of resolution which may be justified at larger datasets and coarser details needed.

In our experiments, the entities comprising singleton Anomalous pattern clusters are frequently erroneous, that is, erroneous values are in some of their features such as, for instance, the human age of 538 years. That means, that Anomalous pattern clustering can be used as a device for capturing errors in data entries.

The iK-means method is flexible with regard to outliers and the "swamp" of inexpressive—normal or ordinary—entities around the grand mean. For example, at its step 4, K-means can be applied to either the entire dataset or to the set from which the smaller APs have been removed. This may depend on the domain: in some problems, such as structuring of a set of settlements for better planning or monitoring or analysis of climate changes, no entity should be dropped out of the consideration, whereas in other problems, such as developing synoptic descriptions for text corpora, some "deviant" texts could be left out of the coverage at all.

In a series of experiments with intermixed Gaussian clusters described by Chiang and Mirkin (2010), iK-means has performed rather well and appeared superior to many other options for choosing K. These options included approaches based on post-processing of results of multiple runs of K-means and then treating them according to either of the following:

(a) Variance based approach: using intuitive or model-based functions of criterion (4.3) which should get extreme or "elbow" values at a correct K such as Hartigan's rule above;
(b) Structural approach: comparing within-cluster cohesion versus between-cluster separation at different K;
(c) Consensus distribution approach: choosing K according to the distribution of the consensus matrix for sets of K-means clusterings at different K.

Some other approaches rely on different ideas for choosing K such as

(d) using results of a divisive or agglomerative clustering procedure or
(e) according to the similarity between K-means clustering results on randomly perturbed or sampled data.

Worked Example 4.5. Iterated Anomalous Patterns in Market/Towns
Applied to the range-standardized Market town data, AP algorithm iterated until no unclustered entities remained, has produced 12 clusters of which 5 are singletons. These singletons have strange patterns of facilities indeed. For example, entity 19 (Liskeard, 7044 residents) has an unusually large number of Hospitals (6) and CABs (2), which makes it a singleton cluster. Lists of seven non-singleton clusters are in Table 4.14, in the order of their extraction from the dataset.

This cluster structure doesn't much change when, according to the iK-means algorithm, Batch K-means is applied to the seven centers (with the five singletons put back into the data). Moreover, similar results have been observed with clustering of the original all-England list of about thirteen hundred Market towns described by a wider list of eighteen characteristics of their development: the number of non-singleton clusters was the same, and with much similar descriptions.

Q.4.18. Formulate a version of K-means to alternatingly maximize criterion (4.6) rather than to minimize (4.3) as the generic version.
Q.4.19. Formulate a version of K-means to alternatingly maximize criterion (4.9) rather than to minimize (4.3) as the generic version (a "parallel" version of the Anomalous Pattern method). Take care of starting from a most distant set of centers.
Q.4.20. Why is the contribution of AP 4 in Table 4.14, 18.6%, greater than that of the preceding AP3, 10.0%?
A. Because of much larger number of entities, 18 against 6 in AP 3. Even if the center of AP 3 is further away from 0 than center of AP 4, which is the cause that AP 3 is extracted first, the contribution takes into account the number of entities as well!

Before substantiating AP algorithm, let us give it a more explicit formulation.

Table 4.14 Iterated AP market town non-singleton clusters

Cluster #	Size	Contents	Contribution, %
1	8	4, 9, 25, 26, 29, 35, 41, 44	35.1
3	6	5, 8, 12, 16, 21, 43	10.0
4	18	2, 6, 7, 10,13, 14, 17, 22, 23, 24, 27, 30, 31, 33, 34, 37, 38, 40	18.6
5	2	3, 32	2.4
6	2	1,11	1.6
8	2	39, 42	1.7
11	2	20,45	1.2

Anomalous Pattern (AP) Algorithm

1. *Pre-processing.* Specify a reference point $a = (a_1, ..., a_V)$ (when in doubt, take a to be the data grand mean) and standardize the original data table by shifting the origin to $a = (a_1, ..., a_V)$ and rescaling if needed.
2. *Initial setting.* Put a tentative center, c, as the entity farthest away from the origin, 0.
3. *Cluster update.* Determine cluster list S around c against the only other "center" 0, so that entity y_i is assigned to S if $d(y_i, c) < d(y_i, 0)$.
4. *Center update.* Calculate the within S mean c' and check whether it differs from the previous center c. If c' and c do differ, update the center by assigning $c \Leftarrow c'$ and go to Step 3. Otherwise, go to 5.
5. *Output.* Output list S and center c, with accompanying interpretation aids (as advised in the next section), as the anomalous pattern.

It is not difficult to prove that, like K-means itself, the Anomalous pattern alternately minimizes a specific version of K-means general criterion $W(S, c)$ (4.3),

$$w(S,c) = \sum_{i \in S} d(y_i, c) + \sum_{i \notin S} d(y_i, 0) \qquad (4.16)$$

where S is a subset of I rather than partition and c its center. Yet AP differs from 2-Means in the following aspect: there is only one center, c, which is updated in AP; the other center, 0, never changes and serves only to attract not-anomalous entities. This is why 2-Means produces two clusters whereas AP—only one, that is farthest away from the reference point, 0.

In fact, criterion (4.16) can be equivalently rephrased using Eqs. (4.8) and (4.9) representing the complimentary criterion $B(S, c)$. When (4.9) applies to the situation of two clusters, one with center in c, the other in 0, it becomes of finding a cluster S maximizing its contribution to the data scatter $T(Y)$:

$$\mu^2 = z^T Y Y^T z / z^T z = c_v^2 |S| = d(0, c) |S| \qquad (4.17)$$

This means that AP algorithm straightforwardly follows the Principal Component Analysis one-by-one extraction strategy extended to binary scoring vectors. That is, the model behind AP is a version of the PCA Eq. (2.17) on p. 124 in which the scoring values z^*_i are but zeros or ones:

$$y_{iv} = \begin{cases} c_v + e_v, i \in S \\ 0 + e_v, i \notin S \end{cases} \qquad (4.18)$$

where S is the cluster list of the anomalous pattern to be found.

In spite of the rather simplistic assumption presented in (4.18), AP clusters fare well with real data. They can be extracted one-by-one, along with their contributions to the data scatter (4.17) showing cluster saliencies. These saliencies can be used to halt the process when the contribution of the next cluster drops decisively, thus leading to an incomplete clustering when needed.

Here are steps of iK-means(t) where t is the cluster resolution threshold—the minimum number of entities in a pattern that can be considered a cluster on its own. In most applications dealing with moderately sized data (up to a few hundred entities) t can be put to be equal to 1.

iK-means(t) Algorithm

0. *Setting.* Preprocess and standardize the data set. Take t as the threshold of resolution. Put $k = 1$ and $I_k = I$, the original entity set.
1. *Anomalous pattern.* Apply AP to I_k to find k-th anomalous pattern S_k and its center c_k.
2. *Test.* Remove S_k from I_k to make $k \Leftarrow k + 1$ and $I_k \Leftarrow I_k - S_k$. If the new I_k is empty, go to Step 3; if not, go to Step 1 again.
3. *Discarding small clusters.* Denote the number of clusters containing t entities or more by K and re-label them so that their centers are $c_1, c_2, \ldots, c_K$.
4. *K-means.* Do Batch K-means using $c_1, c_2, \ldots, c_K$ as initial seeds.

The Stop-condition in this method can be any or all of the following:

(a) All of I has been clustered, $S_k = I_k$, so that there are no unclustered entities left.
(b) Large cumulative contribution. The total contribution of the first k clusters to the data scatter has reached a pre-specified threshold such as 50%.
(c) Small cluster contribution. Contribution of S_k is too small; for example, it is comparable with the average contribution of a single entity, T/N, where T is the data scatter.
(d) Number of clusters, k, has reached its pre-specified value K.

Condition (a) is reasonable if there are "natural" clusters that indeed differ in their contributions to the data scatter. Conditions (b) and (c) can be considered as related to the degrees of granulation at which the user looks at the data. Unlike (d), they appeal to the structure of the data set rather than prior considerations.

Case-Study 4.5. iK-means Clustering of a Normally Distributed 1D Dataset
Let us generate a one dimensional set X of 280 points generated according to Gaussian $N(0, 10)$ distribution (see Fig. 4.12). This data set is printed in the Appendix, section A6. Many would say that this sample constitutes a single, Gaussian, cluster. Yet the idea of applying a clustering algorithm seems attractive as a litmus paper to capture the pattern of clustering embedded in iK-means algorithm.

In spite of the symmetry in the generating model, the sample is slightly biased to the negative side; its mean is -0.89 rather than 0, and its median is about -1.27. Thus, the maximum distance from the mean is at the maximum of 32.02 rather than at the minimum of -30.27.

The Anomalous pattern starting from the furthest away value of maximum comprises 83 entities between the maximum and 4.28. Such a stripping goes along real-world conventional procedures. For example, consider the heights of a sample of young males to be drafted for a military action; the histogram of heights is

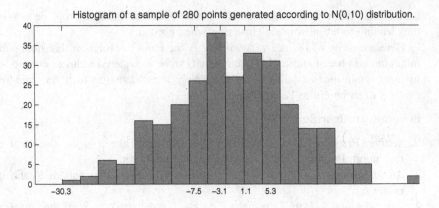

Fig. 4.12 Histogram of the sample of 280 values generated by Matlab's randn command from the Gaussian distribution N(0, 10)

known to be bell-shaped like Gaussian. Those on the either side of the bell-shaped height histogram are not quite fitting for action: those too short cannot accomplish many a specific task whereas those too tall may have problems in closed spaces such as submarines or aircraft.

The iterative Anomalous pattern clustering would sequentially strip the remaining margins off too. The set of fragments of the sorted sequence in Table 4.15 that have been found by the Anomalous pattern clustering algorithm in the order of their forming, including the cluster means and contributions to the data scatter.

The last extracted clusters are all around the mean and, predictably, small in size. One also can see that the contribution of a following cluster can be greater than that

Table 4.15 A summary of the iterative Anomalous pattern clustering results for the sample of Gaussian distribution in Table A5.2. Clusters are shown in the extraction order, along with their sizes, left and right boundary entity indices, means and contributions to the data scatter

Order of extraction	Size	Left index	Right index	Mean	Contrib, %
1	83	198	280	11.35	34.28
2	70	1	70	−14.32	44.03
3	47	71	117	−5.40	4.39
4	41	157	197	2.90	1.11
5	18	118	135	−2.54	0.38
6	10	147	156	0.27	0.002
7	6	136	141	−1.42	0.039
8	2	145	146	−0.49	0.002
9	3	142	144	−0.77	0.006

of the preceding cluster thus reflecting the local nature of the Anomalous pattern algorithm which intends to find the maximally contributing cluster each time. The total contribution of the nine clusters is about 86% to which the last five clusters contribute next to nothing.

Project 4.1. Does PCA Clean the Data Structure Indeed: K-means After PCA
There is a wide-spread opinion that in a situation of many features, the data structure can come less noisy if the features are first "cleaned" by applying PCA and using a few principal components instead of the original features. Although strongly debated by specialists (see, for example, Kettenring 2006), the opinion is wide-spread among the practitioners. One of attacks against this opinion was undertaken by late Kryshtanowski (2008) who provided an example of data structure that, in his words, "becomes less pronounced in the space of principal components".

The example refers to data of two Gaussian clusters, each containing five hundred of 15-dimensional entities. The first cluster can be generated by the following MatLab commands:

> > $b(1:500,1) = 10 * randn(500,1);$
> > $b(1:500,2:15) = repmat(b(1:500,1),1,14) + 20 * randn(500,14);$

The first variable in the cluster is Gaussian with the mean 0 and standard deviation 10, whereas the other fourteen variables add to that another Gaussian variable whose mean and standard deviation are 0 and 20, respectively. That is, this set is a sample from a 15-dimensional Gaussian with a diagonal covariance matrix, whose center is in or near the origin of the space, with the standard deviations of all features at 22.36, the square root of $10^2 + 20^2$, except for the first one that has the standard deviation of 10.

The elements in the second cluster are generated as the next 500 rows in the same matrix in a similar manner:

> > $b(501:1000,1) = 20 + 10 * randn(500,1);$
> > $b(501:1000,2:15) = repmat(b(501:1000,1),1,14) + 20 * randn(500,14) + 10;$

The first variable now is centered at 20, and the other variables, at 30. The standard deviations follow the pattern of the first cluster.

Since the standard deviations by far exceed the distance between centers, these clusters are not easy to distinguish: see Fig. 4.13 illustrating the data cloud, after centering, on the plane of the first principal components.

When applying iK-means to this data, preliminarily centered and range-normalized, the algorithm finds in them indeed much more clusters, 13 altogether, at the discarding threshold $t = 1$. However, when the discarding threshold is set to $t = 200$, to remove any less populated anomalous patterns, the method arrives at just two clusters that differ from those generated by 96 entities (see the very first resulting

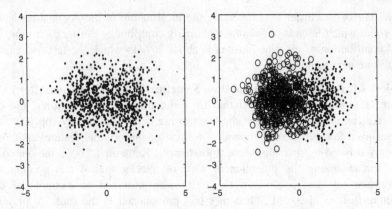

Fig. 4.13 Data of two clusters generated as described above, after centering, are represented by points on the plane of the two first principal components (on the left); the second cluster is represented by circles on the right

Table 4.16 The numbers of errors of iK-means clustering at different data transformations: over the original data differently normalized and over four principal components derived at different data normalizations

Data	Original 15 features		Four principal components		
Normalized by	Range	St. deviation	No normalization	Range	St. deviation
Cluster 1	44	43	37	51	47
Cluster 2	52	56	47	47	45
Total	96	99	84	98	92

column in Table 4.16 presenting the results of the computation) constituting the total error of 9.6%. The same method applied to the data z-score standardized, that is, centered and normalized by the standard deviations, arrives at 99 errors; a rather modest increase, probably due to specifics of the data generation.

Since Kryshtanowski (2008) operated with the four most contributing principal components, we also take the first four principal components, after centering by the means and normalizing data by the range:

```
> > n = 1000; br = (b − repmat(mean(b), n, 1))./repmat(max(b) − min(b), n, 1);
> > [zr, mr, cr] = svd(br);
> > zr4 = zr(:, 1 : 4);
```

These four first components bear 66% of the data scatter. We also derived four principal components from the data non-normalized (yet centered) and from the data normalized by the standard deviations (z-score standardized). The latter is especially important in this context, because Kryshtanowski (2008) used the conventional form of PCA based on the correlation matrix between the variables, which is equivalent to the model-based PCA applied to the data after z-score standardization. The iK-means method applied to each of these data sets at the discarding threshold of 200, has shown rather consistent results (see Table 4.16).

Overall, these results seem to support the idea of better structuring under principal components rather than to refute it. The negative results of using principal components by Kryshtanowski (2008) probably could be attributed to his indiscriminate usage of Batch K-means method with random initializations that failed to find a "right" pair of initial centers, in contrast to iK-means.

Project 4.2. Using Contributions to Determine the Number of Clusters
The question of determining the stopping rule can be addressed with the model (4.18) itself, applied to the cluster contribution values as the raw data (compare to Cangelosi and Goriely 2007). Assume the contributions are sorted in the descending order and denoted by h_k so that $h_1 \geq h_2 \geq \ldots (k = 1, 2, \ldots)$.

If one assumes that the first K values are all approximately equal to each other, whereas the rest approximate zero, then the optimal K can be derived as follows.

Denote the average of the first K contributions as $h(K)$. Then criterion (4.6) to maximize is the product $Kh^2(K)$. The optimal K obviously satisfies inequality $Kh^2(K) > (K + 1)h^2(K + 1)$. Since the average $h(K + 1)$ can be expressed as $h(K + 1) = (K*h(K) + h_{K+1})/(K + 1)$, the inequality can be easily transformed to $h^2(K) - 2h(K)h_{K+1} + h_{K+1}^2/K > 0$ which can be further presented as $(h(K) - h_{K+1})^2 > h_{K+1}^2 (1-1/K)$. Since $h(K) \geq h_{K+1}$, this inequality can be further simplified to $h(K) - h_{K+1} > h_{K+1}\sqrt{(1 - 1/K)}$, that is,

$$h(K) > h_{K+1}(1 + \sqrt{1 - 1/K}) \qquad (4.19)$$

which is, roughly, $h_{K+1} < h(K)/2$. This has an advantage that the threshold is not pre-specified but rather determined according to the structure of gaps between the numbers h_k in their sorted order. The value of K at which (4.19) holds can be considered as a candidate for the right number of clusters or components or, in fact, anything evaluated by contributions.

Similar inequalities can be derived at different models for the chosen contribution values. One may try, for example, the power law assumption that $h(k) = ak^{-b}$ for $k = 1, \ldots, K$ and $h(k) = 0$ for $k > K$.

Method iK-means utilizes a slightly different strategy for choosing the right K. This strategy involves (1) all the anomalous patterns rather than those most contributing, thus involving the patterns close to the reference points too, and (2) a different scoring device—the intuitively clear number of entities rather than a purely geometric contribution whose intuitive value is unclear.

4.4 Cluster Interpretation Aids at Mixed Data Scales

4.4.1 Classical Statistics View at Interpretation of Partitions: Tabular Regression and Correlation Ratio

A set of clusters or just nominal feature values represented by subsets of objects falling in each of them, S_k ($k = 1, 2, ..., K$), can be associated with any quantitative feature y on I with the so-called tabular regression described in Sect. 3.8.3. The tabular regression of quantitative y over categorical x whose values are clusters S_k ($k = 1, 2,..., K$) is a table comprising three columns corresponding to:

(1) Categories/clusters S_k
(2) Within category means of y, $\bar{y}_k$
(3) Within category standard deviations of y, σ_k.

The association between partition $S = \{S_k\}$ and feature y is defined as the correlation ratio, that is the relative drop in the variance of y when it is predicted according to the piece-wise constant model in Sect. 3.8.3. In other words, this is the relative proportion of the explained part of the variance:

$$\eta^2 = 1 - \sigma_w^2 / \sigma^2 \tag{4.20}$$

where

$$\sigma_w^2 = \sum_k p_k \sigma_k^2 \tag{4.21}$$

the average within-cluster variance and $p_k = |S_k|/N$, the proportion of S_k in the entire set and σ_k^2 the variance of y within S_k.

This works as well for categorical features whose categories are represented by 0/1 dummy columns in the data table. Consider a category l whose probability over the whole entity set is p_l, which is the mean of a 0/1 column playing the role of feature y for the category. That means that the variance of the dummy feature is $\sigma_l^2 = p_l (1 - p_l)$. Let us denote the proportion of category l and cluster k co-occurring by p_{kl}. Then the frequency of category l within cluster S_k will be $p(l|k) = p_{kl}/p_k$ where p_k is the proportion of cluster S_k in the entity set. Of course, $p(l|k)$ is the mean of the dummy feature l within cluster S_k, so that its within-cluster variance is $\sigma_{kl}^2 = p(l|k)(1 - p(l|k))$. Then the variance σ_w^2 within partition $S = \{S_k\}$ is equal to

$$\sigma_{lw}^2 = \sum_k p_k p(l|k)(1 - p(l|k)) = \sum_k p_{kl}(1 - p_{kl}/p_k)$$

$$= \sum_k p_{kl} - \sum_k p_{kl}^2/p_k = p_l - \sum_k p_{kl}^2/p_k$$

Then the defining ratio is

$$\frac{\sigma_{lw}^2}{\sigma_l^2} = \frac{p_l - \sum_k p_{kl}^2/p_k}{p_l(1-p_l)} = \frac{1 - \sum_k \frac{p_{kl}^2}{p_k p_l}}{1 - p_l}$$

Therefore, the correlation ratio for the cluster partition S and category l is equal to

$$\eta_l^2 = 1 - \frac{1 - \sum_k \frac{p_{kl}^2}{p_k p_l}}{1 - p_l} = \frac{\sum_k \frac{p_{kl}^2}{p_k p_l} - p_l}{1 - p_l} \tag{4.22}$$

This formula shows a close relation between the Pearson hi-squared contingency table association index and correlation ratio. This can be seen especially clear if one considers a "total correlation ratio", the ratio of summary within cluster variances σ_{lw}^2 and summary variances σ_l^2 over all the categories l of a given nominal feature. The former is equal to the Pearson's chi-squared between the feature and partition S, and the latter is Gini index of the feature.

Worked Example 4.6. Correlation Ratio as an Association Measure
Here are two examples of association between Occupation categories of students and their features, Age, in years, and exam mark over OOProgramming, in a 0–100 scale (Tables 4.17 and 4.18).
Which of the tables manifests a closer association?

Q.4.21. In Table 4.18, there is a positive relation between the Occupation and the OOP mark, with the largest mark, 76.1, going to IT and the smallest mark, 50.7, to AN. There is no such a relation in Table 4.17 in which AN's Age is in the middle between that at the other two groups. Is it that feature of Table 4.18 that leads to a higher correlation ratio?

	Occupation	Age mean	Age StD
Table 4.17 Tabular regression of age over occupation in students data	IT	28.2	5.6
	BA	39.3	7.3
	AN	33.7	8.7
	Total	33.7	8.5

	Occupation	OOP mean	OOP StD
Table 4.18 Tabular regression OOProg/ occupation	IT	76.1	12.9
	BA	56.7	12.3
	AN	50.7	12.4
	Total	61.6	16.5

A. No; the order of means is irrelevant at the tabular regression. The correlation ratio is higher at Table 4.18 than at Table 4.17 because of the tighter boundaries on the quantitative feature within the groups in Table 4.18: the within-group standard deviations in Table 4.18 are consistently lower, by about 25%, than the standard deviation in the whole set, which is not the case in Table 4.17.

Specifically, correlation ratios for the tables computed by using formula (4.20) are:

Occupation/Age	28.1%
Occupation/OOProg	42.3%

The drop in the variance expressed by the correlation ratio is greater at the second table, that is, the correlation between Occupation and OOProg mark is greater than that between Occupation and Age.

4.4.2   *Formulas for Cluster Interpretation Aids at Mixed Scale Data*

According to (4.6) and (4.7), clustering (S, c) decomposes the scatter $T(Y) = \Sigma_{i,v} y_{iv}^2$ of data matrix Y in the explained and unexplained parts, $B(S, c)$ and $W(S, c)$, respectively. The latter is the square-error K-means criterion, whereas the explained part $B(S, c)$ is clustering's contribution to the data scatter, which is equal, according to (4.6), to

$$B(S, c) = \sum_{k=1}^{K} \sum_{v \in V} c_{kv}^2 N_k \qquad (4.23)$$

This is the sum of additive items

$$B_{kv} = N_k c_{kv}^2, \qquad (4.24)$$

each accounting for the contribution of a feature-cluster pair, $v \in V$ and S_k ($k = 1, 2, ..., K$), in the data scatter.

Since the total contribution of feature v to the data scatter is $T_v = \Sigma_{i \in I} y_{iv}^2$, its unexplained part can be expressed as $W_v = T_v - B_{+v}$ where $B_{+v} = \Sigma_{k=1}^{K} B_{kv}$ is feature's v explained part, the total contribution of v to the cluster structure. This can be displayed as a Scatter Decomposition (ScaD) table whose rows correspond to clusters, columns to variables and entries to the contributions B_{kv} (see Table 4.19).

The summary rows, Explained, Unexplained and Total, as well as column Total can be expressed as percentages of the data scatter $T(Y)$. The contributions highlight relative roles of features both at individual clusters and in total.

Feature Cluster	f_1	f_2	f_V	Total
S_1	B_{11}	B_{12}	B_{1V}	B_{1+}
S_2	B_{21}	B_{22}	B_{2V}	B_{2+}
S_K	B_{K1}	B_{K2}	B_{KV}	B_{K+}
Explained	B_{+1}	B_{+2}	B_{+V}	$B(S, c)$
Unexplained	W_{+1}	W_{+2}	W_{+V}	$W(S, c)$
Total	T_1	T_2	T_V	$T(Y)$

Table 4.19 ScaD: Data scatter decomposed over clusters and features using notation introduced above

The explained part $B(S, c)$ is, according to (4.8), the sum of contributions of individual feature-to-cluster pairs $B_{kv} = c_{kv}^2 N_k$ which can be used for interpretation of the clustering results. The sums of B_{kv}'s over features or clusters express total contributions of individual clusters or features into the explanations of clusters.

Summary contributions of individual data features to clustering (S, c) have something to do with statistical measures of association in bivariate data, such as correlation ratio η^2 (3.65) in Sect. 3.8.3 and chi-squared X^2 (3.31) in Sect. 3.6 (Mirkin 2012). In fact, the analysis in Sect. 3.8.2 applies in full to the case when target features are those used for building clustering S.

Specifically, for a quantitative feature v represented by the standardized column y_v, its summary contribution B_{+v} to the data scatter is equal to

$$B_{+v} = N\sigma_v^2\eta_v^2$$

Consider now a nominal feature v represented by a set of binary columns, dummies, corresponding to individual categories $l \in v$. The grand mean of binary column for $l \in v$ is obviously the proportion of this category in the set, p_{+l}. To standardize the column, one needs to subtract the mean, p_{+l}, from all its entries and divide them by the scaling parameter, b_l. After the standardization, the center of cluster S_k can be expressed through co-occurrence proportions too:

$$c_{kl} = (\frac{p_{kl}}{p_k} - p_{+l})/b_l \tag{4.25}$$

where p_{kl} is the proportion of entities falling in both category l and cluster S_k; the other symbols: p_{+l} is the frequency of l, p_k the proportion of entities in S_k, and b_l the normalizing scale parameter.

According to Eqs. (4.24) and (4.25), the summary contribution of all pairs category-cluster (l, k) is equal to

$$B(v/S) = N\sum_{l \in v}\sum_{k=1}^{K}\frac{(p_{kl} - p_k p_{+l})^2}{p_k b_l^2} \tag{4.26}$$

This is akin to several contingency table association measures considered in the literature including Pearson chi-squared X^2 in (3.31) and Gini impurity function, or summary absolute Quetelet index, in (3.35), see also (3.61). To make $B(v/S)$ equal to the chi-squared coefficient, the scaling of binary features must be done by using $b_l = \sqrt{p_{+l}}$, which is the standard deviation of the so-called Poisson probabilistic distribution that randomly throws $p_l N$ unities into an N-dimensional binary vector. To make $B(v/S)$ equal to Gini impurity function, no normalization of the dummies is to be done, or rather the recommended option of normalization by ranges applies since the range of a dummy is 1.

One should not forget the additional normalization of the binary columns by the square root of the number of categories in a nominal feature v, $\sqrt{|v|}$ leading to both the individual contributions B_{kv} in (4.24) and the total contribution $B(v/S)$ in (4.26) divided by the number of categories $|v|$. When applied to Pearson chi-squared, the division by $|v|$ can be considered as another normalization of the coefficient. As mentioned in Sect. 3.3, the maximum of Pearson chi-squared (related to N) is $\min(|v|, K) - 1$. Therefore, when $|v| \leq K$, the division would lead to a normalized index whose values are between 0 and $1 - 1/|v|$. If, however, the number of categories is larger so that $K < |v|$, then the normalized index could be very near 0 indeed. In this regard, it should be of interest to mention that in the literature some other normalizations have been considered. Specifically, Pearson chi-squared is referred to as Cramer coefficient if related to $\min(|v|, K) - 1$, and as Tchouprow coefficient if related to $\sqrt{(|v| - 1)(K - 1)}$ (Kendall and Stewart 1973).

Therefore, the contributions for mixed scale data in tables above are not anything weird, but rather extensions of the classic statistical constructions such as chi-squared contingency and correlation ratio coefficients.

Q.4.22. Prove that, for any cluster k in K-means clustering, $\Sigma_{i \in S_k} y_{iv}^2 = |S_k|(c_{kv}^2 + \sigma_{kv}^2)$.

Q.4.23. How one should interpret the normalization of a category by the $\sqrt{p_v}$? What category gets a greater contribution: that more frequent or that less frequent?

Comment 4.1. Zeros in Contingency Tables

When the chi-squared contingency coefficient or related indexes are applied in the traditional statistics context, the presence of zeros in a contingency table becomes an issue because it contradicts the hypothesis of statistical independence. In the context of data recovery clustering, zeros are treated as any other numbers and create no problems at all because the coefficients are measures of contributions and bear no other statistical meaning in this context.

Comment 4.2. Advantages and Drawbacks of K-means

K-means advantages:

The method

   i   Models typology building activity
  ii Computationally effective both in memory and time
 iii Can be utilized incrementally, "on-line"
 iv Straightforwardly associates feature salience weights with feature scales
  v Applicable to both quantitative and categorical data and mixed data provided that care has been taken of the relative feature scaling
 vi Provides a number of interpretation aids including cluster prototypes and features and entities most contributing to cluster specificity

K-means issues:

  vii Simple convex spherical shape of clusters
 viii Choosing the number of clusters and initial seeds
  ix Instability of results with respect to initial seeds
   x No help in defining feature weights; cannot distinguish between useful and redundant features.

Although conventionally considered as shortcomings, issues vii–ix can be beneficial too. To cope with issue vii, the feature set should be chosen carefully. Then the simple shape of a cluster will provide for a simpler conceptual description of it. To cope with issue viii, the initial seeds should be selected not randomly but rather based on preliminary analysis of the substantive domain or using anomalous approaches Build or AP. Another side of issue ix is that solutions are close to pre-specified centers, which is good when the centers have been chosen carefully. To cope with issue x, a preliminary systems analysis should be carried out to bring forward features of the same granularity level.

Q.4.24. Find ScaD decomposition for the product clusters in Company data.
A. This is in Table 4.20. Table 4.20 shows feature EC as the one most contributing to the Product A cluster, feature MSha to the Product B cluster, and features SupN and Retail to the Product C cluster. The relatively high contribution of MSha to B cluster is not that obvious because that of EC, 0.42, is higher. It becomes clear only on the level of relative contributions relating the absolute values to their respective Exp counterparts, 0.25/0.41 and 0.42/1.88—the former prevails indeed. Clusters A, B, and C can be distinctively described by statements "EC==0", "MSha < 28", and "SupN > 3" (or "Sector is Retail"), respectively, as highlighted in Table 4.20.

Table 4.20 Decomposition of the data scatter over product clusters in Company data; notation is similar to that in Table 4.19; within-row maxima used in Q.4.24 are highlighted in bold

Product	Income	MSha	SupN	EC	Util	Indu	Retail	Total	Total %
A	0.03	0.05	0.04	**1.17**	0.00	0.09	0.06	1.43	24.08
B	0.14	**0.25**	0.15	0.42	0.09	0.00	0.06	1.10	18.56
C	0.06	0.12	**0.50**	0.28	0.09	0.09	**0.38**	1.53	25.66
Exp	0.23	0.41	0.69	1.88	0.18	0.18	0.50	4.06	68.30
Unexp	0.51	0.28	0.20	0.00	0.44	0.44	0.00	1.88	31.70
Total	0.74	0.69	0.89	1.88	0.63	0.63	0.50	5.95	100.00

4.4.3   Data Analysis View of Cluster Interpretation Aids

The practice of using clustering in data analysis for the past decades somewhat extended and enriched the statistics tools. One can distinguish the following four approaches to interpretation of clusters, at least:

(a) Comparing feature values at cluster centers with their grand means (feature averages on the entire data set)
(b) Finding cluster representatives
(c) Estimating cluster-feature contributions to the data scatter to use the largest of them
(d) Conceptually describing clusters.

One should not forget that, under the zero-one coding system for categories, cluster-to-category cross-classification frequencies are, in fact, cluster centers—therefore, (a) includes looking at cross-classifications between the clustering partition and categorical features—this is why we do not consider the cross-classifications as a separate interpretation device—in contrast to conventional approaches.

Consider these in turn.

(a)   Cluster centers versus grand means

These should be utilized in both, original and standardized, formats. The standardized format allows one to see the differences between cluster centers and grand means across all the features expressed in comparable scales. The largest differences show what features are responsible for specifics of individual clusters with respect to the "norm" corresponding to the point of grand means.

To express a standardized center value c_{kv} of feature v in cluster S_k resulting from a K-means run, in the original scale of feature v, one needs to invert the scale transformation by multiplying over rescaling factor b_v with the follow up adding the shift value a_v, so that this becomes $C_{kv} = b_v c_{kv} + a_v$.

Worked Example 4.7. Centers of Market Town Clusters

Let us take a look at centers of the seven clusters of Market towns data both in real and range standardized scales in Table 4.21.

These show some tendencies rather clearly. For instance, the first cluster appears to be a set of larger towns that score 30–50% higher than the average on almost all of the 12 features. Similarly, cluster 3 obviously relates to smaller, than the average, towns. However, in other cases, it is not always clear what feature(s) caused a cluster to separate. On the first glance, both clusters 6 and 7 seem too close to the average to make any real difference at all. However, let us take a look at SP (Swimming Pool) feature, since it makes the maximum standardized deviation from the average in both clusters, -0.24 and 0.25, respectively. The real difference is clearly seen: there is an SP in each member of cluster 7 and no SP in members of cluster 6.

(b) Cluster representative

A cluster is typically characterized by its center consisting of the within-cluster feature means. Sometimes, the means make no sense—like the number of suppliers 4.5 above. In such a case, it is more intuitive to characterize a cluster by its "typical" representative. This is especially appealing when the representative is a well-known object. Such an object can give much better intuition to a cluster than a logical description in situations in which entities are complex and the features are superficial. This is the case, for instance, in mineralogy where a class of minerals can be represented by its "stratotype" mineral, or in art studies where a general concept such as "surrealism" can be represented by an art object such as a painting by S. Dali.

A cluster representative must be the nearest to its cluster's center. An issue is that two different expressions for K-means lead to two different measures. The sum of entity-to-center distances $W(S, c)$ in (4.3) leads to the strategy that can be referred to as "the nearest in distance." The sum of entity-to-center inner products for $B(S, c)$ in (4.10) and (4.23) leads to the strategy "the nearest in inner product". Intuitively, the choice according to the inner product follows tendencies represented in c_k towards the whole of the data expressed in grand mean position whereas the distance follows just c_k itself. These two principles usually lead to similar choices, although not necessarily.

Worked Example 4.8. Representatives of Company Clusters

Consider, for example, A product cluster in Company data as presented in Table 4.22: The nearest to center in distance is Ant and nearest in inner product is Ave.

To see why is that, let us take a closer look at the two companies. Ant and Ave are similar on all four binary features. Each is at odds with the center's tendency on one feature only: Ant is zero on NSup while center is negative, and Ave is negative on Income while center is positive on that. The difference, however, is in feature contributions to the cluster; that of Income is less than that of NSup, which makes

Table 4.21 Patterns of market towns in the cluster structure found with iK-means (see Table 4.15). For each of the clusters, real values are in the first line and the standardized values are in the second line

#	Pop	PS	Do	Ho	Ba	Su	Pe	DIY	SP	PO	CAB	FM
1	18,484	7.63	3.63	1.13	11.63	4.63	4.13	1.00	1.38	4.38	1.25	0.38
	0.51	0.38	0.56	0.36	0.38	0.38	0.30	0.26	0.44	0.47	0.30	0.17
2	5268	2.17	0.83	0.50	4.67	1.83	1.67	0.00	0.50	1.67	0.67	1.00
	-0.10	-0.07	-0.14	0.05	0.02	-0.01	-0.05	-0.07	0.01	-0.12	0.01	0.80
3	2597	1.17	0.50	0.00	1.22	0.61	0.89	0.00	0.06	1.44	0.11	0.00
	-0.22	-0.15	-0.22	-0.20	-0.16	-0.19	-0.17	-0.07	-0.22	-0.15	-0.27	-0.20
4	11,245	3.67	2.00	1.33	5.33	2.33	3.67	0.67	1.00	2.33	1.33	0.00
	0.18	0.05	0.16	0.47	0.05	0.06	0.23	0.15	0.26	-0.04	0.34	-0.20
5	5347	2.50	0.00	1.00	2.00	1.50	2.00	0.00	0.50	1.50	1.00	0.00
	-0.09	-0.04	-0.34	0.30	-0.12	-0.06	-0.01	-0.07	0.01	-0.14	0.18	-0.20
6	8675	3.80	2.00	0.00	3.20	2.00	2.40	0.00	0.00	2.80	0.80	0.00
	0.06	0.06	0.16	-0.20	-0.06	0.01	0.05	-0.07	-0.24	0.02	0.08	-0.20
7	5593	2.00	1.00	0.00	5.00	2.67	2.00	0.00	1.00	2.33	1.00	0.00
	-0.08	-0.09	-0.09	-0.20	0.04	0.10	-0.01	-0.07	0.26	-0.04	0.18	-0.20
GM	7351.4	3.02	1.38	0.40	4.31	1.93	2.04	0.22	0.49	2.62	0.64	0.20

Table 4.22 Standardized entities and center of cluster A in Company data. The nearest to center are: Ant, in distance, and Ave, in inner product (both are in thousand and highlighted in bold)

Cluster	Income	MShar	NSup	EC	Util	Manu	Retail	Distance	InnerPr
Center	0.10	0.12	−0.11	−0.63	0.17	−0.02	−0.14		
Ave	−0.20	0.23	−0.33	−0.63	0.36	−0.22	−0.14	222	**524**
Ant	0.40	0.05	0.00	−0.63	0.36	−0.22	−0.14	**186**	521
Ast	0.08	0.09	0.00	−0.63	−0.22	0.36	−0.14	310	386

Ave to win, as a follower of NSup, over the inner product expressing contributions of entities to the data scatter. With the distance measure, the cluster tendency by itself does not matter at all because it is expressed in the signs of the standardized center.

Q.4.25. Find representatives of Company clusters B and C.

(c) Feature-cluster contributions to the data scatter

To see what features do matter in each of the clusters, the contributions of feature-cluster pairs to the data scatter are to be invoked. The feature-cluster contribution is equal to the product of the squared feature (standardized) center component and the cluster size. In fact, this is proportional to the squared difference between the feature's grand mean and its within-cluster mean: the further away the latter from the former, the greater the contribution! This is illustrated on Fig. 4.14.

Worked Example 4.9. Contributions of Features to Market Town Clusters
The cluster-specific feature contributions are presented in Table 4.23, along with their total contributions to the data scatter in row Total. The intermediate rows Exp and Unexp show the explained and unexplained parts of the totals, with Exp being the sum of all cluster-feature contributions and Unexp the difference between the Total and Exp rows.

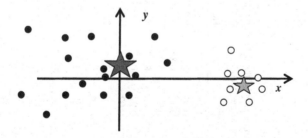

Fig. 4.14 Contributions of features x and y in the group of blank-circled points are proportional to the squared differences between their values at the grand mean (large star) and within-group center (small star)

Table 4.23 Decomposition of the data scatter over clusters and features at market town data; row Exp sums all the cluster contributions, row Total gives the feature contributions to the data scatter, and row Unexp is the difference, Total-Exp; within-row maxima are highlighted in bold

#	P	PS	Do	Ho	Ba	Su	Pe	DIY	SP	PO	CAB	FM	Total	Total, %
1	**2.09**	1.18	**2.53**	1.05	1.19	1.18	0.71	0.54	1.57	1.76	0.73	0.24	14.77	35.13
2	0.06	0.03	0.11	0.01	0.00	0.00	0.02	0.03	0.00	0.09	0.00	**3.84**	4.19	9.97
3	0.86	0.43	0.87	0.72	0.48	0.64	0.49	0.10	0.85	0.39	**1.28**	0.72	7.82	18.60
4	0.10	0.01	0.07	**0.65**	0.01	0.01	0.16	0.07	0.20	0.00	0.36	0.12	1.75	4.17
5	0.02	0.00	**0.24**	0.18	0.03	0.01	0.00	0.01	0.00	0.04	0.06	0.08	0.67	1.59
6	0.02	0.02	0.12	0.20	0.02	0.00	0.01	0.03	**0.30**	0.00	0.03	**0.20**	0.95	2.26
7	0.02	0.02	0.03	0.12	0.00	0.03	0.00	0.02	**0.20**	0.00	0.09	0.12	0.66	1.56
Exp	3.16	1.69	3.96	2.94	1.72	1.88	1.39	0.79	3.11	2.29	2.56	5.33	30.81	73.28
Unexp	0.40	0.59	0.70	0.76	0.62	0.79	1.02	0.96	1.20	0.79	1.52	1.88	11.23	24.72
Total	3.56	2.28	4.66	3.70	2.34	2.67	2.41	1.75	4.31	3.07	4.08	7.20	42.04	100.00

Table 4.24   Centers and feature-to-cluster contributions for product clusters in Company data

Item		Income	MSh	NSup	EC	Util	Manu	Retail	Total
Cluster	A	0.10	0.12	−0.11	−0.63	0.17	−0.02	−0.14	
Centers	B	−0.21	−0.29	−0.22	0.38	−0.02	0.17	−0.14	
Standardized	C	0.18	0.24	0.50	0.38	−0.22	−0.22	0.43	
Cluster	A	0.03	0.05	0.04	**1.17**	0.09	0.00	0.06	1.43
Contributions	B	**0.14**	**0.25**	0.15	0.42	0.00	0.09	0.06	1.10
	C	0.06	0.12	**0.50**	0.28	0.09	**0.09**	**0.38**	1.53
Total	Exp	0.23	0.41	0.69	1.88	0.18	0.18	0.50	4.06
Contributions	Data	0.74	0.69	0.89	1.88	0.63	0.63	0.50	5.95
	Ex%	31.1	59.4	77.5	100.0	28.6	28.6	100.0	68.3
Relative	A	14.7	29.5	18.5	**258.1**	59.9	0.0	49.5	
Contribution	B	101.2	**191.1**	90.2	120.2	0.0	77.7	64.2	
Indexes, %	C	31.7	67.0	**219.5**	58.5	54.7	54.7	**297.0**	
Cluster		24.1	39.23	2.67	**0.00**	0.67	0.33	0.00	
Centers		18.73	**22.37**	2.33	1.00	0.33	0.67	0.00	
Real		25.55	44.10	**4.50**	1.00	0.00	0.00	**1.00**	

The columns on the right show the total contributions of clusters to the data scatter, both as is and per cent. The cluster structure in total accounts for 73.3% of the data scatter, a rather high proportion. Of the total contributions, three first clusters have the largest ones totaling to 24.78, or 87% of the explained part of the data scatter, 30.81. Among the variables, FM gives the maximum contribution to the data scatter, 5.33. This can be attributed to the fact that FM is a binary variable on the data set—binary variables, in general, have the largest total contributions because they are bimodal. Indeed, FM's total contribution to the data scatter is 7.20 so that its explained part amounts to 5.33/7.20 = 0.74 which is not as much as that of, say, the Population resident feature, 3.16/3.56 = 0.89, which means that overall Population resident better explains the clusters than FM.

Worked Example 4.10. Contributions and Relative Contributions of Features at Company Clusters

Consider the clustering of Companies data according to their main product category, A or B or C, to find out what features can be associated with each of the clusters. The cluster centers as well as feature-cluster contributions are presented in Table 4.24. The summary contributions over clusters are presented in the last column of Table 4.24, and over features, in the first line of third row of Table 4.24 termed "Explain". The feature contributions to the data scatter, that is, the sums of squares of the feature's column entries, are in the second line of the third row— these allow us to express the explained feature contributions per cent, in the third line.

Now we can take a look at the most contributing feature-to-cluster pairs. This can be done by considering relative contributions within individual lines (clusters)

or within individual columns (features). For example, in the third row of Table 4.24, each contribution that covers half or more of the explained contribution by the feature is highlighted in bold. Obviously, the within-line maxima do not necessarily match those within columns. The relative contribution indexes in the fourth row of Table 4.24 combine these two perspectives: they are ratios of two relative contributions: the relative explained feature contribution within a cluster to the relative feature contribution to the data scatter. For example, relative con-ꞌ tribution index of feature MarketS to cluster B, 1.911, is found by relating its relative explained contribution 0.25/1.10 to its relative contribution to the data scatter, 0.69/5.95. Those of the relative contribution indexes that are greater than 150%, so that the feature contribution to the cluster structure is at least 50% greater than its contribution to the data scatter, are highlighted.

The most contributing features are those that make the clusters different. To see this, Table 4.24 is supplemented with the real values of the within-cluster feature means, in its last row. The values corresponding to the outstanding contributions are highlighted in bold. Cluster A differs by feature EC—A-listed companies do not use e-commerce; cluster B differs by the relatively low Market Share; and cluster C differs by either the fact that it all falls within Retail sector or the fact that its companies have relatively high numbers of suppliers, 4 or 5. It is easy to see that each of these statements not only points to a tendency but distinctively describes the cluster as a whole.

Case-Study 4.6. 2D Analysis of Most Contributing Features
Consider 2D analysis of the relationship between the Company data partition in three product classes, A, B, and C, and the most contributing of the quantitative features in Table 4.22—the Number of suppliers (77.5%) as illustrated in Table 4.25.

To calculate the correlation ratio of the NSup feature according to formula (4.20), let us first calculate the average within-class variance $\sigma_u^2 = (3 * 0.22 + 3 * .022 + 2 * 0.25)/8 = 0.23$; the correlation ratio then will be equal to $\eta^2 = (\sigma^2 - \sigma_u^2)/\sigma^2 = (1.00 - 0.23)/1.00 = 0.77$.

The case of a nominal feature can be analyzed similarly. Consider contingency table between the product-based partition S and feature Sector in Company data (Table 4.26).

Table 4.25 Tabular regression of NSup feature over the product-based classes in the Company dataset in Table 4.2. (Note that the averaging of the squared deviations is made over N = 8 rather than N−1 = 7.)

Classes	#	NSup mean	NSup standard deviation	NSup variance
A	3	2.67	0.47	0.22
B	3	2.33	0.47	0.22
C	2	4.50	0.5	0.25
Total	8	3.00	1.0	1.00

Table 4.26 Contingency table between the product-based classes and nominal feature Sector in the Company dataset according to Table 4.2

Category class	Utility	Manuf	Retail	Total
A	2	1	0	3
B	1	2	0	3
C	0	0	2	2
Total	3	3	2	8

Table 4.27 Relative frequencies together with absolute and relative Quetélet indexes for contingency Table 4.26

Cat. Class	Utility	Manuf. Retail		Total	Utility	Manuf. Retail		Utility	Manuf. Retail	
	Relative frequencies				Absolute Quetélet ind.			Relative Quetélet ind.		
A	0.25	0.12	0.00	0.37	0.29	−0.04	−0.25	0.78	−0.11	−1.00
B	0.12	0.25	0.00	0.37	−0.04	0.29	−0.25	−0.11	0.78	−1.00
C	0.00	0.00	0.25	0.25	−0.38	−0.38	0.75	−1.00	−1.00	3.00
Total	0.37	0.37	0.25	1.00						

In contrast to the classical statistics perspective, the small and even zero values are not of an issue here.

Table 4.27 presents, on the left, the same data in the relative format; the other two parts present absolute and relative Quetelet indexes as described in Sect. 3.3.

These indexes have something to do with the cluster-feature contributions in Table 4.23. Given that the categories have been normalized by unities as well as the other features, the absolute Quetélet indexes are involved. [To use the relative Quetélet indexes, the categories have to be normalized by the square roots of their frequencies, as explained in Sect. 3.8.2 and p.342.] Their squares multiplied by the cluster cardinalities and additionally divided by the squared rescaling parameter, 3 in this case, are the contributions, according to formula (4.26), as presented in the following Table 4.28.

(d) Conceptual description of clusters

If a contribution is high, then, as can be seen in Fig. 4.14, it is likely that the corresponding feature can be utilized for conceptual description of the corresponding class.

Table 4.28 Absolute Quetélet indexes from Table 4.27 and their squares factored according to formula (4.24) and follow-up; maxima are highlighted in bold

Cat. class	Size	Utility	Manuf	Retail	Utility	Manuf	Retail	Total
		Absolute Quetélet indexes			Contributions			
A	3	0.29	−0.04	−0.25	**0.085**	0.002	0.062	0.149
B	3	−0.04	0.29	−0.25	0.002	**0.085**	0.062	0.149
C	2	−0.38	−0.38	0.75	0.094	0.094	**0.375**	0.563
Total	8				0.181	0.181	0.500	0.862

Worked Example 4.11. Describing Market Town Clusters Conceptually

Consider, for example, Table 4.23 of contributions of the clusters found at Market towns data. Several entries in Table 4.23 are highlighted in bold as those most contributing to the data scatter parts explained by clusters, the columns on the right. Take a look at them, cluster-wise.

Cluster 1 is indeed characterized by its two most contributing features, Population resident (P, contribution 2.09) and the number of doctor surgeries (Do, contribution 2.53). It can be described as a "set of towns with the population resident P not less than 10,200 and number of doctor surgeries Do not less than 3"—this description perfectly fits the cluster with no errors, be it false positive or false negative. Cluster 2 is blessed with an unusually high relative contribution of FM, 3.84 of the total; this may be seen as the driving force of the cluster's separation: it comprises all the towns with a Farmers market that have not been included in cluster 1! Other clusters can be described similarly. Let us note the difference between clusters 6 and 7, underlined by the high contributions of swimming pools (SW) to both, though by different reasons: every town in cluster 7 has a swimming pool whereas any town in cluster 6 has none.

Worked example 4.12. Describing Company clusters conceptually

Conceptual descriptions can be drawn for the product clusters in Company data according to Table 4.24. This Table shows that feature EC is the most contributing to the Product A cluster, feature MSha to the Product B cluster, and features SupN and Retail to the Product C cluster. The relatively high contribution of MSha to B cluster is not that obvious because that of EC, 0.42, is even higher. It becomes clear only on the level of relative contributions when the contributions are related to their respective Total counterparts, 0.25/0.69 and 0.42/1.88—the former prevails indeed. Clusters A, B, and C can be distinctively described by the statements "EC==0", "MSha < 28", and "SupN > 3" (or "Sector is Retail"), respectively.

Unfortunately, high feature contributions not always lead to clear-cut conceptual descriptions. The former are based on the averages, whereas the latter are defined by clear-cut divisions, and division boundaries can be at odds with the averages.

4.4.4 Comparing Within-Cluster and Grand Means Using Bootstrap

As explained in the previous section, comparing the within cluster-central values with the corresponding grand means is at the heart of cluster interpretation. The greater the difference, the more important is the feature to the cluster as an aggregate entity. To extend the comparison result beyond the current dataset, one would want to make a probabilistic assessment of the comparison.

Rather than going into details of mathematics-statistical modeling, we will rely on the computational tool of bootstrapping introduced in Sect. 2.2.3.3. Specifically, given a dataset I of size N and its subset, a cluster S, let us consider a feature x at I. Generate a number, M, of bootstrap trials, each being a sequence of indices from I, and separate its subsequence related to S. Then, for each of the sequences compute the mean of x over it, gx, and the mean of x over the subsequence related to S, sx, and take the difference, $sg = sx - gx$. Then the set of M sg values is used for deriving probabilistic conclusions.

There can be either of three hypotheses: (i) $sg > 0$ (right-tailed testing), (ii) $sg \neq 0$ (two-tailed testing), and (iii) $sg < 0$ (left-tailed testing). Each of these three is against the normalcy hypothesis, which is $sg = 0$. Specifying the 95% level of confidence, these hypotheses are tested by: (i) removing the minimal 5% of the sg values; (ii) removing 2.5% of entities from both the right end and the left end; and (iii) removing the maximal 5% of the sg values. The hypothesis is rejected in either case, if the remaining subsample covers 0.

Worked Example 4.13. Comparing Grand Mean of Sepal Width with its Mean in Taxon 3

Consider the Iris dataset feature Sepal Width. Its means and standard deviations are in Table 4.29. Let us compare the grand mean, gm = 3.057 with the mean in Taxon 3, sm = 2.974. The normalcy hypothesis is that these are equal to each other, up to the sampling errors. I prefer to think that gm > sm, which is to be subject to the right-tailed testing.

Consider 5000-strong bootstrap sample of differences *gm-sm* over bootstrap samples. Its mean and standard deviation are 0.0826 and 0.0443, respectively. The 5% quantile for a Gaussian distribution is, as well-known, 1.645 standard deviations from the mean. Therefore, the difference $0.0826 - 1.645*0.0443 = 0.0097$

Table 4.29 Characteristics of the Sepal Width in the Iris dataset	Set	Mean	Standard deviation
	Taxon 1	3.428	0.379
	Taxon 2	2.770	0.314
	Taxon 3	2.974	0.322
	Total	3.057	0.436

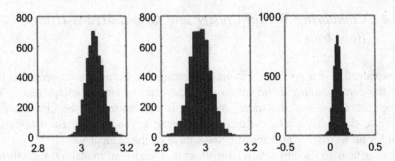

Fig. 4.15 Histograms of bootstrap values for the means of the Sepal Width with respect to the entire Iris dataset, *gm* (on the left), Iris taxon 3, *sm* (in the middle), and the difference *gm−sm* (on the right)

marks the boundary separating the upper 95% from the lower 5% of the Gaussian density function area. The value 0 is not in the confidence 95% interval, meaning that the pivotal bootstrap supports the hypothesis that the grand mean, 3.057, is indeed greater than the within taxon mean, 2.974, up to the 95% confidence (Fig. 4.15).

To apply the non-pivotal bootstrapping, we sort the values of the difference vector, *gm-sm*, in the ascending order and remove the 5% = 250 minimal values. The value number 251 in the sorted list is 0.0107, which is positive, so that this method supports the conclusion, too. Of course, if we go for a 99% confidence level, the data would not support the claim. Indeed, after removal of just 1% = 50 minimal values, the 51st value in the sorted list is negative, −0.0196, so that 0 belongs to the confidence interval and the hypothesis must be rejected.

4.5   Extension of K-means to Different Cluster Structures

So far the clustering was to encode a data set with a number of clusters forming a partition. Yet there can be differing partition-like clustering structures of which, arguably, the most popular are:

   I   *Fuzzy*: Cluster membership of entities may be not necessarily confined to one cluster only but shared among several clusters;

  II   *Probabilistic*: Clusters can be represented by probabilistic distributions rather than manifolds;

 III   *Self-organizing Map (SOM)*: Capturing clusters within cells of a plane grid along with the grid's neighborhood structure.

Further on in this section extensions of K-means to these structures are presented.

4.5.1 Fuzzy K-means' Clustering

A fuzzy cluster is represented by its membership function $z = (z_i)$, $i \in I$, in which z_i ($0 \leq z_i \leq 1$) is interpreted as the degree of membership of entity i to the cluster. This extends the concept of conventional, hard (crisp) cluster, which can be considered a special case of the fuzzy cluster corresponding to membership z_i restricted to only 1 or 0 values.

A conventional (crisp) cluster k ($k = 1, \ldots, K$) can be thought of as a pair consisting of center $c_k = (c_{k1}, \ldots, c_{kv}, \ldots, c_{kV})$ in the V feature space and membership vector $z_k = (z_{1k}, \ldots, c_{ik}, \ldots, c_{Nk})$ over N entities so that $z_{ik} = 1$ means that i belongs to cluster k, and $z_{ik} = 0$ means that i does not. Moreover, clusters form a partition of the entity set so that every i belongs to one and only one cluster if and only if $\Sigma_k z_{ik} = 1$ for every $i \in I$.

These are extended to the case of fuzzy clusters, so that fuzzy cluster k ($k = 1, \ldots, K$) is a pair comprising center $c_k = (c_{k1}, \ldots, c_{kv}, \ldots, c_{kV})$, a point in the feature space, and membership vector $z_k = (z_{1k}, \ldots, c_{ik}, \ldots, c_{Nk})$ such that all its components are between 0 and 1, $0 \leq z_{ik} \leq 1$, expressing the extent of belongingness of i to each of the clusters k. Fuzzy clusters form what is referred to as a fuzzy partition of the entity set, if the summary membership of every entity $i \in I$ is unity, that is, $\Sigma_k z_{ik} = 1$ for each $i \in I$. One may think of the total membership of any entity i as a substance that can be differently distributed among the centers.

These concepts are especially easy to grasp if membership z_{ik} is considered as the probability of belongingness. However, in many cases fuzzy partitions have nothing to do with probabilities. For instance, dividing all people by their height may involve fuzzy categories "short," "medium" and "tall" with fuzzy meanings such as those shown in Fig. 4.16.

Fuzzy clustering can be of interest in applications related with natural fuzziness of cluster boundaries such as image analysis, robot planning, geography, etc.

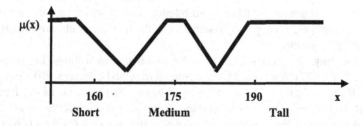

Fig. 4.16 Possible trapezoid fuzzy sets corresponding to fuzzy concepts of man's height: short, regular, and tall

If fuzzy cluster memberships are put into the bilinear PCA model, as K-means crisp memberships have been (see formula (4.14)), they make a rather weird structure in which centers are not average but rather extreme points in their clusters, which can be relaxed in a certain way and make clusters interesting and appealing, if somewhat unusual (Nascimento 2005).

An empirically convenient criterion (4.27) below differently extends that of (4.3) by using the Euclidean squared distance d and factoring in an exponent of the membership, z^α. The value α affects the fuzziness of the optimal solution: at $\alpha = 1$,

$$F(\{c_k, z_k\}) = \sum_{k=1}^{K} \sum_{i=1}^{N} z_{ik}^\alpha d(y_i, c_k) \qquad (4.27)$$

the optimal memberships are proven to be crisp, the larger the α the 'smoother' the membership. Usually α is taken to be $\alpha = 2$.

Globally minimizing criterion (4.27) is a difficult task. Yet the alternating minimization of it appears rather simple. As for Batch K-means, this works in iterations, starting from somehow initialized centers. Each iteration proceeds in two steps: (1) membership update and (2) center update. Specifically, (1): given cluster centers, cluster memberships are updated; (2): given memberships, centers are updated—after which everything is ready for the next iteration. The process stops when the updated centers are close enough to the previous ones. Updating formulas are derived from the first-order optimality conditions. They require the partial derivatives of the criterion (4.27) over the optimized variables to be set to 0.

Membership update formula:

$$z_{ik} = 1 / \sum_{k'=1}^{K} [d(y_i, c_k)/d(y_i, c_{k'})]^{\frac{1}{\alpha-1}} \qquad (4.28)$$

Center update formula:

$$c_{kv} = \sum_{i=1}^{N} z_{ik}^\alpha y_i / \sum_{i'=1}^{N} z_{i'k}^\alpha \qquad (4.29)$$

Since Eqs. (4.28) and (4.29) are the first-order optimality conditions for criterion (4.27) leading to unique solutions, convergence of the method, usually referred to as fuzzy K-means (c-means, too, assuming c is the number of clusters, see Bezdek et al. 1999), is guaranteed.

The meaning of criterion (4.27) can be formulated as follows. Let us present criterion F in (4.27) as $F = \Sigma_i F(i)$, the sum of weighted distances $F(i)$ from points $i \in I$ to all cluster centers. At a stationary solution, $F(i)$ is equal to the harmonic average of the individual memberships if $\alpha = 2$ (see Stanforth et al. 2007). Figure 4.17 presents the indifference contours of the averaged F values versus those of the distances from the nearest centers. The former are much smoother.

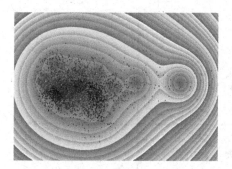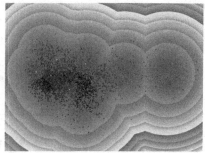

Fig. 4.17 Maps of indifference levels for the membership function $F(i)$ at about 14,000 chemical compounds clustered with iK-means in 41 clusters (**a**); (**b**) scores membership using only the nearest cluster's center

The Anomalous pattern method is applicable as a tool for initializing Fuzzy K-means as well as crisp K-means, leading to reasonable results as reported by Stanforth et al. (2007). Nascimento and Franco (2009) applied this method for segmentation of sea surface temperature maps; found fuzzy clusters closely follow the expert-identified regions of the so-called coastal upwelling, that are relatively cold, and nutrient rich, water masses. In contrast, the conventional fuzzy K-means, with user defined K, under- or over-segments the images.

Q.4.26. Regression-wise clustering. In general, centers c_k can be defined in a space which is different from that of the entity points y_i ($i \in I$). Such is the case of regression-wise clustering. Recall that a regression function $x_V = f(x_1, x_2, ..., x_{V-1})$ may relate a target feature, x_V, to (some of the) other features $x_1, x_2, ..., x_{V-1}$ as, for example, the price of a product to its consumer value and production cost attributes. In regression-wise clustering, entities are grouped together according to the degree of their correspondence to a regression function rather than according to their closeness to the gravity center. That means that regression functions play the role of centers in regression-wise clustering (see Fig. 4.18).

Consider a version of Straight K-means for regression-wise clustering to involve linear regression functions relating standardized variable y_V to other standardized variables, $y_1, y_2, ..., y_{V-1}$, in each cluster. Such a function is defined by the equation $y_V = a_1 y_1 + a_2 y_2 + ... + a_{V-1} y_{V-1} + a_0$ for some coefficients $a_0, a_1, ..., a_{V-1}$. These coefficients form a vector, $a = (a_0, a_1, ..., a_{V-1})$, which can be referred to as a regression-wise center.

When a regression-wise center is given, its distance to an entity point $y_i = (y_{i1}, ..., y_{iV})$ is defined as $r(i, a) = (y_{iV} - a_1 y_{i1} - a_2 y_{i2} - ... - a_{V-1} y_{i,V-1} - a_0)^2$, the squared difference between the observed value of y_V and that calculated from the regression equation. To determine the regression-wise center $a(S)$, given a cluster list $S \subseteq I$, the standard technique of multivariate linear regression analysis is applied, which is but minimizing the within cluster summary residual $\Sigma_{i \in S} r(i, a)$ over all possible a.

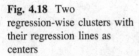

Fig. 4.18 Two
regression-wise clusters with
their regression lines as
centers

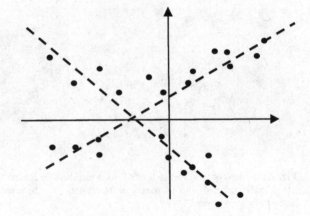

Formulate a version of the Straight K-means for this situation.

Hint: Same as Batch K-means, except that:

(1) centers must be regression-wise centers and
(2) the entity-to-center distance must be $r(i, a)$.

4.5.2   Mixture of Distributions and EM Algorithm

Data of financial transactions or astronomic observations can be considered as a
random sample from a (potentially) infinite population. In such cases, the data
structure can be analyzed with probabilistic approaches of which arguably the most
radical is the mixture of distributions approach.

According to this approach, each of the yet unknown clusters k is modeled by a
density function $f(x, \alpha_k)$ which represents a family of density functions over x defined up to a parameter vector α_k. Consider a one-dimensional density function $f(x)$,
that, for any x and very small change dx, assigns its probability $f(x)dx$ to the interval
between x and $x + dx$, so that the probability of any interval (a, b) is integral
$\int_a^b f(x)dx$, which is the area between x-axis and $f(x)$ within (a, b) as illustrated on
Fig. 4.19 for interval (6, 8). Multidimensional density functions have a similar
interpretation.

Usually, the cluster density $f(x, \alpha_k)$ is considered uni-modal with the mode
corresponding to the cluster standard point. Such is the normal, or Gaussian, density
function defined by α_k consisting of its mean vector m_k and covariance matrix Σ_k:

$$f(x, m_k, \Sigma_k) = \exp(-(x - m_k)^T \Sigma_k^{-1}(x - m_k)/2) \Big/ \sqrt{(2\pi)^V |\Sigma_k|} \qquad (4.30)$$

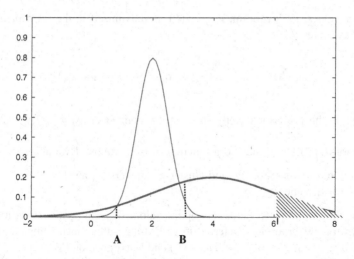

Fig. 4.19 Two Gaussian clusters represented by their density functions drawn with a thin and solid lines. The probability of interval (6, 8) in the solid line cluster is shown by the area with diagonal filling. The interval (A, B) is the only place in which the thin line cluster is more likely than the bold line cluster

The shape of Gaussian clusters is ellipsoidal because any surface at which $f(x, \alpha_k)$ is constant satisfies equation $(x - m_k)^T \Sigma_k^{-1} (x - m_k) = c$, where c is any constant, that defines an ellipsoid. This is why the PCA representation is highly compatible with the assumption of the underlying distribution being Gaussian. The mean vector m_k specifies the k-th cluster's location.

The mixture of distributions clustering model can be set as follows. The row points $y_1, y_2, \ldots, y_N$ are considered a random sample of $|V|$-dimensional observations from a population with density function $f(x)$ which is a mixture of individual cluster density functions $f(x, \alpha_k)$ ($k = 1, 2, \ldots, K$) so that $f(x) = \Sigma_{k=1}^K p_k f(x, \alpha_k)$, where $p_k \geq 0$ are the mixture probabilities such that $\Sigma_{k=1}^K p_k = 1$.

To estimate the individual cluster parameters, the principle of maximum likelihood, one of the main approaches in mathematical statistics, applies. The approach is based on the postulate that the events that have really occurred are those that are most likely. In general, this is not correct—everybody can recall a situation in which a less likely event has occurred. But the principle, applied for parameter estimation, is as much effective as a similarly wrong principle of the maximum parsimony, and even more. In its simplest version, the approach requires to find the mixture probabilities p_k and cluster parameters α_k, $k = 1, 2, \ldots, K$, by maximizing the likelihood of the observed data under the assumption that the observations come independently from a mixture of distributions. It is not difficult to show, under the assumption that the observations come independently of each other, that the likelihood is the product of the density values, $P = \prod_{i=1}^N \Sigma_{k=1}^K p_k f(y_i, a_k)$. To computationally handle the maximization problem for P with respect to the unknown parameter values, its logarithm, $L = log(P)$, is maximized in the form of the following expression:

$$L = \sum_{i=1}^{N} \sum_{k=1}^{K} g_{ik} [\log(p_k) + \log(f(y_i, \alpha_k) - \log(g_{ik})], \qquad (4.31)$$

where g_{ik} is the posterior density of cluster k defined as $g_{ik} = p_k \, f(y_i, \alpha_k)/\Sigma_k \, p_k \, f(y_i, \alpha_k)$.

Criterion L can be considered a function of two groups of variables:

(1) the mixture probabilities p_k and cluster parameters α_k, and
(2) posterior densities g_{ik},

to apply the method of alternating optimization. The alternating maximization algorithm for this criterion is referred to as EM-algorithm since computations are performed as a sequence of Expectation (E) and Maximization (M) steps. As usual, to start the process, the variables must be initialized. Then E-step is executed: Given p_k and α_k, optimal g_{ik} are found. Given g_{ik}, M-step finds the optimal p_k and α_k. This brings the process to an E-step again to follow by an M-step. And so forth. The computation stops when the current parameter values approximately coincide with the previous ones. This algorithm has been developed, in various versions, for Gaussian density functions as well as for some other parametric families of probability distributions. It should be noted that developing a fitting algorithm is not that simple, and not only because there are too many parameters here to estimate. One should take into consideration that there is a tradeoff between the complexity of the probabilistic model and the number of clusters: a more complex model may fit to a smaller number of clusters. To select a better model one can utilize the likelihood criterion penalized for the complexity of the model. A popular penalized log-likelihood criterion is referred to as Bayesian Information Criterion (BIC) and is defined, in this case, as

$$BIC = 2 \, log \, p(X/p_k, a_k) - \lambda \, log(N), \qquad (4.32)$$

where X is the observed data matrix, λ the number of parameters to be fitted, and N the number of observations, that is, rows in X. The greater the value, the better. BIC analysis has been shown to be useful, for example, in assessing the number of clusters K for the mixture of Gaussians model.

The goal of EM algorithm is determining the density functions rather than assigning entities to clusters. If the user needs to see the "actual clusters", the posterior probabilities g_{ik} can be utilized: i is assigned to that k for which g_{ik} is the maximum. Since this "optimal assignment" rule deviates from the distribution of g_{ik}, the proportions of entities in clusters obtained in this way will deviate from the mixture probabilities p_k. This is why it is advisable to consider the relative values of g_{ik} as fuzzy membership values.

The situation, in which all Gaussian clusters have their covariance matrices constant diagonal and equal to each other, so that $\Sigma_k = \sigma^2 E$, where E is identity matrix and σ^2 the variance, is of a theoretical interest. In this case, all clusters have uniformly spherical distributions of the same radius. The maximum likelihood criterion P in this case is equivalent to the criterion of K-means and, moreover, there is a certain homology between the EM and Batch K-means algorithms in this case.

To see what is going on here, consider feature vectors corresponding to entities x_i, $i \in I$, as randomly and independently sampled from the population, with an unknown assignment of the entities to clusters S_k. The likelihood of this sample is determined by the following equation:

$$P = C \prod_{k=1}^{K} \prod_{i \in S_k} \sigma^{-V} \exp\{-(x_i - m_k)^T \sigma^{-2} (x_i - m_k)/2\},$$

because in this case the determinant in (4.30) is equal to $|\Sigma_k| = \sigma^{2V}$ and the inverse covariance matrix is $\sigma^{-2} E$. The logarithm of the likelihood is proportional to

$$L = -2V \log(\sigma) - \sum_{k=1}^{K} \sum_{i \in S_k} (x_i - m_k)^T (x_i - m_k)/\sigma^2.$$

It is not difficult to see from the first-order optimality conditions for L that, given partition $S = \{S_1, S_2, ..., S_K\}$, the optimal values of m_k and σ are determined according to the usual formulas for the mean and the standard deviation. Moreover, given m_k and σ, the partition $S = \{S_1, S_2, ..., S_K\}$ maximizing L will simultaneously minimize the double sum in the right part of its expression above, which is exactly the summary squared Euclidean distance from all entities to their centers, that is, criterion $W(S, m)$ for K-means in (4.3) except for a denotation: the cluster gravity centers are denoted here by m_k rather than by c_k, which is not a big deal after all.

Thus the mixture model leads to the conventional K-means method as a method for fitting the model, under the condition that all clusters have spherical Gaussian distribution of the same variance. This leads some authors to conclude that K-means is applicable only under the assumption of such a model. However, this conclusion is wrong because it involves a logic trap: it is well known that the fact that A implies B does not necessarily mean that B implies A—there are plenty of examples to the opposite. Note however that the K-means data recovery model, also leading to K-means, assumes no restricting hypotheses on the mechanism of data generation. It also implies, through the data scatter decomposition, that useful data standardization options should involve dividing by range or similar range-related indexes rather than by the standard deviation, associated with the spherical Gaussian model. In general, the situation here is similar to that of the linear regression, which is a good method to apply when there is a Gaussian distribution of all variables involved, but it can and should be applied under any other distribution of observations if they tend to lie around a straight line.

4.5.3   Kohonen's Self-Organizing Maps SOM

Kohonen's Self-Organizing Map is an approach to visualize the data cluster structure by explicitly mapping it onto a plane grid (see Kohonen 1995). Typically, the grid is rectangular and its size is determined by the user-specified numbers of its rows and columns, r and c, respectively, so that there are $r \times c$ nodes on the grid. Each of the grid nodes, g_k ($k = 1, 2, ..., rc$), is one-to-one associated with the so-called model, or reference, vector m_k which is of the same dimension as the entity points y_i, $i \in I$.

The grid has a neighborhood structure which is to be set by the user. In a typical case, the neighborhood G_k of node g_k is defined as the set of all the grid nodes whose path distance from g_k is less than a pre-selected threshold value (see Fig. 4.20).

Then each m_k is associated with some data points—a process that can be reiterated. In the end, data points associated at each m_k are visualized at the grid point g_k ($k = 1, ..., rc$) (see Fig. 4.21). Historically, all SOM algorithms have been set in a neural-network-like incremental manner, but later, after some theoretical investigation, straight/batch versions appeared, such as the following.

Initially, vectors m_k are initialized in the data space either randomly or according to an assumption of the data structure such as, for instance, centers of K-means clusters found at $K = rc$. Given vectors m_k, entity points y_i are partitioned into "neighborhood" sets I_k. For each $k = 1, 2, ..., rc$, the neighborhood set I_k is defined as consisting of those y_i that are assigned to m_k according to the Minimum distance rule. Given sets I_k, model vectors m_k are updated as centers of gravity of all entities y_i assigned to grid nodes in the neighborhood of g_k, that is, such y_i that $i \in I_t$ for some $g_t \in G_k$. Then a new iteration of building I_k with the follow-up updating m_k's, is run. The computation stops when new m_k are close enough to the previous ones or after a pre-specified number of iterations.

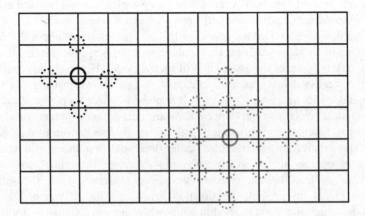

Fig. 4.20 A 7×12 SOM grid on which nodes $g1$ and $g2$ are shown along with their neighborhoods defined by thresholds 1 and 2, respectively

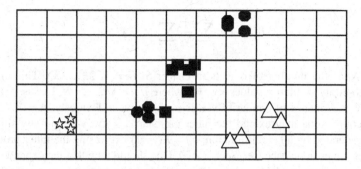

Fig. 4.21 A pattern of final SOM structure using entity labels of geometrical shapes

As one can see, SOM in this version is much similar to Straight/Batch K-means except for the following:

(a) number $K = rc$ of model vectors is large and has nothing to do with the number of final clusters—this comes visually as the number of grid clusters;
(b) data points are averaged over the grid neighbourhood, not the feature space neighborhood;
(c) there are no interpretation rules except according to positioning of points on the grid.

Item (a) results in the fact that many of final I_k's are empty, so that relatively very few of grid nodes are populated, which may create a powerful image of a cluster structure that may go to a deeper—or more interesting—minimum than K-means, because of (b).

4.6   Partitioning for Similarity Data

4.6.1   Extending K-means to Similarity Data

In Sect. 4.3.2, a complementary clustering criterion $B(S, c)$ was derived such that $T(Y) = W(S, c) + B(S, c)$ where $T(Y)$ is the data scatter and $W(S, c)$ the K-means clustering criterion. We recall that, given a data matrix $Y = (y_{iv})$, the data scatter is the sum of all the entries, squared, $T(Y) = \Sigma_{i,v} y_{iv}^2$. Given a partition $S = \{S_1, S_2, ..., S_K\}$ with a set of centers $c = \{c_1, c_2, ..., c_K\}$, over the set of objects, the criterion is the total of distances between objects $i, i = 1,2,..., N$, and their cluster centers c_k ($k = 1, 2, ..., K$). The complementary criterion $B(S, c)$ was derived in several versions, one of which was:

$$B(S,c) = \sum_{k=1}^{K} \frac{1}{N_k} \sum_{i,j\in S_k} <y_i, y_j>,$$

where N_k is the number of objects in cluster S_k ($k = 1, 2, ..., K$). This involves entity-to-entity similarity values $<y_i, y_j>$ where $i, j = 1, 2, ..., N$. Let us denote by A the $N \times N$ matrix of similarity values $a_{ij} = <y_i, y_j>$. Obviously, $A = YY^T$.

Of course, one may extend the inner product $a_{ij} = <y_i, y_j>$ onto any similarity index generated by a kernel function, $a_{ij} = K(y_i, y_j)$ expressing the inner product in a higher dimension space (see Sect. 3.4.3). One should note that the dependence on centers c_k is lost in (4.11). Thus we reformulate the criterion, in a kernel-based form, as follows:

$$g(S) = \sum_{k=1}^{K} \frac{1}{N_k} \sum_{i,j\in S_k} a_{ij} \tag{4.33}$$

The meaning of the criterion (4.33)—the sum of contributions of individual clusters, each being the within-cluster semi-average

$$g(S_k) = \frac{1}{N_k} \sum_{i,j\in S_k} a_{ij} \tag{4.34}$$

that is to be maximized.

Consider the average within-cluster similarity,

$$\lambda(S_k) = \frac{1}{N_k^2} \sum_{i,j\in S_k} a_{ij} \tag{4.35}$$

Then $g(S_k)$ in (4.34) can be expressed as $g(S_k) = N_k\lambda(S_k)$, the product of the number of elements and the average within-cluster similarity; this is why we refer to criterion (4.33) as the semi-average clustering criterion. To maximize $g(S_k)$, both the number of elements and the average within-cluster similarity must be maximized. These two goals, however, usually contradict each other: the greater the number of elements, the smaller the within-cluster similarities. The product, thus, balances them and, in this way leads to relatively tight clusters of reasonable sizes.

A more detailed analysis of $g(S_k)$ (4.34) will be given in the next chapter devoted specifically to the analysis of individual clusters, see Sect. 5.4.

Worked Example 4.14. Similarity Matrix for Normalized Company Data
Consider the 8×7 Company data matrix in its standardized format (Table 4.30 copied for convenience from Table 4.2). Its last two rows show the column contributions to the data scatter, both as are and per cent.

Let us compute two recommended similarity matrices, the matrix of inner products $A = YY^T$ (see in Table 4.31) and the matrix of affinity data $a_{ij} = exp(-d(x_i, x_j)/s)$ found

Table 4.30 Company data matrix standardized by subtraction of feature means and division by feature ranges, with the follow-up further division of columns 5,6, and 7 over the square root of 3

Av	−0.20	0.23	−0.33	−0.63	0.36	−0.22	−0.14
An	0.40	0.05	0	−0.63	0.36	−0.22	−0.14
As	0.08	0.09	0	−0.63	−0.22	0.36	−0.14
Ba	−0.23	−0.15	−0.33	0.38	0.36	−0.22	−0.14
Br	0.19	−0.29	0	0.38	−0.22	0.36	−0.14
Bu	−0.60	−0.42	−0.33	0.38	−0.22	0.36	−0.14
Ci	0.08	−0.10	0.33	0.38	−0.22	−0.22	0.43
Cy	0.27	0.58	0.67	0.38	−0.22	−0.22	0.43
Cnt	0.74	0.69	0.89	1.88	0.62	0.62	0.50
Cnt, %	12.42	11.66	14.95	31.54	10.51	10.51	8.41

Table 4.31 Matrix of row-to-row inner products for matrix of normalized Company data in Table 4.30 (positive entries highlighted in bold)

Entities	Av	An	Ast	Ba	Br	Bu	Ci	Cy
Av	**0.794**	**0.519**	**0.260**	**0.086**	−0.474	−0.237	−0.478	−0.470
An	**0.519**	**0.752**	**0.293**	−0.137	−0.307	−0.629	−0.299	−0.191
As	**0.260**	**0.293**	**0.604**	−0.404	−0.048	−0.126	−0.330	−0.250
Ba	0.086	−0.137	−0.404	**0.527**	**0.005**	**0.320**	−0.069	−0.328
Br	−0.474	−0.307	−0.048	**0.005**	**0.457**	**0.347**	**0.090**	−0.069
Bu	−0.237	−0.629	−0.126	**0.320**	**0.347**	**0.983**	−0.074	−0.583
Ci	−0.478	−0.299	−0.330	−0.069	**0.090**	−0.074	**0.549**	**0.612**
Cy	−0.470	−0.191	−0.250	−0.328	−0.069	−0.583	**0.612**	**1.279**

Table 4.32 Matrix of row-to-row squared Euclidean distance values for the matrix of normalized Company data in Table 4.30 (highlighted in bold are the values less than 1)

Entities	Av	An	Ast	Ba	Br	Bu	Ci	Cy
Av	0.000	**0.508**	**0.877**	1.149	2.199	2.251	2.299	3.012
An	**0.508**	0.000	**0.770**	1.554	1.824	2.994	1.899	2.414
As	**0.877**	**0.770**	0.000	1.939	1.157	1.840	1.814	2.384
Ba	1.149	1.554	1.939	0.000	**0.974**	**0.871**	1.215	2.462
Br	2.199	1.824	1.157	**0.974**	0.000	**0.746**	**0.826**	1.873
Bu	2.251	2.994	1.840	**0.871**	**0.746**	0.000	1.681	3.429
Ci	2.299	1.899	1.814	1.215	**0.826**	1.681	0.000	**0.605**
Cy	3.012	2.414	2.384	2.462	1.873	3.429	**0.605**	0.000

by using the Gaussian kernel transformation, where $d(x_i, x_j)$ is the squared Euclidean distance $d(x_i, x_j)$ and s a scaling parameter. At range-normalized data, our experimentally tested value for s is $s = 1/2$. The former is presented in Table 4.31, the latter in Table 4.33: the intermediate Table 4.32 contains the squared Euclidean distances $d(x_i, x_j)$.

Table 4.33 Matrix of row-to-row Gaussian kernel values for matrix of normalized Company data in Table 4.29. All values that are greater than 0.14 are highlighted in bold

Entities	Av	An	As	Ba	Br	Bu	Ci	Cy
Av	1.000	**0.362**	**0.173**	0.100	0.012	0.011	0.010	0.002
An	**0.362**	1.000	**0.214**	0.045	0.026	0.003	0.022	0.008
As	**0.173**	**0.214**	1.000	0.021	0.099	0.025	0.027	0.008
Ba	0.100	0.045	0.021	1.000	**0.142**	**0.175**	0.088	0.007
Br	0.012	0.026	0.099	**0.142**	1.000	**0.225**	**0.192**	0.024
Bu	0.011	0.003	0.025	**0.175**	**0.225**	1.000	0.035	0.001
Ci	0.010	0.022	0.027	0.088	**0.192**	0.035	1.000	**0.298**
Cy	0.002	0.008	0.008	0.007	0.024	0.001	**0.298**	1.000

In the inner product matrix A (Table 4.31) all positive entries are highlighted in bold. One can see that almost all of them are concentrated within the three product-based clusters {Av, An, As}, {Ba, Br, Bu}, and {Ci, Cy}, except for $c_{14} = 0.186$ and $c_{57} = 0.090$. This, including rather small values of the "oddities" may convince one that the product-based partition is optimal indeed.

Also, one should notice that the diagonal elements are by far the largest in their rows and columns, which is no wonder as the inner product geometrically is the product of vector norms and cosine of the angle between them.

Although rather auxiliary in the context of computation of the Gaussian kernel, the squared Euclidean distance matrix in Table 4.28 may have an independent standing too. Indeed, according to Eq. (4.5) in Q.4.9, these distances stand as the only data involved in an equivalent reformulation of the K-means criterion; the summary within-cluster semi-average value of them must be minimized to achieve the goal of K-means. The distances less than 1 are highlighted in bold in Table 4.31. As one can see, the distances are much consistent with the three product-based clusters: only one between cluster distance, $d(Ba, Ci) = 0.826$, does not follow this pattern.

4.6.2   Algorithms for the Semi-average Clustering Criterion

The semi-average criterion for partitioning does not admit a simple algorithm for its maximization even if the similarity matrix has no negative entries. We consider two local search algorithms here, agglomeration of clusters and moving objects. But we first point out a property which is important further on. The property is that if partition $S = \{S_1, S_2, \ldots, S_K\}$ is a maximizer of the criterion g(S) in (4.33) over similarity matrix A, then the S is a maximizer of (4.33) over symmetric similarity matrix $A^{sym} = (A + A^T)/2$ with entries $a_{ij}^{sym} = (a_{ij} + a_{ji})/2$. The proof easily follows from an obvious fact that the value of g(S) over A coincides with that of g(S) over A^{sym}. Why? Because, every pair i, j adds to g(S) the same value $a_{ij} + a_{ji}$ in both cases.

Before proceeding to the stage of clustering itself, data matrix A should be pre-processed. First of all, it should be converted to a symmetric form by transforming it into $A^{sym} = (A + A^T)/2$. This will not affect the result, according to the equivalence property above. Another pre-processing step is highly recommended too: put zeros in all the diagonal elements to make $a_{ii} = 0$ for all objects $i = 1, 2, ..., N$. Indeed, clusters should reflect the similarity between objects rather than the weighting of objects. A strong diagonal, like that in the Confusion data (see Table 1.8 in Chap. 1), reflecting the strong- and differing-similarity of objects to themselves, may affect the results as illustrated in Worked examples 4.14 and 4.15.

Agglomeration of Clusters Approach

This approach follows a conventional for all agglomerative algorithms scheme. It transforms the original matrix A of object-to-object similarity index values into a binary cluster hierarchy. It starts at N singleton clusters and one-by-one merges chosen pairs of clusters to, conventionally, reach the global universal cluster consisting of all the objects. Here, however, this convention may fail to work. Why?

Take a look at the current partition $S = \{S_1, S_2, ..., S_K\}$, $K \le N$. and the criterion value $g(S)$ (4.33). What the criterion may lose if two clusters, S_k and S_l, are merged together to form another partition, $S(k, l)$, consisting of $K-1$ clusters, of which one is $S_k \cup S_l$ and the others are the other clusters from S. Let us denote

$$A(I_1, I_2) = \sum_{i \in I_1} \sum_{j \in I_2} a_{ij}, \tag{4.36}$$

the summary similarity between sets of objects, I_1 and I_2, which may overlap or even coincide.

We claim that the loss $\Delta(k, l)$ at merging S_k and S_l is equal to

$$\Delta(k, l) = g(S) - g(S(k, l))$$
$$= [N_k A(S_l, S_l)/N_l + N_l A(S_k, S_k)/N_k - 2A(S_k, S_l)]/(N_k + N_l) \tag{4.37}$$

Indeed, the difference involves only those parts (4.29) that relate to S_k, S_l, and $S_k \cup S_l$:

$$\Delta(k, l) = g(S) - g(S(k, l))$$
$$= \frac{1}{N_k} \sum_{i,j \in S_k} a_{ij} + \frac{1}{N_l} \sum_{i,j \in S_l} a_{ij} - \frac{1}{N_k + N_l} \sum_{i,j \in S_k \cup S_l} a_{ij} = \frac{1}{N_k N_l (N_k + N_l)} [N_l(N_k$$
$$+ N_l)A(S_k, S_k) + N_k(N_k + N_l)A(S_l, S_l) - N_k N_l A(S_k \cup S_l, S_k \cup S_l)]$$
$$= \frac{1}{N_k N_l (N_k + N_l)} [N_l^2 A(S_k, S_k) + N_k^2 A(S_l, S_l) - 2N_k N_l A(S_k, S_l)].$$

The last equation follows from the fact that $A(S_k \cup S_l, S_k \cup S_l) = A(S_k, S_k) + 2A(S_k, S_l) + A(S_l, S_l)$ and proves (4.37).

Now we should explain why we refer to the difference $g(S) - g(S(k, l))$ as a loss, assuming that $g(S) > g(S(k, l))$. In the normal circumstances, when elements of

A are inner product or kernel values, the Pythagorean decomposition (4.8) of the data scatter holds, so that the data scatter is the sum of the quadratic error $W(S, c)$ and the "explained part", $B(S, c)$. Merging two clusters may only increase the quadratic error $W(S, c)$ and, thus, only decrease the complementary criterion $B(S, c)$ and, of course, its identical twin, $g(S)$.

The circumstances are not normal when A is not a matrix of inner products anymore, as, for example, if its diagonal is forcibly zeroed, as advised above, or if that is just an observed data case, like that of Eurovision song contest results in Table 1.11. In such a case non-monotonicity of $g(S)$ regarding the merging process can be taken as a stop condition: why one should continue mergers if the increment $\Delta(k, l)$ gets negative? Indeed, the goal is to maximize $g(S)$. This is how the semi-average criterion brings forth a natural stopping condition to the process of mergers, in this way determining the natural number of clusters.

One issue remaining to get discussed is a visual representation of the merger process with a dendrogram. A dendrogram is a rooted tree along with a height function. The height function assigns real numbers to tree nodes in such a monotone way that any parental node's height is greater than that of its children. Heights of leaves usually are taken to be zero.

Agglomerative Clustering Algorithm for the Semi-average Criterion: SA-Agglomeration

1. *Initial setting.* Set $K = N$, put all entities $k \in I$ to singleton clusters $S_k = \{k\}$, with their cardinalities set to unity and heights to zero; form a cluster-to-cluster similarity matrix $AA = (A(k, l))$ between them, with $A(k, l) = a_{kl}$, and a list of maximal clusters M consisting of all the singletons at this stage. Set flag $= 1$ and do the following steps till the flag changes.
2. *Finding minimum.* Using matrix AA, compute all $\Delta(k, l)$ (4.37) and find their maximum $\Delta(k^*, l^*)$ where $k^* < l^*$.
3. *Stop condition.* If $\Delta(k^*, l^*) < 0$, put flag $= 0$. If $K = 2$, put flag $= 0$.
4. *Merge and update.* Merge two maximal clusters, S_{k^*} and S_{l^*}, together to form their parent in the dendrogram, a new maximal cluster $S_{k^* \cup l^*} = S_{k^*} \cup S_{l^*}$. To do this, add the row l^* to the row k^* in AA; after this, add column l^* to column k^* in AA, after which remove l^* row and column from AA; its size now is $K-1$, so change K by subtracting 1 from it. Similarly, add maximal cluster $M(l^*)$ to maximal cluster $M(k^*)$, after which remove the l^* cluster from M. Define the new cluster's cardinality as $N_{k^*} + N_{l^*}$, and put that as the cardinality of cluster k^*. Put the height added to the height of the previous merger equal to $\Delta(k^*, l^*)$ and draw the merger accordingly. (In fact, any sequence of reals monotonically following the order of mergers can be used as a height function.)
5. *Output*: the set of all merged clusters along with their heights. Draw a cluster dendrogram, that is a tree, if $K = 1$, or a forest with K rooted trees if $K > 1$.

Worked Example 4.15. Semi-average Clustering at the Original Confusion Data

Let us take a symmetric version of the Confusion data obtained from the matrix A in Table 1.8 using transformation $(A + A^T)/2$, see Table 4.34.

The values of change of $g(S)$ at consecutive mergers from S to $S(k, l)$ form the following sequence $\Delta(k, l)$: −484.0, −492.5, −605.3, −607.5, −667.6, −726.2, −746.4, −767.3. This corresponds to the idea that each merger decreases the semi-average $g(S)$ because the very strong diagonal similarities are getting mixed with much weaker off-diagonal similarities. Another feature of the sequence is that each time the change is greater. Probably the observed monotonicity reflects the presence of a natural cluster structure in the data. At such a structure, within-cluster similarities are in general greater than those between clusters.

The observed monotonicity can be utilized in drawing the dendrogram. The vertical axis on the right in Fig. 4.22 scores the heights of the interior nodes in the dendrogram; the heights correspond to $-470-\Delta(k, l)$ values.

Let us point to less obvious features of the dendrogram. Clusters 1–4–7 and 3–9–5 quite naturally fit at larger non-diagonal similarities between members of the clusters (see in Table 4.34). On the other hand, a relatively high similarity 122

Table 4.34 A symmetric version of confusion data in Table 1.7

Stimulus	Response									
	1	2	3	4	5	6	7	8	9	0
1	877.0	10.5	18.0	85.5	9.0	20.0	164.5	5.5	14.5	11.0
2	10.5	782.0	38.0	13.0	31.0	30.5	9.0	28.5	18.0	11.0
3	18.0	38.0	681.0	5.5	30.5	3.5	30.5	28.5	131.5	11.0
4	85.5	13.0	5.5	732.0	9.0	11.0	25.5	12.5	43.5	5.5
5	9.0	31.0	30.5	9.0	669.0	88.0	7.0	12.5	104.0	10.5
6	20.0	30.5	3.5	11.0	88.0	633.0	2.0	112.5	11.0	30.5
7	164.5	9.0	30.5	25.5	7.0	2.0	667.0	5.5	12.5	16.0
8	5.5	28.5	28.5	12.5	12.5	112.5	5.5	577.0	74.5	121.5
9	14.5	18.0	131.5	43.5	104.0	11.0	12.5	74.5	550.0	32.0
0	11.0	11.0	11.0	5.5	10.5	30.5	16.0	121.5	32.0	818.0

Fig. 4.22 Dendrogram of the semi-average agglomeration process for similarity data in Table 4.34

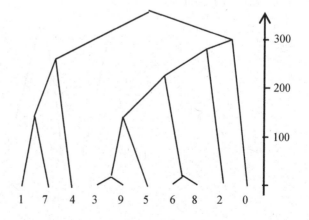

between 0 and 8 has not made it through: 0 is the last to merge into the tree. Why? Perhaps the relatively high diagonal value, 818, for 0, has contributed to this. We try to see to it in the next worked example.

Worked Example 4.16. Semi-average Clustering at the Confusion Data with a Zeroed Diagonal

Let us take the very same Confusion similarity matrix A in Table 4.34 and make all its diagonal entries 0, so that $a_{ii} = 0$ for all $i = 1, 2, ..., 10$. Applying the same Semi-average agglomeration algorithm to this dataset, we obtain the sequence of mergers shown with the dendrogram in Fig. 4.23.

The sequence of changes $\Delta(k, l)$ of the criterion $g(S)$ at consecutive mergers from S to $S(k, l)$ is this: 164.5, 131.5, 121.5, 88.0, 19.2, 11.7, −30.6, −48.1. The sequence is monotone decreasing, like the similar sequence in the Worked example 4.15 above. However, unlike that sequence, this one consists of mostly positive reals. Why is that? Because at the starting point of the agglomeration process here are singleton clusters with zero within-cluster similarities, implying that $g(S) = 0$ at the starting trivial partition. Merging two most similar objects, 1 and 7, adds the average within-cluster similarity 164.5 to the value of $g(S)$. Next merger, 3 and 9, adds to this the similarity, 131.5, between objects forming a new cluster in the dendrogram. Further four mergers add more to this, as reflected in the sequence of the corresponding $\Delta(k, l)$ values: 121.5, 88.0, 19.2, 11.7. Then the maximum change $\Delta(k, l)$ becomes negative, −30.6, at clusters 2–5–6 and 8–10 to be merged. That means that the value $g(S)$ would decrease at the merger and, therefore, no further mergers should be made anymore, as the goal is to maximize $g(S)$.

By comparing $g(S)$ for the obtained partition $S = \{1–7–4, 5–6–2, 8–0, 3–9\}$ with $g(T)$ where $T = \{1–7–4, 3–9–5, 6–8\}$ is that found at Worked example 4.15, one can see that $g(S) = 536.3$ and $g(T) = 473.5$. Therefore, at A with the diagonal zeroed, S in Fig. 4.23 is by far better than T in Fig. 4.22.

Let us introduce the second partitioning algorithm, which implements the idea of iteratively moving objects one-by-one into different clusters.

Fig. 4.23 Dendrogram of the semi-average agglomeration process for similarity data in Table 4.34 with the zeroed diagonal

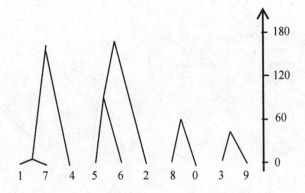

Moving Objects Between Clusters Algorithm: SA-Move

This algorithm takes in a partition S on the set of objects and loops through the objects $i = 1, 2, ..., N$, by testing whether moving i into a different cluster would increase the value of criterion $g(S)$. Because of its additivity, the change of $g(S)$ at moving object i from cluster S_k to cluster S_l can be expressed as

$$\Delta(i, S_k, S_l) = [g(S_k) - 2A(i, S_k)]/(N_k - 1) - [g(S_l) + 2A(i, S_l)]/(N_l + 1) \quad (4.38)$$

where N_k and N_l are cluster cardinalities, $g(S_k)$ and $g(S_l)$ are parts (4.34) of the criterion $g(S)$ related to S_k and S_l, respectively, and $A(i, S)$ the summary similarity between object i and subset S. An important caveat: Eq. (4.38) is derived under assumption that all the diagonal elements a_{ii} are zeros.

An exact formulation of the algorithm of local improvements by moving objects can be different depending on the effort for finding the best version for a move. A most rigorous version would require finding an I and S_l such that $\Delta(i, S_k, S_l)$ in (4.38) is maximized. A most relaxed version would require pre-specifying an order of objects and order of clusters and moving objects, in this order, to the very first cluster S_l at which $\Delta(i, S_k, S_l) > 0$.

Worked Example 4.17. SA-Move at the Confusion Data with a Zeroed Diagonal

Let us apply SA-Move to the symmetric Confusion data with a zeroed diagonal beginning with partition $S = \{1-7-4, 5-6-2, 8-0, 3-9\}$ found in the Worked example 4.16. It is not difficult to see that no move of an object can increase the value $g(S) = 536.3$. For example, moving object 2 into cluster 1–7–4 would result in the value of g-criterion equal to 495.0. Values of g equal to 510.5 and 518.2 would be returned if 2 is moved to cluster 8–0 or 3–9, respectively. Leaving 2 a singleton would return 524.7. These all are less than $g(S) = 536.3$.

Let us then try the SA-Move algorithm beginning with $T = \{1-7-4, 3-9-5, 6-8\}$. Adding object 2, which is outside of the partition T, to either of clusters 1–7–4, 3–9–5, 6–8 returns g-value 443.8, 472.7, 475.3, respectively. Of these, 475.3 is greater than g(T) = 473.5. Should we thus take cluster 6–8–2 instead of 6–8 then? No, this is not advisable because adding 0 to cluster 6–8 gives a by far greater rise of the value of g, to 537.3! Unfortunately, further adding of 2 to either of the three triplet clusters may only decrease the g-value (to 507. 7, 536.5, 528.2, respectively). Now we see that the updated partition $T' = \{1-7-4, 3-9-5, 6-8-0, 2\}$ is better than the partition $S = \{1-7-4, 5-6-2, 8-0, 3-9\}$ found with the SA-Agglomeration algorithm!

Why then the SA-Move failed in improving S? To properly answer to the question, one must note that T' cannot be obtained from S by moving single objects. Two objects, 5 and 6, should be moved simultaneously, to 3–9 and 6–8, respectively, to obtain T' from S. It is not difficult to modify the SA-Move algorithm for moving pairs rather than individual objects; however, so far, no such a version has been proposed and substantiated.

A conclusion: SA-Agglomeration algorithm does not necessarily lead to optimal partitions, even with a zeroed diagonal. Moving individual objects may fail to further optimize the Agglomeration results.

Q.4.27. Take the Eurovision song contest data A from Table 1.11, symmetrize it using transformation $AA = (A + A^T)/2$ and apply SA-Agglomeration method. How many clusters are recommended by the semi-average criterion g(S)?
A. The symmetric data matrix is in Table 4.35.

The order of mergers is illustrated by the dendrogram in Fig. 4.24. The values of the change $\Delta(k, l)$ at mergers form the sequence seq = 10.0, 9.0, 8.6, 8.0, 6.0, 4.6, 4.2, 4.0, 2.8, 2.0, 1.9, 0.5, −0.4, −0.4, −2.2, −6.1, −7.8. The heights of the corresponding nodes correspond to values 11-seq. As one can see, only 12 mergers lead to the increase in g(S) value.

Therefore, the number of clusters recommended by the semi-average criterion is 7. They are:

(1) Az-Ukr-Rus,
(2) Bul-Greece-Serbia,
(3) Belg-Neth,
(4) Ger-UK,
(5) Est-Pol,
(6) Fr-Isr-Switz, and
(7) It-Rom-Port-Spain

The clusters correspond to cultural or ethnic relations between countries. Specifically, cluster (1) is for the former Soviet Union, cluster (2) for Balkans, cluster (3) for Low countries, cluster (5) for Baltics, cluster (7), for Romance languages. Less obvious are clusters (4) and (6).

4.6.3 Summary Similarity Clustering

4.6.3.1 Min-Cut and Ford-Fulkerson Network Flow

Consider a heuristic clustering criterion which is somewhat simpler than the semi-average similarity in (4.33)—the sum of within cluster similarities

$$f(S) = \sum_{k=1}^{K} \sum_{i,j \in S_k} a_{ij} = \sum_{k=1}^{K} A(S_k, S_k) \tag{4.39}$$

where $A(S_k, S_k)$ is the summary similarity (4.36) within cluster S_k.

Table 4.35 The symmetrized version of the average scores by European countries from Table 1.11 (only non-zero entries are shown)

	Az	Be	Bu	Es	Fr	Ge	Gr	Is	It	Ne	Pol	Por	Ro	Ru	Se	Sp	Sw	Ukr	UK
Azerbaijan		1.9	3.4	2.0			5.8	4.9		2.0	4.2		5.1	8.2				10.0	2.1
Belgium	1.9				1.8	2.0	2.0			4.6	2.2	1.8				1.7			2.1
Bulgaria	3.4				2.2		8.6		5.0					2.4	5.6	3.9		2.2	
Estonia	2.0								2.2		2.0			4.4				2.2	1.8
France		1.8	2.2			1.7	2.0	2.8	5.0			2.7			4.0	2.2	2.2		2.0
Germany		2.0			1.7		1.8	1.8		1.9	2.8	2.2			3.5	2.6			4.0
Greece	5.8	2.0	8.6		2.0	1.8							6.1	1.8	7.6	4.4		1.9	1.9
Israel	4.9				2.8	1.8				4.4		2.0	3.3	5.5	2.5	2.1	2.1	3.1	2.2
Italy			5.0	2.2	5.0						4.5	4.0	9.0			3.7	2.4	3.2	2.6
Netherland	2.0	4.6				1.9		4.4							3.5				
Poland	4.2	2.2		2.0		2.8			4.5									7.1	
Portugal		1.8			2.7	2.2		2.0	4.0				2.6	2.4	2.8	4.2		3.7	2.2
Romania	5.1						6.1	3.3	9.0			2.6			2.4	4.0		4.0	1.8
Russia	8.2		2.4	4.4			1.8	5.5				2.4			6.2			8.8	
Serbia			5.6		4.0	3.5	7.6	2.5		3.5		2.8	2.4	6.2		2.4	5.3	6.7	1.8
Spain		1.7	3.9		2.2	2.6	4.4	2.1	3.7			4.2	4.0		2.4		2.0	2.3	
Switzerland					2.2			2.1	2.4						5.3	2.0			2.0
Ukraine	10.0		2.2	2.2			1.9	3.1	3.2		7.1	3.7	4.0	8.8	6.7	2.3			
UK	2.1	2.1		1.8	2.0	4.0	1.9	2.2	2.6			2.2	1.8		1.8		2.0		

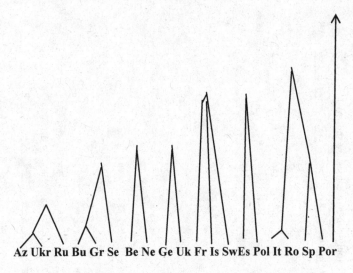

Az Ukr Ru Bu Gr Se Be Ne Ge Uk Fr Is SwEs Pol It Ro Sp Por

Fig. 4.24 Dendrogram of the semi-average agglomeration process for Eurovision song contest data (symmetric)

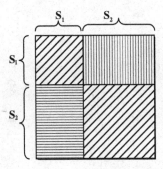

Fig. 4.25 The structure of the similarity matrix regarding partition $\{S_1, S_2\}$ of the entity set, which is assumed sorted so that elements of S_1 stand first. The blocks out of the main diagonal show similarities between S_1 and S_2, whereas those on the main diagonal refer to similarities within the parts

This seems a perfect criterion for clustering—it is simple and intuitive (see Fig. 4.25). The greater the within cluster total similarity, the tighter are the clusters. Maximizing this criterion should lead to a partition consisting of clusters of highest internal similarity.

What is nice about $f(S)$ in (4.39) is that its maximum corresponds to the minimum of the sum of all between cluster similarities

$$h(S) = \sum_{k \neq l} A(S_k, S_l)$$

Obviously, the sum of within-cluster and between-cluster similarities $f(S) + h(S)$ covers all the entries in A; each only once. This implies that $f(S) + h(S) = A(I, I)$ where I is the set of all N objects, so that $A(I, I)$ is constant at a specified matrix A. To illustrate the claim, just take a look at Fig. 4.25: the square representing the similarity matrix there is divided in two parts: that of diagonal strips, the within-cluster similarities, and that with horizontal strips, the between-cluster similarities. Therefore, any S maximizing $f(S)$ simultaneously minimizes $h(S)$.

In the case, when there are only two clusters, $S = \{S_1, S_2\}$, the value of h(S) is referred to as a cut and the problem of minimization of $h(S)$ is referred to as the min-cut problem in the theory of optimization.

At a nonnegative A, $h(S)$ obviously reaches its minimum when the between-cluster blocks in Fig. 4.25 are reduced to just mere one line and column, that one having the minimum sum of its entries. Therefore, an optimal cut is a most unbalanced partition: a singleton and the rest, which is not what should be considered a proper clustering. Such a solution can be avoided if one imposes admissibility constraints on possible solutions. For example, only equal-sized clusters may be admitted—this constraint make the problem very hard. A popular approach is to specify two objects that must be in different clusters. a capacity A flow in a network is arranged between two nodes, a source and a sink. Each link (i, j) is considered as a pipeline capacity of which is limited by the similarity value a_{ij}. A flow is a skew-symmetric function $f(i, j)$ (that is, $f(i, j) = -f(j, i)$ for any pair $i \neq j$) defined over the network so that $f(i, j) \leq a_{ij}$, and, for any i which is neither the source nor the sink, the sum of $f(i, j)$ over all j is zero. The sum of $f(\text{source}, j)$ over all j is referred to as the flow value. The problem is to find a flow of the maximum value. This can be done iteratively by sending a unit of the flow over each path from the source to sink, which has yet a capacity to do so, that is, if $a_{ij} - f(i, j) > 0$ for all edges (i, j) on the path. There is a celebrated Ford-Fulkerson theorem stating that the maximum flow between the source and the sink is equal to the min-cut separating them (see Ahuja et al. 2014 for a more precise exposition). The maximum flow can be used to determine the corresponding min-cut. The cluster containing the source is determined as the set of all nodes reachable from the source by using "unsaturated" paths. Given a maximum flow f, an unsaturated path is a set of neighboring edges (i, j) such that $a_{ij} - f(i, j) > 0$.

Q.4.28. Given a network with similarity matrix A, let us denote $a(s, t)$ the value of min-cut in the network with source s and sink t. Prove that the pair-wise min-cut index $a(s, t)$ is an ultra-similarity. A pair-wise similarity index $a(s, t)$, $s, t = 1, 2, \ldots,$ N, is referred to as an ultra-similarity if inequality

$$a(s,t) \geq \min(a(s,r), a(t,r)) \tag{4.40}$$

holds for all nodes s, t, r.

A. To correctly answer this question, one needs to see that the ultra-similarity condition (4.40) means that at every triplet s, t, r, two of the values $a(s, t)$, $a(s, r)$, and $a(t, r)$, are equal to each other, whereas the third one is either larger or the same. Indeed, if the opposite is true, that is, only one of the values, say $a(s, t)$, is smaller than the others, then of course (4.40) does not hold for $a(s, t)$. Assume now that a min-cut similarity index dos not satisfy (4.40) at some triplet s, t, r, so that $a(s, t) < a(s, r) \leq a(t, r)$. Consider min-cut at source s and sink t; its value is $a(s, t)$. The node r must belong to either of two clusters, say, to the cluster containing s. Then the current partition can be considered as a cut, that is, partition, with source r and sink t such that its value, $a(s, t)$, is smaller than the min-cut $a(t, r) = a(r, t)$. The obtained contradiction proves the statement.

To obtain meaningful clusters, similarity data have to be transformed to sharpen the structure to be found in a manner similar to the transformation of object-to-feature data by standardization. Since all similarities are measured in the same scale, no normalization by rescaling is applied—that would just change the unit of measurement. The number of models of background, to be subtracted from the data, is potentially infinite. So far, two of them made it into the literature,

(i) Uniform scale shift, or soft threshold, in which the background value of similarity indices is considered constant, and
(ii) Modularity function, in which the background is assumed to be generated by random interactions between objects.

These two will be considered in the next two sections.

Subtraction of background makes some A-entries negative, leaving others positive. The problem of minimum cut with negative similarity values becomes computationally intensive, referred to as NP-complete in the theory of combinatorial optimization (see, for example, Johnsonbaugh and Schaefer 2004). This implies that local or approximate algorithms would be a welcome development for the problem.

4.6.3.2   Uniform Similarity Shift Value Subtracted

Given a similarity shift value π, shifted similarity matrix $A - \pi$ has $a_{ij} - \pi$ as its (i, j)-th entry. Therefore, the summary within-cluster criterion (4.39) can be reformulated here as

$$f(S, \pi) = \sum_{k=1}^{K} \sum_{i,j \in S_k} (a_{ij} - \pi) \tag{4.41}$$

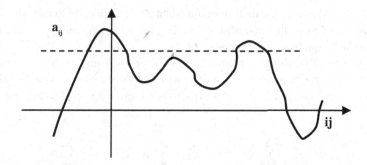

Fig. 4.26 Illustration of the effect of subtraction of a constant background "noise" from the similarity values. The graph shows similarity values (axis y) against some ordering of entity pairs (i, j) over x-axis. At zero noise level, the area of positive similarity values is much larger than that above the dashed line at which the area narrows down to two small high similarity islands

   The summary criterion with the uniform noise subtracted is referred to as uniform clustering criterion (Mirkin 1996, 2012), The criterion leads to a distinction between those pairs (i, j) at which $a_{ij} > \pi$ and those at which $a_{ij} \leq \pi$. In the former case, i and j should be put in a same cluster in S, whereas i and j should be separated in S, in the latter case, because the goal is to maximize (4.41). Thus, π is not only the similarity scale shift, but also a threshold. Unlike the concept of threshold used in the definition of threshold graphs that are obtained from weighted graphs by zeroing those weights which are smaller than the threshold (see Sect. 5.3.1), this threshold is not hard, it is soft: a pair i, j can be put into the same cluster, even for $a_{ij} \leq \pi$, if there are other objects whose similarities to both i and j are great enough. This meaning of π is illustrated by Fig. 4.26.
   To give this meaning a more precise formulation, let us introduce the concept of average similarity between clusters:

$$a(S_k, S_l) = \frac{1}{N_k N_l} \sum_{i \in S_k} \sum_{j \in S_l} a_{ij} = \frac{1}{N_k N_l} A(S_k, S_l)$$

where N_k and N_l are cardinalities of S_k and S_l, respectively. Of course, at $k = l$, the value $a(S_k, S_k)$ is the average similarity within cluster S_k.
   The following statement is true: if S is an optimal partition according to criterion (4.41), then for any its clusters S_k and S_l, $a(S_k, S_k) \geq \pi \geq a(S_k, S_l)$.
   Indeed, those parts of matrix $A - \pi$ that relate to clusters S_k and S_l produce sums $\sigma(S_k, S_l) = A(S_k, S_l) - \pi N_k N_l$ and $\sigma(S_k, S_k) = A(S_k, S_k) - \pi N_k^2$ whose signs are different in an optimal partition. Specifically, if S is optimal, then for any S_k and S_l $\sigma(S_k, S_k) \geq 0$ and $\sigma(S_k, S_l) \leq 0$ if $k \neq l$. Indeed, admitting that $\sigma(S_k, S_k) < 0$ for some k, one can easily improve the partition S by scattering all elements of S_k into singletons, thus adding 0 to criterion (4.41) rather than a negative quantity (let us remind that the diagonal entries are assumed zero, $a_{ii} = 0$). Similarly, admitting that $\sigma(S_k, S_l) > 0$ for some $k \neq l$ leads to a simple way for improving S by merging

clusters S_k and S_l, since the merger would add a positive quantity to the value of the criterion. This proves the inequalities. Dividing them by $N_k N_l$ and N_k^2, respectively, leads to the inequalities under consideration.

Yet one more meaning of π emerges from a simple transformation of criterion (4.41). If we open the parentheses in the criterion, then we will see that the value of π on its own is summed N_k^2 times within k-th cluster S_k, so that the following equation holds:

$$f(S, \pi) = \sum_{k=1}^{K} \sum_{i,j \in S_k} \left(a_{ij} - \pi\right) = \sum_{k=1}^{K} A(S_k, S_k) - \pi \sum_{k=1}^{K} N_k^2 \qquad (4.42)$$

This equation show that criterion f(S, π) in (4.41) is effectively the very same within-cluster summary criterion (4.39) minus a regularizer, the item on the right,

$$\varepsilon(S) = \sum_{k=1}^{K} N_k^2 \qquad (4.43)$$

at which π is just a factor.

Regularizer, in general, is an item to be added to an objective function to improve an aspect of solutions to be found, even at the price of losing some effectivity. What aspect is to be improved by adding $-\pi\varepsilon(S)$? According to the formula in (4.42), the smaller the $\varepsilon(S)$, the greater the $f(S, \pi)$. As is well known, the minimum of $\varepsilon(S)$ is reached at the case at which all the clusters are equal-sized, $N_1 = N_2 = \ldots = N_K$. Therefore, this regularizer tries to make a more balanced partition, and π is just the weight of this additional goal, while the original goal, maximization of $f(S)$ has its weight equal to 1.

The concept of regularizer is known in data analysis for quite a while already, but recently it became ubiquitous. Nowadays, regularizers are applied by the users any time when a solution should be more comprehensible and smoother (see, for example, Efron and Hastie 2016), Here we see a regularizer emerging quite naturally from just the scale shifting. It should be mentioned that, in fact, subtraction of $\pi\varepsilon(S)$ is equivalent to addition of the Gini index (see formula (2.14) in Sect. 2.3.1) multiplied by πN^2 (see Q.4.30).

Q.4.29. Prove that $\varepsilon(S)$ in (4.43) can be expressed through Gini index as $\varepsilon(S) = N^2 - N^2 g(S)$.
A. Obviously follows from the definition of Gini index g(S) in formula (2.14), Sect. 2.3.1.
In fact, the claim that $\varepsilon(S)$ regularizes the balance in cluster sizes can be formulated more precisely.
Q.4.30. Assume that S and S' are optimal partitions according to criterion (4.41) at shifts π and π', respectively. Let, for example, $\pi < \pi'$. Then, necessarily $\varepsilon(S') \leq \varepsilon(S)$. Prove that.

A. To prove the statement, let us consider inequalities $f(S, \pi') \leq f(S', \pi')$ and $f(S', \pi) \leq f(S, \pi)$. These follow from the fact that S' and S are optimal at corresponding values of the shift. Then the item (4.37) in the left and right parts of the inequalities can be equivalently removed, so that the inequalities could be reformulated as $-\pi'\varepsilon(S) \leq -\pi'\varepsilon(S')$ and $-\pi\varepsilon(S') \leq -\pi\varepsilon(S)$. These can be turned in $0 \leq \pi'[\varepsilon(S) - \varepsilon(S')]$ and $\pi[\varepsilon(S) - \varepsilon(S')] \leq 0$, so that $\pi[\varepsilon(S) - \varepsilon(S')] \leq \pi'[\varepsilon(S) - \varepsilon(S')]$. This implies that $(\pi' - \pi)[\varepsilon(S) - \varepsilon(S')] \geq 0$. Dividing the last inequality by the positive difference $(\pi' - \pi) > 0$ proves the statement.

Let us consider an agglomerative algorithm for uniform partitioning. It is quite similar to SA-Agglomeration above, except for the formula of the criterion change at a merger. It is easy to prove that merging clusters S_k and S_l changes $f(S, \pi)$ by

$$\Delta(k,l) = 2A(S_k, S_l) - 2\pi N_k N_l,$$

which is as based on inter-cluster summary similarities $A(S_k, S_l)$ as changes in the Semi-average criterion (4.35), although the relation is somewhat simpler.

This can be made even simpler if matrix A is preliminarily transformed by subtraction π from all its entries, $A \Leftarrow A - \pi$. With thus transformed A, the change of the criterion $f(S, \pi)$ after a merger can be expressed as

$$\Delta(k,l) = 2A(S_k, S_l). \tag{4.42'}$$

Therefore, an agglomerative algorithm for uniform partitioning can be formulated as follows.

Agglomerative Clustering Algorithm for Uniform Partitioning: SU-Agglomeration

1. *Initial setting.* Set $K = N$, put all entities $k \in I$ to singleton clusters $S_k = \{k\}$, with their cardinalities set to unity and heights to zero; form a cluster-to-cluster similarity matrix $AA = (A(k, l))$ between them, with $A(k, l) = a_{kl} - \pi$, and a list of maximal clusters M consisting of all the singletons at this stage. Set flag $= 1$ and do the following steps till flag changes.
2. *Finding maximum.* Using matrix AA, compute all $\Delta(k, l)$ (4.42) and find their maximum $\Delta(k^*, l^*)$ where $k^* < l^*$.
3. *Stop condition.* If $\Delta(k^*, l^*) < 0$, put flag $= 0$. If $K = 2$, put flag $= 0$.
4. *Merge and update.* Merge two maximal clusters, S_{k^*} and S_{l^*}, together to form their parent in the dendrogram, a new maximal cluster $S_{k^* \cup l^*} = S_{k^*} \cup S_{l^*}$. To do this, add the row l^* to the row k^* in AA; after this, add column l^* to column k^* in AA, after which remove l^* row and column from AA—its size now is $K-1$, so change K by subtracting 1 from it. Similarly, add maximal cluster $M(l^*)$ to maximal cluster $M(k^*)$, after which remove the l^* cluster from M. Define the new cluster's cardinality as $N_{k^*} + N_{l^*}$, and put that as the cardinality of cluster k^*. Put the height added to the height of the previous merger equal to $\Delta(k^*, l^*)$ and draw the merger accordingly. (In fact, any sequence of reals monotonically following the order of mergers can be used as a height function.)

5. *Output*: the set of all merged clusters along with their heights. Draw a cluster dendrogram, that is a tree, if $K = 1$, or a forest with K rooted trees if $K > 1$.

To complete the general description, let us briefly discuss the issue of choice of the scale shift in criterion $f(S, \pi)$. This can be considered a device in the user's hands to control the granularity of the clustering by using a different perspective. So far, we encountered two ways for regulation of the granularity of the analysis:

(i) Directly, by specifying a global parameter of an admissible partition, the number of clusters K, as it is in K-means clustering;
(ii) Less directly, by specifying a local parameter of an admissible partition, the minimum cluster size, as it is in iK-means clustering.

Here is one more parameter, which involves no partition but rather the data under analysis. What similarity level is insignificant? That lesser than threshold π. An example of a situation in which that can be useful: let entries a_{ij} reflect the level of interaction between two control units at an industrial enterprise, so that

mark 1 scores "opinion exchange";
mark 5, "giving/receiving helpful advices";
mark 15, "giving/receiving orders";
mark 20, "giving/receiving resources".

Then the user may establish $\pi = 10$ if the goal of the analysis is to analyse the linear control structure, or $\pi = 1$ if the goal of the analysis is to cover all types of working interactions.

In a general case, when setting threshold value is of an issue, it is advisable setting π on the average similarity level.

Worked Example 4.18. Uniform Clustering of Confusion Dataset with a Zeroed Diagonal

Let us take the Confusion similarity matrix A in Table 4.34 with all its diagonal entries set to 0, so that $a_{ii} = 0$ for all $i = 1, 2, \ldots, 10$. Find the average non-diagonal value by dividing the sum of all the entries over 90. That produces $\pi = 33.46$.

Applying the SU-agglomeration algorithm to this dataset with this π, we obtain the sequence of mergers shown with the dendrogram in Fig. 4.27.

The sequence s of changes $\Delta(k, l)$ (4.42) of the criterion $f(S, \pi)$ at consecutive mergers from S to $S(k, l)$ is this: 262.1, 196.1, 176.1, 152.2, 135.2, 88.2. No further mergers are made, because next mergers would decrease the value of the uniform partitioning criterion: next best merger, item 2 into cluster 3–9–5, would make the criterion value to go down by adding −26.7. The sequence is monotone decreasing, like in the other similar sequences. Therefore, this process brings forward four natural confusion clusters: 2, 5–3–9, 0–6–8, and 1–4–7. The interior point heights in the dendrogram follow the rule 280 - s. This set of clusters accords well with other clustering results over the dataset.

Fig. 4.27 Dendrogram of the
SU-agglomeration process for
similarity data in Table 4.31
with the zeroed diagonal at
the scale shift $\pi = 33.46$

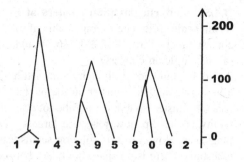

Q.4.31. Uniform partition clusters at Eurovision song contest data.
Take the Eurovision song contest data A from Table 4.35 and apply
SU-Agglomeration method at $\pi = 2$. How many clusters are found by the method?
A. The order of mergers and other relevant information is provided in Table 4.36.

These mergers provide no monotonic sequence of changes Δ, unlike in the
previous examples in Sect. 4.6.2. There are four non-singleton clusters which
embrace 16 objects out of 19—Est, Pol, and Switz have not made it to the clusters.
Moreover, the current clusters somewhat differ from those in Fig. 4.24, as previously found clusters (1) and (2), Former Soviet Union and Baltics, form a cumulative cluster in line 7 of Table 4.36. The previous "less obvious" cluster (6) has
disappeared altogether while leaving the Romance languages cluster intact (in line
12 of Table 4.35).

Table 4.36 Mergers of Eurovision song contest data according to the uniform partitioning
criterion at $\pi = 2$

Merger	Cluster S_k	Cluster S_l	$\Delta(k, l)$
1	Az	Ukr	16.1
2	Az, Ukr	Rus	25.9
3	Az, Ukr, Rus	Isr	15.0
4	Az, Ukr, Rus, Isr	Serb	14.8
5	Az, Ukr, Rus, Isr, Serb	Gre	14.3
6	Az, Ukr, Rus, Isr, Serb, Gre	Rom	21.4
7	Az, Ukr, Rus, Isr, Serb, Gre, Rom	Bulg	16.5
8	Fra	Ita	6.1
9	Fra, Ita	Por	5.5
10	Belg	Neth	5.3
11	Ger	UK	4.1
12	Fra, Ita, Por	Spain	3.9

Q.4.32. Uniform partition clusters at Eurovision song contest data at a differing scale shift. The average value of off-diagonal entries of the Eurovision song contest data in Table 4.35 is 1.786. Would using this threshold increase the number of non-singleton clusters?

A. No, it would not. Although in this setting Pol and Est merge at 13th step, to form a new non-singleton cluster, at the next 14th step Fra-Ita-Por cluster merges with Ger-UK cluster, so that the number of non-singleton clusters, four, remain. Another oddity is that Sp now joins the largest cluster rather than that of the Romance languages. However, this solution seems less convincing than that in the previous Question. There is no sharp boundary between "wanted" and "unwanted" mergers here: the two last mergers added mere 0.33 and 0.07, respectively, to the value of the criterion. This contrasts the much larger value 4.53 added at the previous, 12th merger (compare with the last positive change of the criterion, 3.9, in the setting of Q.4.31 at $\pi = 2$).

Case-Study 4.7. Summary Clustering at Flat Networks

Consider graphs in Fig. 4.28. They can be considered as representing 0–1 similarity matrices whose ij-th entry is 1 if i and j are linked in the corresponding graph, and 0 if i, j nodes are not connected. Similarity matrices corresponding to these graphs are in Tables 4.36 and 4.37.

Before applying SU-Agglomeration to the Cockroach graph on Fig. 4.28a take a look at its structure with a naked eye. The structure seems a bit difficult for clustering, although sets 1–2–3, 7–8–9 and the rest seem good candidates for clusters.

The average value in the matrix (Table 4.37) is 0.26. After subtracting that, the unity entry changes for 0.74 and the zero entry, for −0.26.

How can one apply the maximum similarity rule in the algorithm in a case like this: all positive entries are the same? Well, the MatLab routine for computing the maximum value takes care of that. It keeps data of the very first maximum value if there are several of them. Therefore, the SU-Agglomeration algorithm first merges doubletons 1–2, 3–4, 5–6, 7–8, 9–10, and 11–12. After this, only one positive nondiagonal entry at AA matrix remains, 0.96, between 5–6 and 11–12. This

Fig. 4.28 Illustrative ordinary graphs: "Cockroach" in (**a**) and "Tail" in (**b**)

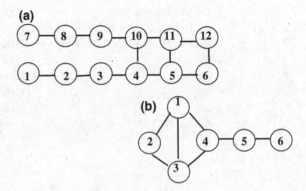

Table 4.37 Similarity matrix of cockroach graph, Fig. 4.28a; empty places stand for 0

#	1	2	3	4	5	6	7	8	9	10	11	12
1		1										
2	1		1									
3		1		1								
4			1		1					1		
5				1		1					1	
6					1							1
7								1				
8							1		1			
9								1		1		
10				1					1		1	
11					1					1		1
12						1					1	

Table 4.38 Similarity matrix for the tail graph, Fig. 4.28b; empty places stand for 0

#	1	2	3	4	5	6
1		1	1	1		
2	1		1			
3	1	1		1		
4	1		1		1	
5				1		1
6					1	

completes the action with five clusters: 1–2, 3–4, 7–8, 9–10, and 5–6–11–12. At the average level of resolution, only the Cockroach's head makes it into a cluster; the others form doubletons, according to the summary criterion (4.41).

Similarly, SU-Agglomeration at the Tail graph data in Table 4.38 with both the average π value of 0.39 and the mid-interval 0.5 value leads to uniform clustering clusters 1–2–3–4 and 5–6. Almost all—except for (4,5)th entry—the similarities between these two clusters are negative. However, further increase in the scale shift value, say, to $\pi = 2/3$, may break this. Indeed, at this value of π, joining node 4 into a tight cluster 1–2–3 becomes too costly: it brings in two positive entries of 1/3 and two negative values of $-2/3$, totalling to $-2/3$. Therefore, at $\pi = 2/3$, uniform non-singleton clusters are 1–2–3 and 4–5.

4.6.3.3 Modularity Clustering

A rather popular approach to clustering, notably, to discovering communities in networks is referred to as "Modularity clustering". This approach is based on a probabilistic interpretation of the similarities. Any similarity index is considered as emerging from interactions between the objects. According to this interpretation, each object i is assigned with a probability of interaction, equal to the proportion of the summary similarity in the i-th row in the whole summary volume of the similarities. Therefore, random interactions between two entities will occur with the probability equal to the product of their respective probabilities. To clear the scored similarity index values from these random interactions, the latter should be subtracted from the former. Then the summary clustering criterion applies (see Newman 2006).

Let us give a more precise formulation. In this approach, similarity matrix A is considered as a contingency table. Then relative frequencies are derived. Specifically, define the interaction summary values $a_{i+} = \Sigma_{j \in I}\, a_{ij}$ at any of the objects, i. The total sum is $a_{++} = \Sigma_{i,j \in I}\, a_{ij}$. Under the assumption that the random interaction between entities i and j is proportional to the interaction summary values, the background similarity is defined as the product $k_{ij} = a_{i+}a_{j+}/a_{++}$; the denominator is added to return the product to the original unit of measurement of similarities in A. The within-cluster summary similarity criterion (4.39) is applied to the residual similarity values found by subtracting the "background" similarity from A. This criterion is referred to as the modularity criterion:

$$m(S) = \sum_{k=1}^{K} \sum_{i,j \in S_k} \left(a_{ij} - k_{ij}\right) = \sum_{k=1}^{K} \sum_{i,j \in S_k} \left(a_{ij} - a_{i+}a_{+j}/a_{++}\right)$$

Obviously, this criterion is the within-cluster summary criterion $f(S)$ (4.39) applied to matrix $C = (c_{ij})$ where $c_{ij} = a_{ij} - a_{i+}a_{j+}/a_{++}$.

Worked Example 4.19. Modularity Clustering for Confusion Dataset
To apply the modularity transformation, let us take the original Confusion data matrix in Table 1.5 and compute for it the matrix of random interactions (k_{ij}). This matrix is subtracted from the data matrix, the result of which is made symmetric with its transpose added and the result halved. Then the diagonal is zeroed. The resulting matrix is in Table 4.39.

The SU-Agglomeration algorithm in this case is modified at the very first step: the preliminary transformation now is $a_{ij} \Leftarrow a_{ij} - k_{ij}$ (modularity), rather than $a_{ij} \Leftarrow a_{ij} - \pi$ (uniform).

The agglomeration process is illustrated with the dendrogram in Fig. 4.29; the heights correspond to the sequence of changes in the value of the criterion at the mergers, 259.1, 176.0, 175.6, 150.7, 130.5, 121.7, 44.1, subtracted from 280.

Further mergers would decrease the criterion value; thus three clusters, 1–7–4, 3–9–5, and 8–0–6–2, are output. These much resemble clusters found with the same algorithm at the data preprocessed by subtraction of a uniform scale shift value. The only difference is in the allocation of digit 2. In contrast to joining the cluster of "round" figures here, that forms a cluster of its own at the uniform scale shift value, which is probably about right. The difference is caused by the fact that similarities

Table 4.39 Confusion data after subtraction of random interactions and symmetrization

0	−12.70	−16.96	57.70	−26.93	−18.89	129.54	−41.18	−35.74	−12.20
−12.70	0	19.50	0.22	12.28	11.56	−7.62	3.42	−9.71	−5.26
−16.96	19.50	0	−14.90	0.97	−26.56	4.06	−10.95	88.01	−14.05
57.70	0.22	−14.90	0	−11.59	−9.56	7.55	−15.23	12.72	−13.22
−26.93	12.28	0.97	−11.59	0	57.61	−19.71	−27.50	59.87	−15.07
−18.89	11.56	−26.56	−9.56	57.61	0	−24.80	71.71	−34.14	3.66
129.54	−7.62	4.06	7.55	−19.71	−24.80	0	−30.40	−27.29	−7.88
−41.18	3.42	−10.95	−15.23	−27.50	71.71	−30.40	0	15.69	87.82
−35.74	−9.71	88.01	12.72	59.87	−34.14	−27.29	15.69	0	−4.70
−12.20	−5.26	−14.05	−13.22	−15.07	3.66	−7.88	87.82	−4.70	0

Fig. 4.29 Dendrogram of the SU-agglomeration process for similarity data in Table 4.39

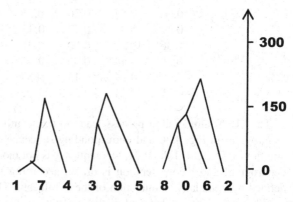

of digit 2 with the rest are relatively small. This separates 2 from the rest at the uniform shift value. This, however, makes its random interactions smaller as well, so that the residual values remain positive leading to the fact that 2 becomes a welcome member of the cluster 8–0–6.

Worked Example 4.20. Modularity Clustering of Tail Dataset

For matrix A of the Tail graph dataset (Table 4.38), the matrix of random interactions $a_{i+}a_{+j}/a_{++}$ is as follows:

$$
\begin{array}{cccccc}
0.64 & 0.43 & 0.64 & 0.64 & 0.43 & 0.21 \\
0.43 & 0.29 & 0.43 & 0.43 & 0.29 & 0.14 \\
0.64 & 0.43 & 0.64 & 0.64 & 0.43 & 0.21 \\
0.64 & 0.43 & 0.64 & 0.64 & 0.43 & 0.21 \\
0.43 & 0.29 & 0.43 & 0.43 & 0.29 & 0.14 \\
0.21 & 0.14 & 0.21 & 0.21 & 0.14 & 0.07 \\
\end{array}
$$

By subtracting that from A and zeroing the diagonal entries, one obtains the residual similarity matrix R:

Fig. 4.30 Dendrogram of the SU-agglomeration process for the residual Tail matrix R above

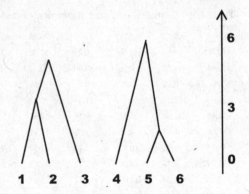

$$R = \begin{matrix} 0.00 & 0.57 & 0.36 & 0.36 & -0.43 & -0.21 \\ 0.57 & 0.00 & 0.57 & -0.43 & -0.29 & -0.14 \\ 0.36 & 0.57 & 0.00 & 0.36 & -0.43 & -0.21 \\ 0.36 & -0.43 & 0.36 & 0.00 & 0.57 & -0.21 \\ -0.43 & -0.29 & -0.43 & 0.57 & 0.00 & 0.86 \\ -0.21 & -0.14 & -0.21 & -0.21 & 0.86 & 0.00 \end{matrix}$$

The SU-Agglomeration process leads to consecutive mergers illustrated in the dendrogram in Fig. 4.30 and corresponding sequence of changes of the summary criterion, 1.71, 1.14, 1.86, 0.71, this sequence is not monotone again. Therefore, the dendrogram under consideration is drawn according to a proper height function defined by cumulative summation of these values: 1.71, 2.86, 4.71, 5.43.

What seems somewhat odd at these results in the view of graph structure in Fig. 4.28b: (1) rather peripheral items 5 and 6 merge first; (2) item 4 joins not to the triangle 1–2–3, with which it is connected by 2 links, but rather to a weaker doubleton 5–6. This is explained by the structure of positive entries in matrix R, which is, in turn, can be explained by the very same phenomenon, as in the previous example. The residual similarities between 4, 5, and 6 are greater because weaker interactions are subtracted from the 0/1 values in the original data matrix. The modularity approach makes weaker links comparatively larger.

Q.4.33. What clusters are brought in by the modularity approach to the Cockroach data in Table 4.37?
A. Finally, there are three modularity clusters: 1–2–3, 7–8–9, and 4–5–10–11–6–12. They differ from those found with uniform partitioning in Case-Study 4.7, but probably may be considered as natural as those found there, 1–2, 3–4, 7–8, 9–10, and 5–6–11–12.

4.7 Consensus Partitioning

To introduce the subject, let us consider the Digits data from Sect. 1.2: these are ten numerals for which we have both a naïve digital system for putting them on screen, and a matrix of confusion between them by a human eye. The naïve digital drawings are represented by seven binary features corresponding to presence/absence of a corresponding segment in the drawing (see Table 4.40).

. Different algorithms or subjective opinions may bring forward a number of partitions of them, such a those presented in Table 4.41.

If one takes a close look at the partitions R1, R2, ..., R11 in Table 4.41, one can see that they are not incompatible. Say, digits 1 and 7, as well as 8 and 0, belong in the same part in almost all eleven partitions, except for one.

Moreover, partitions R1, R3 and R9 coincide. Figure 4.31 illustrates the partitions: it shows, by closed pointed lines, parts/clusters in partitions R1 (see part (a)), R7 (see part (b)), and R4 (see part (c)).

There can be several situations at which the user encounters an ensemble of partitions such as that in Table 4.39. Two most usual are:

- Partitions are results of different clustering algorithms or even of different runs of the same clustering algorithm, such as K-means at different initial settings;
- Partitions represent various nominal features of the objects.

The ensemble in Table 4.41 has been obtained by runs of iK-means algorithm over reduced data tables obtained from that in Table 4.40 by removing two columns selected randomly. This may be considered an imitation of the situation at which no data holder has a full dataset.

A partition which "accommodates" most of common characteristics of the partition ensemble under consideration is referred to as a consensus partition. A most popular approach to building a consensus partition is through defining it as a central partition according to some partition-to-partition association or distance measure. If such a measure is selected, then the central partition for a given ensemble of partitions $R_1, R_2, \ldots, R_m$ of a finite set I is defined as such an S that minimizes the summary distance

$$D(S) = \sum_{t=1}^{m} d(S, R_t) \tag{4.44}$$

where $d(S, R_t)$ is the selected distance measure between partitions S and R_t.

In Chap. 3, we defined two distance measures between partitions, the mismatch distance (3.54) and dummy matrix regression distance (3.41). Let us recall them in this order.

The mismatch distance can be defined using binary $N \times N$ matrices of partitions. Given a partition R on set I, its binary matrix $\mathbf{r} = (r_{ij})$ is defined as:

Table 4.40 The table of digit-to-edge features e1 to e7 according to Fig. 1.2 in Sect. 1.2.1

Numeral	e1	e2	e3	e4	e5	e6	e7
1	0	0	1	0	0	1	0
2	1	0	1	1	1	0	1
3	1	0	1	1	0	1	1
4	0	1	1	1	0	1	0
5	1	1	0	1	0	1	1
6	1	1	0	1	1	1	1
7	1	0	1	0	0	1	0
8	1	1	1	1	1	1	1
9	1	1	1	1	0	1	1
0	1	1	1	0	1	1	1

Table 4.41 Eleven partitions of the set of 10 numerals, each represented by a column assigning entities to the corresponding cluster numbers

Digit	R1	R2	R3	R4	R5	R6	R7	R8	R9	R10	R11
1	1	1	1	2	1	1	1	1	1	2	1
2	2	2	2	2	3	3	3	2	2	3	2
3	3	3	3	2	3	3	4	3	3	2	3
4	3	1	1	3	3	1	1	1	3	2	1
5	3	3	3	1	2	2	2	3	3	1	3
6	2	2	2	1	2	2	2	2	2	1	2
7	1	1	1	2	1	3	1	1	1	2	1
8	2	2	2	3	3	4	3	2	2	3	2
9	3	3	3	3	3	3	4	3	3	3	3
0	2	2	2	2	3	4	3	2	2	3	2

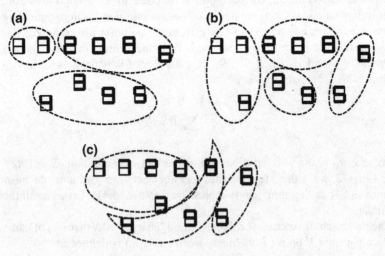

Fig. 4.31 Three replicas of segmented numerals from Fig. 1.2 in Sect. 1.2.3. Each shows a partition from Table 4.41, specifically, R1 (in (a)), R7 (in (b)), and R4 (in (c))

Table 4.42 Binary 10×10 matrices of partitions R1 and R7 from Table 4.41 (zeros are presented by blank spaces)

(a)

rl	1	2	3	4	5	6	7	8	9	0
1	1						1			
2		1				1		1		1
3			1	1	1				1	
4			1	1	1				1	
5			1	1	1				1	
6		1				1	1			1
7	1						1			
8		1				1		1		1
9			1	1	1				1	
0		1				1	1			1

(b)

r7	1	2	3	4	5	6	7	8	9	0
1	1		1				1			
2		1						1		1
3			1					1		
4	1			1			1			
5					1	1				
6					1	1				
7	1			1			1			
8		1						1		1
9			1						1	
0		1						1		1

$$r_{ij} = \begin{cases} 1 \ \text{if} \, (i,j) \in \rho \\ 0 \ \text{if} \, (i,j) \notin \rho \end{cases}$$

so that $r_{ij} = 1$ if and only if i and j belong in the same part of R. For example, matrix of partition R1 in Table 4.41 is in Table 4.42a, whereas matrix of partition R7 in Table 4.41 is in Table 4.42b.

The mismatch distance between partitions is the mismatch (Hamming) distance between their matrices:

$$d(\boldsymbol{R}, \boldsymbol{R}') = \sum_{i,j \in I} \left| r_{ij} - r'_{ij} \right| = \sum_{i,j \in I} \left(r_{ij} - r'_{ij} \right)^2$$

The right-hand equality follows from the fact that $\left| r_{ij} - r'_{ij} \right|$ is either 0 or 1. Therefore, $d(r_1, r_7) = 1 + 1 + 2 + 5 + 4 + 4 + 1 + 1 + 2 + 1 = 22$. For the sake of convenience, the total sums contributions of all rows.

The second distance is a bit more complex. First of all, it is asymmetric: one needs to decide which is taken to be a predictor for the second. Let it be R7 as that with a greater number of parts. Then one needs the orthogonal projector P_X for the predictor. Rather than using $N \times N$ similarity matrices, this approach uses partition incidence matrices. A dummy incidence matrix Y for R1 is a binary $N \times 3$ matrix whose columns correspond to individual parts, and (i, t)-th entry is 1 if object i belongs to t-th part, and 0, if not. The distance is measured by the squared differences between Y and its projection $P_X Y$ onto the linear subspace span by columns of X, a similar object-to-part binary matrix for the predictor, R7. All these matrices are presented in Table 4.43 including the difference $Y - P_X Y$. The sum of the squared entries of this latter matrix, is equal to 2.333 (Table 4.43).

It appears, the central partition according to each of these two distances involves the so-called consensus matrix. This is a $N \times N$ similarity matrix $C = (c_{ij})$. Given an ensemble of m partitions on I, for every pair of objects, i and j, its entry c_{ij} is defined as the number of partitions at which the objects belong in the same cluster.

The consensus matrix for the eleven partitions in Table 4.41 is presented in Table 4.44. How its elements are calculated? Let us illustrate with some entries in 3-d row:

- $c_{33} = 11$ because 3 is in a cluster in all the eleven partitions;
- $c_{34} = 4$ because digits 3 and 4 are in the same cluster in four partitions: R1, R5, R9, and R10;
- $c_{37} = 3$ because digits 3 and 7 are in the same cluster in three partitions: R4, R6, and R7;
- $c_{39} = 9$ because digits 3 and 9 are in different clusters in only two partitions out of 11: R4 and R10.

What this has to do with the concept of consensus central partition in (4.44)? The following statements are true.

Given an ensemble of m partitions and the corresponding consensus matrix $C = (c_{ij})$, partition $S = \{S_1, S_2, \ldots, S_K\}$ is central for the mismatch distance if and only if it maximizes the summary within-cluster criterion (4.40)

$$f(S, \pi) = \sum_{k=1}^{K} \sum_{i,j \in S_k} (c_{ij} - \pi) \tag{4.40'}$$

at $\pi = m/2$.

Given an ensemble of m partitions and the corresponding consensus matrix $C = (c_{ij})$, partition $S = \{S_1, S_2, \ldots, S_K\}$ is central for the dummy matrix regression distance if and only if it maximizes the semi-average within-cluster similarity criterion (4.31):

$$g(S) = \sum_{k=1}^{K} \frac{1}{N_k} \sum_{i,j \in S_k} c_{ij} \tag{4.31}$$

Let us prove these statements. Let us begin with the mismatch distance consensus.

Table 4.43 Data of partition incidence matrices and their comparison, with X taken predictor and Y target

#	Y			X				P_X									P_XY			$Y - P_XY$		
1	1	0	0	1	0	0	0	0.33	0	0.33	0	0.33	0	0	0	0	0.67	0	0.33	0.33	0	−0.33
2	0	1	0	0	1	0	0	0	0.33	0	0	0	0	0.33	0	0.33	0	1.0	0	0	0	0
3	0	0	1	1	0	0	0	0.33	0	0.33	0	0.33	0	0	0	0	0.67	0	0.33	−0.67	0	0.67
4	0	0	1	0	0	1	0	0	0	0	0.5	0	0	0	0.5	0	0	0.5	0.5	0	−0.5	0.5
5	1	0	0	1	0	0	0	0.33	0	0.33	0	0.33	0	0	0	0	0.67	0	0.33	0.33	0	−0.33
6	0	0	1	0	0	0	1	0	0	0	0	0	1.0	0	0	0	0	0	1.0	0	0	0
7	0	1	0	0	1	0	0	0	0.33	0	0	0	0	0.33	0	0.33	0	1.0	0	0	0	0
8	0	1	0	0	0	1	0	0	0	0	0.5	0	0	0	0.5	0	0	0.5	0.5	0	0.5	−0.5
9	0	1	0	0	1	0	0	0	0.33	0	0	0	0	0.33	0	0.33	0	1.0	0	0	0	0

Table 4.44 Consensus matrix for the eleven partitions in Table 4.41

	1	2	3	4	5	6	7	8	9	0
1	11	1	2	7	0	0	10	0	0	1
2	1	11	3	1	0	6	2	9	3	10
3	2	3	11	4	6	0	3	1	9	2
4	7	1	4	11	2	0	6	2	4	1
5	0	0	6	2	11	5	0	0	6	0
6	0	6	0	0	5	11	0	6	0	6
7	10	2	3	6	0	0	11	0	1	1
8	0	9	1	2	0	6	0	11	3	10
9	0	3	9	4	6	0	1	3	11	2
0	1	10	2	1	0	6	1	10	2	11

Consider the summary mismatch distance in (4.44) from a partition S to be found and given partitions $R_1, R_2, \ldots, R_m$ on I (with the corresponding square binary matrices (s_{ij}) and (r_{ij}^t)):

$$D(S) = \sum_{t=1}^{m} d(S, R_t) = \sum_{t=1}^{m} \sum_{i,j \in I} (s_{ij} - r_{ij}^t)^2 = \sum_{t=1}^{m} \sum_{i,j \in I} (s_{ij}$$

$$+ r_{ij}^t - 2s_{ij}r_{ij}^t) = \sum_{i,j \in I} (ms_{ij} + \sum_{t=1}^{m} r_{ij}^t - 2s_{ij} \sum_{t=1}^{m} r_{ij}^t) = \sum_{i,j \in I} (ms_{ij}$$

$$+ c_{ij} - 2s_{ij}c_{ij}) = \sum_{i,j \in I} c_{ij} - 2 \sum_{i,j \in I} (c_{ij} - m/2)s_{ij}.$$

These equations use little algebra and, most importantly, the fact that the consensus matrix C is but the sum of square binary matrices of individual partitions (r_{ij}^t) over $t = 1, 2, \ldots, m$. Since the first item in the last expression is constant, we may conclude that the task of minimization of $D(S)$ (4.44) with the mismatch distance is equivalent to the task of maximization of

$$\sum_{i,j \in I} (c_{ij} - m/2)s_{ij}.$$

Of course, the last expression is obviously equal to the criterion in (4.40) at $\pi = m/2$. This proves the first statement.

To prove the second statement, let us turn to the incidence matrices Y and X^t of partitions S and R^t, respectively. These binary matrices mark by 1 the belongingness of objects (rows) to clusters. The square error criterion can be reformulated in terms of similarities between entities as well. Let us denote the total number of clusters in all the partitions $(t = 1, 2, \ldots, m)$ by L and form $N \times L$ matrix $X = (X^1 X^2 \ldots X^m)$ consisting of all the columns in these matrices. The columns of this matrix correspond to all the clusters in partitions $R^1, R^2, \ldots, R^m$. Then the criterion (4.44) for the dummy matrix regression distance can be expressed as $D(Y) = \|X - P_Y X\|^2$, or equivalently, as $D(Y) = Tr((X - P_Y X)(X - P_Y X)^T)$ where Tr denotes the trace of a square matrix, that is, the sum of its diagonal elements. By opening the parentheses in the latter expression, one can see that $D(Y) = \mathrm{Tr}(XX^T - P_Y XX^T - XX^T P_Y + P_Y XX^T P_Y) = \mathrm{Tr}(XX^T - P_Y XX^T)$. Indeed, the operation Tr is commutative, which implies equations $\mathrm{Tr}(P_Y XX^T) = \mathrm{Tr}(XX^T P_Y)$ and $\mathrm{Tr}(P_Y XX^T P_Y) = \mathrm{Tr}(P_Y P_Y XX^T) = \mathrm{Tr}(P_Y XX^T)$; the last equation follows from the fact that the projection operator P_Y satisfies the so-called idempotence property, that is, its consecutive application changes nothing, $P_Y P_Y = P_Y$. It remains to notice that matrix XX^T is the consensus matrix C: indeed, the (i, j)-th entry is the sum products $x_i^l x_j^l$, $l = 1,\ldots, L$. Each of the products is either 1 or 0; it is 1 if both i and j belong to the l-th cluster. Obviously, $c_{ii} = L$ for all $i \in I$, so that $\mathrm{Tr}(XX^T) = NL$. On the other hand, the (i, i)-th diagonal element of matrix $P_Y C$ equals to the sum of products $p_{ij} c_{ji}$ where p_{ij} is either 0, if i and j are in different S-clusters, or $1/N_k$ if j belongs to the cluster S_k that contains i. This completes the proof.

It should be noted that the criterion (4.31) is akin to the K-means criterion for the total binary matrix X whose columns correspond to all individual clusters; moreover, it can be reformulated in terms of association measures between the partition to be built and ensemble partitions (see Mirkin 2012 and Sects. 3.6 and 3.7).

The proven facts point to both semi-average and uniform partitioning criteria as those implementing the concepts of consensus partition under consideration. Both of the criteria, (4.40) and (4.31), involve the same consensus matrix C, although the former seems somewhat less sensitive to the data structure, as its threshold, $\pi = m/2$, depends on just the number of partitions, not their structure. Indeed, unlike the latter criterion, it does not pass the so-called Muchnik test (Mirkin 2012) which is described in the next Question, Q.4.34.

Q.4.34. Muchnik's test

Consider a partition, $R = \{R_k\}$ with K parts R_k, $k = 1, 2, \ldots, K$, on I. This R can be decomposed in K bisection partitions R^k. Each R^k is defined as a two-part partition to comprise two parts, R_k and its complement, $I - R_k$ ($k = 1, \ldots, K$). Our intuition tells us that the pre-specified R should be the consensus for the set of bisected partitions R^k. Is R a central partition according to either distance concept above? If yes indeed, the distance passes the test; if not, then not.
A. Unfortunately, the mismatch distance fails this test, whereas the dummy matrix regression distance passes it.

Indeed, let us take a look at the consensus matrix $C = (c_{ij})$ at the test. Obviously, if i and j belong to the same part R_k of R, they belong to the same part in every bisected partition R^t, so that $c_{ij} = K$ in this case. If, in contrast, $i \in R_k$ and $j \in R_l$ at $k \neq l$, the objects will belong to different parts in the corresponding bisected partitions R^k and R^l, whereas they would belong to the same class in all the other bisected partitions. This would make $c_{ij} = K-2$ in the case when i and j belong to different parts in R.

Then all items $c_{ij} - \pi$ in criterion (4.40) will be positive at $K \geq 4$. Indeed, in this case $c_{ij} - \pi = K - 2 - K/2 = (K - 4)/2$. This indeed is non-negative; that is, equals 0 at $K = 4$ and greater than 0 at $K > 4$. That means the universal partition consisting just of one part coinciding with all the set I gives the maximum to the criterion, not R. That is a "no pass" for the mismatch distance central partition.

In contrast, the central partition based on the dummy matrix regression distance does pass the test.

Apply the equivalent consensus criterion (4.31) to $S = R$. Obviously, $g(S) = KN$ in this case because $g(S_k) = KN_k$ at each $k = 1, 2, ..., K$. Any allocation of objects, not consistent with R, will bring in lesser values of similarity c_{ij} within clusters which would decrease the criterion value. On the other hand, breaking a part R_k, say, in two would not change the criterion value; it would still be KN. Therefore, R is the maximal consensus partition according to criterion (4.31) in this setting.

The last issue to address in this section is of zeroing the diagonal of consensus matrix. Due to the fact that all the diagonal entries are equal to the same maximum consensus similarity value m, the zeroing is equivalent to subtraction of m from all the N diagonal cells. How is this reflected in the value of criterion $g(S)$ in (4.31)? It is not difficult to see that the value $g(S_k) = C(S_k, S_k)/N_k$ decreases by m as well. Therefore, when m is subtracted from all the diagonal elements of C, the summary value $g(S)$ at K-part partition S decreases by Km. Then the change $\Delta(k, l)$ at a merger of two clusters, S_k and S_l, will be changed by $+m$. Why? That is the difference between the change, $-Km$, before the merger, and $-(K - 1)m$ after the merger. Otherwise, the SA-Agglomeration process remains the same.

Worked Example 4.21. Consensus Partition at Digits Data

Let us apply the SA-Agglomeration process to the consensus matrix in Table 4.44 obtained from 11 partitions of the 10 numerals in Table 4.41. This brings forth a consensus partition according to the dummy matrix regression distance. The sequence of mergers according to this algorithm is reflected in the dendrogram drawn in Fig. 4.32.

The changes $\Delta(k, l)$ of the criterion at the diagonal $m = 11$ are: $-1.0, -1.0,$ $-1.67, -2.0, -5.67, -6.0, -6.83, -20.33$. Those at the zeroed diagonal are: $10.0,$ $10.0, 9.33, 9.0, 5.33, 5.0, 4.17, -9.33$. This latter sequence is equal to the former one plus 11, exactly as proven above since $m = 11$ here. The negative value in the end signals the end of mergers, so that the consensus partition according to this method is $S = \{1\text{–}7\text{–}4, 2\text{–}0\text{–}8\text{–}6, 3\text{–}9\text{–}5\}$. The heights on the dendrogram in Fig. 4.32 are defined as the positive part of the latter sequence of changes subtracted from 11: 1, 1, 1.67, 2, 5.67, 6, 6.83.

The dendrogram differs from analogous dendrograms in Figs. 4.23, 4.26, and 4.28 by this: 2 merges into cluster containing 0 very early here, following the pattern of similarities in the consensus matrix in Table 4.44.

Worked Example 4.22. Consensus Clustering for Eurovision Song Contest Data

Cluster labels of 15 partitions of the Eurovision song contest countries from Table 1.11 are presented in columns of Table 4.46. These partitions contain from 3 to 5 clusters found by the iK-means method over data tables containing 12 randomly sampled columns out of 19 in Table 1.11.

Fig. 4.32 Dendrogram of the
SA-Agglomeration mergers
for the consensus similarity
matrix in Table 4.44

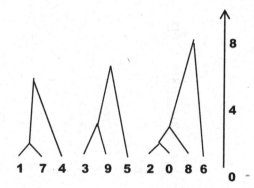

To find a consensus partition by minimizing the summary distance to the fifteen partitions in Table 4.45, one should use the consensus matrix, whichever distance between partitions one chooses, either the mismatch distance or the dummy matrix regression distance. The consensus matrix is in Table 4.46. Its (i, j) entry counts the number of the partitions at which i and j are in the same cluster. Say, Sw and Fr are in the same cluster in each of the partitions leading to (Sw, Fr) entry equal to 15; in contrast, Fr and Bu happen to be in the same cluster in only one partition, R13, thus leading to (Bu, Fr) entry of 1.

The results of SA-Agglomeration process applied to the consensus matrix with its diagonal zeroed are presented as a dendrogram in Fig. 4.33. There are five clusters over there; they are:

(1) Belgium-UK
(2) Italy-Portugal-Spain
(3) France-Germany-Greece-Netherlands-Switzerland
(4) Bulgaria-Romania-Russia-Ukraine
(5) Azerbaijan-Estonia-Israel-Poland-Serbia

This picture somewhat differs from that in answers to **Q.4.27**, see Fig. 4.24, and **Q.4.31**, see Table 4.35, which should not be taken too seriously, as the partition ensemble data here is of a purely illustrative character.

Project 4.3. Median Ranking According to the Mismatch Distance and Muchnik Test

Let us extend the concept of central partition to ordered partitions representing tied rankings (see Sect. 3.7); this will closely follow Mirkin and Fenner (2019).

Given a set of tied rankings R^1, R^2, ..., R^m, let us define its median as a tied ranking R minimizing the sum of mismatch distances between R and R^t ($t = 1, 2, ..., m$). Let us define a ranking consensus matrix $C = (c_{ij})$ as follows: for any pair (i, j), c_{ij} is the number of those rankings R^t, $1 \leq t \leq m$, in which i either precedes j or is indifferent to j.

Table 4.45 Data of 15 partitions over the Eurovision song contest data

Country	R1	R2	R3	R4	R5	R6	R7	R8	R9	R10	R11	R12	R13	R14	R15
Azerbaijan	3	3	2	2	3	3	2	4	2	2	3	2	1	2	1
Belgium	3	2	2	1	3	3	3	4	1	4	2	2	2	4	2
Bulgaria	3	3	2	3	3	4	2	5	2	4	3	2	2	3	1
Estonia	3	3	2	2	3	3	2	4	2	2	1	2	2	2	1
France	2	2	1	1	2	2	3	3	1	3	2	1	2	1	2
Germany	2	2	1	1	2	2	3	3	1	3	2	1	2	4	2
Greece	2	1	1	1	2	2	1	1	1	4	1	1	2	1	2
Israel	3	3	2	2	3	3	2	1	2	2	1	2	1	3	1
Italy	1	2	2	2	1	1	1	4	1	1	2	2	1	2	2
Netherland	2	2	2	1	2	2	3	3	1	4	2	1	2	1	1
Poland	3	1	2	2	3	4	2	3	1	2	1	2	1	3	1
Portugal	1	2	1	1	1	1	3	1	1	1	2	1	2	1	1
Romania	2	3	2	3	2	4	2	5	2	4	3	2	2	3	1
Russia	2	1	1	3	3	4	2	1	2	4	1	1	2	3	1
Serbia	3	3	2	3	3	3	1	4	2	2	3	2	2	2	2
Spain	1	2	2	1	1	1	1	2	1	1	3	1	1	1	2
Switzerland	2	3	1	1	2	2	3	3	2	3	2	1	2	1	2
Ukraine	3	2	1	2	3	4	2	1	1	4	1	2	2	3	1
UK	3	2	1	1	3	3	3	4	1	4	2	2	2	4	2
Cluster #	3	3	2	3	3	4	3	5	2	4	3	2	2	4	2

Table 4.46 Consensus country-to-country matrix for the 15 partitions in Table 4.45

Country	Az	Be	Bu	Es	Fr	Ge	Gr	Is	It	Ne	Pol	Por	Ro	Ru	Se	Sp	Sw	Uk	UK
Azerbai	15	6	9	13	0	0	0	12	6	0	9	2	7	3	11	3	0	8	5
Belgium	6	15	6	7	7	8	5	5	7	8	5	5	4	2	8	5	7	5	14
Bulgaria	9	6	15	9	1	1	2	9	2	2	8	1	13	8	9	2	1	11	5
Estonia	13	7	9	15	1	1	2	12	5	1	9	1	7	5	11	1	1	10	6
France	0	7	1	1	15	14	10	0	4	14	1	8	3	5	2	6	15	2	8
Germany	0	8	1	1	14	15	9	0	4	13	1	7	3	5	2	5	14	2	9
Greece	0	5	2	2	10	9	15	2	3	11	3	5	4	9	3	6	10	5	6
Israel	12	5	9	12	0	0	2	15	4	0	11	2	7	6	8	2	0	11	4
Italy	6	7	2	5	4	4	3	4	15	4	5	8	2	0	6	10	4	2	6
Netherl	0	8	2	1	14	13	11	0	4	15	1	8	4	6	2	6	14	3	9
Poland	9	5	8	9	1	1	3	11	5	1	15	4	6	6	5	4	1	9	4
Portugal	2	5	1	1	8	7	5	2	8	8	4	15	1	3	0	11	8	2	6
Romania	7	4	13	7	3	3	4	7	2	4	6	1	15	10	7	2	3	9	3
Russia	3	2	8	5	5	5	9	6	0	6	6	3	10	15	3	1	5	10	3
Serbia	11	8	9	11	2	2	3	8	6	2	5	0	7	3	15	4	2	6	7
Spain	3	5	2	1	6	5	6	2	10	6	4	11	2	1	4	15	6	0	4
Switzerl	0	7	1	1	15	14	10	0	4	14	1	8	3	5	2	6	15	2	8
Ukraine	8	5	11	10	2	2	5	11	2	3	9	2	9	10	6	0	2	15	6
UK	5	14	5	6	8	9	6	4	6	9	4	6	3	3	7	4	8	6	15

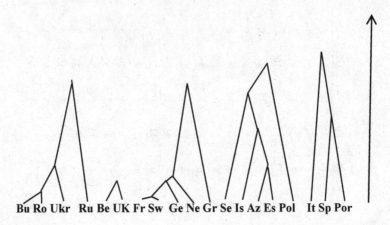

Bu Ro Ukr   Ru Be UK Fr Sw   Ge Ne Gr Se Is Az Es Pol   It Sp Por

Fig. 4.33 Dendrogram of the SU-Agglomeration process for Eurovision song contest consensus partitioning

This means that $C = \sum_t r^t$, where r^t is the binary matrix of the binary relation ρ^t corresponding to R^t, for $1 \le t \le m$. The very same analysis as that provided on p. 392 proves that a tied ranking $R = (R_1, R_2, ..., R_p)$ is a median of the set of tied rankings $R^1, R^2, ..., R^m$ if and only if it maximizes

$$F(R) = \sum_{i,j \in I} \left(c_{ij} - \frac{m}{2}\right) r_{ij} = \sum_{s=1}^{m} \sum_{t \ge s} \sum_{i \in R_s} \sum_{j \in R_t} \left(c_{ij} - \frac{m}{2}\right) \quad (4.45)$$

with respect to a pre-specified set of admissible rankings.

It is not difficult to prove that maximizing $F(R)$ in (4.45) is equivalent to maximizing

$$G(R, \pi) = \sum_{s=1}^{m} \sum_{t \ge s} \sum_{i \in R_s} \sum_{j \in R_t} c_{ij} - \pi \sum_{s=1}^{m} N_s^2, \quad (4.46)$$

where $\pi = m/4$. This latter criterion can be used as a genuine criterion for finding a median for the ensemble of m ordered partitions.

We can also see the very same regularizer $\varepsilon(R) = \sum_{s=1}^{p} N_s^2$ as that in the uniform partitioning criterion in Sect. 4.6.3.2 and the central partition criterion (see Sect. 4.7). The value of the coefficient $\pi = m/4$ here is as insensitive to the structure of the ensemble as that, $\pi = m/2$, for the consensus partition.

This makes us wonder if the Muchnik's test can be applied here. Unfortunately, in the original formulation, not, because it is valid for non-ordered partitions only. However, the test can be easily extended to the concept of median, thanks to the concept of Likert scale developed for psychological measurements (see Allen and Seaman 2007). "This scale is applied when an individual cannot reproduce an entire ranking. A psychologist then specifies a number of attributes, each of which splits the ordering in question into two complementary fragments—the beginning and the end. For each object or observation, the psychologist asks the individual, in respect of a specific attribute, whether the object falls within the beginning or end part of the scale (Mirkin and Fenner 2019)." Asking this as many times as there are boundaries between the scale parts can exactly recover the position of the object on the scale.

In our setting, this can be formulated as follows. There is a ranking with K tied parts, $R = (R_1, R_2, ..., R_K)$. This ranking leads to producing $K-1$ binary rankings $R^1, R^2, ..., R^{K-1}$. Each binary ranking R^t consists of just two parts, U_t, and V_t, in this order, for $t = 1, 2, ..., K-1$. The part U_t is the set-theoretic union of the first t parts of R, and the part V_t is the union of the remaining parts of R. These binary rankings completely determine the underlying ranking R. Then a question arises whether R is the consensus ranking, or median, for the ensemble of binary rankings $R^1, R^2, ..., R^{K-1}$.

Consider, for example, a consensus matrix on Fig. 4.34 and its partition in blocks corresponding to a seven-part ranking over the set of its rows/columns from Mirkin and Fenner (2019).

Parts (a) and (b) in Fig. 4.34 show two cases representing the ranking consensus matrix for the set of binary rankings corresponding to a ranking with $K = 7$ tied parts, with the value $(K-1)/2 = 3$ subtracted from each of the entries. Rather than showing individual elements of the consensus matrix, the figure shows just the block structure of the matrix, where the (s, t)-th block, $1 \leq s, t \leq K$, corresponds to the entries of constant value $c_{ij} - (K-1)/2$. There are only 6 negative blocks in the matrix. The number of negative entries (i, j) clearly depends on the part sizes of R.

Let us consider, for the sake of simplicity, a case in which each part $R_1, R_2, ..., R_k$ of R contains the same number of elements. To maximize the criterion (4.46), we just need to minimize the sum of the entries below the diagonal. Obviously, the parts of R cannot be split in an optimal ranking S. Therefore, a median S can be obtained by merging some parts of R, that is, S should be coarser than R. The best option would, if it were possible, be to merge the parts in such a way that these negative blocks of entries, and only they, are excluded from the upper part of the matrix. However, this is not possible, because, however the parts are aggregated, some positive entries must be present below the diagonal—more precisely, below a borderline delineating the aggregated parts. This is shown in Fig. 4.34a for the case in which the candidate ranking has three parts obtained by merging (i) R_1 and R_2, (ii) R_3, R_4 and R_5, and (iii) R_6 and R_7. The borderline between the entries included in and excluded from the sum in (4.46) is shown in bold. We can see there are 6 positive entries below the borderline, which almost cancel out the negative values. In this sense, the binary ranking R^4 that merges R_1, R_2, R_3 and R_4 into the first part of S, and R_5, R_6 and R_7 into the second part of S, as shown in Fig. 4.34b, is better as it excludes all the negative entries and only 3 positive entries. It is not difficult to prove that this ranking is optimal, and that another optimal ranking is the binary ranking R^3. Similarly, in the general case with a different number of equal-size parts

(a)

3	3	3	3	3	3	3
2	3	3	3	3	3	3
1	2	3	3	3	3	3
0	1	2	3	3	3	3
-1	0	1	2	3	3	3
-2	-1	0	1	2	3	3
-3	-2	-1	0	1	2	3

(b)

3	3	3	3	3	3	3
2	3	3	3	3	3	3
1	2	3	3	3	3	3
0	1	2	3	3	3	3
-1	0	1	2	3	3	3
-2	-1	0	1	2	3	3
-3	-2	-1	0	1	2	3

Fig. 4.34 Values $c_{ij} - (K - 1)/2$ for the blocks of the consensus matrix for the Likert scale consensus problem at $K = 7$. The boundary separating entries summed in criterion (4.46) is shown by the bold multi-line. Part (a) shows the case in which the seven parts are merged into three aggregate parts: (i) the first two parts, (ii) the next three parts, and (iii) the last two parts. Part (b) shows the case in which the seven parts are merged into two aggregate parts: (i) the first four parts, and (ii) the last three parts

in the underlying ranking R, a binary ranking that splits R into two equal-size parts (or as equal as possible) is an optimal ranking.

Therefore, the median rule for the Likert scale cannot reproduce the original ranking R when $K > 2$. The median S in this case is just a coarser binary version of R, that is a rather rough model of the original.

4.8 Summary

This Chapter presents K-means, the most popular clustering method. The method partitions the entity set into clusters along with centers representing them. It is very intuitive and usually does not require that much space to get presented, except of course its various versions such as incremental or nature inspired or medoid based algorithms.

This text includes rarely mentioned theoretical underpinnings of the method. K-means relates to an extension of the Principal Component Analysis model onto binary 1/0 vectors corresponding to clusters. Within this framework, a Pythagorean decomposition of the data scatter is derived. In this decomposition, criterion of the method represents unexplained part of the data scatter, whereas the part "explained" by clusters has meaning of a complementary clustering criterion. This complementary criterion shows that K-means is, in fact, about finding big anomalous clusters. It also brings forth useful tools for interpretation of clusters as well as for initialization—the choice of K and location of centers by using what is referred to as Anomalous and intelligent clustering. The decomposition allows for using a conventional determinacy value for choosing the number of clusters. The intelligent clustering involves a different granularity parameter, the minimum cluster size, which may be more intuitive, in some contexts, that the number of clusters or the explained data scatter proportion. Our attempts at combining the Anomalous cluster approach with such seemingly relevant concepts as Affinity Propagation clustering (Frey and Dueck 2007) and Silhouette Width (Rousseeuw 1987) so far brought no success; thus AP and SW approaches glossed over here, as well as many other popular approaches.

Three extensions of K-means onto different cluster structures are presented too. These are: Fuzzy K-means for finding fuzzy clusters, Expectation-Maximization (EM) for finding probabilistic clusters as items of a mixture of distributions, and Kohonen's self-organizing maps (SOM) that tie up the sought clusters to a visually comfortable two-dimensional grid.

The second part of the chapter is devoted to partitioning over similarity data. First of all, the complementary K-means criterion is equivalently reformulated as the so-called semi-average similarity criterion. This criterion is maximized with a consecutive merger process referred to SA-Agglomeration clustering. This latter method leads to provably tight, on average, clusters, and, most importantly, stops merging clusters when the criterion does not increase anymore if the data has been pre-processed by zeroing the similarities of the objects to themselves. A similar process is considered for another natural criterion, the summary within-cluster similarity, for which two pre-processing options are considered. These are: a popular "modularity" clustering option, based on removal of random interactions, and "uniform" partitioning, based on a scale shift, a.k.a. soft thresholding. The summary clustering also leads to an automated determination of the number of clusters, which involves one more granularity parameter, the scale shift labeling the level of significance of an individual interaction. Examples show the two different data pre-processing options for the summary clustering criterion should be applied in different contexts: the uniform criterion is better when the level of similarity is uniform across the data table, whereas the modularity criterion works better at differently scaled similarities. Another useful feature of these constructions is the role played by zeroing the diagonal entries: this brings forth a natural criterion for halting mergers in the agglomeration process and, therefore, automatic determination of the right number of clusters.

The chapter concludes with a section on consensus clustering, a concept getting popular currently. In the context of central partition for a given ensemble of partitions, two distance-between-partitions measures apply, both leading to the usage of the so-called consensus matrix; with consensus similarity defined, for any two objects, by the number of clusters in the ensemble to which both objects belong. This brings the issue into the context of similarity clustering.

References

R.K. Ahuja, T.L. Magnanti, J.B. Orlin, *Network Flows* (Pearson Education, London, 2014)

J. Bezdek, J. Keller, R. Krisnapuram, M. Pal, *Fuzzy Models and Algorithms for Pattern Recognition and Image Processing* (Kluwer Academic Publishers, 1999)

B. Efron, T. Hastie, *Computer Age Statistical Inference* (Cambridge University Press, Cambridge, 2016)

S.B. Green, N.J. Salkind, *Using SPSS for the Windows and Mackintosh: Analyzing and Understanding Data* (Prentice Hall, Upper Saddle River, 2003)

J.A. Hartigan, *Clustering Algorithms* (Wiley, Hoboken, 1975)

C. Hennig, M. Meila, F. Murtagh, R. Rocci (eds.), *Handbook of Cluster Analysis* (CRC Press, Boca Raton, 2015)

R. Johnsonbaugh, M. Schaefer, *Algorithms* (Pearson Prentice Hall, 2004). ISBN 0-13-122853-6

L. Kaufman, P. Rousseeuw, *Finding Groups in Data: An Introduction to Cluster Analysis* (Wiley, Hoboken, 1990)

M.G. Kendall, A. Stewart, *Advanced Statistics: Inference and Relationship*, 3rd edn. (Griffin, London, 1973). ISBN 0852642154

T. Kohonen, *Self-organizing Maps* (Springer-Verlag, Berlin, 1995)

A. Kryshtanowski, *Analysis of Sociology Data with SPSS* (Higher School of Economics Publishers, Moscow, 2008). (in Russian)

B. Mirkin, *Mathematical Classification and Clustering* (Kluwer Academic Press, 1996)

B. Mirkin, *Clustering: A Data Recovery Approach* (Chapman & Hall/CRC Press, 2012)

S. Nascimento, *Fuzzy Clustering via Proportional Membership Model* (ISO Press, 2005)

B. Polyak, *Introduction to Optimization* (Optimization Software, Los Angeles, 1987). ISBN 0911575144

Articles

I.E. Allen, C.A. Seaman, Likert scales and data analyses. Qual. Prog. **40**(7), 64 (2007)

S. Bandyopadhyay, U. Maulik, An evolutionary technique based on K-means algorithm for optimal clustering in R^N. Inf. Sci. **146**, 221–237 (2002)

R. Cangelosi, A. Goriely, Component retention in principal component analysis with application to cDNA microarray data. Biol. Direct **2**, 2 (2007). http://www.biolgy-direct.com/con-tent/2/1/2

B.J. Frey, D. Dueck, Clustering by passing messages between data points. Science **315**(5814), 972–976 (2007)

M. Girolami, Mercer kernel-based clustering in feature space. IEEE Trans. Neural Netw. **13**(3), 780–784 (2002)

J. Kettenring, The practice of cluster analysis. J. Classif. **23**, 3–30 (2006)

Y. Lu, S. Lu, F. Fotouhi, Y. Deng, S. Brown, Incremental genetic algorithm and its application in gene expression data analysis. BMC Bioinformatics **5**, 172 (2004)

M. Ming-Tso Chiang, B. Mirkin, Intelligent choice of the number of clusters in K-means clustering: an experimental study with different cluster spreads. J. Classif. **27**(1), 3–40 (2010)

B. Mirkin, T.I. Fenner, Distance and consensus for preference relations corresponding to ordered partitions. J. Classif. **36**(2) (2019) (see also https://publications.hse.ru/en/preprints/210704630, HSE Working Paper, 2016, no. 8)

C.A. Murthy, N. Chowdhury, In search of optimal clusters using genetic algorithm, Pattern Recognit. Lett., **17**(8), 825–832 (1996)

S. Nascimento, P. Franco, Unsupervised fuzzy clustering for the segmentation and annotation of upwelling regions in sea surface temperature images, in *Discovery Science*, ed. by J. Gama. LNCS, vol. 5808 (Springer-Verlag, 2009), pp. 212–224

M.E. Newman, Modularity and community structure in networks. Proc. Natl. Acad. Sci. U.S.A. **103**(23), 8577–8582 (2006)

S. Paterlini, T. Krink, Differential evolution and PSO in partitional clustering. Comput. Stat. Data Anal. **50**, 1220–1247 (2006)

P.J. Rousseeuw, Silhouettes: a graphical aid to the interpretation and validation of cluster analysis. J. Comput. Appl. Math. **20**, 53–65 (1987)

R. Stanforth, B. Mirkin, E. Kolossov, A measure of domain of applicability for QSAR modelling based on intelligent K-means clustering. QSAR Comb. Sci. **26**(7), 837–844 (2007)

D. Steinley, M.J. Brusco, Initializing K-means batch clustering: a critical evaluation of several techniques. J. Classif. **24**(1), 99–121 (2007)

Chapter 5
Divisive and Separate Cluster Structures

Abstract This Chapter is about dividing a dataset or its subset in two parts. If both parts are to be clusters, this is referred to as divisive clustering. If just one part is to be a cluster, this will be referred to as separative clustering. Iterative application of divisive clustering builds a binary hierarchy of which we will be interested at a partition of the dataset. Iterative application of separative clustering builds a set of clusters, possibly overlapping. The first three sections introduce three different approaches in divisive clustering: Ward clustering, Spectral clustering and Single link clustering. Ward clustering is an extension of K-means clustering dominated by the so-called Ward distance between clusters; also, this is a natural niche for conceptual clustering in which every division is made over a single feature to attain immediate interpretability of the hierarchy branches and clusters. Spectral clustering gained popularity with the so-called Normalized Cut approach to divisive clustering. A relaxation of this combinatorial problem appears to be equivalent to optimizing the Rayleigh quotient for a Laplacian transformation of the similarity matrix under consideration. In fact, other approaches under consideration, such as uniform clustering and semi-average clustering, also may be treated within the spectral approach. Single link clustering formalizes the nearest neighbor approach and is much related to graph-theoretic concepts: components and maximum spanning trees. One may think of divisive clustering as a process for building a binary hierarchy, which goes "top-down" in contrast to agglomerative clustering (in Sect. 4.6), which builds a binary hierarchy "bottom-up". Two remaining sections describe two separative clustering approaches as extensions of popular approaches to the case. One tries to find a cluster with maximum inner summary similarity at a similarity matrix preprocessed according to the uniform and modularity approaches considered in Sect. 4.6.3 The other applies the encoder-decoder least-squares approach to modeling data by a one-cluster structure. It appears, criteria emerging within the latter approach are much akin to those described earlier, the summary and semi-average similarities, although parameters now can be adjusted according to the least-squares approach. This applies to a distinct direction, the so-called additive clustering approach, which can be usefully applied to the analysis of similarity data.

© Springer Nature Switzerland AG 2019
B. Mirkin, *Core Data Analysis: Summarization, Correlation,*
and Visualization, Undergraduate Topics in Computer Science,
https://doi.org/10.1007/978-3-030-00271-8_5

5.1 Ward Divisive Clustering

5.1.1 *Hierarchy and Dendrogram*

Term hierarchy can mean different things in different contexts. Here it is a dendrogram, that is a decision tree-like nested structure drawn like that on Fig. 5.1 below (see also Figures in Sect. 4.6). Such a hierarchy may relate to mental or real processes such as conceptual structures (taxonomy, ontology) or inheritance structures (genealogy, phylogeny).

The top node, referred to as the root, represents all the entity set I under consideration. Every interior node of the dendrogram has a number of children nodes representing division of the subset—or cluster—represented by the node into smaller clusters. The terminal nodes, that have no children, are referred to as leaves and usually correspond to singletons. A hierarchical structure should be annotated to reflect the correspondence between the nodes and entity sets. Such an annotation, according to bases of division is utilized in decision trees. In clustering, another annotation is frequently used—that imposed by the leaf contents. Every node of the dendrogram corresponds to cluster of those entities that annotate the leaves descending from the node.

On the right of the hierarchy on Fig. 5.1, there is a y-axis to represent the node heights. The node height is a useful device for positioning nodes in layers. Typically, all leaves have zero heights whereas the root is assigned with the maximum height, usually taken as unity or 100%. Some hierarchies are naturally assigned with node heights, e.g., the molecular clock in evolutionary trees, some not, e.g. the decimal classification of library subjects. But to draw a dendrogram as a figure, one needs to define positions for each node, thus its height as well, even if implicitly.

Nodes may be linked by using what is called edges. Only one edge ascends from each node—this is a defining property of nested hierarchies that each node, except

Fig. 5.1 A cluster hierarchy of Company data entities: nested node clusters, each comprising a set of leaves. Cutting the tree at a certain height leads to a partition of the three product clusters here

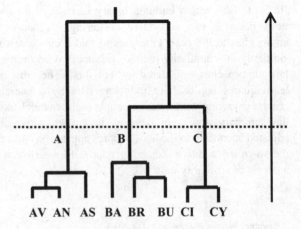

for the root, has one and only one parent. Each hierarchy node, or its parental edge, represents cluster of all leaves descending from the node; such are the edges labeled by product names A, B, and C on Fig. 5.1—they represent the corresponding clusters. These clusters have a very special pattern of overlapping: for any two clusters of a hierarchy, their intersection is either empty or coincides with one of them—this is one more characteristic property of a nested hierarchy.

The tree on Fig. 5.1 has one more specific property—it is binary: each interior node in the tree has exactly two children, that is, split in two parts. Most clustering algorithms, including those presented below, do produce binary trees, along with node heights.

Q.5.1. Given a binary hierarchy H with leaf set I, prove that the number of edges in the hierarchy is $2(|I| - 1)$.

Q.5.2. Consider a binary hierarchy H with node set J and height function $h(j)$, $j \varepsilon J$, such that $h(j) = 0$ at each leaf j. Assume that $h(j)$ is monotone, that is, the closer the node to the root the greater the value of $h(j)$. Define the distance $u(i1, i2)$ between each pair of leaves $i1, i2 \in I$ as the height of the least cluster node $j(i1,i2)$ such that both $i1$ and $i2$ are among its descendants, $u(i1, i2) = h(j(i1, i2))$. Prove that the distance u is an ultrametric, that is, it is not only symmetric, $u(i1, i2) = u(i2, i1)$, and reflexive, $u(i1, i1) = 0$, but also satisfies ultrametric inequality

$$u(i1, i2) \leq \max[u(i1, i3), u(i2, i3)] \tag{5.1}$$

for every triplet of leaves $i1$, $i2$, and $i3$.

Q.5.3. Prove that if distance u is ultrametric then, for each three entities, the three distances between them satisfy the following property: those two larger ones are equal to each other. This can be rephrased as follows: under an ultrametric, every triangle is isosceles.

Q.5.4. Define Baire distance $b(x,y)$ between non-coinciding real numbers x and y, both located in interval [0,1], as follows. Consider their decimal digits, $x = 0.x_1x_2$ ... and $y = 0.y_1y_2$ ..., and set $b(x,y) = 2^{-n}$ where n is the very first digit at which $x_n \neq y_n$. If, for example $x = 0.125$, $y = 0.128$ and $z = 0.250$, then $b(x,y) = 2^{-3}$ and $b(x,z) = 2^{-1}$ (Murtagh et al. 2008). Prove that Baire distance is ultrametric and, moreover, every finite ultrametric can be represented as a Baire metric.

Methods for hierarchical clustering are divided in two classes:

- *Divisive* methods: they build a cluster hierarchy by proceeding top-down, starting from the entire data set and recursively splitting clusters into parts; and
- *Agglomerative* methods: they build a cluster hierarchy by proceeding bottom-up, starting from the least clusters available, usually singletons, and merging those nearest to each other at each step.

5.1.2   Square-Error Criterion and Ward Distance

Consider a partition $S = \{S_1, S_2, \ldots, S_K\}$ on set I, together with centers $c = \{c_1, c_2, \ldots, c_K\}$, and the square error criterion $W(S, c) = \sum_{k=1}^{K} \sum_{i \in S_k} d(i, c_k)$ of K-means. Here $d(i, c_k)$ is the squared Euclidean distance between vector representing object i and the center c_k. Let two of the clusters, S_k, S_l, be merged so that the resulting partition is $S(k, l)$ coinciding with S except for the merged cluster $S_k \cup S_l$; the new center obviously being $c_{k \cup l} = (N_k c_k + N_l c_l)/(N_k + N_l)$, where N_k and N_l are cardinalities of clusters S_k and S_l, respectively. As proven previously—and rather evident indeed (see Fig. 5.2)—the value of square error criterion on partition $S(k, l)$ is greater than $W(S, c)$. But how much greater? The answer is:

$$W(S_k \cup S_l, c_{k \cup l}) - W(S, c) = \frac{N_k N_l}{N_k + N_l} \sum_{v \in V} (c_{kv} - c_{lv})^2 = \frac{N_k N_l}{N_k + N_l} d(c_k, c_l), \quad (5.2)$$

the squared Euclidean distance between centers of the merged clusters S_k and S_l weighted by a factor proportional to the product of cardinalities of the merged clusters (Ward 1963).

To prove this, let us follow the definition and do some elementary transformations. First, we notice that the distances within unchanged clusters do not change in the partition $S(k, l)$ so that the difference between the values of criterion W after and before the merger is

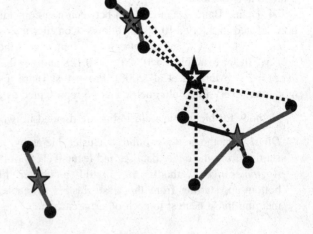

Fig. 5.2 The distances in criterion $W(S, c)$ before (solid lines) and after the merger (dashed lines) of two clusters on the upper right. The numbers of dashed and solid lines coincide, but the dashed-line distances are longer overall

$$\sum_{i\in S_k\cup S_l}\sum_{v\in V}\left(y_{iv}-c_{k\cup l,v}\right)^2-\sum_{i\in S_k}\sum_{v\in V}\left(y_{iv}-c_{kv}\right)^2-\sum_{i\in S_l}\sum_{v\in V}\left(y_{iv}-c_{lv}\right)^2$$

$$=\sum_{i\in S_k}\sum_{v\in V}\left(y_{iv}-c_{k\cup l,v}\right)^2-\sum_{i\in S_k}\sum_{v\in V}\left(y_{iv}-c_{kv}\right)^2+\sum_{i\in S_l}\sum_{v\in V}\left(y_{iv}-c_{k\cup l,v}\right)^2$$

$$-\sum_{i\in S_l}\sum_{v\in V}\left(y_{iv}-c_{lv}\right)^2.$$

Since

$c_{k\cup l,v}=c_{kv}+N_l(c_{lv}-c_{kv})/(N_k+N_l)\ \ =\ \ c_{lv}+N_k(c_{kv}-c_{lv})/(N_k+N_l)$ and the binomial rule $(a+b)^2=a^2+b^2+2ab$, the sum $\sum_{i\in S_f}\sum_{v\in V}(y_{iv}-c_{k\cup l,v})^2$ can be presented as

$$\sum_{i\in S_k}\sum_{v\in V}\left(y_{iv}-c_{kv}\right)^2+\sum_{i\in S_k}\sum_{v\in V}\left(\frac{N_l}{N_k+N_l}\right)^2\left(c_{kv}-c_{lv}\right)^2+2\sum_{i\in S_k}\sum_{v\in V}\frac{N_l}{N_k+N_l}\left(y_{iv}-c_{kv}\right)\left(c_{kv}-c_{lv}\right)$$

where the right-hand item is equal to zero, because $\sum_{i\in S_k}(y_{iv}-c_{kv})=0$. A similar decomposition holds for the sum $\sum_{i\in S_l}\sum_{v\in V}(y_{iv}-c_{k\cup l,v})^2$. These two combined make the difference $W(S(k,l),\ c(k,l))-W(S,c)$ equal to

$$\sum_{v\in V}N_k\left(\frac{N_l}{N_k+N_l}\right)^2\left(c_{kv}-c_{lv}\right)^2+\sum_{v\in V}N_l\left(\frac{N_k}{N_k+N_l}\right)^2\left(c_{lv}-c_{kv}\right)^2$$

$$=\frac{N_kN_l}{N_k+N_l}\sum_{v\in V}\left(c_{kv}-c_{lv}\right)^2$$

which completes the proof of Eq. (5.2).

The weighted distance found

$$wd(S_k,S_l)=\frac{N_kN_l}{N_k+N_l}d(c_k,c_l)\tag{5.3}$$

is referred to as Ward distance between clusters. Its weight coefficient highly depends on the distribution of entities between clusters being merged. This may affect the results of agglomerative or divisive algorithms that utilize Ward distance. Indeed, in an agglomerative process, the Ward distance between clusters to be merged must be as small as possible—which favors merging big and small clusters. On the other hand, in divisive clustering, when splitting, the Ward distance between split parts must be as large as possible, which favors splitting large clusters into relatively equal-sized parts. It is the effect of this weighting that underlies the odd behavior of the square error K-means criterion in the discussion of issues of K-means clustering (see Case-Study 4.2 and Fig. 4.7).

Given a cluster S with its center c, let us denote the square error within S by $W(S)=\sum_{i\in S}d(y_i,c)$. Using this, Eq. (5.2) can be rewritten as

$$W(S_k \cup S_l) = W(S_k) + W(S_l) + wd(S_k, S_l). \qquad (5.2')$$

This explains the additive properties of the square error $W(S)$ when used as the height index in drawing the clustering tree. According to this equation, the height of the parent is equal to the sum of heights of its children plus Ward distance between them. This warrants a specific heights distribution over the tree: the closer to the root, the longer the edges!

Q.5.5. Prove that Ward distance after a merger can be recursively calculated from the distances before the merger according to the following formula (see also Q.5.23 and Table 5.17):

$$
\begin{aligned}
wd(S_{k\cup l}, S_t) = [(N_k + N_t)wd(S_k, S_t) + (N_l + N_t)wd(S_l, S_t) \\
- N_t wd(S_k, S_l)]/(N_k + N_l + N_t).
\end{aligned}
\qquad (5.4)
$$

Q.5.6. Prove that the square error of cluster, $W(S_k) = \sum_{i \in S_k} d(y_i, c_k)$, can be expressed using within cluster distances only:

$$W(S_k) = \sum_{i,j \in S_k} d(y_i, y_j)/N_k. \qquad (5.5)$$

where d is the squared Euclidean distance and N_k is the number of entities in S_k. Hint: Use Eq. (4.5) in Q.4.10.

Equations (5.4) and (5.5) allow carrying Ward's agglomeration process by using only the distances, so that the cluster centers are involved neither for calculating Ward distances nor for the cluster's square errors.

5.1.3 Divisive Clustering

A divisive method works in a top-down manner, starting from the entire data set and splitting each cluster in two, which is reflected in drawing the split cluster as a parental node with two children corresponding to split parts. The splitting process goes on in such a way that each time a leaf cluster is split, two children nodes are sprung from the leaf which thus becomes an internal node.

To specify a method for divisive clustering, one should define the following:

(i)   splitting criterion—how one decides which split is better;
(ii) splitting method—how the splitting is actually done;
(iii) choice of cluster—which of the current leaf clusters is to be split;
(iv) stopping criterion—at what point one decides to stop the splitting.

Let us cover some options that can be recommended based on some theoretical and/or experimental evidence, in this sequence.

(i) **Splitting criterion**

The only splitting criterion that is considered here is K-means criterion of the summary square error which is implemented as Ward criterion, that is the maximum possible reduction in the total squared error caused by the split: the greater the better.

When applied to categorical features represented by their categories enveloped into the corresponding binary features, this criterion can be reinterpreted in terms of what is referred to a goodness-of-split criterion, which usually measures the improvements in the predictability of the categories, from the split partition. The "predictability" can be measured differently, most frequently by involving the general concept of "uncertainty" in the distribution of possible feature values that is captured by the concepts like Gini index, entropy, variance, etc. defined above.

Three popular goodness-of-split criteria that are compatible with the least-squares data recovery framework are: (a) impurity function (Breiman et al. 1984), (b) category utility function (Fisher 1987), and (c) the summary Pearson chi-squared coefficient. The category utility function, in fact, is the sum of impurity functions over all categories in the data, related to the number of clusters in the partition being built. All the three can be expressed in terms of the cluster-category contributions to the data scatter and, thus, can be considered special cases of the square-error clustering criterion at the conventional zero-one coding of categories along with different normalizations of the data, as explained in Sects. 3.6.1 and 3.8.2.

(ii) **Splitting method**

For Ward's criterion, we consider two splitting approaches:

IIA. *K-means at K = 2*, or Bisecting K-means; this leads, typically, to good results if care is taken to find good initial centers. Since the criterion of Bisecting K-means is equivalent to the criterion of maximizing Ward distance between split parts, the divisive algorithm utilizing Bisecting K-means is referred to as Ward divisive algorithm here.

IIB. *Conceptual clustering*—in this, just one of the features is involved in each of the splits, which leads to a straightforward conceptual interpretation of all the clusters. Conceptual clustering builds a cluster hierarchy by sequentially splitting clusters, as all divisive algorithms do, yet here each of the splits uses a single attribute in contrast to the classic clustering that utilizes distances involving all features. The criteria such as summary impurity function or summary Pearson chi-squared are part of the Ward divisive clustering algorithm under a corresponding normalization option (see Sect. 3.8). In this aspect, conceptual clustering should be equated to building classification trees over a multiple target feature set—the only difference is that the very same target features are simultaneously input features!

(iii) **Choice of cluster to split**

The order of splitting conventionally is not considered important: if the set is divided all the way down to singletons, then the order does not much matter indeed. If, however, the goal is to produce a partition by finishing after just a few splits, then Ward's criterion gives the following guiding principle: after each split, all leaf clusters S_k are supplied with their square errors $W(S_k)$. The square-error is the contribution to the unexplained data scatter, that is, the sum of Euclidean squared distances between cluster's entities and its center, which is proportional to the cluster summary variance weighted by its size. To minimize the unexplained part, that cluster whose square-error is maximum is to be split first.

(iv) **Stopping criterion**

Conventionally, the divisions stop when there remains nothing that can be split, that is, when all the leaves are singletons. Yet for the Ward's criterion one can specify a threshold on the value of the square-error at a cluster, the level of "noise" reached by $W(S_k)$ at which a cluster is considered next to noise and not split anymore because of that. This threshold can be set as a proportion of the data scatter, say 5%. Another criterion of course can be just the cluster size—say, clusters whose cardinality is less than 1% of the original data size are not to be split anymore. Tasoulis et al. (2010) propose stopping the splitting process when the density of the cluster points projections to the first principal component has no minima inside the range.

The process of divisive clustering is much like that of building a classification decision tree—the only difference is the criterion. It is correlation to a target variable in the latter case and it is a summary correlation with all the features forming the clustering space, in the former case, even if it is expressed as maximizing Ward distance between split parts. The equivalence between K-means criterion at $K = 2$ and the criterion of maximization of Ward distance justifies the following algorithm.

Ward divisive clustering algorithm

1. **Start**
   Put $J \Leftarrow I$ and draw tree root as a node corresponding to I at the height of $W(I)$ which is the data scatter, by itself or 100%.
2. **Split**
   Split J in two parts, S_1 and S_2, to maximize Ward distance $wd(S_1, S_2)$.
3. **Draw**
   In the dendrogram drawing, add two children nodes corresponding to S_1 and S_2 at the parent node corresponding to J, their heights being their square errors.
4. **Update**
   Find the cluster of maximum height among the leaves of the current cluster hierarchy and make it J.
5. **Stop-Condition**
   Check the stopping condition as described below. If it holds, halt and output the hierarchy and possible interpretation aids; otherwise, go to Step 2.

Developing a good splitting algorithm at Step 2 in divisive clustering can be an issue. We consider two versions: Bisecting K-means and C-splitting. The latter option will be described in the next section.

Bisecting K-means

1. **Initialization**
   Given J, specify initial seeds of its split parts, c_1 and c_2.
2. **Batch 2-Means**.
   Apply Bisecting K-means with initial seeds specified at step 1 and the squared Euclidean distance.
3. **Output**
   Output:

   (a) final split parts S_1 and S_2;
   (b) their centers c_1 and c_2;
   (c) their heights, $h(S_1)$ and $h(S_2)$;
   (d) contribution of the split which is Ward distance $dw(S_1, S_2)$.

To specify two initial seeds in Bisecting K-means, either option can be applied:

   (1a) random selection;
   (1b) maximally distant entities;
   (1c) centers of two Anomalous pattern clusters derived on J as described in Sect. 6.1.5.
   (1d) two centers derived with algorithm Build in Sect. 6.1.5.

Random selection must be repeated many times to get a reasonable solution for any sizeable dataset. Maximally distant entities not necessarily reflect the structure of a good split (see Case-Study 5.1). Therefore, two latter options should be preferred.

Case-Study 5.1. Divisive Clustering of Companies with Bisecting K-means
Consider the Ward divisive clustering method for the Company data, range standardized with the follow-up rescaling the dummy variables corresponding to the three Sector categories in Table 4.2. Using Bisecting K-means algorithm may produce a rather poorly resolved picture if the most distant entities, 6 and 8 according to the distance matrix in Table 5.1, are taken as the initial seeds. Then step 2 would produce tentative clusters {1,3,4,5, 6} and {2,7,8} because 2 is nearer to 8 than to 6 as easily seen in Table 5.1. This partition is at odds with the product clusters.

Unfortunately, no further iterations can change that. This shows that the choice of the initial seeds at the two farthest entities can be not that good an option that it may seem to be. Usage of Build or Anomalous pattern algorithms, explained in Sect. 6.1.5, could lead to better results. Because of our reluctance in using the original Company dataset in this Chapter, let us use Build algorithm which relies on distances only.

Table 5.1 Distances between standardized Company entities (copied from Table 4.10). For the sake of convenience, those smaller than 1, are highlighted in bold

Entities	Av	An	Ast	Ba	Br	Bu	Ci	Cy
1 Av	**0.00**	**0.51**	**0.88**	1.15	2.20	2.25	2.30	3.01
2 An	**0.51**	**0.00**	**0.77**	1.55	1.82	2.99	1.90	2.41
3 As	**0.88**	**0.77**	**0.00**	1.94	1.16	1.84	1.81	2.38
4 Ba	1.15	1.55	1.94	**0.00**	**0.97**	**0.87**	1.22	2.46
5 Br	2.20	1.82	1.16	**0.97**	**0.00**	**0.75**	**0.83**	1.87
6 Bu	2.25	2.99	1.84	**0.87**	**0.75**	**0.00**	1.68	3.43
7 Ci	2.30	1.90	1.81	1.22	**0.83**	1.68	**0.00**	**0.61**
8 Cy	3.01	2.41	2.38	2.46	1.87	3.43	**0.61**	**0.00**

According to distance data in Table 5.1, entity 5 Br is a medoid there since the sum of its distances to the rest, 9.6, is the minimum. Now we form a cluster around each entity to consist of those that are nearer to the entity than to medoid 5: these will be 2 and 3 around 1, 1 and 3 around 2, 1 and 2 around 3, 7 and 8 around each other, and clusters for entities 4,5,6 are singleton themselves. This would give an edge to entity 1 as the next seed, because it is further away from 5 and surrounded by two. Indeed the summary E value for 1, $2.20 + (1.82 - 0.51) + (1.16 - 0.88) = 3.79$ is by far the greatest. Using the entities 5 and 1 as seeds, indeed brings the bisecting K-means to a desired split {1,2,3} versus {4,5,6,7,8}. The hierarchy on Fig. 5.1 then will be found with further splits. The node heights are the same—within cluster squared errors that are scaled as percentages of the pre-processed data scatter.

5.1.4 Conceptual Clustering

This process applies when a hierarchy is needed with immediate interpretation of all the splits and clusters. In this case, at each step of the divisive clustering process divisions are made over a single feature chosen according to a maximum association criterion. Here such an approach is described.

C-splitting (Conceptual clustering with binary splits)

1. **Initial setting**
   Set J to consist of the universal cluster, the entire entity set I.
2. **Evaluation**
   In a loop over all leaf clusters J and variables $v \in V$, for each J and v, consider all possible splits y_{lv} of J over v in two parts. If v is quantitative or ordinal, J-splits are defined by splits of its range in two parts: one part consists of entities at which feature v is less than or equal to y_{iv} and the other of those at which feature v is greater than y_{lv}. If v is nominal, J is split over each of v's categories l in "yes" and "no" parts. This amounts to using quantitative dummy variables for each category.

3. **Split**

Select that triplet (J, v, y_{Jv}) which received the highest score and perform the binary split of J, thus generating two its offspring nodes S_1 and S_2.

4. **Output**

This is the same as in the previous versions plus the variable v and split values, for each of the splits, either $y_{iv} > a$ and $y_{iv} \le a$ for a quantitative feature or $y_{iv} = a$ and $y_{iv} \ne a$ for a categorical feature.

Case-Study 5.2. Conceptual Clustering of Digit Data as Related to Ward Clustering

As shown in Sect. 3.8, divisions over individual features—the essence of conceptual clustering procedures—are governed by the square error criterion if conventional measures for scoring association between dataset features and the partition—impurity index or Pearson chi-squared—are applied.

Let us take the set of 10 styled digits presented in Fig. 5.3 and turn the figure into a dataset by considering each of the seven rectangle edges a feature with two categories, "Present" and "Absent" (see Fig. 5.3 and Table 5.2).

To produce a classification tree leading to a partition S, we use summary Gini index (impurity function) $G(v1/S) + G(v2/S) + \cdots + G(v7/S)$ as the criterion to maximize. Start by trying each of the features as the split base to select the best of them. Consider, for example, partition $S = \{S_1, S_2\}$ of the Digit set according to

Fig. 5.3 Digit rectangle edges as features

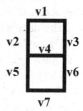

Table 5.2 Digit dataset

	v1	v2	v3	v4	v5	v6	v7
1	0	0	1	0	0	1	0
2	1	0	1	1	1	0	1
3	1	0	1	1	0	1	1
4	0	1	1	1	0	1	0
5	1	1	0	1	0	1	1
6	1	1	0	1	1	1	1
7	1	0	1	0	0	1	0
8	1	1	1	1	1	1	1
9	1	1	1	1	0	1	1
0	1	1	1	0	1	1	1
Total	8	6	8	7	4	9	7

Table 5.3 Cross-classification of S = v2 and v7 on digit dataset

	S_1	S_2	Total
v7 = 1	5	2	7
v7 = 0	1	2	3
Total	6	4	10

attribute v2 which is present at S_1= {4,5, 6, 8, 9, 0} and is absent at S_2= {1,2,3,7}. Cross-classification of S and v7 (see Table 5.3) yields G(v7/S) = 0.053.

To see what this has to do with the setting in which Ward's criterion applies, let us pre-process the Digit data matrix by subtracting the column averages without rescaling them (because the scaling coefficients must be all unity to make Ward's criterion equivalent to the summary Gini index, see Sects. 3.6.1 and 3.8.2). However, the data in Table 5.4 is not exactly the data matrix Y considered theoretically in Sect. 4.3.1. Indeed, the theoretical data matrix in Eq. (4.8) comprises columns corresponding to all of the categories, whereas the data matrices in Tables 5.2 and 5.4 reflect only half of the categories—those of the presence of edges v1–v7, never an absence. Indeed, a column corresponding to an "Absent" category is a mirror of the column corresponding to the "Presence" category, with all ones made zero and, vice versa, all zeros made ones. After the centering, the lacking half of the data table would be the Table 5.4 negated, that is, multiplied by −1. The data scatter is formed by squares of the entries which are the same. That means that this lacking part can be taken into account by just doubling the contributions accounted for with Table 5.4.

The data scatter of matrix in Table 5.4 is the summary column variance times $N = 10$, which is 13.1. However, to get the data scatter in the left-hand side of (4.8), this must be doubled to 26.2 to reflect the "missing half" of the virtual data matrix Y.

Let us now calculate the within class averages c_{kv} of each of the variables, $v = v1, ..., v7$, in clusters $k = 1,2$ and take contributions $B_{kv} = N_{kv} \, c_{kv}^2$ summed over clusters S_1 and S_2. This is done in Table 5.5, the last line of which contains contributions of all features to the explained part of the data scatter.

Table 5.4 Digit dataset pre-processed by centering its columns

Digit	v1	v2	v3	v4	v5	v6	v7
1	−0.8	−0.6	0.2	−0.7	−0.4	0.1	−0.7
2	0.2	−0.6	0.2	0.3	0.6	−0.9	0.3
3	0.2	−0.6	0.2	0.3	−0.4	0.1	0.3
4	−0.8	0.4	0.2	0.3	−0.4	0.1	−0.7
5	0.2	0.4	−0.8	0.3	−0.4	0.1	0.3
6	0.2	0.4	−0.8	0.3	0.6	0.1	0.3
7	0.2	−0.6	0.2	−0.7	−0.4	0.1	−0.7
8	0.2	0.4	0.2	0.3	0.6	0.1	0.3
9	0.2	0.4	0.2	0.3	−0.4	0.1	0.3
0	0.2	0.4	0.2	−0.7	0.6	0.1	0.3

The last item, 0.267, is the contribution of $v7$. Has it anything to do with the reported value of impurity function $G(v7/S) = 0.053$? Yes, it does. There are two reasons to make these two quantities different. First, to get to the contribution from $G(v7/S)$, it must be multiplied by $N = 10$, which would make it 0.533. Second, the 0.267 value is the contribution to the data scatter of matrix Y obtained after enveloping of all 14 categories—not just 7 present in Table 5.5. After the contribution 0.267 is properly doubled, the quantities do coincide. Similar calculations made for the other six attributes, $v1$, $v2$, $v3$, $v4$, $v5$, and $v6$, would lead to the total contribution of S to the data scatter equal $10\Sigma_f G(vf/S) = 5.03$ which is 26.8% of the scatter 26.2.

To find out which of the features is to be used for the first split, all pair-wise Gini index values have been computed and presented in Table 5.6. According to these, feature $v7$ supplies the maximum summary contribution $10 * 0.963 = 9.63$ which is 36.8% of the total data scatter.

Therefore, the first split must be done according to $v7$. Two more splits are due $v5$, contributing 3.90, and $v1$, contributing 3.33, resulting in a four-cluster partition $S = \{1\text{–}4\text{–}7,\ 3\text{–}5\text{–}9,\ 2,\ 6\text{–}8\text{–}0\}$. This partition contributes $9.63 + 3.90 + 3.33 = 16.87 = 64.4\%$ to the total data scatter. The next partition step would contribute less than 10% of the data scatter, which is less than the contribution of one entity on average—a good signal to stop the splitting. The classification tree, or conceptual tree, produced with the splits is presented on Fig. 5.4 along with a visualization of the set of tree-making features on the rectangle base of the Digit data.

What is nice about the tree is that the clusters are well matching those found by using the data on Confusion between the digits in a psychological experiment (see Sects. 4.5, 5.3 and 5.4). This should lead to further analysis of possible importance of features $v7$, $v5$, $v1$ for the human judgment on similarity between the digits.

Table 5.5 Feature contributions to digit clusters according to v2

	v1	v2	v3	v4	v5	v6	v7
v2 = 1	0.007	0.960	0.107	0.107	0.060	0.060	0.107
v2 = 0	0.010	1.440	0.160	0.160	0.090	0.090	0.160
Total	0.017	2.400	0.267	0.267	0.150	0.150	0.267

Table 5.6 Pairwise Gini indexes for all 7 features in digit dataset

	v1	v2	v3	v4	v5	v6	v7
v1	0.320	0.003	0.020	0.015	0.053	0.009	0.187
v2	0.005	0.480	0.080	0.061	0.030	0.080	0.061
v3	0.020	0.053	0.320	0.034	0.003	0.009	0.034
v4	0.020	0.053	0.045	0.420	0.003	0.020	0.115
v5	0.080	0.030	0.005	0.004	0.480	0.080	0.137
v6	0.005	0.030	0.005	0.009	0.030	0.180	0.009
v7	0.245	0.053	0.045	0.115	0.120	0.020	0.420
Total	0.695	0.703	0.520	0.658	0.720	0.398	0.963

Fig. 5.4 Conceptual clustering of Digit dataset and features involved in the splits

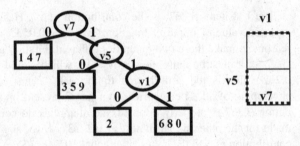

Q.5.7. The appropriateness of using Bisecting K-means as a splitting device in Ward divisive clustering.

On the first glance, the Ward criterion for dividing an entity set in two clusters—maximize Ward distance between the split parts—has nothing to do with that of K-means. In fact, given a parental cluster $J \subseteq I$, the K-means criterion, minimizing $W(S, c) = W(J, c) = \sum_{i \in S_1} d(i, c_1) + \sum_{i \in S_2} d(i, c_2)$ where S_1 and S_2 are the split parts of J, c_1 and c_2 their respective centers and d squared Euclidean distance, is equivalent to the Ward criterion. That means that Ward divisive clustering is adequately served with Bisecting K-means, or Bisecting K-means. Prove it.

A. Let us remind of the complementary to criterion W part of the data scatter according to the Pythagorean decomposition in Eq. (4.8). This complementary criterion is to maximize $B(S, c) = N_1 \langle c_1, c_1 \rangle + N_2 \langle c_2, c_2 \rangle$ where N_1 and N_2 are respective cardinalities of clusters S_1 and S_2.

Let us prove that Ward distance between the two clusters, $dw(S_1, S_2) = \frac{N_1 N_2}{N_1 + N_2} d(c_1, c_2)$, is just that. To proceed, we need two equations. The first just expresses the squared Euclidean distance through inner products, $d(c_1, c_2) = (\langle c_1, c_1 \rangle - \langle c_1, c_2 \rangle) + (\langle c_2, c_2 \rangle - \langle c_1, c_2 \rangle)$. The second is relation between the cluster centers and the parental cluster center c: $N_1 c_1 + N_2 c_2 = (N_1 + N_2)c$. Since c is not involved in $dw(S_1, S_2)$ and, in fact, is irrelevant to it, we may take it to be $c = 0$. Then the latter equation implies that $c_2 = -(N_1/N_2)c_1$. This can be put into $\langle c_1, c_1 \rangle - \langle c_1, c_2 \rangle = \langle c_1, c_1 \rangle + N_1/N_2 \langle c_1, c_1 \rangle = (N_1 + N_2)/N_2 \langle c_1, c_1 \rangle$. Similarly, equation $\langle c_2, c_2 \rangle - \langle c_1, c_2 \rangle = (N_1 + N_2)/N_1 \langle c_2, c_2 \rangle$ is obtained. Substituting these through the first equation in Ward distance, we find $dw(S_1, S_2) = \frac{N_1 N_2}{N_1 + N_2} d(c_1, c_2) = \frac{N_1 N_2}{N_1 + N_2} ((N_1 + N_2)/N_2 \langle c_1, c_1 \rangle + (N_1 + N_2)/N_1 \langle c_2, c_2 \rangle) = B(S, c)$, which proves the statement.

Q.5.8. Equivalence of the square-error criterion and the summary Gini index.

Show that, in the situation in which all the features are categorical, maximizing the summary Gini index $\sum_{v \in V} G(v/J)$ is equivalent to minimizing the least squares criterion.

A. Assume that data matrix Y in this case is drawn by putting a dummy variable for each of the categories with a follow up centering it with the mean which is the category frequency. Then, according to Statement 3.5.3.1(c) in Sect. 3.5.3, Gini index $G(v/S)$, multiplied by the number of entities, is the contribution of the partition of J to the summary scatter of the dummies corresponding to categories of feature v—this, in fact, easily follows from Eq. (3.40). This implies that the summary Gini index, multiplied by the number of entities, is the contribution of the partition to the summary scatter of all the dummy variables, that is, the data scatter of matrix Y. That means that maximum of the summary Gini index is reached at a partition minimizing the total unexplained contribution which is exactly the square error criterion. The statement is proved.

Q.5.9. What data standardization should be applied if one wants to build a conceptual clustering tree maximizing the summary Pearson chi-squared by using Ward distance maximization?

A. Each category is to be represented by a dummy variable which then should be centered by subtracting its frequency and normalized by the square root of the frequency (see Statement 3.6.2.2(c)).

Comment 5.1. Greediness of Divisive Clustering

Some may argue that the framework of divisive clustering is deliberately set as a greedy optimization procedure: at each local splitting step the best solution is taken which is not necessarily the best if one considers summary results of several sequential steps. The greedy-wise nature of the setting is true. Yet it is not easy to formulate a holistic optimization problem for divisive clustering. If for example, the process of splitting goes all the way down to singleton clusters, then perhaps the greedy-wise setting is most natural. When there is a stopping condition such as a pre-specified number K of terminal clusters, then the problem becomes of globally minimizing K-means square-error criterion. One should remember that the K-means criterion has some innate drawbacks related to its rigidity in putting the goal of getting split parts as uniform as possible. This means that achieving the global minimum is not necessarily beneficial from the point of view of data analysis.

Q.5.10. Formulate a version of agglomerative clustering for Ward criterion using the definition of Ward distance with centers.

5.2   Normalized Cut, Laplace Standardization and Spectral Clustering

5.2.1   Normalized Cut and Laplace Transformation

The concept of normalized cut is a relatively recent development started by Shi and Malik (2000). It belongs to a series of graph cutting criteria that balance the cluster sizes by normalizing the sums of within cluster similarities by the size-dependent values. Given a similarity matrix A, let us take the volume of $S \subseteq I$, the summary similarity in rows $i \in S$, $a(S) = \Sigma_{i \in S} a_{i+}$. Then the normalized cut over a partition $\{S_1, S_2\}$ is defined as $a(S_1,S_2)/a(S_1) + a(S_1, S_2)/a(S_2)$. This is to be minimized over all splits $\{S_1, S_2\}$ of set I. An equivalent criterion would maximize the sum of normalized within cluster similarities, $a(S_1, S_1)/a(S_1) + a(S_2, S_2)/a(S_2)$—the two criteria sum to 2.

The normalized cut concept brought forward a less intuitive type of data pre-processing, the Laplace transformation of similarity matrices. In its normalized form, this transformation normalizes every similarity w_{ij} by dividing it by the square root of the product of i and j volumes, $w_{ij}/\sqrt{w_{i+}w_{j+}}$, and then subtracts the resulting matrix from the identity matrix which has all its entries zero except for unities on the diagonal. With this transformation we are into the realm of spectral clustering. The Laplacian matrix L is proven to have all the eigenvalues non-negative. Besides, L has a specific minimum eigenvalue—the zero. Yet the next minimum eigenvalue and the corresponding eigenvector provide for a relaxation of the minimum of normalized cut problem, reformulated in terms of the Rayleigh quotient for the Laplacian matrix. Then split $\{S_1, S_2\}$ can be found by using this second minimum eigenvector so that S_1 is defined by indices of the positive components and S_2, of the negative components.

Let us put this a bit more precisely.

Given a symmetric similarity matrix $A = (a_{ij})$ on set I, consider the issue of dividing I in two parts, S_1 and S_2, in such a way that the similarity between S_1 and S_2 is minimum while it is maximum within the parts. Denote by $A(S_f, S_g)$ the summary similarity "between" S_f and S_g so that $A(S_f, S_g) = \sum_{i \in S_f} \sum_{j \in S_g} a_{ij}$.

The normalized cut utilizes the summary similarities $a_{i+} = A(i,I)$. Denote

$$a(S_k) = \sum_{i \in S_k} a_{i+}$$

Obviously, $a(S_1) = A(S_1, S_1) + A(S_1, S_2)$; a similar equation holds for $a(S_2)$. The normalized cut is defined as

$$nc(S) = \frac{A(S_1, S_2)}{a(S_1)} + \frac{A(S_2, S_1)}{a(S_2)} \tag{5.6}$$

to be minimized.

It should be noted that the minimized cut (5.6), in fact, includes the requirement of maximization of the within-cluster similarities. Indeed consider the normalized within-cluster similarity

$$nt(S) = \frac{A(S_1, S_1)}{a(S_1)} + \frac{A(S_2, S_2)}{a(S_2)},\qquad(5.7)$$

scoring the tightness of clusters. These two measures are highly related: $nc(S) + nt$ $(S) = 2$ (see Q.5.11). This latter equation warrants that minimizing the normalized cut simultaneously maximizes the normalized tightness.

It appears, the criterion of minimizing $nc(S)$ can be expressed in terms of a corresponding Rayleigh quotient—for the so-called Laplacian. Given a (pre-processed) similarity matrix $W = (w_{ij})$, let us denote its row sums, as usual, by $w_{i+} = \sum_{j\in I} w_{ij}$ $(i \in I)$ and introduce diagonal matrix D in which all entries are zero except for diagonal elements (i,i) that hold w_{i+} for each $i\epsilon I$. The so-called (normalized) Laplacian is defined as $L = E - D^{-1/2}WD^{-1/2}$ where E is identity matrix and $D^{-1/2}$ is a diagonal matrix with (i,i)-th entry equal to $1/\sqrt{w_{i+}}$. That means that L's (i,j)-th entry is $\delta_{ij} - w_{ij}/\sqrt{w_{i+}w_{j+}}$ where δ_{ij} is 1 if $i = j$ and 0, otherwise. It is not difficult to prove that $Lf_0 = 0$ where $f_0 = (\sqrt{w_{i+}}) = D^{1/2}1_N$ where 1_N is N-dimensional vector whose all entries are unity. That means that 0 is an eigenvalue of L with f_0 being the corresponding eigenvector.

Moreover, for any N-dimensional f, the following equation holds:

$$f^T Lf = \frac{1}{2}\sum_{i,j\in I} w_{ij}\left(\frac{f_i}{\sqrt{w_{i+}}} - \frac{f_j}{\sqrt{w_{j+}}}\right)^2\qquad(5.8)$$

This equation is easy to prove by opening the parentheses and doing little algebra over the obtained expression. It implies, among other things, that matrix L is semi-positive definite, which means that product $f^T LF$ is not negative for any vector f. As is well known in matrix theory, any semi-positive definite matrix has all its eigenvalues non-negative reals, so that 0 is the minimum eigenvalue.

Given a bisecting partition $S = \{S_1, S_2\}$ of I, let us define vector s by condition $s_i = \sqrt{w_{i+}w(S_2)/w(S_1)}$ for $i\epsilon S_1$ and $s_i = -\sqrt{w_{i+}w(S_1)/w(S_2)}$ for $i\epsilon S_2$. Obviously, the squared norm of this vector is constant, $\|s\|^2 = \sum_{i\in I} s_i^2 = w(S_2) + w$ $(S_1) = w_{++}$. Moreover, s is orthogonal to the trivial eigenvector $f_0 = D^{1/2}1_N$ corresponding to the 0 eigenvalue of L. Indeed, the product of i-th components of these vectors has w_{i+} as its factor multiplied by a value which is constant within clusters. Then summation of these components over S_1 will produce $w(S_1)\sqrt{w(S_2)/w(S_1)} = \sqrt{w(S_1)w(S_2)}$ and summation over S_2, $-w(S_2)\sqrt{w(S_1)/w(S_2)} = -\sqrt{w(S_1)w(S_2)}$. The sum of these two is 0, which proves the statement.

It remains to prove that minimization of the normalized cut (5.6) is equivalent to minimization of $s^T L s / s^T s$ for thus defined s. Indeed, at $f = s$, the squared item in (5.8) is equal to 0 for i,j from the same set, S_1 or S_2. When i and j belong to different classes of S, the squared item is equal to $w(S_1)/w(S_2) + w(S_2)/w(S_1) + 2 = [w_{++} - w(S_2)]/w(S_2) + [w_{++} - w(S_1)]/w(S_1) + 2 = w_{++}/w(S_2) + w_{++}/w(S_1)$. That means that to make the sum (5.8), the obtained quantity is multiplied by the number of non-zero items, $2W(S_1, S_2)$, so that $s^T L s = 2w_{++}nc(S)$, that is, $2nc(S) = s^T L s / s^T s$. Therefore, indeed. minimization of $nc(S)$ is equivalent to minimization of the Rayleigh quotient for L.

We have proven that the normalized cut minimizes the Rayleigh quotient for Laplacian matrix L over specially defined vectors s that are orthogonal to the eigenvector $f_0 = (w_{i+}^{1/2})$ corresponding to the minimum eigenvalue 0 of L.

Therefore, one may consider the problem of finding the minimum non-zero eigenvalue for L along with the corresponding eigenvector as a proper relaxation of the normalized cut problem. That means that the spectral clustering approach in this case would be to grab that eigenvector and approximate it with an s-like binary vector. The simplest way to do that would be by putting all plus components to S_1 and all negative to S_2.

It remains to define the pseudo-inverse Laplacian transformation, Lapin for short, for a symmetric matrix W. Consider all non-zero eigenvalues $\lambda_1, \lambda_2, \ldots, \lambda_r$ of matrix L and corresponding eigenvectors $f_1, f_2, \ldots, f_r$. The following spectral decomposition equation is known to hold:

$$L = \lambda_1 f_1 f_1^T + \lambda_2 f_2 f_2^T + \ldots + \lambda_r f_r f_r^T \tag{5.9}$$

The pseudo-inverse is defined by leaving the same eigenvectors but reversing the eigenvalues, which causes no problems since they are all non-zero:

$$L^+ = \frac{1}{\lambda_1} f_1 f_1^T + \frac{1}{\lambda_2} f_2 f_2^T + \ldots + \frac{1}{\lambda_r} f_r f_r^T \tag{5.10}$$

Q.5.11. Prove that $nc(S) + nt(S) = 2$ where the constituents are defined by Eqs. (5.6) and (5.7).

Q.5.12. Prove that a one cluster extension of the normalized cut criterion, maximize $ng(S) = a(S,S)/a(S)$, does not work at nonnegative similarity data because the maximum is always reached at the universal cluster $S = I$.

A. Indeed, $ng(I) = 1$, whereas $ng(S) < 1$ at all other S unless there are all zeros outside of $A(S,S)$.

Q.5.13. What's wrong with the idea of expressing the summary similarity criterion as a Rayleigh quotient?

Given a non-negative matrix W with none of its rows summing up to 0, its Laplacian can be found with these MatLab commands:

```
≫W=(W + W')/2; % to warrant the symmetry
≫wr=sum(W);
≫D=diag(wr);
≫D=sqrt(D);
≫Di=inv(D);
≫L=eye(size(W)) – Di*W*Di;
```

Then the pseudo-inverse transformation can work like this.

```
≫L=(L + L')/2;
≫[Z,M]=eig(L);
≫ee=diag(M);
≫ind=find(ee ~=0); % indices of non-zero eigenvalues;
≫Zn=Z(ind,ind);
≫Mn=M(ind.ind);
≫Mi=inv(Mn);
≫Lapin=Zn*Mi*Zn';
```

Worked Example 5.1. Normalized Cut for Company Data: Laplace and Lapin Matrices

To show how this works, consider the affinity data for Company data set and its Laplacian matrix in Table 5.7. The minimum eigenvalue of the Laplacian matrix is 0 whereas the second minimum eigenvalue is 0.32. The eigenvector corresponding to the latter, as expected, well separates the first three entities, A-product companies.

Although the result is natural, there is no way to see it from the Laplacian matrix by itself. Unlike the original affinity matrix, the visible structure of the Laplacian gives no useful indications on the cluster structure underlying its entries. To make the structure visible, the Laplacian should be further transformed. The Laplacian Pseudo Inverse (Lapin, for short) transformation takes the spectral decomposition of the Laplacian, inverses the non-zero eigenvalues λ into $1/\lambda$, and returns a pseudo-inverse Laplacian which is presented in the lower triangle of Table 5.8.

Table 5.7 Affinity similarities between eight companies in the Company data in Table 4.2 (upper triangle) and the result of the normalized Laplace transformation (lower triangle)

	1	2	3	4	5	6	7	8
1	1	**0.3623**	**0.1730**	0.1005	0.0123	0.0111	0.0101	0.0024
2	−0.5360	1	**0.2143**	0.0447	0.0261	0.0025	0.0224	0.0080
3	−0.2803	−0.3450	1	0.0207	0.0989	0.0252	0.0266	0.0085
4	−0.1611	−0.0712	−0.0361	1	0.1424	**0.1752**	0.0880	0.0073
5	−0.0177	−0.0372	−0.1548	−0.2207	1	**0.2248**	**0.1918**	0.0236
6	−0.0196	−0.0044	−0.0486	−0.3343	−0.3845	1	0.0347	0.0011
7	−0.0150	−0.0332	−0.0431	−0.1411	−0.2759	−0.0614	1	**0.2982**
8	0.0050	−0.0164	−0.0191	−0.0162	−0.0471	−0.0026	−0.6158	1

Table 5.8 Affinity similarities between eight companies as in Table 5.7 (upper triangle; entries larger than 0.15 are highlighted in bold; those not fitting in the structure underlined) and the result of Lapin transformation (lower triangle, positive entries highlighted in bold, that not fitting underlined)

	1	2	3	4	5	6	7	8
1		**0.3623**	**0.1730**	0.1005	0.0123	0.0111	0.0101	0.0024
2	**0.4734**		**0.2143**	0.0447	0.0261	0.0025	0.0224	0.0080
3	**0.2221**	**0.2677**		0.0207	0.0989	0.0252	0.0266	0.0085
4	−0.2073	−0.2719	−0.2405		<u>0.1424</u>	**0.1752**	0.0880	0.0073
5	−0.4281	−0.4314	−0.2577	**0.0190**		**0.2248**	<u>**0.1918**</u>	0.0236
6	−0.3534	−0.3839	−0.2638	**0.1563**	**0.1984**		<u>0.0347</u>	0.0011
7	−0.5430	−0.5337	−0.4076	−0.1213	<u>**0.0457**</u>	−0.1020		**0.2982**
8	−0.4440	−0.4316	−0.3385	−0.1650	−0.0504	−0.1478	**0.6003**	

Table 5.9 Reciprocal non-zero eigenvalues of the Laplacian and Lapin matrices corresponding to the same eigenvectors

Eigenvalue labels	I	II	III	IV	V	VI
Normalized Laplacian	0.32	0.59	1.11	1.35	1.40	1.55
Lapin	3.08	1.70	0.90	0.74	0.71	0.65

One can easily see that the cluster structure is more pronounced in the Lapin matrix than it is in the original affinity matrix. First, there is no need for guessing a right threshold value to subtract: it is 0 here. Second, there is only one not-fitting entry here, (7,5), but it would not make any difference anyway because it is rather small in comparison with the other negative entries (7,6) and (7,4) linking item 7 to B-product cluster, or entry (8,5) linking item 5 to C-product cluster.

One more result of the Lapin transformation is that the eigenvalue to look for is the maximum one, and it is much better separated from the rest because of the inversion (see Table 5.9). The corresponding eigenvector does not change.

Indeed the three product based clusters, {1, 2, 3}, {4, 5, 6}, {7, 8}, are found with both the summary clustering criterion and spectral approach applied to the Lapin transformed Company affinity data.

Case-Study 5.3. Circular Cluster Exposed by Lapin Transformation

To further demonstrate the formidable ability of the Lapin transformation in manifesting clusters according to human intuition, let us consider the 2D set presented in Fig. 5.5.

This set has been generated as follows. Three 100×2 data matrices, $a1$, $a2$ and $a3$, were generated from Gaussian distribution $N(0,1)$. Then matrix $a2$ was normed row-wise into b, so that each row in b is a 2D normed vector, after which matrix c has been defined as $c = 0.5 * a3 + 8 * b$. Its rows form a ring-wise shape on the Cartesian plane, while rows of $a1$ fall into a heap in the circle's center as presented on Fig. 5.5. Then $a1$ and c are merged into a 200×2 matrix X, in which $a1$ takes the first 100 rows and c the next 100 rows.

Fig. 5.5 Two intuitively obvious clusters that are difficult to separate using conventional approaches: stars in the heap and dots in the ring

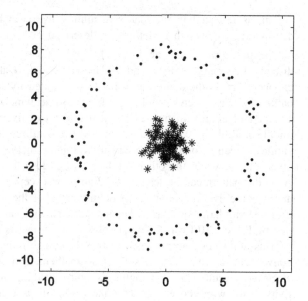

The conventional data standardization methods would not change the picture, and conventional clustering procedures like K-means clustering would not be able to separate the ring as a whole. The single link clustering will be able to separate these two clusters, which would once again remind us of a rift between the data approximation clustering and graph theoretic approaches. Yet the Laplace transformation allows us to put this dataset into the data approximation context too.

The data is first transformed into a 200 × 200 affinity similarity matrix, which is then Lapin transformed into a final similarity matrix. This final matrix shows a clear-cut pattern: all similarities between the first hundred and the second hundred of points are negative whereas all the Lapin similarities within these sets are positive. Such a structure clearly separates the two clusters with any reasonable algorithm, the summary criterion based AddRem and the spectral approach included.

Take a look, for example, at a randomly selected 5 × 2 fragment from matrix X concerning 2 rows from $a1$ and 3 rows from c in the left-hand part of Table 5.10. It is not easy to cluster points 3, 4, 5 together because of the great distances between them.

Table 5.10 Five points from Fig. 5.5, on the left, the affinity similarities between them, in the middle, and Lapin similarities, on the right (those positive are highlighted by bold font)

	x-axis	y-axis	2	3	4	5	2	3	4	5
1	−1.1465	0.3274	0.63	0.00	0.02	0.01	**0.04**	−0.12	−0.14	−0.10
2	0.8956	0.5529		0.00	0.00	0.00		−0.12	−0.15	−0.11
3	0.3086	5.9059			0.00	0.03			**0.16**	**0.40**
4	−5.1827	0.0625				0.01				**0.46**
5	−5.0025	5.8504								

This is reflected in the Gaussian affinity matrix A between 200 rows of the data matrix too, which is defined as described in Sect. 5.1 according to formula $a_{ij} = \exp(-d(x_i, x_j)/s)$ where $d(x, y) = \sum_{v \in V} (x_v - y_v)^2$ is the squared Euclidean distance between vectors x and y. The value of s relates to the denominator of exponent $2\sigma^2$ in the definition of the Gaussian density so that if one takes σ to be half of the range, then s should be about the same, which leads to s = 9 in this case. The part of affinity matrix A related to the set of five points is presented in the middle of Table 5.10. One can see indeed a high affinity value between the first two entities, which belong to the heap in the middle of Fig. 5.5 and are close to each other indeed, while the other similarities are close to zero—no visible structure. A similar pattern can be seen on the Laplacian except that all non-diagonal entries are negative there because of the definition. After the Lapin transformation, however, the similarity structure, once again, becomes clear-cut, as shown on the right part of Table 5.10 for the 5-point subset, and in fact is true for the entire dataset.

This ability of Lapin transformation in transforming elongated structures into convex clusters has been a subject of mathematical scrutiny. An analogy with electricity circuits has been found. Roughly speaking, if w_{ij} measures the conductivity of the wire between nodes i and j in a "linear electricity network", then the corresponding element of a Lapin matrix expresses the "effective resistance" between i and j in the circuit (Klein and Randic 1993). However, there can be cases of elongated structures, as shown in Worked Example 5.2, at which Lapin transformation does not work at all.

Worked Example 5.2. Failure of the Spectral Clustering at Cockroach Network

Lapin matrix for Cockroach network analyzed in Chap. 4, see Fig. 4.28a, is presented in Table 5.11. It manifests a rather clear-cut cluster structure embracing three clusters, {1, 2, 3, 4}, {7, 8, 9, 10}, {5, 6, 11, 12}. Indeed, the positive entries are those within the clusters, except for two positive—but rather small—entries, at (4, 5) and (10, 11).

Yet the first eigenvector reflects none of that; it cuts through by separating six nodes {1, 2, 3, 4, 5, 6} (negative components) from the rest (positive components). This is an example of a situation in which the spectral approach fails: the normalized cut criterion at the partition separating the first 6 nodes from the other 6 nodes is equal to 0.46, whereas its value at cluster {5, 6, 11, 12} cut from the rest is 0.32. The same value of the criterion, 0.32, is attained at cluster {4, 5, 6, 10, 11, 12} cut from the rest. These two cuts are optimal according to the criterion, and the spectral cut is not.

5.2.2 Minimum Cut and Straight Spectral Clustering

It should be mentioned that the matrix spectrum can be used not only for the normalized cut approach but for the summary and semi-average criteria as well.

Table 5.11 Lapin similarity data between nodes of the Cockroach network in Fig. 4.27; the positive entries are highlighted in bold

	2	3	4	5	6	7	8	9	10	11	12
1	**2.43**	**1.18**	**0.05**	-0.52	-0.58	-0.69	-0.92	-0.75	-0.59	-0.69	-0.63
2		**1.75**	**0.16**	-0.64	-0.74	-0.92	-1.22	-0.99	-0.74	-0.88	-0.81
3			**0.44**	-0.36	-0.51	-0.75	-0.99	-0.76	-0.46	-0.60	-0.58
4				**0.14**	-0.15	-0.59	-0.74	-0.46	0.02	-0.16	-0.24
5					**0.68**	-0.69	-0.88	-0.60	-0.16	**0.46**	**0.44**
6						-0.63	-0.81	-0.58	-0.24	**0.44**	**0.94**
7							**2.43**	**1.18**	**0.05**	-0.52	-0.58
8								**1.75**	**0.16**	-0.64	-0.74
9									**0.44**	-0.36	-0.51
10										**0.14**	-0.15
11											**0.68**

Here this is elaborated for the min-cut criterion, that is of minimization of the sum of between-cluster similarities. Indeed, define N-dimensional vector $z = (z_i)$ such that $z_i = 1$ if $i \epsilon S_1$ and $z_i = -1$ if $i \epsilon S_2$. Obviously, $z_i^2 = 1$ for any $i \epsilon I$ so that $z^T z = N$ which is constant at any given entity set I. On the other hand, $z^T A z = A(S_1, S_1) + A(S_2, S_2) - 2A(S_1, S_2) = 2(A(S_1, S_1) + A(S_2, S_2)) - a(I) = a(I) - 4A(S_1, S_2)$, which means that criterion (4.39) is maximized when $z^T A z$ is maximized, that is, the problem of finding a minimum cut is equivalent to the problem of maximization of Rayleigh quotient

$$g(z) = \frac{z^T W z}{z^T z} \tag{5.11}$$

with respect to the unknown N-dimensional z whose components are either 1 or -1. Matrix W is A pre-processed into either B, with subtraction of a threshold, or C, with subtraction of the random interactions (see Sect. 4.6.3.3 for more detail) or using a different transformation.

As is well known, the maximum of (5.11) with respect to arbitrary z is equal to the maximum eigenvalue of W and it is reached at the corresponding eigenvector referred to as the first eigenvector. This brings forth the idea that is referred to as spectral clustering: Find the first eigenvector as the best solution and then approximate it with a $(1, -1)$-vector by putting 1 for positive components and -1 for non-positive components—then produce S_1 as the set of entities corresponding to 1, and S_2, corresponding to -1.

To find the maximum eigenvalue and corresponding eigenvector for a symmetric similarity matrix W, MatLab command [Z,L] = eig(W) should be executed. Resulting L is a diagonal matrix with eigenvalues located on the diagonal in the ascending order, so that the last one is the maximum eigenvalue. Accordingly, the last column is the corresponding, "first", normed eigenvector. Its positive components correspond to one cluster, and the non-positive components to the other. Here is a sequence of commands to determine the split parts S_1 and S_2:

```
≫[Z,L]=eig(W);
≫[n,n]=size(L);
≫z=Z(:,n);
≫S{1}=find(z>0); S{2}=find(z<=0);
```

If W is non-negative then the first eigenvector is proven to be not negative either—no partition of I can emerge in such a situation.

Although not necessarily an optimal partition, this is a practical and, in most cases, good solution.

Unlike the normalized cut approach, no weird data transformation is required here; this is why we refer to this approach as a "straight spectral approach."

Table 5.12 First eigenvectors according to the modularity and uniform data preprocessing options

Modularity				Uniform, current mean subtracted			
Set	0–9	2 3 5 6 8 9 0	2 3 5 9	0–9	2 3 5 6 8 9 0	2 3 5 9	3 5 9
1	−0.57			0.52			
2	0.08	0.06	0.5	−0.08	0.07	0.46	
3	0.07	0.51	0.5	−0.12	0.50	−0.50	0.74
4	−0.27			0.25			
5	0.21	0.22	0.5	−0.24	0.22	−0.29	−0.58
6	0.26	−0.36		−0.29	−0.36		
7	−0.53			0.49			
8	0.34	−0.43		−0.38	−0.44		
9	0.19	0.43	0.5	−0.25	0.43	−0.68	0.35
0	0.21	−0.43		−0.23	−0.44		
λ	703.7	189.6	0.0	355.7	189.6	83.8	−46.4

Worked Example 5.3. Straight Spectral Clusters for Confusion Dataset
Table 5.12 presents results of sequential cuts according to the first eigenvectors on the sets resulting from the previous cuts. As before, the modularity transformation leads to three discernible clusters of numerals, {1, 4, 7}, {6, 8, 0} and {2, 3, 5, 9}, whereas the uniform data transformation, by subtracting the mean of the similarities on the current set, convincingly separates 2 from cluster {2, 3, 5, 9}—the remaining set {3,5,9} cannot be further divided because the maximum eigenvalue at that is negative, thus no positive value of the criterion at the division.

Worked Example 5.4. Straight Spectral Clusters for Cockroach Network
Table 5.13 presents results of the first two cuts according to the spectral clusters derived at the modularity and uniform data transformations. They differ on nodes 4 and 10 at which they are ether merged with the thicker end of the network, at the uniform clustering, or not,—at the modularity clustering. At the second cut, they go to different parts, according to the network topology on Fig. 4.28.

Q.5.14. Consider an agglomerative clustering algorithm, in which the similarity between clusters is the sum of between cluster similarities. Prove that:

(a) this algorithm (locally) maximizes the summary within cluster similarity criterion;

(b) the algorithm stops when all between cluster summary similarities are negative (which will happen if the similarity matrix has been preprocessed with either the modularity or uniform transformation).

Modularity			Uniform, current mean subtracted	
Set	1–12	1–4 7–10	1–12	1–3 7–9
1	0.21	−0.30	−0.22	0.3536
2	0.32	−0.46	−0.26	0.5000
3	0.24	−0.41	−0.10	0.3536
4	0.00	−0.16	0.22	
5	−0.37		0.46	
6	−0.40		0.34	
7	0.21	0.30	−0.22	−0.3536
8	0.32	0.46	−0.26	0.5000
9	0.24	0.41	−0.10	−0.3536
10	0.00	0.16	0.22	
11	−0.37		0.46	
12	−0.40		0.34	
λ	1.71	1.53	1.88	1.41

Worked Example 5.5. Straight Spectral clustering of Affinity Data The spectral
clustering approach is much successful on the affinity data for the Company dataset
—the three clusters corresponding to the three products are recovered well on both
data transformation options, the uniform and the modularity (see Table 5.14). The
uniform version does not divide the B product cluster {4, 5, 6} in smaller parts
because all components of the first eigenvector here have the same sign.

Table 5.14 First eigenvectors according to the modularity and uniform data preprocessing
options at Company affinity data set

Modularity				Uniform, current mean subtracted		
Set	1–8	4–8	4–6	1–8	4–8	4–6
1	0.50			0.53		
2	0.51			0.55		
3	0.32			0.35		
4	−0.13	0.35	0.58	−0.09	0.36	−0.40
5	−0.29	0.25	0.58	−0.23	0.27	−0.62
6	−0.24	0.48	0.58	−0.20	0.49	−0.68
7	−0.38	−0.45		−0.33	−0.43	
8	−0.3	−0.62		−0.29	−0.61	
λ	0.41	0.21	0.00	0.41	0.21	0.01

5.3   Threshold Graph, Connected Component, Single Linkage Clustering, and Maximum Spanning Tree MST

5.3.1   Threshold Graphs

Given a weighted graph with the node set $I = \{1,2,...,N\}$, or, equivalently, a similarity matrix $A = (a_{ij})$, $i,j \epsilon I$, it can be simplified into the so-called threshold graph. Given a real t, a t-threshold graph is defined as a flat graph with nodes in I so that i and j from I are linked if and only if $a_{ij} > t$. In this way, all weights that are less than t are considered as equally insignificant, while those greater than t are considered equally significant. This may seem an over-simplification sometimes—some insist on leaving the significant links in a t-threshold graph as they are, without flattening them into the same unity weight. In this latter case, one encounters a data pre-processing option. This option is powerful but applied quite rarely, probably because of difficulties in substantiation of such a transformation.

Having a flat t-threshold graph, natural graph-theoretic cluster concepts are applicable, first of all, a clique and a component. A clique is a maximal subset of nodes, all of which are mutually linked in the graph. Finding a clique of maximum size is a computationally hard task. A component is a maximal subset of nodes all of which are connected by a path. Finding a component is computationally easy by iteratively joining in all the nodes linked to any already existing member.

Worked Example 5.6. Flat Graphs for Eurovision Song Contest Data and Components

Consider the symmetrized similarity data on the Eurovision song contest (from Table 1.11 in Chap. 1).

Let us take its t-threshold graph at $t = 6$ so that all the weights that are less than 6 are zeroed. It appears, there is only one non-trivial component in this graph (see Fig. 5.6).

The component in Fig. 5.6 shows no separate clusters but rather a transition from East to West.

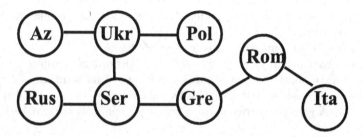

Fig. 5.6 The only non-trivial component of t-threshold graph for the similarity data in Table 5.15, $t = 6$

Q.5.15. Demonstrate that any t-threshold graph for data in Table 5.15 may have only one nontrivial connected component. How many nodes the component contains?

Consider a similarity, rather than dissimilarity, matrix, for a change. All the contents of this section apply to dissimilarity data as well with the only change—of taking maximum for taking minimum.

Weighted graphs, or networks, is a natural way for representing similarity matrices such as those in Tables 1.7 and 1.5. Single link clustering method applies to symmetric matrices, such as that presented in Table 5.16—a symmetric version of the Confusion data Table 1.7.

It is obtained by a most conventional way: given a possibly non-symmetric matrix A, take its transpose A^T and define $\tilde{A} = (A + A^T)/2$. This is a technical way to express the idea that every symmetric pair of non-coinciding entries such as 7 in position (1,3) and 29 in position (3,1) should be substituted by their half-sum:. $36/2 = 18$. To obtain data in Table 5.16, the result was rounded up to the nearest larger integer.

The similarity matrix in Table 5.16 can be represented by a graph whose nodes correspond to the entities $i \epsilon I$ and edge weights to the similarity values. Frequently, a threshold t applies so that only those edges $\{i,j\}$ are put in the graph for which the similarity values are greater than the threshold.

For the threshold $t = 20$, this graph is presented on Fig. 5.7.

Here we present some mathematical properties relating the concepts of Maximum Spanning Tree, connected component of a graph and Nearest Neighbor clustering, which is also referred to as the Single linkage clustering.

5.3.2 Maximum Spanning Trees and Connected Components

In graph theory, a number of concepts have been developed to reflect the structure of weighted graphs of which one of the most popular is the concept of Maximum Spanning Tree (MST). A tree is a graph with no cycles, and a spanning tree is a tree over all the entities under consideration as its nodes. The length of a spanning tree is defined as the sum of weights of all its edges. An MST is a spanning tree whose length is maximum.

Let us first recall some definitions from graph theory.

A weighted (similarity) graph $\Gamma = (I,G,A)$ is defined as a triplet of: (i) an N-element set of nodes I; (ii) set of edges, that is, two-element subsets of I, G; and (iii) edge weight function represented by a symmetric matrix $A = (a_{ij})$ so that $a_{ij} = 0$ if $\{i,j\} \notin G$. A graph is referred to as an ordinary graph if its nonzero weights are all unities.

A path between nodes i and j in Γ is a sequence of nodes $i_1, i_2, ..., i_n$ such that $\{i_m, i_{m+1}\} \epsilon G$ for each $m = 1, 2, ..., n-1$ and $i_1 = i$, $i_n = j$. A path is referred to as a

Table 5.15 The symmetrized version of the average scores by European countries from Table 1.11

	Az	Be	Bu	Es	Fr	Ge	Gr	Is	It	Ne	Pol	Por	Ro	Ru	Se	Sp	Sw	Ukr	UK
Azerbaijan		1.9	3.4	2.0			5.8	4.9		2.0	4.2		5.1	5.2				10.0	
Belgium	1.9				1.8	2.0	2.0			4.6	2.2	1.8				1.7			2.1
Bulgaria	3.4				2.2		5.6		5.0				2.4	5.6	3.9			2.2	
Estonia	2.0										2.0	2.0		4.4				2.2	1.8
France		1.8	2.2			1.7	2.0	2.8	5.0		2.7			4.0		2.2			2.0
Germany		2.0			1.7		1.8	1.8		1.9	2.8	2.2			3.5	2.6			4.0
Greece	5.8	2.0	5.6		2.0	1.8					6.1	1.8	7.6	4.4				1.9	1.9
Israel	4.9				2.8	1.8				4.4			2.0	3.3	5.5	2.5	2.1	3.1	2.2
Italy			5.0		5.0						4.5		9.0		3.7	2.4		3.2	2.6
Netherland	2.0	4.6				1.9		4.4						3.5					
Poland	4.2	2.2		2.0	2.7	2.8	6.1		4.5									7.1	
Portugal		1.8		2.0		2.2	1.8						2.6		2.8	2.4		3.7	2.2
Romania	5.1		2.4				7.6	2.0	9.0			2.6			4.0	4.0			1.8
Russia	5.2	5.6		4.4	4.0		4.4	3.3		3.5					6.2		5.8	6.7	5.8
Serbia			3.9			3.5		5.5	3.7			2.8	4.0	6.2			5.3	6.7	1.8
Spain		1.7			2.2	2.6		2.5	2.4			2.4	4.0				2.0	2.3	2.0
Switzerland								2.1						5.8	5.3	2.0			
Ukraine	10.0		2.2	2.2			1.9	3.1	3.2		7.1	3.7		6.7	6.7	2.3			2.0
UK		2.1		1.8	2.0	4.0	1.9	2.2	2.6			2.2	1.8	5.8	1.8	2.0		2.0	

Table 5.16 A symmetric version of Confusion data in Table 1.8

Stimulus	Response									
	1	2	3	4	5	6	7	8	9	0
1	877	11	18	86	9	20	165	6	15	11
2	11	782	38	13	31	31	9	29	18	11
3	18	38	681	6	31	4	31	29	132	11
4	86	13	6	732	9	11	26	13	44	6
5	9	31	31	9	669	88	7	13	104	11
6	20	31	4	11	88	633	2	113	11	31
7	165	9	31	26	7	2	667	6	13	16
8	6	29	29	13	13	113	6	577	75	122
9	15	18	132	44	104	11	13	75	550	32
0	11	11	11	6	11	31	16	122	32	818

Fig. 5.7 Threshold graph of connections corresponding to similarity weights of 21 or greater in matrix of Table 5.16

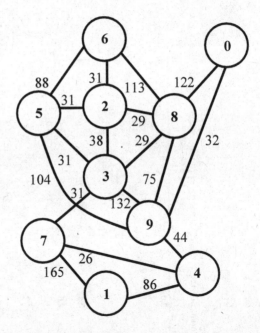

cycle if $i_1 = i_n$. A subset of nodes S is referred to as a connected component if there is a path within S between each pair of nodes in S, and S is maximal in this sense so that addition of any supplementary node to S breaks the property. Graph Γ is called connected if it consists of just one connected component.

Given a connected weighted graph $\Gamma = (I, G, A)$, a connected weighted graph $T = (J, H, B)$, with no cycles, is referred to as its spanning tree if $J = I$, $H \subset G$, and B is A restricted to H, so that $b_{ij} = a_{ij}$ for $\{i,j\} \in H$ and $b_{ij} = 0$ for $\{i,j\} \notin H$. A characteristic property of a spanning tree T is that it has exactly $N - 1$ edges:

if there are more edges than that, T must contain a cycle, and if there are less edges than that, T cannot span the entire set I and, therefore, it would consist of several connected components.

The weight of a spanning tree $T = (I,H,B)$ is defined as the total weight of its edges, that is, the sum of all elements of weight matrix B. A spanning tree of maximum weight is referred to as a Maximum Spanning Tree, MST.

Given a weighted graph $\Gamma = (I,G,A)$ and a real t, an ordinary graph $\Gamma_t = (I,G_t,A_t)$ is referred to as a threshold graph if $G_t = \{\{i,j\}: a_{ij} > t\}$. Given a spanning tree T and a threshold t, its threshold graph can be found by cutting out those edges whose weights are smaller than or equal to t. It appears T bears a lot of structural information of the corresponding graph $\Gamma = (I,G,A)$.

In particular, connected components of an MST found by cutting those links from MST that are less than t one-to-one correspond to connected components of the threshold graph Γ_t.

Indeed, consider a component S of MST T obtained by cutting all edges of the T whose weights are less than t. We need to prove that for each pair $i,j \in S$ there is a path between i and j such that it all belongs to S, so that the weight of each edge in the path is greater than t, and, moreover, for all i,k such that $i \in S$ and $k \notin S$, if $\{i,k\} \in G$, then $a_{ik} \leq t$. But the former obviously follows from the fact that S is a connected component of T in which all weights are greater than t since the others have been cut out. The latter is not difficult to prove either: assumption that $a_{ik} > t$ for some $i \in S$ and $k \notin s$ such that $\{i,k\} \in G$ would contradict the assumption that T is an MST, that is, that the weight of T is maximal, because by substituting the edge connecting S and the component containing k by edge $\{i,k\}$, one would obtain a spanning tree of a greater weight. Assume now that an S is a connected component of the threshold graph (at threshold t), and prove that S is a component of the threshold graph, at the same threshold t, for any MST. Indeed, if S overlaps two components, $S1$ and $S2$, of the threshold graph of some MST T, then there must be a pair i,j in S such that $i \in S1$ and $j \in S2$ and $a_{ij} > t$, which again contradicts the fact that $S1$ and $S2$ are not connected in the threshold graph of T. This completes the proof.

Building a Maximum Spanning Tree

To find an MST, several "greedy" approaches can be undertaken. One of them, by Kruskal (1956), finds an MST by picking up edges; the other, by Prim (1957), picks up nodes. Prim's algorithm builds an MST T from an arbitrary node by finding the weakest link to the tree from outside and adding it to tree at each step. An exact formulation is this.

Prim's algorithm

1. **Initialization.**
   Start with tree T consisting of an arbitrary node $i \in I$ with no edges.
2. **Tree update.**
   Find $j \in I\text{-}T$ maxiimizing a_{ij} over all $i \in T$ and $j \in I - T$. Add j and edge $\{i,j\}$ with the maximal a_{ij} to T.

3. **Stop-condition.**
    If $I{-}T = \varnothing$, halt and output tree T. Otherwise, go to 2.

To build a computationally effective procedure for the algorithm may be a cumbersome issue, depending on how maxima are found, to which a lot of work has been devoted. A simple pre-processing step can be quite useful: in the beginning, find a nearest neighbor for each of the entities; only they may go to MST. At each step, update the neighbors of all elements in $I{-}T$ so that they lead to elements of T (Murtagh 1985). The claim that the algorithm builds an MST indeed can be proven using inductive statement that T at each step is part of an MST.

Worked Example 5.7. Concept of MST
Consider graph of Fig. 5.7. Figure 5.8 highlights two of its spanning trees. The length of that on the left is $165\{1{-}7\} + 31\{7{-}3\} + 44\{4{-}9\} + 132\{9{-}3\} + 31\{3{-}5\} + 38\{3{-}2\} + 29\{3{-}8\} + 31\{2{-}6\} + 122\{8{-}0\} = 623$; here, curly braces' contents correspond to the edges in the tree. The length of that on the right is $86\{1{-}4\} + 165\{1{-}7\} + 44\{4{-}9\} + 132\{3{-}9\} + 104\{9{-}5\} + 88\{5{-}6\} + 113\{6{-}8\} + 38\{3{-}2\} + 122\{8{-}0\} = 892$, which is much greater. In fact the latter is a Maximum Spanning Tree.

Given a weighted graph, or similarity matrix, an MST T can be built by using Prim's algorithm which collects T step by step starting from a singleton node, in fact any of the nodes, and then adding a maximum outside link to T one by one.

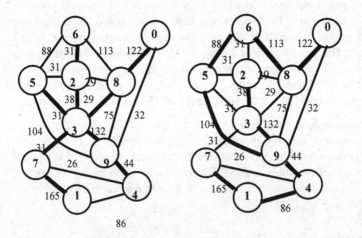

Fig. 5.8 Two spanning trees on the graph of Fig. 5.7 are highlighted by bold edges. The length of the tree on the left is 623, and that on the right, 892. Which one is an MST?

Worked Example 5.8. Building an MST on Confusion Data

Let us build a Maximum Spanning Tree for the network in Fig. 5.7 by using Prim's algorithm. Start, for example, with $T = \{0\}$ and add to T that link which is maximum, that is, obviously $122\{0-8\}$. Since T has two nodes now, we check external links of each of them to find a maximum external link from T to the rest, $113\{8-6\}$, thus getting three nodes, 0, 8, 6 and two links, $\{0-8\}$ and $\{8-6\}$, in T. The maximum external link now is $88\{6-5\}$ bringing 5 and link $\{6-5\}$ into T. Next maximum links are $104\{5-9\}$, $132\{9-3\}$ and $44\{9-4\}$ bringing 9, 3 and 4, respectively, into T. Of the three remaining nodes outside T, 2, 7 and 1, the maximum link is 86 $\{4-1\}$ followed by $165\{1-7\}$. Node 2's maximum connection is $38\{2-3\}$ thus completing the MST drawn on the right hand side of Fig. 5.8.

Prim's algorithm is what is called greedy—it works node-by-node and picks up the best solution at the given step, paying no attention to what happens next. The MST problem is one of a very few combinatorial problems that can be solved indeed by a greedy algorithm. On the other hand, one should not be overly optimistic about performances of the algorithm because it finds, at each step a maximum of a number of elements, on average—half the number of entities, and one should not forget that finding maximum is a rather expensive operation.

Another potential drawback, related to the weight data size, which is quadratic over the number of entities, is not that bad. Specifically, if the similarities are computed from data in the entity-to-feature format, the difference between the data sizes can grow fast indeed: say 500 entities over 5 features take about 2500 numbers, whereas the corresponding similarity matrix will have about 250,000 numbers—a hundred times greater. Yet if the number of entities grows 20 times to 10,000—a rather modest size nowadays—the raw data table will take 50,000 numbers whereas the similarity matrix, of the order of 100,000,000, a hundred million, which is two thousand times greater! Yet it is possible to organize the computation of an MST by storing and updating lists of nearest neighbors in the network in such a manner that the quadratic size increase is not necessary, because almost all necessary similarities can be calculated from the raw data when needed.

5.3.3 Single Link Hierarchical Clustering by Using MST

Single link clustering is, primarily, a hierarchical clustering method in which the similarity between two clusters, S_1 and S_2, is defined according to nearest neighbor rule as the maximum similarity between elements of these clusters, $a(S_1, S_2) = \max_{i \in S_1, j \in S_2} a_{ij}$—the fact that the between-cluster similarity is defined by just one link underlies the name of the method.

In an agglomerative process, there is no need to revise all the maximum similarities after every merger step. New similarities can be revised dynamically in the
agglomeration process according to the following rule:

$$a(S, S_1 \cup S_2) = \max[a(S, S_1), a(S, S_2)],$$

where $S_1 \cup S_2$ is the result of the agglomeration step.

Thus, the only intensive computation is finding the maximum in the newly
formed column $a(S, S_1 \cup S_2)$ of the similarity matrix over all current clusters S.

There is another way to proceed, however—by building an MST first. All the
merger steps can be made then according to the MST topology. First, the $N-1$
edges of the tree are to be sorted in the descending order. Then the following
recursive steps apply. On the first step, take any maximum similarity edge $\{i,j\}$,
combine its nodes into a cluster and merge i and j by removing the edge. On the
general step, take any remaining similarity edge $\{i,j\}$ of the maximum similarity
value (among those left) and combine clusters containing i and j nodes into a
merged cluster. Halt, when no edges remain in the sorted order.

This operation is legitimate because the following property holds: clusters found
in the process of mergers according to the sorted list of MST edges are clusters
obtained in the agglomerative Single Link clustering procedure.

There is no straightforward divisive version of the Single Link method as
originally defined. However, it is rather easy to do if an MST is built first. Then
cutting the tree over any of its weakest, that is, minimum, links produces the first
single link division. Each of the split parts is divided in the same way—by cutting
out one of the weakest links.

**Worked Example 5.9. A Divisive MST based Single Link hierarchy for
Confusion data**

Let us use the MST found in Worked Example 5.8 for Single Link divisive clustering. We start from cutting the MST in two parts over the weakest link, in this
case, edge linking 2 and 3 with the weight 38. This division separates digit 2 from
the rest. Next weakest link is 44 between 4 and 9, thus cutting cluster 1–4–7 out.
Next divisions can be followed by the dendrogram on the right in Fig. 5.9: it shows
not only cuts but, also, the cut link values (in the bottom of the cluster boxes),
which are used as the height function, as well. Since they monotonically increase
from the root, not decrease, the height axis is turned upside down.

Q.5.16. Let us refer to a similarity matrix A as ultra-similarity if it satisfies the
following property: for any triplet $i,j,k \in I$, $a_{ij} \geq min\ (a_{ik}, a_{kj})$, that is, two of the
values a_{ij}, a_{ik}, a_{kj} are equal, whereas the third one may be greater than that. Given

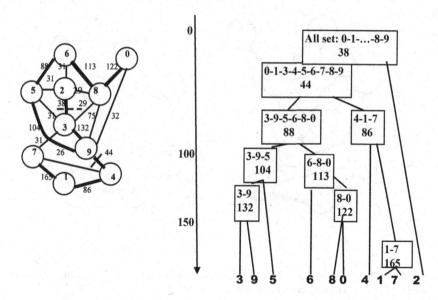

Fig. 5.9 An MST for the graph of Fig. 5.7 is on the left, and, on the right, a dendrogram for the results of a Single Link divisive clustering procedure by sequentially cutting the weakest links. The node heights correspond to the split values

an MST T, define a new similarity measure between any nodes i and j by using the unique path $T(i,j)$ connecting them in T: $at(i,j) = min_{k,l \in T(i,j)} a_{kl}$. Prove that:

(i)   Similarity $at(i,j)$ coincides with that defined by the agglomerative hierarchy built according to the Single Link algorithm;

(ii) Similarity $at(i,j)$ is an ultra-similarity;

(iii) Similarity $at(i,j)$ is the minimum ultra-similarity satisfying condition $at(i, j) \geq a_{ij}$ for all $i,j \in I$. (Hint: use similar properties of ultrametric.)

Worked Example 5.10. MST and Connected Components

Let us sort all the edges in MST found at the graph on Fig. 5.8 in the ascending order: 38{3–2}, 44{4–9}, 86{1–4}, 88{5–6}, 104{9–5}, 113{6–8}, 122{8–0}, 132 {9–3}, 165{1–7}. Given a threshold t, say $t = 50$, cut the 2 edges in the tree that are less than the threshold, 3–2 and 4–9—the tree will be partitioned in $2 + 1 = 3$ fragments corresponding to connected components of the corresponding threshold graph (see the cuts on the left in Fig. 5.9).

Figure 5.10 presents, on the left, a threshold graph at $t = 50$, along with clearly seen components consisting of subsets {1, 4, 7}, {2}, and {3, 5, 6, 8, 9, 0}. The same subsets are seen on the right where the two weakest links are cut out of the MST. The fact that the threshold graph components are MST fragments is not a coincidence but rather a mathematically proven property of the MSTs.

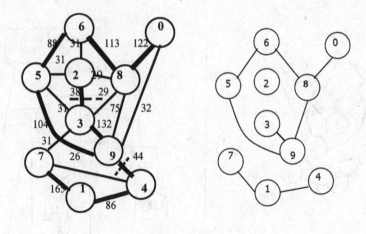

Fig. 5.10 Threshold graph at t = 50 for the graph of Fig. 5.7 is on the right, and the MST with 2 weakest links shown using crossing dashed lines is on the left

Worked Example 5.11. MST and Single Linkage Clusters for Company Dataset

This example illustrates a property of the MST based clustering. The results depend on the link between just two objects, which can be at odds with the data structure sometimes. Let us consider a Minimum Spanning Tree built on distances between Companies in Table 5.1. Starting from Av, we add minimum link An(0.51)Av to tree T being built, then we add to T the minimum distance link As(0.77)An. Then the minimum distance is 1.15 between Ba and Av, which brings next links Bu(0.87) Ba, Br(0.75)Bu, followed by Ci(0.83)Br and Cy(0.61)Ci. (Note that all row-wise minimum distances highlighted on Table 5.1 have been brought in the tree T. These minimum distances can be used, in fact, in a different method for building an MST —Boruvka's algorithm (1926), arguably the very first clustering method!). This MST T, which is in fact a path, is presented on the left side of Fig. 5.11.

This path goes along the product clusters so that B companies are all between A and C companies. Yet the single link clusters shown on the right side of Fig. 5.11 do not reflect the structure of the set but separate a distant company Ba and mix together products B and C—all this just because the right-to-remove link Br-Ci (0.83) appears a bit smaller than wrong-to-remove link Ba-Bu (0.87).

Case-Study 5.4. Difference Between K-means and Single Link Clustering

Consider a set of 2D points presented on Fig. 5.12. Those on the left have been clustered by using the single link approach, whereas those on the right, by using the square error criterion of K-means.

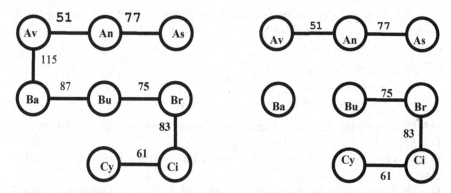

Fig. 5.11 Minimum Spanning Tree for company dataset in Table 5.1: the structure of company products is reflected on the tree and lost on the clusters because of wrong cuts. (For the sake of convenience, the distances are multiplied by 100.)

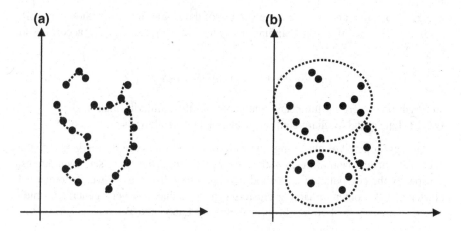

Fig. 5.12 A 2D point set: clustered with the single link method (**a**) and K-means (**b**)

Overall, this example demonstrates the major difference between conventional clustering and single link clustering: the latter finds elongated structures whereas the former cuts out convex parts. Sometimes, especially in the analysis of results of physical processes or experiments over real-world particles, the elongated structures do capture the essence of the data and are of great interest. In other cases, especially when entities/features have no intuitive geometric meaning—think of bank customers or internet users, for example, convex clusters make much more sense as groupings around their centers.

An interesting property of the single linkage method is that it involves just $N-1$ similarity entries occurring in an MST rather than all $N(N-1)/2$ entries in A. This results in a threefold effect:

(1) a nice mathematical theory,
(2) fast computations, and
(3) poor application capability.

Q.5.17. Explain the sequence of splits in a divisive algorithm according to tree of Fig. 5.9.

Q.5.18. How many edges in an MST are to be cut if the user wants to find 5 clusters?

Q.5.19. Prove that the number of edges in an MST is always the number of entities short one.

Q.5.20. Apply Prim's algorithm to Amino acid similarity data in Table 1.9.

Q.5.21. Prove that the MST would not change if the similarities are transformed with a monotone transformation, that is, a function $\varphi(x)$ such that $\varphi(x1) > \varphi(x2)$ if $x1 > x2$. Hint: Because the sequence of events in Prim's algorithm does not change.

Q.5.22. Prove that an agglomerative version of the single linkage method can work recursively by modifying the similarities, after every merger $S_1 \cup S_2$, according to formula

$$a(S, S_1 \cup S_2) = max[a(S, S_1), a(S, S_2)],$$

and each time merging the nearest neighbors in the similarity matrix.

Q.5.23. Lance and Williams family of agglomerative algorithms.

An agglomerative clustering algorithm works one-by-one mergers of the "nearest" clusters beginning from the trivial partition of the entity set in singletons. To specify the computation, a method for computing distances between the merged cluster and the other remaining clusters can be defined. A very general formula covering many a method has been proposed by Lance and Williams (1967):

$$d(k, f \cup g) = \alpha_f d(k, f) + \alpha_g d(k, g) + \beta d(f, g) \\ + \gamma |d(k, f) - d(k, g)|. \tag{5.12}$$

Values of the coefficients in (5.12) for some popular methods are presented in Table 5.17. One of the methods, popular in bioinformatics, is referred to as UPGMA (Unweighted Pair Group Method with Arithmetic Means): the dissimilarity between two clusters is defined as the average distance between all entities of the two. Prove that coefficients in Table 5.17 are correct.

Table 5.17 Lance-Williams coefficients for some popular agglomerative clustering methods

Method	α_f	α_g	β	γ
Single linkage	½	½	0	$-½$
Complete linkage	½	½	0	½
UPGMA	$N_f/(N_f+N_g)$	$N_g/(N_f+N_g)$	0	0
Ward	$(N_f+N_k)/$ $(N_f+N_g+N_k)$	$(N_g+N_k)/$ $(N_f+N_g+N_k)$	$(N_f-N_k)/$ $(N_f+N_g+N_k)$	0

5.4 Separate Clusters

Separation of one cluster from the rest is akin to divisive clustering; just criteria differ, In divisive clustering, the two parts obtained have equal "statuses" as clusters. In separate clustering, just one cluster is sought, whatever relations between other objects are. This is a frequent situation in data analysis when the user is interested at separating just one cluster from the rest, be it a cluster of endemic territories or a subset of would-be fraudsters. We already encountered this when considering Anomalous clusters at K-means. This is why the case of entity-to-feature data will be considered in brief only. Most attention will be given to the case of similarity data.

5.4.1 Anomalous Cluster

Separation of a cluster S from the rest may follow the model considered in Chap. 4 (see Eqs. (4.18)). Specifically, a cluster is defined by its center $c = (c_v)$ so that every object in S coincides with c, up to errors that are to be minimized. A decoder-based model is this:

$$y_{iv} = \begin{cases} c_v + e_v, & i \in S \\ 0 + e_v, & i \notin S \end{cases} \tag{5.13}$$

The least squares approach requires to minimize the summary squared error, equivalently represented as

$$W(S,c) = \sum_{i \in S} d(y_i, c) + \sum_{i \notin S} d(y_i, 0) \tag{5.14}$$

where S is the cluster to be found and c its center. This criterion much reminds the K-means criterion at K = 2, except that here one of the centers, 0, is unvaried. Obviously, minimizing (5.14) requires the cluster S' center, c, to be as far away from 0 as possible. An appropriate version of K-means can be formulated as follows.

Anomalous Cluster (AC) algorithm

1. *Pre-processing.* Specify a reference point $a = (a_1, ..., a_V)$ (when in doubt, take a to be the data grand mean point) and standardize the original data table by shifting the origin to $a = (a_1, ..., a_V)$—by subtracting a from every object point vector.
2. *Initial setting.* Put a tentative center, c, as the entity farthest away from the origin, 0.
3. *Cluster update.* Determine cluster list S around c against 0, so that entity y_i is assigned to S if $d(y_i, c) < d(y_i, 0)$.
4. *Center update.* Calculate the within S mean c' and check whether c' differs from the previous center c. If c' and c do differ (up to a pre-specified precision error), update the center by assigning $c \Leftarrow c'$ and go to Step 3. Otherwise, go to 5.
5. *Output.* Output list S and center c, with accompanying interpretation aids (as advised in Chap. 4).

Examples of application of the Anomalous clustering were given in Chap, 4. Here, we are just going to illustrate the difference between the separation of the Anomalous cluster and Ward division method.

Case-Study 5.5. Anomalous Cluster Versus K-means Bisecting
Consider 280 values generated according to the one-dimensional Gaussian distribution N(0,10) with zero mean and standard deviation equal to 10 (see Table A.6), presented on Fig. 5.13 and try divide it in two clusters. When it is done with a splitting criterion, the division goes just over the middle, cutting the bell-shaped curve in two equal halves. When one takes Anomalous clusters, though, the divisions are much different: first goes a quarter of the entities on the right (to the right of point A on Fig. 5.13), because the right end in this individual sample is a bit farther from the mean than the left one; then a similar chunk to the left of point B, etc. Yet if one uses the anomalous clusters in iK-means, just as an initial centers generator, things differ. With the discarding threshold of 60, only two major Anomalous patterns found in the beginning remain. Further 2-means iterations bring a rather symmetric solution reflected in the leftmost part of Table 5.18. The right-hand part of the table shows results found with an incremental version of Bisecting K-means method.

This leads to a slightly different partition, with three entities swapping their membership, which is slightly better, achieving 65.4% of the explained scatter

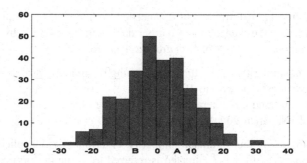

Fig. 5.13 Histogram of the one dimensional sample of 280 entities from N(0,10) distribution. Points A and B denote the boundaries of the right and left anomalous fragments found with the Anomalous pattern algorithm

Table 5.18 Two-class partitions found using different strategies

Cluster	iK-means with t = 60			Incremental Bisecting K-means		
	Size	Mean	Explained, %	Size	Mean	Explained, %
1	139	5.63	32.6	136	5.82	26.8
2	141	−9.30	32.1	144	−9.12	35.6

A better result by the incremental Bisecting K-means should be attributed to its one-by-one entity moving procedure

versus 64.8% at iK-means. This once again demonstrates that the incrementally processing individual entities is a more precise option than the all-as-one switching in iK-means.

5.4.2 One-Cluster Summary Criterion and Its Properties

Given a cluster S, its within-cluster similarities can be characterized by the summary value

$$A(S, S) = \sum_{i,j \in S} a_{ij}. \tag{5.15}$$

Obviously, the greater the sum (5.15), the better the cluster S.

Given a non-negative matrix A, the maximum of $A(S, S)$ is obviously reached at the universal cluster $S = I$, because then the sum (5.15) is the greatest possible. Provided that all rows/columns have at least one positive entry, $S = I$ is the only maximizer of (5.15). Does it mean that the summary criterion should be discarded as leading to no nontrivial clusters as is conventionally suggested?

Not at all! Just a transformation of A is needed to sharpen the portrait of a cluster structure hidden in the data, A most natural data pre-processing is by subtracting some background. Two types of background data are:

(i)   a constant similarity level π that has meaning of similarity scale shift or "soft" similarity threshold (Mirkin 1987)
(ii) an effect of "random" interactions based on the relative "strength" of entities involved (Newman 2006).

The maximum of the summary similarity criterion (5.15) applied to matrix A after a similarity shift π is subtracted is referred to as the uniform criterion:

$$f(S, \pi) = \sum_{i,j \in S} (a_{ij} - \pi) \tag{5.16}$$

Obviously, this criterion is the same as $A(S, S)$ (5.15) applied to matrix $A' = (a_{ij}')$ with $a_{ij}' = a_{ij} - \pi$.

As mentioned in Chap. 4, pair $\{i, j\}$ should be put in cluster S if $a_{ij} > \pi$, and rather not if $a_{ij} < \pi$. The value of threshold π can be defined using external information (see, for example, Mirkin et al. 2010).

The background similarity in the case (ii) needs no external information. In this approach, matrix A is treated as a contingency table. Consider the summary values $a_{i+} = \sum_{j \in I} a_{ij}$, and $a_{++} = \sum_{i,j \in I} a_{ij}$. Under the assumption that there is random interaction between entities i and j, which is proportional to these summary values, the background similarity is defined as the product $k_{ij} = a_{i+}a_{j+}/a_{++}$; the denominator is added to return the product to the original scaling of similarities in A. The within-cluster summary similarity criterion (5.15) applied to matrix A after the "background" similarity is subtracted is the modularity criterion extended to one-cluster case:

$$m(S) = \sum_{i,j \in S} (a_{ij} - k_{ij}) = \sum_{i,j \in S} (a_{ij} - a_{i+}a_{j+}/a_{++}) \tag{5.17}$$

Obviously, this criterion is the same as $A''(S, S)$ (5.15) applied to matrix $A'' = (a_{ij}'')$ where $a_{ij}'' = a_{ij} - a_{i+}a_{j+}/a_{++}$.

Let us briefly analyze some properties of these versions of the summary criterion. For the case of $f(S, \pi)$ in (5.16), let us focus on the case when the diagonal entries are not considered so that $i \neq j$ in (5.15):

$$f(S, \pi) = \sum_{\substack{i,j \in S \\ i \neq j}} (a_{ij} - \pi) = \sum_{\substack{i,j \in S \\ i \neq j}} a_{ij} - \pi |S|(|S| - 1) \tag{5.18}$$

where $|S|$ denotes the number of elements in S. When the diagonal elements are present, the right-hand item in (5.18) would be $\pi |S|^2$ rather than $\pi |S|(|S| - 1)$.

An irregular structure of similarities may prevent the threshold π to be a separator between all within-cluster and out-of-cluster similarities, but it certainly works on average. Indeed, let us denote the average similarity between $i{\in}I$ and $S{\subseteq}I$ by $a(i,S)$—this may be referred to as the uniform attraction of i to S. Obviously, $a(i,S) = \Sigma_{j\in S}a_{ij}/(|S| - 1)$ if $i{\in}s$, and $a(i,S) = S_{j\in S}a_{ij}/|S|$ if $i \notin S$, because of the assumption that the diagonal similarities a_{ii} are not considered.

Let us refer to cluster S as uniformly π-tight if, for any entity $i{\in}I$, its uniform attraction to S is greater than or equal to π if $i{\in}S$ and it is less than π, otherwise. Then the following statement, in support of the claim that an optimal cluster S should be tight, holds.

If S maximizes criterion $f(S, \pi)$ in (5.18) then S as uniformly π-tight, that is, $a(i,S) \geq \pi$ for all $i{\in}S$, and $a(i,S) \leq \pi$ for all $i \notin S$.

To prove it, let us change the state of an entity i^* with respect to cluster S, that is, add i^* to S if it does not belong to S or remove it from S if it does. Now take the difference between $f(S, \pi)$ and the result of the state change, that is, $f(S - i^*,\pi)$ if $i^*{\in}S$, or $f(S + i^*, \pi)$ if $i^* \notin S$ where $S - i^*$ and $S + i^*$ denote S with i^* removed or added, respectively:

$$
\begin{aligned}
f(S, \pi) - f(S - i*, \pi) &= 2\big(\Sigma_{j\in S}a_{i*j} - \pi(|S| - 1)\big), \\
f(S, \pi) - f(S + i*, \pi) &= 2\big(-\Sigma_{j\in S}a_{i*j} + \pi(|S|)\big),
\end{aligned}
\tag{5.19}
$$

Equations (5.19) are rather obvious if one consults Fig. 5.14: all the differences between $f(S, \pi)$ and its value after the change of state of i^* come from the boxed fragments of i^*th row and i^*th column. Since S is assumed to be optimal, both of the differences in (5.19) are to be non-negative. Take, for example, the case $i^* \notin S$. Then $-\Sigma_{j\in S}a_{i*j} + \pi|S| \geq 0$. This implies $\pi \geq \Sigma_{j\in S}a_{i*j}/|S| = a(i*, S)$—the statement is proven for $i^* \notin S$. The case of $i^* \in S$ is treated similarly, which completes the proof.

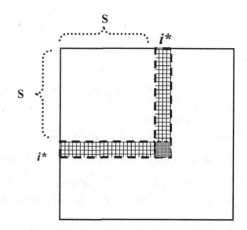

Fig. 5.14 A schematic representation of the similarities with respect to cluster S under the assumption that entities are sorted so that elements of S are followed by i^* followed by the rest; then entries related to entity i^* are in the boxed row and column on S's margin

In fact, a wider statement is proven. Let us refer to S as being locally optimal if, for any entity $i \in I$, $f(S, \pi)$ does not decrease under the change of i's state with respect to S. The proof warrants that any locally optimal cluster is uniformly π-tight.

A similar statement can be proven for the modularity criterion $m(S)$ in (5.17). For the sake of simplicity, assume that the diagonal entries are all zeros. Let us consider the summary similarity of S with all objects,

$$a(S) = \sum_{i \in S} \sum_{j \in I} a_{ij} = \sum_{i \in S} a_{i+}$$

and refer to it as the volume of S. In particular, an entity i's volume will be $a(i) = a_{i+}$ and the universal cluster I's volume, $a(I) = A(I,I) = a_{++}$. Then the modularity criterion (5.17) can be rewritten as

$$m(S) = a(S, S) - a(S)^2 / a(I). \tag{5.17'}$$

Let us introduce the modularity attraction of an entity $i \in I$ to S, $m(i, S) = S_{i \in S} a_{ij} / a_{i+}$, and the relative volume of S in I, $v(S) = a(S)/a(I)$. Then the relative volume of an entity i would be $v(i) = a_{i+}/a_{++}$. Let us refer to a cluster S as being modularity tight if, for any entity $i \in I$, its modularity attraction to S is greater than or equal to the relative volume of S, minus a half of $v(i)$, if $i \in S$, and it is less than the relative volume of S plus a half of $v(i)$, otherwise. That is, S is modularity tight if $m(i, S) \geq a(S)/a(I) - v(i)/2$ for all $i \in S$, and $m(i, S) \leq a(S)/a(I) + v(i)/2$ for all $i \notin S$. Then the following statement is true. If S is a local maximizer of criterion $m(S)$ in (5.17) then S is modularity tight.

To prove the statement, let us take $i*$ and change its state with respect to S. Then the increment of criterion $m(S)$ expressed in terms of $c_{ij} = a_{ij} - a_{i+}a_{j+}/a_{++}$ will be equal to

$$m(S \pm i^*) - m(S) = \pm 2 \sum_{j \in S} c_{i*j} + c_{i*i*}$$

$$= \pm 2 a_{i*+} \left(\sum_{j \in S} a_{i*j}/a_{i*+} - a(S)/a(I) \right) \mp a_{i*+} / 2a(I) \right).$$

That is,

$$m(S \pm i^*) - m(S) = \pm 2 a_{i*+} (m(i^*, S) - v(S) \mp v(i^*)/2). \tag{5.20}$$

The proof follows from (5.20) and the fact that the increment must be non-positive at a locally optimal S.

Local algorithms for one-cluster summary criterion

At a preprocessed, by subtracting background similarities, similarity matrix $A = (a_{ij})$ the summary criterion is rather easy to (locally) optimize by adding entities one-by-one starting, say, from a most linked couple i and j, and at each step adding just one entity $i*$—that one which is most similar to S. The computation stops when the summary similarity stops increasing, which will be the case if many of A entries are negative. There are two issues about this algorithm:

- Starting configuration—the pair of entities of the maximum similarity. This may not work in some cases such as the case of a flat graph matrix in which all nonzero entries are the same. Also, this choice is not flexible and may lead to a clearly suboptimal cluster and missing larger subsets whose elements are well connected but with similarity levels slightly smaller than the maximum.
- Addition with no removals. This can be of an issue because at a later stage of collecting a cluster some entities, picked up in the very beginning, can be far away from the later arrivals and should be removed at later stages.

The following algorithm AddRem tackles both of these as follows. To not get stuck in a wrong place, it runs as many times as there are entities, each time starting from another singleton $S = \{i\}$. To have an opportunity to remove a wrong element, at each step the algorithm considers the increment of the criterion caused by the change of state of every entity with respect to the current cluster. To do so, N-dimensional $1/-1$ vector $z = (z_i)$ is maintained such that $z_i = 1$ if i belongs to the current cluster and $z_i = -1$ if not. Then the change of state of $i \in I$ with respect to the cluster is equivalent to changing the sign of z_i. The change of the criterion value because of this can be expressed as follows. Denote by z the vector at current cluster S and by $z(i)$ the result of change of sign of z_i in it, so that $S(i) = S - i$ if $z_i = 1$ and $S(i) = S + i$ if $z_i = -1$. This makes the operations of addition or removal of an entity to or from the current cluster computationally similar. The increment of the summary criterion after the change is equal to

$$\Delta(i) = -2z_i \sum_{j \in S} a_{ij} + \delta a_{ii} \qquad (5.21)$$

where $\delta = 1$ if the diagonal entries are taken into account and $\delta = 0$, otherwise.

AddRem(k) algorithm

Input: matrix $A = (a_{ij})$; output: cluster S related to entity k and value of the summary criterion.

1. *Initialization.* Set N-dimensional z to have all its entries equal to -1 except for $z_k = 1$, the summary similarity equal to δa_{ii}.
2. *General step.* For each entity $i \in I$, compute the value $\Delta(i)$ according to (5.21) and find $i*$ maximizing it.
3. *Test.* If $\Delta(i*) > 0$, change the sign of z_{i*} in vector z, $z_{i*} \Leftarrow -z_{i*}$, after which recalculate the sum by adding $\Delta(i)$ to it. (In the case of large data, computing the summary values in (5.21) can be costly. Therefore, a vector of these values should be maintained and dynamically changed after each addition/removal step.), and go to 2. Otherwise, go to 4.

4. *Output*. S and the summary criterion value.

A tightness property of the resulting cluster S depending on the pre-processing step, at any starting $i \in I$, holds as established in above in this section because S is locally optimal.

Algorithm AddRem(k) utilizes no ad hoc parameters, except for the k of course, so the cluster sizes are determined by the process of clustering itself. Multiple runs of AddRem(k) at different starting points k allow to (a) find a better cluster S maximizing the summary similarity criterion over the runs, and (b) explore the cluster structure of the dataset by analyzing both differing and overlapping clusters.

Let us consider now application of this method, under each of the two background removal options—modularity and uniform, to instances of the three data types: genuine similarity data, flat graphs and affinity data.

Worked Example 5.12. Summary Similarity Clusters at a Genuine Similarity Dataset

Consider a similarity data set such as Confusion between numerals in Table 1.8, already analyzed in Sect. 5.4.

A symmetric version of the Confusion data is presented in Table 5.19: the sum of $A + A^T$ without further dividing it by 2, for the sake of wholeness of the entries. In this table, care has been taken of the main diagonal. The diagonal entries are by far the largest and considerably differ among themselves, which may highly affect further computations. Since we are interested in patterns of confusion between different numerals, this would be an unwanted effect so that the diagonal entries should be made to bear no effect on the clustering process. They are changed for zeros in Table 5.19.

The results of the clustering algorithm AddRem(k) applied, at each k, in the two different settings—modularity and uniform—are presented in Table 5.20. The arbitrariness of choosing an entity to start has no effect in this case. Not too many

Table 5.19 Confusion data from Table 1.7 summed with the transpose after the diagonal elements removed

	1	2	3	4	5	6	7	8	9	0
1	0	21	36	171	18	40	329	11	29	22
2	21	0	76	26	62	61	18	57	36	22
3	36	76	0	11	61	7	61	57	263	22
4	171	26	11	0	18	22	51	25	87	11
5	18	62	61	18	0	176	14	25	208	21
6	40	61	7	22	176	0	4	225	22	61
7	329	18	61	51	14	4	0	11	25	32
8	11	57	57	25	25	225	11	0	149	243
9	29	36	263	87	208	22	25	149	0	64
0	22	22	22	11	21	61	32	243	64	0
Total	677	379	594	422	603	618	545	803	883	498

Table 5.20 One-cluster structures found with the summary criterion at symmetric Confusion data in Table 5.19

Modularity		Uniform, π = Mean = 66.91		Uniform, π = 100	
Cluster	Criterion	Cluster	Criterion	Cluster	Criterion
2 3 5 6 8 9 10	1135.2	3 5 6 8 9 10	1200.7	1 4 7	502
1 4 7	805.2	1 4 7	700.5	3 5 9	464
				6 8 10	458
				2	

clusters have been found anyway. The modularity criterion is capable of separating the cluster $\{1, 4, 7\}$ from the rest, albeit with a somewhat lesser criterion value, but the rest also appears to be a cluster, in fact a tighter one. The uniform criterion at the threshold subtracted at the average level of 66.91 with no diagonal entries considered, achieves a similar fit, though it loses digit 2 from the "rest" cluster—which is good because this digit keeps a company of its own being very rarely confused for anything else. Yet at a larger threshold value of $a = 100$, the uniform criterion leads to four high density clusters—exactly those produced in Sect. 5.3 by the conceptual clustering applied to the styled numerals' images. This would be a success story provided that the user knew beforehand the right threshold value, which is a rather bold hypothesis.

Results reported in Table 5.20 lead to the following idea. Would the structure be revealed in a more uniform way if clusters are taken sequentially, so that once clustered entities are removed from the set, the remainder is considered as a new set to cluster. That is, a new random interaction or average similarity data on the remaining set is compiled and AddRem is applied after that—doing the removals again and again. There may be a problem with this approach, which can be clearly seen in Table 5.20: the cluster to remove should be the set of seven numerals rather than the remainder consisting of three numerals, 1, 4 and 5. To tackle the issue, each part, both the remainder and cluster, should be clustered again (see Case-Study 5.6).

Case-Study 5.6. Repeated One-Cluster Clustering with Repeated Removal of Background

Let us, after each clustering step, consider the unclustered part as a fresh data set, a ground set, to perform the background similarity removal again. The results of this approach are presented in Table 5.21 in such a way that each cluster that has appeared on the right, in its column, has been clustered again. All the three modularity clusters have produced themselves as their modularity subclusters. On the contrary, at the uniform criterion, each of the three-element clusters has produced a proper subcluster as shown in the further rows of the right-hand part of the table.

Table 5.21 Partitions found at the symmetric Confusion data by sequentially extracting clusters one by one, recomputing the background similarities at each subset to be analyzed

Modularity, set adjusted		Uniform, mean set adjusted	
Ground set	Cluster	Ground set	Cluster
0–9	1 4 7	0–9	1 4 7
0 2 3 5 6 8 9	2 3 5 9	0 2 3 5 6 8 9	3 5 9
0 6 8	0 6 8	0 2 6 8	0 6 8
		1 4 7	1 7
		3 5 9	3 9
		0 6 8	0 8

To explain this phenomenon, let us take a closer look, say, at cluster $\{1,4,7\}$. Table 5.22 presents the original within cluster similarities as well as those found by subtracting the average similarity, for the uniform clustering, or the random interactions, for the modularity clustering.

The total sum of similarities in set $\{1,4,7\}$, the volume, according to the left part of Table 5.22 is 1102 = 2 * 551 (factor 2 applies to make up for the absent lower triangle of the similarity matrix), of which entity 1 takes 45.4%, entity 4, 20.1%, and entity 7, 34.5%. The volume of entity 4, 20.1%, is by far the smallest of the three, which straightforwardly translates to the level of its random interactions: they are smaller than those of the others so that the subtracted part of entity 4's similarities is relatively small. This is why the summary similarity of 4 in the right-hand part of the table is positive, $c_{41} + c_{47} = 70.3 - 25.6 = 44.7 > 0$, making 4 a welcome member of the cluster according to the modularity criterion. This is not so according to the uniform criterion: the summary similarity of 4 with two others is negative, $b_{41} + b_{47} = -12.7 - 132.7 = -145.4 < 0$, setting 4 apart from the rest. A similar effect is at work with entity 2: 2 is rather remote from anything else so that its similarities become negative when the average similarity is subtracted, which is not the case with the random interactions because the latter are by far smaller at 2 than those at other entities.

The analysis reported in Case-Study 5.6 shows that the two criteria—or, better to say, the same criterion at the two different data pre-processing formulas—should be applied in different contexts: the uniform criterion is better when the meaning of

Table 5.22 Similarities between numerals 1, 4 and 7 according to Table 5.19 and, also, after subtraction of the background according to each, uniform and modularity, criterion

	Raw similarities		Mean subtracted similarities		Random interactions subtracted similarities	
	4	7	4	7	4	7
1	171	329	−12.7	145.3	70.3	156.6
4		51		−132.7		− 25.6

similarity is uniform across the table, whereas the modularity criterion works better when the similarities should be scaled depending on the individual entities.

Case-Study 5.7. Summary Criterion Clusters at Ordinary Network Data
Consider two network graphs on Fig. 5.15a and b. The former's cluster structure is rather simple—it consists of two connected components. There is no visible cluster structure in the graph (b). The latter graph consists of just one component—but can the cluster structure hidden in it be discovered using a less rigid instrument than the concept of connected component?

An even less structured is a "cockroach" graph on Fig. 4.28 in Sect. 4.6.3.2, taken from Guattery and Miller (1998) as an example of a structure that is difficult for clustering (Luxburg 2007).

The results of AddRem clustering algorithm runs starting from every node for the modularity criterion at the cockroach network of Fig. 4.28 are given in the left part of Table 5.23.

There are three highly overlapping clusters, two of them reflecting the topology of the graph with the winning cluster embracing four nodes in the right-hand side of graph in Fig. 4.28. The second column reflects an attempt at finding a partition using one-by-one clustering: after first cluster is found, its entities are removed, and the method is applied to the remaining part of the data matrix, with the random interactions readjusted to the topology of the ground set to be analyzed.

Similar attempts, but with the uniform criterion with the noise threshold set at the average similarity value, are presented in the right part of the table. The clusters demonstrate five patterns of which the lead, 4–5–6–10–11–12, embracing the right-hand half of the graph, concurs with the human view of the topology (see Luxburg 2007). After removal of this cluster, the algorithm finds remaining

Fig. 5.15 Two graphs on a set of eight entities; that on the left consists of two components whereas that on the right has a few additional edges to make it just a component

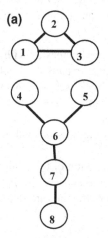

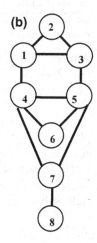

Table 5.23 One cluster multiple solutions using the summary criteria at Cockroach network data

Modularity as is and sequentially adjusted				Uniform with the mean subtracted once			
Cluster	Criterion	Ground set	Cluster	Cluster	Criterion	Ground set	Cluster
5 6 11 12	5.15	1–12	5 6 11 12	4 5 6 10 11 12	5.09	1–12	4 5 6 10 11 12
1 2 3 4 5 6	4.69	1–4 7–10	1 2 3 4 10	2 3 4 5 10 11	6.09	1–3 7–9	1 2 3
3 4 7 8 9 10	4.69	7 8 9	7 8 9	4 5 8 9 10 11	6.09	7 8 9	7 8 9
				3 4 5 9 10 11	6.09		
				1 2 3 4 5 6	4.09		

connected components—see the clusters presented in the right-hand column of Table 5.23. In contrast to the modularity criterion, the value of threshold subtracted from the data is kept the same through all the iterations because of both flat values of similarities and the thrust of the uniform criterion towards to a unified scale across the entire network.

Good clustering results found here with the uniform criterion are not easy to match with other clustering methods, which supports the view that the ordinary graphs, that is, flat networks, could be a natural niche at which the uniform criterion, with a flat value subtracted, can produce good results.

Affinity data are similarities between entities in an entity-to-feature table. They are usually defined by a kernel function depending on entity-to-entity distances such as a Gaussian kernel function $G(x, y) = e^{-d(x,y)/2\sigma^2}$ where $d(x,y)$ is the squared Euclidean distance between x and y if $x \neq y$. The denominator $2\sigma^2$ may greatly affect results and is subject to the user's choice. In our experiences, consistent results are obtained with $2\sigma^2 = 1/2$ corresponding to $\sigma = 1/2$ after each feature has been normalized by its range.

One more parameter at defining the affinity data is the distance threshold, R, such that the similarity between entities is defined as 0 if the distance between them is greater than R. The usage of this parameter appears highly successful in such areas as image analysis (Shi and Malik 2000).

Worked Example 5.13. Similarity Clusters at Affinity Data

The affinity data for eight entities in Company data table (range normalized with the last three columns further divided by $\sqrt{3}$, see Sect. 4.1) are presented in Table 5.24. Similarity values that are greater than 0.15 are highlighted in bold. The two affinity values that are at odds with the three-product cluster structure $S = \{\{1,2,3\}, \{4,5,6\}, \{7,8\}\}$ are underlined: the absent within-cluster link (4,5) and the unwanted between-cluster link (6,7).

This similarity matrix after subtraction of the random interactions background is presented in Table 5.25; the positive entries are highlighted in bold and those at odds with the three-product cluster structure are underlined.

Table 5.24 Affinity similarities between eight companies in the Company dataset

	2	3	4	5	6	7	8
1	**0.3623**	0.1730	0.1005	0.0123	0.0111	0.0101	0.0024
2		**0.2143**	0.0447	0.0261	0.0025	0.0224	0.0080
3			0.0207	0.0989	0.0252	0.0266	0.0085
4				0.1424	**0.1752**	0.0880	0.0073
5					**0.2248**	**0.1918**	0.0236
6						0.0347	0.0011
7							**0.2982**

Those greater than 0.15 are highlighted by bold

Table 5.25 Table 5.24 data after subtraction of the background of the random interactions; highlighted in bold are positive values; those at odds with the product-based clusters are underlined

	2	3	4	5	6	7	8
1	**0.2654**	0.0922	**0.0180**	−0.0903	−0.0565	−0.0857	−0.0473
2		**0.1325**	−0.0389	−0.0779	−0.0660	−0.0745	−0.0424
3			−0.0490	**0.0123**	−0.0319	−0.0543	−0.0335
4				**0.0540**	**0.1169**	**0.0055**	−0.0356
5					**0.1523**	**0.0892**	−0.0297
6						−0.0330	−0.0341
7							**0.2484**

Table 5.26 One-cluster and one-by-one partition structures found at the Company affinity data in Table 5.25

Modularity				Uniform, π = Mean			
One-cluster		One-by-one clustering		One cluster		One-by-one clustering	
Cluster	Criterion	Ground set	Cluster	Clu	Criterion	Ground set	Cluster
4 5 6 7 8	1.068	1–8	1 2 3	1 2 3	1.057	1–8	1 2 3
1 2 3	0.980	4–8	4 5 6	4 5 6 7 8	0.901	4–8	4 5 6 7
		7 8	7 8			4–7	4 5 6

The results of AddRem clustering for the affinity data are presented in Table 5.26. This time, both—modularity and uniform—criteria give similar results: two clusters only, with cluster of product A separated from the rest. The only difference is that the modularity criterion assigns a larger value to the combined cluster of B and C products, whereas the uniform criterion with the subtracted average affinity value gives a larger value to the cluster of A product.

Consider now the case at which one-by-one clustering process applies with the background adjusted at the set at which methods apply. In contrast to the previous case, it is the modularity function that finds a good solution, whereas the uniform criterion cannot find the cluster of two product C companies, making each of them a singleton.

5.4.3 Approximate Cluster

Given a similarity matrix $W = (w_{ij})$, we are going to develop a machinery for finding an approximate cluster, that is a subset of objects S assigned with an intensity weight λ. The assumption is that objects in S are all mutually related to each other with the intensity λ. According to this view, other objects have no relation to S whatsoever. We represent S by a binary N-dimensional zero-one vector of the membership values $s = (s_i)$ where $s_i = 1$ for $i \in S$ and $s_i = 0$ for $i \notin s$. Therefore, we turn to a simplest version of approximation model for one cluster:

$$w_{ij} = \lambda s_i s_j + e_{ij}, \tag{5.22}$$

where w_{ij} are not necessarily the original similarities but rather pre-processed similarities including the scale shifted a_{ij}, or a_{ij} after subtraction of random inter-actions. This can be a_{ij} after application the Laplace transformation to it. The meaning of the model: w_{ij} is λ at any pair i and j from S, or 0, otherwise, up to small residuals e_{ij}. To warrant that the residuals are small indeed, the square error criterion

$$L^2(\lambda, s) = \sum_{i,j \in I} (w_{ij} - \lambda s_i s_j)^2 \tag{5.23}$$

is to be minimized with respect to unknown S and, possibly, λ.

We first note that, with no loss of generality, the similarity matrix W can always be considered symmetric, because otherwise W can be equivalently changed for a symmetric matrix $\hat{W} = (W + W^T)/2$: the solution will not change.

Indeed, the part of criterion (5.23) related to a particular pair $i, j \in I$ is $(w_{ij} - \lambda s_i s_j)^2 + (w_{ji} - \lambda s_j s_i)^2$ which is equal to $w_{ij}^2 + w_{ji}^2 - 2\lambda(w_{ij} + w_{ji})s_i s_j + 2\lambda^2 s_i s_j$. The $s_i s_j$ on the right are not squared because they are 0 or 1, thus do not change under this operation. The same part at matrix $\hat{W} = (\hat{w}_{ij})$ reads as $(w_{ij}^2 + w_{ji}^2 + 2w_{ij}w_{ji})/2 - 2\lambda$ $(w_{ij} + w_{ji})\, s_i s_j + 2\lambda^2 s_i s_j$ so that the only parts affected are constant while those depending on the cluster to be found are identical, which proves the statement. Thus, the assumption that the similarity matrix is symmetric does not change anything: it can always be transformed to a symmetric form $\hat{W} = (W + W^T)/2$.

For the sake of simplicity, we assume that the matrix W comes with diagonal entries, usually all set to zero.

Let us take a look at criterion (5.23) under each of two assumptions (Mirkin et al. 2010):

(a) Cluster intensity λ is pre-specified by the user
(b) Cluster intensity λ is to be found according to the criterion.

We first analyze the case (a) of λ pre-specified. Let us slightly rewrite criterion (5.23):

$$L^2(\lambda, s) = \sum_{i,j\in I}(w_{ij} - \lambda s_i s_j)^2 = \sum_{i,j\in I} w_{ij}^2 - 2\lambda \sum_{i,j\in I}\left(w_{ij} - \frac{\lambda}{2}\right)s_i s_j \qquad (5.23')$$

Assume that λ is positive. Then minimizing (5.23) is equivalent to maximizing the sum on the right, which is just the summary uniform criterion (5.16) at $\pi = \lambda/2$ that has been described and utilized in Sect. 5.4.2. Indeed, the equation $\sum_{i,j\in I}(w_{ij} - \lambda/2)s_i s_j = \sum_{i,j\in S}(w_{ij} - \lambda/2)$ easily follows from the fact that $s_i = 1$ if and only if $i\in S$. That means that the algorithm AddRem from Sect. 5.4.2 is applicable here to produce $\lambda/2$-tight clusters.

Consider now case (b), when intensity λ in (5.23) is to be adjusted to further minimize the criterion. It is easy to prove that, given an S, the optimal λ is just the average of within cluster similarities, $\lambda = \lambda(S)$, where

$$\lambda(S) = \frac{\sum_{\substack{i,j\in I \\ i\neq j}} w_{ij}s_i s_j}{\sum_{i,j\in I} s_i s_j} = \frac{\sum_{\substack{i,j\in S \\ i\neq j}} w_{ij}}{|S|^2} \qquad (5.24)$$

as it is always the case for the least-squares approximation of a series of numbers by a central value (see Sect. 2.2).

That means that, again, the criterion is equivalent to the summary uniform criterion (5.16), but this time with a variable value of the threshold $\pi = \lambda(S)/2$ that depends on S. In particular, a locally optimal cluster is $\lambda(S)/2$-tight: the average similarities of entities $i\in I$ to S are greater than $\lambda(S)/2$ for those i in S and smaller than $\lambda(S)/2$ for i's out of S.

If one puts the optimal $\lambda = \lambda(S)$ in (5.23), the least squares criterion is decomposed as follows

$$L^2(\lambda(S), s) = \sum_{i,j\in I}(w_{ij} - \lambda(S)s_i s_j)^2 = \sum_{i,j\in I} w_{ij}^2 - \lambda^2(S)\sum_{\substack{i,j\in I \\ i\neq j}} s_i s_j = T - \lambda^2(S)|S|^2$$

where T is the data scatter, the sum of all the similarities squared, so that a Pythagorean decomposition of the data scatter holds:

$$T = \sum_{i,j\in I} w_{ij}^2 = \lambda(S)^2|S|^2 + L^2 \tag{5.25}$$

where L^2 is the unexplained minimized part (5.23) whereas the item in the middle is the explained part of the data scatter.

The explained part in (5.25), which is to be maximized to minimize L^2 because the scatter T is constant, is

$$g(S)^2 = [\lambda(S)|S|]^2 = \left[\frac{s^T W s}{s^T s}\right]^2 \tag{5.26}$$

which is but the square of the Rayleigh quotient

$$g(S) = \frac{s^T W s}{s^T s} = \lambda(S)|S| \tag{5.27}$$

Since it is assumed that at least some of the similarities in A are positive, the maximum of (5.27) over all binary s's is positive as well. Indeed, take a positive w_{ij} and a vector s with all components equal to zero except for just i-th and j-th components that are unities. Obviously (5.27) is positive on that, the more so the maximum. If, however, all the similarities between entities are negative, then no non-singleton cluster can make (5.27) positive—that is, no nontrivial cluster can come up with the criterion.

That means that a version of AddRem(k) algorithm with a variant threshold π, AddRemAdd(k) in Sect. 5.5.3, in fact (locally) optimizes the Rayleigh quotient (5.27).

5.5 Decomposing a Similarity Matrix Over Additive Clusters

5.5.1 General

The concept of additive clustering was proposed independently by Shepard and Arabie (1979) and Mirkin (1987); see these references for further history comments. The idea behind additive clustering is this. Since the raw data are similarities measuring relations between entities, let us decode a cluster in the same relational format. That is, let us make a cluster S to assign every two entities, i and j, a similarity value: say unity if they belong to the cluster or 0 if at least one of them does not. This cluster similarity matrix s plays the role of a dummy variable—in the format of a similarity matrix. Consider, for example, subset $S = \{1,3,4\}$ of $I = \{1,2,3,4,5,6\}$: its corresponding matrices s, $2s$, and $2s - 1$ are in Table 5.27.

Table 5.27 Binary matrices for cluster $S = \{1,3,4\}$ in a 6-element set

Matrix s							Matrix $2s$							Matrix $2s - 1$					
	1	2	3	4	5	6		1	2	3	4	5	6		1	2	3	4	5	6
1	1	0	1	1	0	0		2	0	2	2	0	0		1	-1	1	1	-1	-1
2	0	1	0	0	0	0		0	2	0	0	0	0		-1	1	-1	-1	-1	-1
3	1	0	1	1	0	0		2	0	2	2	0	0		1	-1	1	1	-1	-1
4	1	0	1	1	0	0		2	0	2	2	0	0		1	-1	1	1	-1	-1
5	0	0	0	0	1	0		0	0	0	0	2	0		-1	-1	-1	-1	1	-1
6	0	0	0	0	0	1		0	0	0	0	0	2		-1	-1	-1	-1	-1	1

Table 5.28 Part of matrix in Table 1.9 related to amino acids A, C, D, E, F: Original on the left, rearranged in the middle, and with 4 added to all entries on the right

	A	C	D	E	F	C	D	E	F	C	D	E	F
A	4	0	-2	-1	-2	0	-2	-1	-2	4	2	3	2
C	0	9	-3	-4	-2		-3	-4	-2		1	0	2
D	-2	-3	6	2	-3			2	-3			6	1
E	-1	-4	2	5	-3				-3				1
F	-2	-2	-3	-3	6								

Therefore, it is reasonable to think that a similarity matrix may reflect a number of attribute-based similarity matrices possibly taken with different weights. Consider, for example, the matrix of similarities between first five amino acids in Table 1.9 (B is omitted from the list because it is synonymous to D, see Table 1.10) as presented in Table 5.28; that on the right is obtained by subtracting the minimum, −4, from all entries to make it non-negative.

Table 5.29 presents similarity matrices between these amino-acids according to attributes from Table 5.32 in Project 5.1 further on. The attributes on the left in Table 5.29, reflect popular molecular properties of amino acids related to their size (Small or not), electricity charge (Polar or not) and the propensity to keep inside of the molecules (Hydrophobic or not). The matrix on the right represents a weighted sum of the three with an added intercept to mimic the matrix of similarities from BLOSUM62 in Table 1.9.

The two similarity matrices are compared in Table 5.30. Overall, the result does not look too bad: there are only two significant differences, in similarities between amino acids A and E, and C and D, that probably require taking into account more attributes. If we go for regression of the observed similarity over attribute-based similarity we could get slightly better results. This idea is pursued further on in Project 5.1 on the whole set of amino acids.

There are situations, though, in which the user prefers to find clusters underlying the observed similarities, according to the additive model, by the matrix itself, without much bothering of trying to obtain related attributes. This is the realm of the additive clustering model in (5.28) analyzed further in Project 5.1. This model can be considered as an extension of the spectral decomposition of similarity matrices to the case when the vectors to be found are constrained to be 1/0 binary. Assuming the conventional least-squares criterion for this specification of the summarization problem, a natural idea coming to mind is to mimic the one-by-one approach of the Principal Component Analysis. The other idea, just working on all clusters in parallel, is not considered in this text.

Yet even at the restricted, one cluster, model (5.23), there can be a number of different approaches to minimizing the least-squares criterion or, equivalently, maximizing the Rayleigh quotient. Two of the approaches, tried at the maximum tightness criteria in Sect. 5.4, should be considered first:

Table 5.29 Similarity matrices between five amino acids according to attributes Small, Polar and Hydrophobic from Table 5.32 in Project 5.1 further on

	Sm	Po	Hy
A	+		+
C	+	+	
D	+	+	
E		+	
F			+

Sm:

	C	D	E	F
A	1	1	0	0
C		1	0	0
D			0	0
E				0

Po:

	C	D	E	F
A	0	0	0	0
C		0	0	0
D			1	0
E				0

Hy:

	C	D	E	F
A	0	0	0	0
C		0	0	1
D			0	0
E				0

$2Sm + 6Po + 2Hy + 1$:

	C	D	E	F
A	3	3	1	1
C		3	1	2
D			6	1
E				1

Table 5.30 Comparison of two similarity matrices between five amino acids, one taken from observations (Table 5.28), and the other additively composed using attribute clusters (Table 5.29)

	BLOSUM62				2Sm + 6Po + 2Hy + 1				Difference			
	C	D	E	F	C	D	E	F	C	D	E	F
A	4	2	3	2	3	3	1	1	1	−1	2	0
C		1	0	2		3	1	2		−2	−1	0
D			6	1			6	1			0	0
E				1				1				0

(i) Spectral approach

Let us drop the constraint of vectors being binary and find the optimal solution among arbitrary vectors, that is, the maximum eigenvalue and corresponding eigenvector, and then adjust somehow its components to the zero-one setting. It seems reasonable that the larger components of the eigenvector are to be changed for unity while those smaller ones are changed for zero. If true, this would drastically reduce computation.

(ii) Hill-climb clustering

The strategy of finding a cluster by adding/removing entities in a best possible way implemented, for the summary similarity criterion, in Sect. 5.4.2 can be applied here too. At least, it leads to provably tight clusters. This is the strategy pursued further in this text with AddRem algorithm from Sect. 5.4.2.

The one-cluster model assumes, rather boldly, that all observed similarities can be explained by a summary action of just two constant-level causes and noise.

This is a much-simplified model, but it brings in a nice clustering criterion to implement the least-squares approach: the underlying cluster S must maximize the criterion in (5.27), the product of the average within-cluster similarity $a(S)$ and the number of elements in S, $|S|$: $g(S) = |S|a(S)$. The greater the within-cluster similarity, the better, and the larger the cluster, the better too. These two objectives do not necessarily go along. In fact, they are at odds in most cases: the greater the number of elements in in a cluster, the smaller the within-cluster similarities are. That is, criterion $g(S)$ is a compromise between the two. When S is small, an increase in its size would dominate the unavoidable fall in similarities. But later in the addition process, when S becomes larger, the relative size change diminishes and cannot dominate the fall in within-cluster similarities—the process of generating S stops. This is further elaborated in Sect. 5.5.2 in which an analogue to the AddRem(k) algorithm (see Sect. 5.4.2) is formulated for the semi-average clustering criterion (5.27). This analogue, referred to as AddRemAdd(k), outputs a cluster which is provably tight in the following sense. The attraction of entity i to S, its average similarity with elements in S, is greater than half the within cluster average if i belongs in S; and it is less than that if i does not belong in S.

Worked Example 5.14. Additive Clusters at Confusion Dataset
Consider the symmetrized Confusion data set in Table 5.16 and apply algorithm AddRem at different levels of similarity shifts starting at different entities (Table 5.31).

Table 5.31 presents each approximate cluster with all three characteristics implied by the additive clustering model:

(1) The cluster list S of its entities;
(2) The cluster-specific intensity $\lambda = a(S)$, the average within cluster similarity;
(3) The cluster contribution to the data scatter, $g^2(S) = \lambda^2 |S|^2$.

The cluster sizes decrease when the similarity threshold grows, as illustrated on Fig. 4.25 and stated in Q.5.26. The corresponding intensity changes reflect the ever increasing shift values subtracted from the similarities. The table also shows that there is no point in making the similarity shift values greater than the average similarity value. In fact, setting the similarity shift value equal to the average can be seen as a step of the one-by-one cluster extracting strategy: subtracting the average from all the similarities is equivalent to extracting the universal cluster with its optimal intensity value—provided the cluster is considered on its own, without the presence of other clusters. At the similarity shift equal to the average, cluster $\{1,4,7\}$ loses digit 4 because of its weak connections. The results best matching those of Fig. 1.24 in Sect. 1.3.3 are found at the similarity shift equal to Av/2.

Table 5.31 Non-singleton clusters at symmetrized, no diagonal, Confusion matrix found at different similarity shift values; the average out-of-diagonal similarity value is Av = 33.46

Similarity shift	Cluster lists	Intensity	Contribution
0	(i) 2 3 5 8 10	45.67	35.14
	(ii) 1 4 7	91.83	21.46
Av/2 = 16.72	(i) 1 4 7	75.11	21.11
	(ii) 3 5 9	71.94	19.37
	(iii) 6 8 0	71.44	19.10
Av = 33.46	(i) 1 7	131.04	25.42
	(ii) 3 5 9	55.21	13.54
	(iii) 6 8 0	54.71	13.29
3Av/2 = 50.18	(i) 1 7	114.32	16.31
	(ii) 3 9	81.32	5.25
	(iii) 6 8 0	35.98	5.40
2Av = 66.91	(i) 1 7	95.59	5.08
	(ii) 3 9	64.59	3.54
	(iii) 8 0	54.59	2.53
	(iv) 6 8	45.59	1.76

Project 5.1. Analysis of Structure of Amino Acid Substitution Rates

Let us consider the data of substitution between amino acids in Table 1.9 and try explaining them in terms of properties of amino acids. An amino acid molecule can be considered as consisting of three groups of atoms: (i) an amine group, (ii) a carboxylic acid group, and (iii) a side chain. The side chain varies between different amino acids, thus affecting their biochemical properties. Among important features of side chains are the size and polarity, the latter affecting the interaction of proteins with solutions in which the life processes act: the polar amino acids tend to be on protein surfaces, i.e., hydrophilic, whereas other amino acids hide within membranes (hydrophobicity). There are also so-called aromatic amino acids, containing a stable ring, and aliphatic amino acids whose side chains contain only hydrogen or carbon atoms. These are presented in Table 5.32. As can be easily seen, these five attributes cover all amino acids but only once or twice.

A natural idea would be to check what relation these features have to the substitutions between amino acids. To explore the idea one needs to represent the features in the format of the matrix of substitutions, that is, in the similarity matrix format. Such a format is readily available as the adjacency matrix format. That is, a feature, say, "Small" corresponds to a subset S of entities, amino acids, that fall in it. The subset generates a binary relation "i and j belong to S" expressed by the Cartesian product $S \times S$ or, equivalently, by the $N \times N$ binary entity-to-entity

Table 5.32 Attributes of twenty amino acids

Amino acid		Small	Polar	Hydrophobic	Aliphatic	Aromatic
A	Ala	+			+	
C	Cys	+		+		
D	Asp	+	+			
E	Glu		+			
F	Phe			+		+
G	Gly	+			+	
H	His					+
I	Ile			+	+	
K	Lys		+			
L	Leu			+	+	
M	Met			+		
N	Asn	+	+			
P	Pro	+				
Q	Gln		+			
R	Arg		+			
S	Ser	+				
T	Thr	+				
V	Val			+	+	
W	Trp			+		+
Y	Tyr					+

similarity matrix $s = (s_{ij})$ such that $s_{ij} = 1$ if both i and j belong to S, and $s_{ij} = 0$, otherwise. For example, on the set of first five entities $I = \{A, C, D, E, F\}$ in Table 5.32, the binary similarity matrices for attributes Small, Polar and Hydrophobic are presented in Table 5.29.

To analyze contributions of the attributes to the substitution rate data A one can use a linear regression model (see Sect. 3.3)

$$A = \lambda_1 Sm + \lambda_2 Po + \lambda_3 Hy + \lambda_4 Al + \lambda_5 Ar + \lambda_0$$

which in this context suggests that the similarity matrix A (after the intercept λ_0 is subtracted from it) can be decomposed, up to a minimized residual matrix, according to features in such a way that each coefficient $\lambda_1, \ldots, \lambda_5$, expresses the intensity level supplied by it to the overall similarity. The intercept λ_0, as usual, sums shifts in the individual attribute similarity scales.

To fit the regression model, let us utilize upper parts of the matrices only. In this way, we

(i) take into consideration the similarity symmetry and
(ii) make the diagonal substitution rates, that is, similarity to itself, not affecting the results.

As one can see from Table 5.33, the estimates of the slope regression coefficients are all positive, giving them the meaning of the weights or similarity intensities indeed, of which dummies representing categories Small, Polar, and Aromatic are the most contributing, according to the last line in Table 5.33. The intercept, though, is negative.

Unfortunately, the five attributes cannot explain the pattern of amino acid substitution: the determinacy coefficient is just 35.3%, less than a half. That means one needs to find different attributes for explaining the amino acid substitution patterns.

Then the idea of additive clustering comes. Why cannot we find attributes to fit in the similarity matrix from the matrix itself rather than by trying to search the amino acid feature databases? That is, let us consider unknown subsets $S_1, S_2, \ldots, S_K$ of the entity set along with the corresponding binary membership vectors $s_1, s_2, \ldots, s_K$ such that $s_{ik} = 1$ if $i \in S_k$, and $s_{ik} = 0$, otherwise, $k = 1, 2, \ldots, K$, and find them according to model

$$a_{ij} = \lambda_1 s_{i1} s_{j1} + \lambda_2 s_{i2} s_{j2} + \ldots + \lambda_K s_{iK} s_{jK} + \lambda_0 + e_{ij} \qquad (5.28)$$

Table 5.33 Least-squares regression results

	Sm	Po	Hy	Al	Ar	Intercept
Intensity λ	2.46	1.48	1.02	0.81	2.65	−2.06
Standard deviation	0.27	0.31	0.36	0.22	2.06	
Standardized intensities	0.66	0.47	0.36	0.18	0.46	

The last line entries (standardized intensities) are products of the corresponding entries in the first and second lines

According to this model, each of the similarities a_{ij} is equal to a weighted sum of the corresponding cluster similarities s_{ik} s_{jk}, up to small residuals, e_{ij} $(i,j\in I)$.

Unfortunately, there are too many items to find, given the similarity matrix $A = (a_{ij})$: the number of clusters K, the clusters S_1, S_2,..., S_K themselves as well as their intensity weights, λ_1, λ_2, ..., λ_K, and the intercept, λ_0. This makes the solution much dependent on the starting point, as it is with the general mixture of distributions model.

If, however, we rewrite the model by moving the intercept to the left as

$$a_{ij} - \lambda_0 = \lambda_1 s_{i1} s_{j1} + \lambda_2 s_{i2} s_{j2} + \ldots + \lambda_K s_{iK} s_{jK} + e_{ij}, \tag{5.29}$$

the model reminds the equation for the Principal Component Analysis very much, especially as expressed in terms of the square matrices—the $a_{ij} - \lambda_0$ plays the role of the covariance values, s_{ik}, the role of the loading/values, that is, k-th eigenvector, and λ_k, the role of the k-th eigenvalue, the only difference being that the binarity constraints are imposed on the values s_{ik} that must be either 1 or 0.

In (5.28), the intercept value λ_0 is the intensity of the universal cluster $S_0 = I$ which is assumed to be part of the solution. In (5.29), however, this is just a similarity shift, with the shifted similarity matrix $A^s = (a_{ij}^s)$ defined by $a_{ij}^s = a_{ij} - \lambda_0$ which is akin to the uniform data transformation in Sect. 4.6.3. Most important is that the value of λ_0 in model (5.29) ought to come from external considerations rather than from inside of the model as it is in (5.28).

The machinery for identifying additive clusters one-by-one developed further on leads to the following clusters found at different scale shift value λ_0 (see Table 5.34).

Table 5.34 Non-singleton clusters at Amino acid substitution data found at different similarity shift values; the average out-of-diagonal similarity value is Av = -1.43

Similarity shift	Cluster lists	Intensity	Contribution
0	(i) ILMV	1.67	2.04
	(ii) FWY	2	1.47
	(iii) EKQR	1.17	1
	(iv) DEQ	1.33	0.65
	(v) AST	0.67	0.16
Av/2 = -0.71	(i) ILMV	2.38	6.47
	(ii) DEKNQRS	1.05	4.38
	(iii) FWY	2.71	4.21
	(iv) AST	1.38	1.09
Av = -1.43	(i) DEHKNQRS	1.6	16.83
	(ii) FILMVY	1.96	13.44
	(iii) FWY	3.43	5.22

At the similarity shift equal to the average, there are three clusters covering 35.5% of the variance of the data. These concern three features of those considered above: Polar (cluster i), Hydrophobic (cluster ii), and Aromatic (cluster iii). The clusters slightly differ from those presented in Table 5.32, which can be well justified by the physical and chemical properties of amino acids. In particular, cluster (i) adds to Polar group two more amino acids: H (Histidine) and S (Serine). These two, in fact, are frequently considered polar too. Cluster (ii) differs from the Hydrophobic group by the absence of C (Cysteine) and W (Tryptophan) and the presence of Y (Tyrosine). This corresponds to a specific aspect of hydrophobicity, the so-called octanol scale that does exclude C and include Y (for some most recent measurements, see, for example, http://blanco.biomol.uci.edu). The absence of Tryptophan from the cluster is probably due to the fact that it is not easily substituted by the others because it is by far the most hydrophobic of the pack. Cluster (iii) consists of hydrophobic aromatic amino acids which excludes F (Phenylalanine) because it is not hydrophobic.

5.5.2 Additive Clusters One-by-One

Let us reformulate the additive cluster model from Project 5.1, keeping the same equation labels. Let I be a set of entities under consideration and $A = (a_{ij})$ a symmetric similarity matrix $i,j \in I$. The additive clustering model assumes that the similarities in A are generated by a set of additive clusters $S_k \subseteq I$ together with their intensities λ_k ($k = 0, 1, \ldots, K$) in such a way that each a_{ij} is approximated by the sum of the intensities of those clusters that contain both i and j:

$$a_{ij} = \lambda_1 s_{i1} s_{j1} + \lambda_2 s_{i2} s_{j2} + \ldots + \lambda_K s_{iK} s_{jK} + \lambda_0 + e_{ij} \qquad (5.28)$$

where $s_k = (s_{ik})$ are the membership vectors of unknown clusters S_k, and λ_k are their positive intensity values, $k = 1, 2, \ldots, K$. Residuals e_{ij} are to be minimized.

The zero's cluster S_0 is assumed to coincide with the entire set I so that its intensity λ_0 is the intercept in (5.28). On the other hand, λ_0 has a meaning of the similarity shift, with the shifted similarity matrix $A' = (a'_{ij})$ defined by $a'_{ij} = a_{ij} - l_0$. Equation (5.28) for the shifted model can be rewritten as

$$a'_{ij} = \lambda_1 s_{i1} s_{j1} + \lambda_2 s_{i2} s_{j2} + \ldots + \lambda_K s_{iK} s_{jK} + e_{ij}, \qquad (5.29)$$

so the shifted similarity matrix $a'_{ij} = a_{ij} - \lambda_0$ is the sum of cluster binary matrices weighted by their intensities. The role of the intercept λ_0 in (5.29) as a "soft" similarity threshold is of a special interest when λ_0 is user specified, because the shifted similarity matrix a'_{ij} may lead to different clusters at different λ_0 values, as Fig. 4.25 and Q.4.31 clearly demonstrate.

Model (5.29) can be considered using two different assumptions of the underlying cluster structure:

A. Overlapping additive clusters
B. Non-overlapping clusters

In the latter case, the summation in model (5.28) and (5.29) hides the fact that no
summation of intensities goes on. Every similarity a'_{ij} is assumed to be approximately equal to the intensity value of that cluster that contains both i and j, or 0 if no
cluster contains both of the entities.

The equations in (5.29) coincide with those in the spectral decomposition (5.25)
(see (A.13) in Sect. A.3.2) up to the condition that vector s_k's in (5.29) is bound to
be 1/0 binary, whereas no constraint is imposed on c's in (A.13). That means that
the additive clustering model is an extension of the spectral decomposition onto the
case when vectors are binary. This type of decomposition, with additional constraints such as say non-negativity of the elements of the solution is becoming
increasingly popular in data analysis. Assuming the conventional least-squares
criterion for this specification of the summarization problem, a natural idea coming
to mind is to imitate the one-by-one approach of the Principal Component Analysis.
The other idea, just working on all clusters in parallel, is not considered in this text.

Now we can return to the case of the original model with multiple clusters. The
situation will slightly differ depending on whether clusters are assumed
non-overlapping (A) or possibly overlapping (B).

Consider, first, the case of model (5.29) with the restriction that clusters to be
found may not overlap. The fact that clusters S_f and S_g do not overlap can be
equivalently stated in terms of their binary membership vectors s_f and s_g: these must
be orthogonal so that $\langle s_f, s_g \rangle = 0$. This implies that the shifted data scatter admits
the following decomposition:

$$\langle A', A' \rangle = \sum_{k=1}^{K} \left[s_k^T A s_k / s_k^T s_k \right]^2 + \langle E, E \rangle \tag{5.30}$$

which extends Eq. (5.25) to the multiple cluster case. In (5.30), the inner products
$\langle A', A' \rangle$ and $\langle E, E \rangle$ denote the sums of the squared elements of the corresponding
matrices. To derive (5.30), one can take the inner product of Eq. (5.29) by itself,
considering all matrices as $N \times N$ vectors, and taking into account the fact that
matrices $s_k s_k^T$ and $s_l s_l^T$ are orthogonal as $N \times N$ vectors at $k \neq l$, because the corresponding vectors s_k and s_l are orthogonal.

Equation (5.30) means that each of the optimal non-overlapping clusters indeed
contributes the squared Rayleigh quotient (5.26) to the shifted data scatter, and,
moreover, the optimal intensity value λ_k of cluster S_k is, in fact, the within cluster
average $\lambda_k = \lambda(S_k)$. The sum in the middle represents the part of the data scatter
$\langle A', A' \rangle$ "explained" by the model, whereas $\langle E, E \rangle$ relates to the "unexplained"
part. Both can be expressed in percentages of the data scatter. Obviously, the greater
the explained part the better the fit.

Assuming that the cluster contributions differ significantly, one can apply the one-by-one principal component analysis strategy to the cluster case as well—though, in this case, the process does not necessarily lead to an optimal solution. This strategy can be put as follows. First, a cluster S is found at the entire data set to maximize the Rayleigh quotient (5.27). It is denoted by S_1 along with its intensity value $\lambda_1 = \lambda(S_1)$ and the contribution $g^2(S_1)$ in (5.30) and removed from the entity set I. The next cluster S_2 is found in the same way over the remaining entity set, and removed as well. The process of sequential extraction iterates until no positive entries in A' over the remaining entities can be found. This would mean the remaining entities are all to remain singletons. In general, the process yields sub-optimal, not necessarily optimal, clusters.

Let us turn now to the case of overlapping clusters B.

To fit the model (5.29), the one-by-one cluster extracting strategy will require maximizing, at each step $k = 1, 2, \ldots, K$, the criterion (5.26) applied to a corresponding residual similarity matrix A_k (Mirkin 2012). Specifically, A_1 is taken to coincide with the shifted similarity matrix, $A_1 = A'$. At k-th step, a (locally) optimal cluster maximizing $g^2(S)$ in (5.26) over $W = A_k$ is found to be set as S_k along with its intensity value λ_k, equal to the average of the residual similarities within S_k. Its contribution to the data scatter is equal to the optimized criterion (5.26). The residual similarities are updated after each step k by subtracting the found $\lambda_k s_{ik} s_{jk}$:

$$a_{ij,k+1} = a_{ij,k} - \lambda_k s_{ik} s_{jk}. \tag{5.31}$$

In spite of the fact that thus found clusters may and frequently do overlap, this one-by-one strategy leads to an additive decomposition of the data scatter into the contributions of the extracted clusters (S_k, λ_k) and the minimized residual square error

$$\langle A', A' \rangle = \sum_{k=1}^{K} \left[s_k^T A_k s_k / s_k^T s_k \right]^2 + \langle E, E \rangle \tag{5.32}$$

except that it is the residual similarity matrix A_k to stand in the middle rather than the original matrix A'.

To prove (5.32), one needs just the Eq. (5.27) applied to $W = A_k$,

$$\sum_{i,j \in I} a_{ij,k}^2 = \left(s_k^T A_k s_k / s_k^T s_k \right)^2 + L_k^2 \tag{5.33}$$

Since $L_k^2 = \sum_{i,j \in I} a_{ij,k+1}^2$, (5.32) can be obtained by summing all the Eqs. (5.33) over all $k = 1, 2, \ldots, K$.

Q.5.24. What happens if $\lambda < 0$ in criterion (5.23)?

A. According to formula (5.23′), that would mean that the summary uniform similarity $\sum_{i,j \in S} (w_{ij} - \lambda/2)$ must be minimized rather than maximized. An optimal set S would consist of most dissimilar entities. Such a set sometimes is referred to as an anti-cluster.

Q.5.25. Can you think of a real-world problem that would amount to the goal of finding anti-clusters rather than clusters?
A. This can be any domain in which a parsimonious representation of a given set of entities is required. One such domain is bioinformatics at which such a subset of genes or gene products is required that express as differently as possible.
Q.5.26. Consider the uniform summary criterion $f(S, \pi)$ in (5.16) and two values of threshold, $\pi_1 > \pi_2$. Prove that the size of optimal cluster at π_1 cannot be greater than that at π_2.
A. Let $S1$ be an optimal cluster at π_1, and $S2$, at π_2. Then $f(S2, \pi_1) \leq f(S1, \pi_1)$ and $f(S1, \pi_2) \leq f(S2, \pi_2)$ because of the optimality. With little algebra, the former inequality can be reformulated as

$$\sum_{i,j \in S2} a_{ij} = \sum_{i,j \in S1} a_{ij} \leq \pi_1[|S2|(|S2| - 1) - |S1|(|S1| - 1)]$$

and the latter inequality, as

$$\sum_{i,j \in S2} a_{ij} = \sum_{i,j \in S1} a_{ij} \leq \pi_2[|S2|(|S2| - 1) - |S1|(|S1| - 1)]$$

where $|S|$ stands for the number of elements in S. Combining these leads to

$$\pi_2[|S2|(|S2| - 1) - |S1|(|S1| - 1)] \leq \pi_1[|S2|(|S2| - 1) - |S1|(|S1| - 1)],$$

that is,

$$(\pi_1 - \pi_2)[|S2|(|S2| - 1) - |S1|(|S1| - 1)] \geq 0.$$

Since $\pi_1 - \pi_2 > 0$, this implies that $|S2|(|S2| - 1) - |S1|(|S1| - 1) \geq 0$, which is only possible when $|S2| \geq |S1|$, thus proving the claim.

In this derivation, an implicit assumption was that the diagonal elements a_{ii} are not present, so that there are only $|S|(|S| - 1)$ pairs to be considered in S. However, the same derivation holds for the case at which diagonal elements are present, so that all $|S|^2$ pairs are to be considered. The only change would be in using the $|S|^2$ item rather than $|S|(|S| - 1)$, which would not affect the conclusion.

5.5.3   Finding (Sub)Optimal Additive Clusters

Before starting computation of additive clusters, the similarity matrix should be made symmetric, by averaging it with its transpose, and shifted by a scale shift value λ_0 which is to be user defined. A default value for λ_0 can be the average value of the similarity matrix if similarity values vary across the matrix or $\lambda_0 = 1/2$ if the similarity matrix is the flat zero-one matrix of an ordinary graph. For the sake of simplicity, it is assumed here that the diagonal elements a_{ii} are present. In real-world computations, it is advisable always put the diagonal elements equal to 0, except in the cases at which the "individual strength" of entities is to be taken into account while looking for clusters.

We consider here only one cluster based additive clustering algorithms.

Given a matrix $W = (w_{ij})$, consider an additive clustering analogue to AddRem (k) algorithm from Sect. 5.4.2 to adjust the cluster intensity, that is, the threshold value, according to criterion (5.27). Again, vector $z = 2s - 1$ is used to hold the information of cluster S being built. Its components are: $z_i = 1$ if $i \in S$ and $z_i = -1$, otherwise. This allows for the same action of changing the sign of z_i to express both addition of i into S if $i \notin S$ and removal of i from S if $i \in S$.

AddRemAdd(k) algorithm

Input: matrix $W = (w_{ij})$ and initial element in the cluster, k; output: cluster S, its intensity λ and contribution g^2 to the original A' matrix scatter.

1. *Initialization.* Set N-dimensional z to have all its entries equal to -1 except for $z_k = 1$, the number of elements in it $n = 1$, intensity $\lambda = 0$, and contribution $g^2 = 0$. For each entity $i \in I$, define its average similarity to S, $w(i,S) = w_{ij}$.
2. *Selection.* Find i^* maximizing $w(i,S)$ over all $i \in I$.
3. *Loop*

   1. While $-z_i * [w(i^*,S) - \lambda/2] > 0$

      a. Update:
         $g \Leftarrow n[g - z_i * 2w(i^*,S)]/(n - z_{i*})$ (the criterion value),
         $w(i,S) \Leftarrow [nw(i,S) - z_{i*}w_{ii*}]/n$ (the average similarities of all entities i to S),
         $n \Leftarrow n - z_{i*}$ (the number of elements in S),
         $\lambda \Leftarrow g/n$ (the intensity of S),
         $g^2 = g^2$ (the contribution), and go to 2.
      b. Change the sign of z_{i*} in vector z, $z_{i*} \Leftarrow -z_{i*}$ (either add i^* to S or remove i^* from S)

   2. Output S, λ and g^2.

The general step is justified by the fact that indeed maximizing $w(i,S)$ over all $i \in I$ does maximize the increment of $g(S)$ among all sets that can be obtained from S by either adding an entity to S or removing an entity from S. Updating formulas can be derived from the definitions of the concepts involved.

One such formula is for updating $g(S)$:

$$g(S \mp i^*) = \frac{|S|g(S) - z_{i^*}2|S|w(i^*, S) + w_{i^*i^*}}{|S| - z_{i^*}} \qquad (5.34)$$

so that the change of the criterion is

$$g(S \mp i^*) - g(S) = \frac{z_{i^*}[g(S) - 2|S|w(i^*, S)] + w_{i^*i^*}}{|S| - z_{i^*}}.$$

It should be mentioned that these two formulas are derived under the assumption that any diagonal values may be present in matrix W. The action of changing the state of i^* is executed if and only if the numerator in the latter formula is positive. This explains the condition of the loop in AddRemAdd(k) above (under the assumption that the diagonal is zero). The proof of the former formula closely follows that of formula (5.19) in Sect. 5.4.2, based on using the image in Fig. 5.14 and, thus, is omitted. The latter formula is easily derived from the former.

The rule for updating the average similarities between entities and cluster S is based on formula:

$$w(i, S \mp i^*) = \frac{|S|w(i, S) - z_{i^*}w_{ii^*}}{|S| - z_{i^*}},$$

Which is easily derived from the definition of the average similarity between $i \in I$ and S,

$$w(i, S) = \frac{\sum_{j \in S} w_{ij}}{|S|}.$$

The algorithm AddRemAdd(k) utilizes no ad hoc parameters, except for the similarity shift value, so the cluster sizes are determined by the process of clustering itself. Yet, changing the similarity shift λ_0 may affect the clustering results indeed, which can be of an advantage when one needs to contrast within- and between-cluster similarities.

To use AddRemAdd algorithm for the case of non-overlapping clusters, one needs to perform a set of repetitive steps arranged as the algorithm ADN (ADditive clusters Non-overlapping) as follows.

ADN algorithm
Input: matrix $A' = (a'_{ij})$; Output: a set of non-overlapping clusters S_1, S_2, ..., S_K where

(i) number of clusters K is not pre-specified and
(ii) they do not necessarily cover all the entity set,

together with their intensities λ_k and contributions g_k^2 to the A' matrix scatter.

0. *Initialization.* Set $k = 1$, $I_k= I$ and $A_k= A'$.
1. *Stopping test.* Check whether I_k contains more than one entity and whether A_k contains positive values. If either is not true, the computation stops and those clusters found so far are output.
2. *Cluster.* Apply AddRemAdd(j) for every $j{\in}I_k$. Select that of the results maximizing the contribution and put is as S_k along with the corresponding intensity λ_k and contribution g_k^2.
3. *Update.* Set $I_k = I_k - S_k$, $k = k + 1$, and A_k the part of matrix A' related to elements of I_k only.

The number of clusters is not pre-specified by the user with ADN nor the subset of entities remaining unclustered. Yet both are predetermined by the choice of the scale shift parameter λ_0 leading to matrix A'. This choice, in fact, defines the granularity of clustering.

A similar algorithm, ADO (ADditive clusters Overlapping) can be drawn for the case when clusters are not necessarily non-overlapping.

ADO algorithm
Input: matrix $A' = (a'_{ij})$ and parameters for halting the computation: (i) threshold of contribution of individual clusters ς, say $\varsigma = 5\%$, (ii) threshold of explained contribution η, say $\eta = 50\%$; Output: a set of possibly overlapping clusters S_1, S_2, ..., S_K where

  (i)   number of clusters K is not pre-specified,
  (ii) they do not necessarily cover all the entity set, and
  (iii) they may overlap,

together with their intensities λ_k and contributions g_k^2 to the A' matrix scatter.

0. *Initialization.* Compute the data scatter $D = \langle A',A' \rangle$. Set $k = 1$ and $A_k= A'$.
1. *Cluster.* Apply AddRemAdd(j) to A_k for every $j{\in}I$. Select the result maximizing the contribution and put is as S_k along with the corresponding intensity λ_k and contribution g_k^2.
2. *Stopping test.* Check whether $g_k^2/D > \varsigma$ and $\Sigma_{f \leq k}g_k^2/D \leq \eta$. If either is not true, the computation stops and only clusters found at the previous iterations are output.
3. *Update.* Set $A_k = A_k - \lambda_k s_k s_k^T$, $k = k + 1$.
4. *Similarity positivity test.* Check whether A_k contains positive values. If yes, go to 1. If not, the computation stops and all clusters found so far are output.

Algorithm ADO extracts clusters from the similarity matrix one by one so that the residual elements are getting smaller at each step overall (Mirkin 1996). A drawback of ADO is that any cluster, once extracted, is never updated, so that a version of the algorithm should be developed with an inbuilt mechanism for updating the extracted clusters.

5.6   Summary

This Chapter is about dividing a dataset or its subset in two parts in either divisive or separative manner. In the former, both parts are to be clusters; in the latter, just one.

The first three sections introduce three different approaches in divisive clustering: Ward clustering, Spectral clustering and Single link clustering. Ward clustering is an extension of K-means clustering dominated by the so-called Ward distance between clusters. Within this approach, an important direction is the so-called conceptual clustering in which every division is made over a single feature to attain immediate interpretability of the hierarchy branches and nodes, that are clusters. Spectral clustering gained popularity with the so-called Normalized Cut approach to divisive clustering. A matrix algebra relaxation of this combinatorial problem appears to be equivalent to optimizing the Rayleigh quotient for Laplacian and pseudo-inverse Laplacian transformation of the similarity matrix under consideration. This amounts to using maximum and minimum eigenvalues and corresponding eigenvectors. In fact, other approaches under consideration also may be treated within the spectral approach, so a few examples of that, applied within the summary similarity clustering framework, are presented in the end of the section. Single link clustering is a popular method for extraction of elongated structures from the data. It formalizes the nearest neighbor approach and is much related to graph-theoretic concepts: components and Maximum Spanning Trees including Prim's algorithm for MST building.

Two remaining sections describe two separative clustering approaches as extensions of popular partitional approaches to the case. One tries to find a cluster with maximum inner summary similarity at a similarity matrix preprocessed according to the uniform or modularity approaches considered in Sect. 4.6. Both of the approaches are effective; they do find good clusters, although there are specifics such as, for example, that the uniform criterion is better fitting to flat ordinary graph structures while the modularity criterion is better fitting at the data reflecting the diversity of intensity in establishing interrelations by individual entities.

The other approach applies the encoder-decoder least-squares approach to modeling data by a one-cluster structure. It appears, criteria emerging within the latter approach are much akin to those described earlier, the summary and semi-average similarities, although parameters now can be adjusted according to the least-squares criterion. This applies to the so-called additive clustering mechanism, an encoder-decoder based model for similarity data, which can be usefully applied to the analysis of similarity data.

References

L. Breiman, J.H. Friedman, R.A. Olshen, C.J. Stone, *Classification and Regression Trees* (Wadswarth, Belmont, Ca, 1984)

B. Mirkin, *Mathematical Classification and Clustering* (Kluwer Academic Press, 1996)

B. Mirkin, *Clustering: A Data Recovery Approach* (Chapman & Hall/CRC, 2012)

F. Murtagh, *Multidimensional Clustering Algorithms* (Physica-Verlag, Vienna, 1985)

Articles

O. Boruvka, Příspěvek k řešení otázky ekonomické stavby elektrovodních sítí (Contribution to the solution of a problem of economical construction of electrical networks)" (in Czech). Elektronický Obzor **15**, 153–154 (1926)

D.H. Fisher, Knowledge acquisition via incremental conceptual clustering. Mach. Learn. **2**, 139–172 (1987)

S. Guattery, G. Miller, On the quality of spectral separators. SIAM J. Matrix Anal. Appl. **19**(3), 701–719 (1998)

C. Klein, M. Randic, Resistance distance. J. Math. Chem. **12**, 81–95 (1993)

J.B. Kruskal, On the shortest spanning subtree of a graph and the traveling salesman problem. Proc. Am. Math. Soc. **7**(1), 48–50 (1956)

G.N. Lance, W.T. Williams, A general theory of classificatory sorting strategies: 1. Hierarchical Systems. Comput. J. **9**, 373–380 (1967)

U. Luxburg, A tutorial on spectral clustering. Stat. Comput. **17**, 395–416 (2007)

B. Mirkin, Additive clustering and qualitative factor analysis methods for similarity matrices. J. Classif. **4**, 7–31 (1987); Erratum **6**, 271–272 (1989)

B. Mirkin, R. Camargo, T. Fenner, G. Loizou, P. Kellam, Similarity clustering of proteins using substantive knowledge and reconstruction of evolutionary gene histories in herpesvirus. Theor. Chem. Acc.: Theory, Comput., Model. **125**(3–6), 569–582 (2010)

F. Murtagh, G. Downs, P. Contreras, Hierarchical clustering of massive, high dimensional data sets by exploiting ultrametric embedding. SIAM J. Sci. Comput. **30**, 707–730 (2008)

M.E.J. Newman, Modularity and community structure in networks. PNAS **103**(23), 8577–8582 (2006)

R.C. Prim, Shortest connection networks and some generalizations. Bell Syst. Tech. J. **36**, 1389–1401 (1957)

J. Shi, J. Malik, Normalized cuts and image segmentation. IEEE Trans. Pattern Anal. Mach. Intell. **22**(8), 888–905 (2000)

R.N. Shepard, P. Arabie, Additive clustering: Representation of similarities as combinations of discrete overlapping properties. Psychol. Rev. **86**, 87–123 (1979)

S.K. Tasoulis, D.K. Tasoulis, V.P. Plagianakos, Enhancing principal direction divisive clustering. Pattern Recogn. **43**, 3391–3411 (2010)

J.H. Ward Jr., Hierarchical grouping to optimize an objective function. J. Am. Stat. Assoc. **58**, 236–244 (1963)

Appendix

Boris Mirkin

Department of Data Analysis and Machine Intelligence, Higher School of Economics, 20 Miasnitskaya Street, Moscow 101000 RF
Department of Computer Science and Information Systems, Birkbeck University of London, Malet Street, London WC1E 7HX UK

Abstract This material consists of six sections. Four sections are to help an interested reader in getting acquainted with

- Basic linear algebra and data approximation
- Basic optimization
- Basic MatLab

Section A5 lists MatLab codes for some of the methods, in addition to the codes supplied in computational parts of the text. These are:

cm.m	Evolutionary method for finding Minkowski's center of a series
plan.m	A set of modules for fitting power law regression by using both evolutionary method and linearization; includes a module for saving results in an ascii file (can be used as a template for saving results)
nnn.m	Learning a neural network with one hidden layer
clatree.m	Building binary classification trees using Gini or Pearson chi-squared or Information gain criterion.

In last Sect. A6, presents two randomly generated samples: three samples from different distributions 50-strong each—the file short.dat, and a 280 strong sample from N(0,10).

© Springer Nature Switzerland AG 2019

B. Mirkin, *Core Data Analysis: Summarization, Correlation, and Visualization*, Undergraduate Topics in Computer Science,
https://doi.org/10.1007/978-3-030-00271-8

A1 Basic Linear Algebra

A1.1 Matrix

Table A.1 presents data matrix from Table 4.2. It has 8 rows and 7 columns, that is, its data contents is a 8×7 matrix X below.

Matrix X is a set of elements x_{iv} labeled by two indices, i related to its rows, and v, related to its columns. It is a common convention that the first of the two indices refers to row and the second to column. The pair, number of rows N and the number of columns V, forms the dimension of matrix X, $N \times V$, or in this case, 8×7.

$$
X = \begin{matrix}
-0.20 & 0.23 & -0.33 & -0.63 & 0.36 & -0.22 & -0.14 \\
0.40 & 0.05 & 0 & -0.63 & 0.36 & -0.22 & -0.14 \\
0.08 & 0.09 & 0 & -0.63 & -0.22 & 0.36 & -0.14 \\
-0.23 & -0.15 & -0.33 & 0.38 & 0.36 & -0.22 & -0.14 \\
0.19 & -0.29 & 0 & 0.38 & -0.22 & 0.36 & -0.14 \\
-0.60 & -0.42 & -0.33 & 0.38 & -0.22 & 0.36 & -0.14 \\
0.08 & -0.10 & 0.33 & 0.38 & -0.22 & -0.22 & 0.43 \\
0.27 & 0.58 & 0.67 & 0.38 & -0.22 & -0.22 & 0.43
\end{matrix}
$$

A1.2 Inner Product and Distance

Every row in data matrix Table A.1 represents an entity as a 7-dimensional vector, or point, such as e1 = (−0.20, 0.23, −0.33, −0.63, 0.36, −0.22, −0.14) which is simultaneously a 1×7 matrix. Similarly, every column represents a feature or category as an 8-dimensional vector, or a 8×1 matrix, such as

Table A.1 Company data standardized

	v1	v2	v3	v4	v5	v6	v7
e1	−0.20	0.23	−0.33	−0.63	0.36	−0.22	−0.14
e2	0.40	0.05	0	−0.63	0.36	−0.22	−0.14
e3	0.08	0.09	0	−0.63	−0.22	0.36	−0.14
e4	−0.23	−0.15	−0.33	0.38	0.36	−0.22	−0.14
e5	0.19	−0.29	0	0.38	−0.22	0.36	−0.14
e6	−0.60	−0.42	−0.33	0.38	−0.22	0.36	−0.14
e7	0.08	−0.10	0.33	0.38	−0.22	−0.22	0.43
e8	0.27	0.58	0.67	0.38	−0.22	−0.22	0.43

$$
\begin{array}{c}
v1 \\
-0.20 \\
0.40 \\
0.08 \\
-0.23 \\
0.19 \\
-0.60 \\
0.08 \\
0.27
\end{array}
$$

or, its transpose, a $1{\times}8$ row.

$$v1^{T} = (-0.20, \quad 0.40, \quad 0.08, \quad -0.23, \quad 0.19, \quad -0.60, \quad 0.08, \quad 0.27)^{T}.$$

Elements of vectors are referred to as their components. Each component has an index which is an integer running from 1 to 8, in this case. In general case, the index runs from 1 to N, the number of components. In Computer Sciences, sometimes indices run from 0 to N − 1. The index is the base for matching different vectors.

Operations of summation and subtraction are defined component-wise:

$$e1 = (-0.20, 0.23, -0.33, -0.63, 0.36, -0.22, -0.14)$$
$$+$$
$$e2 = (0.40, 0.05, \quad 0, \quad -0.63, 0.36, -0.22, -0.14)$$
$$e1 + e2 = (0.20, 0.28, -0.33, -1.26, 0.72, -0.44, -0.28)$$

and

$$e1 = (-0.20, 0.23, -0.33, -0.63, 0.36, -0.22, -0.14)$$
$$-$$
$$e2 = (0.40, 0.05, \quad 0, \quad -0.63, 0.36, -0.22, -0.14)$$
$$e1 - e2 = (-0.60, 0.18, -0.33, \quad 0, \quad 0, \quad 0, \quad 0)$$

The second important operation is multiplication of a vector by a real defined as multiplication of all components simultaneously:

$$3 * e1 = (-0.60, 0.69, -0.99, -1.89, 1.08, -0.66, -0.42),$$
$$10 * e1 = (-2.00, 2.30, -3.30, -6.30, 3.60, -2.20, -1.40)$$

To get some intuition, let us consider Cartesian plane representation of 2D vectors obtained by removing all components of the rows except for the first two (Fig. A.1a).

Figure A.1 illustrates two geometric facts: (a) the sum of two vectors $e1 + e2$ sits in the fourth node of the parallelogram formed by connecting 0 and the vectors;

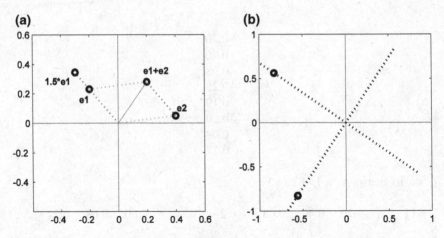

Fig. A.1 Plane geometry representation of 2D vectors e1, e2, e1 + e2, and 1.5 * e1 on (**a**) and eigenvector lines for symmetric matrix A (**b**)

(b) given vector x, all vectors ax at a the constant a taking any value between minus infinity and plus infinity, form a line through the origin 0 and x.

The third important operation over vectors is the inner product. The inner, or scalar, product is defined for every pair of vectors x and y of the same dimension and it is equal to—not a vector—but just a number, the sum of the products of corresponding components and denoted by $<x.y>$. For example, for 2D parts of vectors $e1 = (-0.20, 0.23)$ and $e2 = (0.40, 0.05)$, the inner product is $<e1, e2> = -0.20 * 0.40 + 0.23 * 0.05 = -0.08 + 0.01 = -0.07$. A full computation of $<e1,e2>$: $= \text{sum}(e1.* e2)$ is below:

$$
\begin{array}{rlllllll}
e1 = & (-0.20, & 0.23, & -0.33, & -0.63, & 0.36, & -0.22, & -0.14) \\
e2 = & (0.40, & 0.05, & 0, & -0.63, & 0.36, & -0.22, & -0.14) \\
e1.*e2 = & (-0.08, & 0.01, & 0, & 0.39, & 0.13, & 0.05, & 0.02) \\
\langle e1, e2 \rangle = \text{sum}(e1*e2) & -0.08+ & 0.01+ & 0+ & 0.39+ & 0.13+ & 0.05+ & 0.02 = 0.52
\end{array}
$$

Here we denote $x.*y$ an operation over $n \times 1$ vectors x and y resulting in a vector $x * y = (x_i y_i)$ whose components are products of components of the same index in both x, and y, $x_i y_i$.

The inner product is a linear operation so that, for example, $<e1, 2 * e1 + 3 * e2> = 2 * <e1, e1> + 3 * <e1, e2>$, which can be proven in this case straightforwardly by computation.

The inner square, that is, the product of a vector by itself, like $<e1,e1> = -0.20 * (-0.20) + 0.23 * 0.23 = 0.040 + 0.053 = 0.093$, is the sum of squares of its components, which is the square length of the line connecting the origin 0 and the point on Cartesian plane such as Fig. A.1a. This follows from the Pythagoras theorem illustrated on Fig. A.2. The theorem states that the square of hypothenuse's length in any right-angled triangle is equal to the sum of squares of the sides'

Fig. A.2 Pythagoras' theorem: the squared Euclidean distance between x1 and x2 is $d(x1,x2) = (x_{11} - x_{21})^2 + (x_{12} - x_{22})^2$

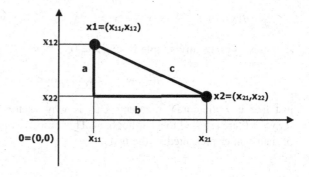

lengths, $c^2 = a^2 + b^2$. By extending this property to multidimensional points and vectors, the square root of the inner square $\langle x,x \rangle$ is referred to as the norm of x and denoted $\|x\|$.

This allows us to introduce Euclidean distance between any two vectors/points x and y as the norm of their difference, $r(x,y) = \|x - y\|$. In MatLab, this can be expressed as r(x,y) = sqrt(sum((x − y).*(x − y))). For example, the distance between e1 and e2 as rows of Table A.1 can be computed as follows:

e1 − e2 = $\quad\quad\quad\quad\quad (-0.60, 0.18, -0.33, 0, 0, 0, 0)$
(e1 − e2). ∗ (e1 − e2) = $\quad (\;\; 0.36, 0.03, \quad 0.11, 0, 0, 0, 0)$
d(e1, e2) = sum((e1 − e2). ∗ (e1 − e2)) = 0.36 + 0.03 + 0.11 + 0 + 0 + 0 + 0 = .50

$$r(e1, e2) = sqrt(d(e1, e2)) = sqrt(0.50) = 0.71$$

An important function in this computation is the squared Euclidean distance $d(e1,e2)$—this is the base of the least-squares approach in data analysis.

Some other concepts of distance are popular too. Among them: Manhattan or City-block distance defined as $m(x1,x2) = |x_{11} - x_{21}| + |x_{12} - x_{22}| + \cdots + |x_{1V} - x_{2V}|$ and Chebyshev or L_∞ distance defined as $c(x1,x2) = max(|x_{11} - x_{21}|, |x_{12} - x_{22}|, ..., |x_{1V} - x_{2V}|)$. A popular exercise in getting intuition about the distances is drawing sets of points that are equidistant to origin 0: this is a circle in the case of Euclidean distance, rhomb in the case of city-block distance, and square in the case of Chebyshev distance. All these are special cases of more general, so-called, Minkowski distance, $lp(x1,x2) = (|x_{11} - x_{21}|^p + |x_{12} - x_{22}|^p + \cdots + |x_{1V} - x_{2V}|^p)^{1/p}$, where p is any positive real number. Obviously, p = 1 corresponds to Machattan distance, p = 2, to Euclidean distance, and at p tending to ∞, to Chebyshev distance.

An important relation between the (Euclidean squared) distance and inner product is this:

$$d(x,y) = <x-y, x-y> \ = \ <x,x> \ + \ <y,y> \ -2<x,y>,$$

which is especially simple if $<x,y> = 0$:

$$d(x,y) = \ <x,x> \ + \ <y,y>$$

just like in Pythagoras' theorem. This is why vectors/points x and y satisfying $<x,y> = 0$ are referred to as orthogonal. This property underlies the decompositions of data scatter presented in the text.

A1.3 Matrix Operations

A general notation for a matrix A is like this:

$$A = \begin{pmatrix} a_{11} & a_{12} & \cdots & a_{1V} \\ \vdots & \ddots & & \vdots \\ a_{N1} & a_{N2} & \cdots & a_{NV} \end{pmatrix}$$

so that A has N rows and V columns and the $N \times V$ size, and a common element is a_{iv} ($i = 1,...,N$, $v = 1,...,V$)—the row's index always goes first. The transpose A^T of matrix A is defined by switching the rows and columns so that $A^T = (a_{vi})$ is of $V \times N$ size:

$$A^T = \begin{pmatrix} a_{11} & a_{21} & \cdots & a_{N1} \\ \vdots & \ddots & & \vdots \\ a_{1V} & a_{2V} & \cdots & a_{NV} \end{pmatrix}$$

A matrix of $N \times V$ size is referred to as a square matrix if $N = V$. A square matrix A is referred to as symmetric if $A = A^T$. The set of elements a_{ii} with coinciding indices is referred to as the diagonal of matrix A. The symmetry then literally is over the diagonal.

Operations of summation, subtraction and multiplication by a number are defined for matrices component-wise exactly as they are for vectors. Here is an example:

-0.20 0.23			0.19 -0.29			0.18 -0.35
-0.40 0.05	+	2*	0.60 -0.42	=		0.80 -0.79
0.08 0.09			0.08 -0.10			0.24 -0.11

Matrices of different sizes cannot be summed with or subtracted from each other. The rules are more complicated for the operation of matrix multiplication.

An $N \times V$ matrix A can be multiplied by a column vector b of the size $V \times 1$ to produce an $N \times 1$ vector $c = Ab$—note that the number of components in b must be equal to the number of columns in A. The result is just the sum of A columns weighted by the corresponding components of b (hence is the rule that the size V of b is equal to the number of columns in A). Here is an example:

$$\begin{pmatrix} -0.20 & 0.23 \\ 0.40 & 0.05 \\ 0.08 & 0.09 \\ -0.23 & -0.15 \end{pmatrix} \begin{pmatrix} 3 \\ 2 \end{pmatrix} = 3 \begin{pmatrix} -0.20 \\ 0.40 \\ 0.08 \\ -0.23 \end{pmatrix} + 2 \begin{pmatrix} 0.23 \\ 0.05 \\ 0.09 \\ -0.15 \end{pmatrix} = \begin{pmatrix} -0.14 \\ 1.30 \\ 0.42 \\ -0.99 \end{pmatrix}$$

This definition can be reformulated using the inner product: in fact, each component of Ab is the inner product of the corresponding row of A and b. Using the same example,

$$\begin{pmatrix} -0.20 & 0.23 \\ 0.40 & 0.05 \\ 0.08 & 0.09 \\ -0.23 & -0.15 \end{pmatrix} \begin{pmatrix} 3 \\ 2 \end{pmatrix} = \begin{pmatrix} <(-0.20 & 0.23), & (3\ 2)> \\ <(0.40 & 0.05), & (3\ 2)> \\ <(0.08 & 0.09), & (3\ 2)> \\ <(-0.23 & -0.15), & (3\ 2)> \end{pmatrix} = \begin{pmatrix} -0.14 \\ 1.30 \\ 0.42 \\ -0.99 \end{pmatrix}$$

Consider, as an example, the operation of multiplication of a row-vector $a^T = (a_1, a_2, \ldots, a_n)$ which is a $1 \times n$ matrix by its transpose, $(a^T)^T = a$, which is an $n \times 1$ matrix. Obviously, $a^T a = a_1^2 + a_2^2 + \cdots + a_n^2 = <a, a> = \|a\|^2$. On the other hand, $a a^T$ is an $n \times n$ matrix whose (i,j)-th entry is $a_i a_j$.

It should be noted that the multiplication is a linear operation. That means that taking a linear combination $\alpha x + \beta y$ of two V-dimensional vectors, x and y, where α and β are arbitrary reals, the product $A(\alpha x + \beta y) = \alpha A x + \beta A y$, the same linear combination of products. Consider the span $L(A)$ of matrix A, that is the set of vectors $c = Ab$ for all possible V-dimensional vectors b. Obviously, $L(A)$ is a linear space, that is, such an algebraic structure that for any $x, y \in L(A)$, any linear combination also belongs to $L(A)$, $\alpha x + \beta y \in L(A)$. A geometric expression of this is that lines are mapped into lines by any matrix.

Consider, for example, $A = \begin{bmatrix} 6 & 7 \\ 5 & 8 \\ 8 & 9 \end{bmatrix}$ and $b = \begin{bmatrix} 2 \\ 3 \end{bmatrix}$; then $c = Ab = \begin{bmatrix} 33 \\ 34 \\ 43 \end{bmatrix}$.

The line through 0 and b consists of points αb at all possible real α (see Fig. A.3 on the left), and its image over mapping A is line formed by points αc (see Fig. A.3 on the right).

Based on the above, matrix product AB is defined for matrices A of size $N \times V$ and B of size $V \times M$ (note coinciding V number!) as a matrix of size $N \times M$ whose columns are products of A and corresponding columns of B. Let us extend our example to this case:

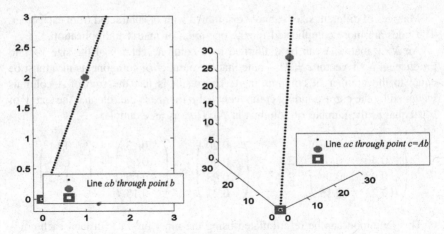

Fig. A.3 Line through 0 and point b, on the left, and line through 0 and point c = Ab, on the right

$$
\begin{pmatrix}
-0.20 & 0.23 \\
0.40 & 0.05 \\
0.08 & 0.09 \\
-0.23 & -0.15
\end{pmatrix}
\begin{pmatrix}
3 & 1 \\
2 & 0
\end{pmatrix}
=
\begin{pmatrix}
-0.14 & -0.20 \\
1.30 & 0.40 \\
0.42 & 0.08 \\
-0.99 & -0.23
\end{pmatrix}
$$

One may consider an $N \times V$ matrix A as a mapping from R^N, the space of N-dimensional vectors x to the space R^V of V-dimensional vectors $y = Ax$. Then the matrix product AB for $N \times V$ matrix A and $V \times M$ matrix B is but the product of the corresponding mappings, a mapping from R^N to R^M, that is the result of the consecutive application of mapping B after mapping A. Then one can see it only natural, that the transpose of the product obeys the following rule $(AB)^T = B^T A^T$.

Given a square $n \times n$ matrix A and an $n \times 1$ vector b, the product $c = Ab$ is again an $n \times 1$ vector. That means that $n \times n$ matrices can be considered as mappings in the space R^n of n-dimensional vectors. Given a square matrix A, matrix A^{-1} is referred to as inverse of A if $AA^{-1} = A^{-1}A = E$, where E is the so-called identity matrix, the diagonal entries of which are unities and the rest are zeros:

$$
e_{ij} = \begin{cases} 1, i = j \\ 0, i \neq j. \end{cases}
$$

A square matrix is referred to as a diagonal one if its (i,j)-entries are all equal to 0 at $i \neq j$. Of course, the identity matrix is diagonal.

Obviously, $AE = EA = A$ for each A. The inverse exists if and only if the mapping of R^n represented by matrix A is injective, that is, one-to-one. In such a case, A is referred to as a non-singular matrix; otherwise, singular. It can be proven that A is non-singular if and only if for any $b \neq 0$, $Ab \neq 0$ as well. Here 0 is an n-dimensional vector in which all components are zeros; this vector is called the origin of the space sometimes. If $n \times n$ A is non-singular, then given a vector $b \in R^n$, equation $Ax = b$ can be easily solved as $x = A^{-1}b$.

A1.4. Spectral Theory

A vector b is of a special interest if $c = Ab$ lies on the line drawn through 0 and b, that is, if equation $Ab = \lambda b$ holds for some number λ. Such a number is referred to as an eigenvalue of A and b the corresponding eigenvector of A. The set of eigenvalues is not too large—the number of eigenvalues cannot exceed the matrix size n. If A is symmetric, then all its eigenvalues are real numbers, and the eigenvectors corresponding to different eigenvalues are orthogonal to each other. In data analysis, it is usually assumed that all the non-zero eigenvalues are different indeed since the matrices are based on observations of quantitative variables because of random errors. We also adhere to this convention. Then the eigenvectors of A represent "inner" directions for Cartesian space axes that follow the structure of A. Geometrically, matrix multiplication transforms lines into lines. Then it would be correct to say that A transforms axes of the Cartesian space into its inner axes specified by the eigenvectors. Figure A.1b represents the eigenvector-defined axes for matrix $A = e + e^T$ where e is the matrix composed of two-dimensional row-vectors $e1$ and $e2$ considered above.

Let A be an $n \times n$ square symmetric matrix with non-zero, necessarily real, eigenvalues $\lambda_1 > \lambda_2 > \cdots > \lambda_r$ and corresponding eigenvectors $c_1, c_2, \ldots, c_r$, where $r \leq n$. Equality $n = r$ holds if and only if A is non-singular. The number r is what is referred to as the rank of A. The rank of a matrix is defined as the maximum size of its square submatrix at which the submatrix is non-singular. The set of eigenvalues is referred to as the spectrum of A. Let us denote by Λ the diagonal $n \times n$ matrix whose diagonal elements are the eigenvalues, so that $\lambda_{11} = \lambda_1, \lambda_{22} = \lambda_2, \ldots, \lambda_{rr} = \lambda_r$, and $\lambda_{pp} = 0$ for all $p > r$. Assume that all the eigenvectors are normed, so that the sum of squares of components in each c_k, $k = 1, \ldots, r$, is unity. Then the so-called spectral decomposition holds:

$$A = C \Lambda C^T, \qquad (A.1)$$

where $C = (c_1, c_2, \ldots, c_r)$ is $n \times r$ matrix in which $c_1, c_2, \ldots, c_r$ are columns. This matrix equation can be formulated element-wise,

$$a_{ij} = \sum_{k=1}^{r} \lambda_k c_{ik} c_{jk} \qquad (A.1')$$

There can be an intermediate notation used as well:

$$A = \sum_{k=1}^{r} \lambda_k c_k c_k^T \qquad (A.1'')$$

These all equations mean that the eigenvalues and eigenvectors provide for a decomposition of all the entries of A in sums of products $c_{ik}c_{jk}$. The spectral decomposition captures the structure of matrices, which is important in data analysis. In data analysis, usually the eigenvalues are numbered so that $\lambda_1 > \lambda_2 > ... > \lambda_r$, assuming that in any real-world dataset, it is highly unlikely that (some of) eigenvalues can be equal to each other.

Since the eigenvectors in C are mutually orthogonal and normed, $C^T C = E$ where E is the identity matrix. Multiplying (A.1) by itself, we obtain $AA = (C\Lambda C^T)$ $(C\Lambda C^T)$, so that $A^2 = C\Lambda\Lambda C^T = C\Lambda^2 C^T$. Since Λ^2 is the diagonal matrix at which λ_k^2 stands on the diagonal rather than λ_k, the latter equation, in format of (A.1''), is

$$A^2 = \sum_{k=1}^{r} \lambda_k^2 c_k c_k^T \qquad (A.2)$$

Of course, this holds for every integer power p, not only $p = 2$. In the case when A is not singular, a similar equation holds for the inverse matrix A^{-1}:

$$A^{-1} = \sum_{k=1}^{n} \frac{1}{\lambda_k} c_k c_k^T \qquad (A.3)$$

differing from (A.1) by only the reciprocals $1/\lambda_k$ used instead of the eigenvalues λ_k. Equation (A.3) can be extended to the case of singular A at which the rank r is smaller than n. In such a case, the equation is

$$A^{+} = \sum_{k=1}^{r} \frac{1}{\lambda_k} c_k c_k^T \qquad (A.3')$$

This equation defines, and provides an algorithm for computing of, the concept of pseudo-inverse matrix.

Multiplying (A.1) by C on the right, we obtain thus $AC = C\Lambda$ (check that this is the defining condition for eigenvalues!). Multiplying this by C^T from the left, we obtain

$$\Lambda = C^T A C, \qquad (A.4)$$

that is,

$$\lambda_k = c_k^T A c_k,$$

under the condition that c_k is a normed vector, which brings us to the so-called Rayleigh quotient,

$$q(x) = \frac{x^T A x}{x^T x} \tag{A.5}$$

This gives an independent way for deriving eigenvalues: the Rayleigh quotient $q(x)$ reaches its maximum over all possible x at the first eigenvector c_1 and its value is $q(c_1) = \lambda_1$. Moreover, if we maximize $q(x)$ at only those x that are orthogonal to the first $k - 1$ eigenvectors $c_1, c_2, ..., c_{k-1}$, then the maximum is λ_k and it is reached at $x = c_k$.

A1.5. Singular Values and SVD

Given an $n \times m$ matrix X, its singlular triplet is defined as a set (μ, u, v) in which μ is a non-negative real, u an m-dimensional normed vector $u \in R^m$ and v n-dimensional normed vector $v \in R^n$, such that

$$\begin{cases} Xu = \mu v \\ X^T v = \mu u \end{cases} \tag{A.6}$$

The number μ is referred to as a singular value and vectors u, v the corresponding singular vectors. Geometrically, u represents a direction $(0, u)$ from the origin in R^m, whereas v a direction $(0, v)$ from the origin in R^n (as shown on Fig. 5.3 with $u = b$ and $v = c$). Equations (A.6) claim that the direction $(0, v)$ is the image of $(0, u)$ under mapping X, while the direction $(0, u)$ is the image of $(0, v)$ under mapping X^T. Geometrically, X^T represents the inverse mapping to X. The situation much resembles that in the concept of eigenvector. The difference is that the eigenvectors are arguments and images of the same mapping in the same space, whereas singular vectors are in different spaces and under somewhat different mappings. Can we somehow put this into a larger space in which both mappings, X and X^T, act simultaneously? Yes, we can.

Given a rectangular $n \times m$ matrix X, define a square $(n + m) \times (n + m)$ matrix X^*, that consists of four blocks two of which, the diagonal $n \times n$ and $m \times m$ blocks, are all zeros, whereas two others are X and X^T:

$$X^* = \begin{pmatrix} 0 & X \\ X^T & 0 \end{pmatrix}$$

Note that X^* is symmetric, which implies that all its eigenvalues are real. Consider an $(n + m)$—dimensional vector $w = (u, v)$ in which u is its m-dimensional part and v n-dimensional part. The condition that w is an eigenvector of X^* is $X * w = \lambda w$. Because of the special structure of X^*, this can be reformulated by using the constituent parts of w, u and v, as $0u + Xv = \lambda u$, $X^T u + 0v = \lambda v$, which is exactly equations (A.6) except that λ here plays the role of μ. What remains to be seen is whether the norms of parts u and v are necessarily the same in $w = (u,v)$ or not. If this is true, then both vectors can be taken as normed and, in this way, satisfying the definition. Let us multiply the first of the equations (A.6) by v^T from the left and the second equation by u^T from the left. This will produce

$$\begin{cases} v^T X u = \mu v^T v = \mu ||v||^2, \\ u^T X^T v = \mu u^T u = \mu ||u||^2. \end{cases}$$

But $v^T X u$ and $u^T X^T v$ coincide because $v^T X u$ is of 1×1 size, that is, a real number and, therefore, is equal to its transpose $(v^T X u)^T = u^T X^T v$. The following statement is proven.

Pair λ, $w = (u,v)$ combines a non-negative eigenvalue and the corresponding eigenvector of X^* if and only if the triplet (λ,u,v) is singular for X.

This leaves yet an issue of the negative eigenvalues of X^*, which appears to be resolved in a quite satisfactory way. Indeed, it is easy to prove, just by using negations in equations (A.6) that a pair λ, $w = (u,v)$ is an eigenpair for X if and only if so is the pair $-\lambda$, $w^- = (-u,v)$. Therefore, one may only take into account the non-negative eigenvalues of X^*; the negative ones are just replicas of the positive ones.

Another, much more popular, geometrical view of singular values as eigenvalues, comes from consecutively applying the mapping X and inverse mapping X^T so that the resulting mapping works in only one of the spaces, R^m or R^n. Let us multiply the first equation in (A.6) by X^T from the left. We obtain $X^T X u = X^T \mu v = \mu X^T v$. The last equation follows from the fact that product of a matrix and a number is commutative. Now we put $X^T v = \mu u$, from the second equation in (A.6), into the expression on the right. This produces $X^T X u = \mu^2 u$. The $n \times n$ matrix $A = X^T X$ is what is referred to as Gram matrix; its elements are inner products of the columns of a matrix, X, in this case. Of course, $X^T X$ is symmetric. The obtained equation $X^T X u = \mu^2 u$ implies that u is an eigenvector of $X^T X$ corresponding to its eigenvalue μ^2 which is non-negative. Therefore, eigenvalues of a Gram matrix are non-negative reals. This implies that a Gram matrix $A = X^T X$ is semi-definite positive, that is, for any m-dimensional x, the product $x^T A x$ is non-negative so that $x^T A x \geq 0$ for any x.

Similarly, one can prove that $X X^T v = \mu^2 v$. Allover, the triplet (μ, u, v) is a singular triplet for matrix X if and only if is an eigen value of matrices $X^T X$ and $X X^T$ and u is the corresponding normed eigenvector of matrix $X^T X$ and v, of matrix $X X^T$. This implies that singular m-dimensional vectors $u_1, u_2,..., u_r$ are mutually orthogonal, as well as singular n-dimensional vectors $v_1, v_2,..., v_r$.

The number r of positive singular values of X is equal to the rank of X. For rectangular matrices, the rank is defined in the very same way as for the square ones. The rank of X is the maximum size of such its square submatrix, that is, non-singular. Obviously, the rank is less than or equal to the minimum of m and n.

Let $\mu_1 > \mu_2 > \cdots > \mu_r > 0$ be positive singular values and $u_1, u_2,..., u_r$ and $v_1, v_2,..., v_r$ the corresponding singular vectors for $n \times m$ matrix X. The so-called singular value decomposition (SVD) of elements x_{ij} of X holds:

$$x_{ij} = \sum_{k=1}^{r} \mu_k \sum_{i=1}^{m} \sum_{j=1}^{n} u_{ik} v_{jk} \qquad (A.7)$$

Let us consider $m \times r$ matrix $U = (u_{ik})$ whose columns are singular vectors u_k ($i = 1,...,m$, $k = 1,..., r$) and $n \times r$ matrix $V = (v_{jk})$ whose columns are singular vectors v_k ($j = 1,...,n$, $k = 1,..., r$). Let us denote the $r \times r$ diagonal matrix of singular values μ_1, μ_2, ...,μ_r by M. Then the SVD in (5.7) can be reformulated as a matrix equation:

$$X = UMV^T, \qquad (A.7')$$

or, a bit less cryptically,

$$X = \sum_{k=1}^{r} \mu_k v_k u_k^T \qquad (A.7'')$$

A2 Basic Optimization

Given a function $f(x)$ for $x \in X$, it is natural to look for points x in X at which $f(x)$ takes extreme values, ether maximum or minimum, hence is the problem of optimization, that is, finding a point that either minimizes f or maximizes it. Let us focus on minimization, for certainty. There are two approaches to optimization: one is the classical one and the other of nature-inspired computational intelligence.

The classical approach is informed by calculus.

This approach has been first developed for one-dimensional functions $f(x)$ like the one whose graph is on Fig. A.4. In the point of minimum, like A or D, or maximum, like C, or change in the orientation of convexity, like B, the first derivative $f'(x)$, which expresses the tangent of the curve $f(x)$ in the point, is 0—this is what is referred to as the first-order necessary condition of minimum. It is possible to separate the minima from the rest by using the second order derivatives, but there is no way to tell one local minimum from the other unless reaching each of them, and to add to the misery, there is not much usually known of how to find them all or just the global minimum either. Sometimes the calculus is not of much help—a case at hand is the curve on Fig. A.5: its global minimum is at the very left point of the graph, and the first-order condition cannot help because it is valid only in interior points of the admissible set X.

Yet to reach a local minimum satisfying the first-order minimum condition, a most universal method is of steepest descent. This method relies on the derivative of the function in any given point. This shows the direction of the steepest ascent over the optimized function, so that the opposite direction makes it steepest descent.

Fig. A.4 Graph of a typical multi-optimum function

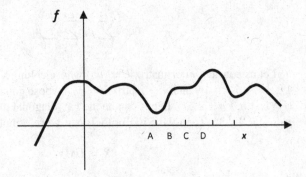

Fig. A.5 New point taken in the direction opposite to the tangent

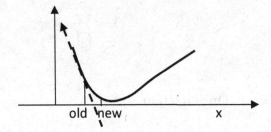

Given an x and values $f(x)$ and $f'(x)$, this method finds another point x_{new} by subtracting the derivative scaled by a step factor, $x_{new} = x - \mu f'(x)$, where μ is the step factor. The closer the point to the minimum, the smaller is the value of the derivative, thus the smaller the change. Of course, the method can converge to a local minimum point, not necessarily the global minimum.

The situation when x is multidimensional is even more complex. The mathematics have made a good progress on the theory of optimization when only one minimum can exist—such is the case of so called convex optimisation when both, function $f(x)$ and set of admissible points X, are convex. In a more general situation, though, the steepest descent frequently remains the only tool available, even in spite of the fact it finds a local minimum with no estimates on the global one. Here, however, the concept of gradient is involved rather that of the derivative. For a function of n-dimensional vectors, $f(x_1, x_2, \ldots, x_n)$, the gradient is an n-dimensional vector $\text{grad}(f(x))$ whose k-th component is partial derivative $\frac{\partial y}{\partial x_k}$ ($k = 1, \ldots, n$). The different term is used because there are examples of functions that have all the partial derivatives at some points but still have no gradients in those points. The gradient, in n-dimensional space, shows direction of the steepest ascent. So, by taking the opposite direction, the process is supposed to go in the direction of the steepest descent. That makes the method of steepest descent to work in iterations.

Each iteration takes in a point $x = (x_1,x_2,...,x_n)$ and outputs a new point in the direction opposite to the gradient:

$$x(new) = x(old) - \mu * grad(f(x(old)))$$

where μ is the step size. This new point is taken then by the next iteration. By changing the step-size from iteration to iteration, one may achieve a better rate of convergence.

In the case when the set of arguments can be naturally partitioned in two or more parts such that the function is easy to minimize over each part taken separately, an iterative process applies to involve steps optimizing each part at pre-specified values of the other parts. This process is referred to as alternating minimization. Consider that $x = (y,z)$ so that $f(x) = f(y,z)$ and, at any given $y*$ and $z*$, the minimum of $f(y*,z)$ with respect to z can be found easily, as well as minimum $f(y,z*)$ over y. Then, starting from some y^0 the alternating minimization process would produce a sequence $y^0, z^1, y^1, z^2, y^2,...$ in which z^t is a minimizer of $f(y^{t-1},z)$ and y^t a minimizer of $f(y,z^t)$ at each $t = 1, 2,....$ This sequence would provide for an ever decreasing sequence of values $f(y^t,z^t)$. In a situation when there is a bound on them from below, this would warrant that the sequence converges to a local minimum. If either y or z can have only a finite number of values, the process of alternating minimization would converge in a finite number of steps.

Q.A.1. What is gradient of function: (i) $f(x_1,x_2) = x_1^2 + x_2^2$, (ii) $f(x_1,x_2) = (x_1 - 1)^2 + 3 * (x_2 - 4)^2$, (iii) $f(z_1,z_2) = 3 * z_1^2 + (1 - z_2)^4$?

A: (i) $(2x_1, 2x_2)$, (ii) $[2 * (x_1 - 1), 3 * (x_2 - 4)]$, (iii) $(6 *z_1, - 4 * (1-z_2)^3)$.

In contrast to classical approaches, a nature inspired optimization approach does not try to reach a minimum by improving and updating a single solution point. Just the opposite. According to this approach, a population of admissible solutions is thrown in randomly and all the attention is given not to an individual solution but the population as a whole. Probabilistic rules are defined to generate the next generation of the population, usually in the same numbers, so that a process of evolution of the population from generation to generation is defined and executed computationally. Because of its probabilistic rules, each instance of the process may differ from the others. To warrant that the population improves in the process of evolution, a special "elite maintenance" policy is defined so that the elite—which is the best solution or a set of best solutions reached so far in the process—is used as an improvement device. The presence of the probabilistic component is considered an important device to warrant that the population does not stuck in a local optimum but rather covers the entire area of admissible solutions.

A3 Approximate Data

A3.1 Orthogonal Projection and Projector

Consider an $n \times m$ matrix X. The set $L(X)$ of all n-dimensional vectors c such that $c = Xb$ for some b is referred to as the L-span of X or just X-span. Given an n-dimensional vector u, not necessarily belonging to $L(X)$, the problem is: find a vector $\hat{u} \in L(X)$ such that the Euclidean distance from u to $\hat{u}$ is at its minimum over all the vectors in $L(X)$. The global solution to this problem is well-known: it is provided by a matrix P_X applied to u:

$$\hat{u} = P_X u \tag{A.8}$$

where P_X is the so-called orthogonal projection operator, or orthogonal projector, an $n \times n$ matrix defined as:

$$P_X = X \left(X^T X\right)^{-1} X^T \tag{A.9}$$

so that

$$\hat{u} = X \left(X^T X\right)^{-1} X^T u \text{ or } \hat{u} = Xb \text{ at } b = \left(X^T X\right)^{-1} X^T u. \tag{A.10}$$

Matrix P_X projects every n-dimensional vector u to its nearest match in the n-dimensional X-span space. This matrix depends only on the space $L(X)$ itself, not on the specifics of matrix X. An $n \times n$ matrix P is an orthogonal projector if and only if it is symmetric, $P = P^T$ and idempotent; that is, its repeated application does not change the image, $P^2 = P$.

The inverse $(X^T X)^{-1}$ in (A.9) does not exist if the rank of X, as it may happen, is less than the number of columns in X, m, that is, if matrix $X^T X$ is singular or, equivalently, the dimension of X-span is less than m. In this case, the so-called pseudo-inverse matrix $(X^T X)^+$ can be used as well. This is not a big deal computationally: for example, in MatLab one just puts command pinv($X^T X$) instead of inv ($X^T X$), see formulas (A.3) and (A.3') above.

A3.2. SVD and Data Approximation

Let us extend the problem of projection of a vector to data matrices by relaxing both conditions, that the projection space is pre-specified and that the data is just a single point. Given an $n \times m$ data matrix X and a relatively small integer $p < r$ where r is the rank of X, consider the problem of finding such a linear space L of the dimensionality p that is the nearest to X. One can formalize the problem as follows. Find an $n \times p$ matrix Z consisting of p mutually orthogonal columns such that the projection $P_Z X$ of X into the Z-span $L(Z)$ is as near X as possible with respect to all possible p-dimensional spaces.

To measure the distance between matrices, one usually uses a straightforward extension of the Euclidean vector norm onto matrices. Given two $n \times m$ matrices X and Y, the norm $\|X - Y\|$ is defined, via its square, as

$$\|X - Y\|^2 = \sum_{i=1}^{n} \sum_{j=1}^{m} (x_{ij} - y_{ij})^2.$$

This can be expressed in terms of matrix operations by using the notion of trace of a square matrix. Given a square matrix A, its trace is defined as the sum of its diagonal elements, $Tr(A) = \sum_{i=1}^{n} a_{ii}$. The trace is equal to the sum of all eigenvalues of A.

It appears, $\|A\|^2 = Tr(A^T A) = Tr(AA^T)$. Therefore, the problem is to find such an $n \times p$ matrix Z that minimizes

$$\|X - P_Z X\|^2 = Tr\left([X - P_Z X]^T [X - P_Z X]\right)$$

under the orthonormality constraint

$$Z^T Z = E.$$

The function to minimize can be simplified to

$$\|X - P_Z X\|^2 = Tr(X^T X) - Tr(Z^T X X^T Z).$$

The first item on the right, $T(X) = Tr(X^T X) = \sum_{i=1}^{n} \sum_{j=1}^{m} x_{ij}^2$ is but the sum of squared distances from all the row-vectors (or column-vectors) to 0 and referred to as the data scatter. Since $T(X)$ is constant, the problem can be reformulated as of finding an $n \times p$ matrix Z satisfying equation $Z^T Z = E$ and maximizing $Tr(Z^T X X^T Z)$. This is much reminiscent of equations (5.4) for the spectrum of $A = XX^T$ and $C = Z$. Indeed, consider Z consisting of the p eigenvectors $v_1, v_2, ..., v_p$, of $A = XX^T$ corresponding to its maximal p eigenvalues $\lambda_1 > \lambda_2 > ... > \lambda_p$, respectively. The (globally) optimal projection $\tilde{X}(p) = P_Z X$ in this case is part of (A.7'') comprising the first p singular triplets of X:

$$\tilde{X}(p) = \sum_{k=1}^{p} \mu_k v_k u_k^T \tag{A.11}$$

where $\mu_1 > \mu_2 > ... > \mu_p > 0$ are the maximal singular values of X. Note that $\mu_k = \lambda_k^{1/2}$. Readers who have read the previous part of this section attentively, will not be surprised to learn that this solution satisfies a Pythagorean equation decomposing the data scatter in two parts, that taken into account by the approximation, $\tilde{X}(p)$, and that residual, $\left\| X - \tilde{X}(p) \right\|^2$:

$$T(X) = \|X\|^2 = \sum_{i,j} x_{ij}^2 = \mu_1^2 + \mu_2^2 + \cdots + \mu_p^2 + \left\| X - \tilde{X}(p) \right\|^2 \qquad (A.12)$$

The part taken into account is itself decomposed in contributions of individual singular triplets, μ_k^2. The residual part is equal to the sum of contributions of singular triplets corresponding to the smaller singular values,

$$\left\| X - \tilde{x}(p)^2 \right\| = \sum_{k=p+1}^{r} \mu_k^2$$

The case of $p = 1$ is of a special interest both from theoretical and practical perspectives. Let μ be the maximum singular value of X and $u = (u_j)$, $v = (v_i)$ corresponding normed singular vectors. Then the approximating matrix is $\tilde{X}$ in which (i,j)-th entry is the product $v_i u_j$. Any matrix of rank 1 has a similar structure generated by two vectors, one from R^n the other from R^m; all its rows are proportional to each other and all its columns are proportional to each other.

This all can be extended to any square $n \times n$ matrix A that is not necessarily XX^T. The first p items in equation (A.1″) form a matrix $A(p)$ which is the best at approximating A among all the square matrices of rank p. More precisely, let us consider matrix

$$A(p) = \sum_{k=1}^{p} \lambda_k c_k c_k^T$$

and the difference $E = A - A'$ where A' is any $n \times n$ matrix of rank $p < n$. It appears, the minimum of $\|E\|^2 = \|A - A'\|^2$, with respect to all matrices A' of rank p, is reached at $A' = A(p)$. Therefore, equation

$$A = \sum_{k=1}^{p} \lambda_k c_k c_k^T + E \qquad (A.13)$$

decomposes any square matrix A into the best approximating matrix of rank $p < n$ and the residual E.

There is a special method, the so-called power method, for finding the maximum μ and corresponding singular values. The method can be presented in various forms. Consider, say, matrix $A = XX^T$ where X is an $n \times m$ matrix. To compute its first eigenvector, take an arbitrary positive $x0 \in R^n$ and apply A to it to compute $x1$ by dividing the result by its norm, $x1 = A * x0 / \|A * x0\|$. Reiterate to obtain $x2$: $x2 = A * x1 / \|A * x1\|$. Continuing this process, we obtain a sequence $x0, x1, x2, \ldots, xs$

which converges to the first eigenvector v while the norm $\|A * xs\|$, to the maximum eigenvalue λ.

For example, take matrix

$$X = \begin{pmatrix} 0.3 & 0.4 & 0.1 & 0.1 \\ 0.5 & 0.6 & 0.2 & 0.0 \\ 0.4 & 0.7 & 0.1 & 0.2 \end{pmatrix}$$

and compute

$$A = XX^T = \begin{pmatrix} 0.27 & 0.41 & 0.43 \\ 0.41 & 0.65 & 0.64 \\ 0.43 & 0.64 & 0.70 \end{pmatrix}$$

The following Table A.2 presents results of a few iterations of the power method applied to A.

To explain the convergence, note that s iterations of the power method are equivalent to one-time application of the s-th power of A, A^s, up to rescaling normalization steps. According to equations (A.2) extended to the s-th power of A,

$$A^s = \sum_{k=1}^{r} \lambda_k^s c_k c_k^T = \lambda_1^s \left[c_1 c_1^T + \sum_{k=2}^{r} \left(\frac{\lambda_k}{\lambda_1}\right)^s c_k c_k^T \right].$$

At growing s, the fractions in power s tend to 0 since λ_1 is strictly greater than any λ_k at $k = 2,3,\dots r$. The factor λ_1^s at the square brackets is taken care of by the normalizing steps. Therefore, the result of application of s iterations to any initial vector will approach $c_1 c_1^T x0$ which is proportional to c_1, the eigenvector of A for the eigenvalue λ_1.

Moreover, the equation above warrants that A^s is approximated by matrix $c_1 c_1^T$ multiplied by a number, so that all the columns of matrix A^3 itself are versions of the corresponding eigenvector c_1. Indeed, in our case,

Table A.2 Results of three iterations of the power method for matrix A. The sequence of obtained vectors in the top row and corresponding norms in the bottom

X0	X1	X2	X3
1	0.4121	0.4120	0.4120
1	0.6311	0.6311	0.6311
1	0.6571	0.6572	0.6572
$\|A * X0\|$	$\|A * X1\|$	$\|A * X2\|$	$\|A * X3\|$
2.6935	1.5841	1.5841	1.5841

$$A^3 = \begin{pmatrix} 0.6747 & 1.0336 & 1.0783 \\ 1.0336 & 1.5834 & 1.6489 \\ 1.0763 & 1.6489 & 1.7171 \end{pmatrix}$$

The norms of columns of A^3 are, in respect, 1.6376, 2.5088, 2.6126. Relating the columns to their norms will produce the very same normed vector c_1.

It should be noted that the very same power method can be applied to finding the maximum eigenvalue and corresponding eigenvector even in the case when A is not symmetric, provided that the maximum eigenvalue is known to be real and positive! Such is the case at which all entries of A are positive. In this case the so-called Perron-Frobenius theorem is true. The theorem states that the power method for such an asymmetric A converges to the first eigenvector c_1 and A^s, to a matrix whose columns are collinear to c_1.

Returning to $A = XX^T$, we may proceed to singular triplets for X. This eigenvector c_1 is the singular vector $v \in R^n$ for matrix X, corresponding to its maximum singular value $\mu_1 = \lambda_1^{1/2}$. The other singular vector, $u \in R^m$, can be found from the defining equation $X^T v = \mu u$: $u = X^T v / \mu$.

Applying this to our example, we have the maximum singular value of X equal to $(1.584)^{1/2} = 1.2586$. One singular vector is

$$v = \begin{matrix} 0.4120 \\ 0.6311 \\ 0.6572, \end{matrix}$$

And the other singular vector is

$$u = X^T v / \mu = \begin{matrix} 0.5578 \\ 0.7973 \\ 0.1852 \\ 0.1372 \end{matrix}$$

This allows us to compute the best approximating 1-rank matrix $\tilde{X}$ as well as the residual matrix $X - \tilde{X}$, all presented in Table A.3.

According to the Pythagorean decomposition (A.12), the approximating matrix $\tilde{X}$ contributes 1.5841 to the data scatter $Tr(XX^T) = Tr(A) = 1.62$, that is, 97.78% of the latter. That means that the residual matrix accounts for 2.22% of the data scatter

Table A.3 Original matrix X, approximating matrix of rank one $\tilde{X}$, and the residual matrix $X - \tilde{X}$

Matrix X	One-rank matrix $\tilde{X}$	Residual matrix $X - \tilde{X}$
0.3 0.4 0.1 0.1	0.289 0.413 0.096 0.071	0.011 −0.013 0.004 0.029
0.5 0.6 0.2 0.0	0.443 0.633 0.147 0.109	0.057 −0.033 0.053 −0.109
0.4 0.7 0.1 0.2	0.461 0.660 0.153 0.114	−0.061 0.040 −0.053 0.086

only, although some of the residuals may look rather large (such as the (2,4)-entry, -0.109 in Table A.3).

One may now compute the second largest eigenvalue and its corresponding eigenvector for matrix $A = XX^T$ by applying the same process to the residual matrix $X - \tilde{X}$. Then the same process applied to the next residual matrix will compute the third largest eigenvalue and corresponding eigenvector. This process of one-by-one extraction may continue till the total contribution of the found SVD parts reaches an acceptable threshold which may be sometimes, in social sciences for example, as small as 40–50% of the data scatter.

A3.3 Canonical Correlation

Given two data matrices for the same set of objects, say $n{\times}m_1$ matrix X and $n{\times}m_2$ matrix Y, let us try scoring correlation between them. A rather straightforward formulation may be this: let us consider X-span and Y-span spaces $L(X)$ and $L(Y)$ and find such $x{\in}L(X)$ and $y{\in}L(Y)$, that is, $x = Xa$ and $y = Yb$, that the squared Euclidean distance $\|x - y\|^2$ between them is at its minimum. The intersection $L(X) \cap L(Y)$ of the spaces is not empty, as $0{\in} L(X) \cap L(Y)$. To avoid such a trivial solution, assume that both x and y belong to a unit sphere in its space, that is, $\|x\| = \|y\| = 1$.

Therefore, the problem can be formulated as follows. Find $a \in R^{m_1}$ and $b \in R^{m_2}$ minimizing $\|Xa - Xb\|^2$ with regard to normalization constraints $(Xa)^T(Xa) = 1$ and $(Yb)^T(Yb) = 1$. Since $\|Xa - Xb\|^2 = (Xa)^T(Xa) + (Yb)^T(Yb) - 2(Xa)^T(Yb) = 2 - 2(Xa)^T(Yb)$, the problem can be equivalently reformulated as of finding $a \in R^{m_1}$ and $b \in R^{m_2}$ maximizing the inner product $<x,y> = (Xa)^T(Yb) = a^T X^T Y b$ with regard to the normalization conditions above. The solution to this problem satisfies equations following from the first order optimality conditions:

$$P_X y = \lambda x, \quad P_Y x = \lambda y,$$

where P_X and P_Y are orthogonal projectors to $L(X)$ and $L(Y)$, respectively, and λ the maximum value of the inner product $<x,y>$. By substituting the equations into one another, one obtains final equations for the solution:

$$P_X P_Y x = \lambda^2 x, \quad P_Y P_X y = \lambda^2 y. \tag{A.14}$$

These show that the sought vectors $x = Xa$ and $y = Yb$ are eigenvectors of the matrices $P_X P_Y$, $P_Y P_X$, respectively, corresponding to their maximum eigenvalue λ^2. Spectra of the matrices $P_X P_Y$ and $P_Y P_X$ coincide; the square roots of the eigenvalues are referred to as canonical correlations, and corresponding eigenvectors x and y, canonical vectors. The canonical correlations are non-negative and are smaller than or equal to 1. The number r of positive canonical correlations is equal to the minimum of ranks of X and Y. The canonical vectors x and y corresponding to the same eigenvalue are mutually orthogonal. To find the vectors a and b, let us consider matrices $A = (X^TX)^{-1}X^TY$ and $B = (Y^TY)^{-1}Y^TX$. It appears, a and b are eigenvectors of AB and BA, respectively, corresponding to the same eigenvalue λ^2:

$$ABa = \lambda^2 a, \quad BAb = \lambda^2 b. \tag{A.15}$$

Multiplying either of the equations in (A.15) from the left by B or A, respectively, one can see that Ba in fact coincides, up to normalization, with b and, vice versum, Ab coincides, up to normalization, with a. In practical computations, one should use that of the matrices AB, BA, P_XP_Y, P_YP_X, which has the smallest dimensions.

Given a relatively small integer $p < r$, matrices $\tilde{X}(p)$ and $\tilde{Y}(p)$ consisting of the canonical vectors x_1, x_2, ...,x_p and y_1, y_2,...,y_p, respectively, corresponding to the first p maximal eigenvalues $\lambda_1^2 > \lambda_2^2 > \cdots > \lambda_p^2$ of matrices AB, BA, P_XP_Y, P_YP_X, solve the following problem. Find in the spaces $L(X)$ and $L(Y)$ such sets of p mutually orthogonal normed vectors represented, respectively, by matrices Up and Vp that the distance between them $||Up - Vp||^2$ is as small as possible, or the summary canonical correlations $<x_k, y_k> = \lambda_k$ ($k = 1,2,...,p$) are as large as possible. This is why the canonical correlations are frequently used for scoring the level of similarity between the spaces $L(X)$ and $L(Y)$ or between the data matrices X and Y themselves.

A4 Basic MatLab

A4.1 Introduction

The working place within a processor's memory is up to the user. A recommended option:

- a folder with user-made MatLab codes, termed say Code, and two or more subfolders, Data and Result, in which data and results, respectively, are to be stored.

MatLab's icon then is clicked on, after which MatLab opens as a three-part window, of which that on the right is working area referred to as Command Window, and the two parts on the left are auxiliary. MatLab can be brought in to the working folder/directory with traditional MSDOS or UNIX based commands such as: cd <Path_To_Working_Directory> in its Command Window. MatLab remembers then this path; and it is available to the user in a tiny window on top of the Command Window.

MatLab is organized as a set of packages, each in its own directory, consisting of program files with extension .m each. Character '%' symbolizes a comment, usually made for humans till the end of the line.

Help can be invoked Windows-wise or within the working area. In the latter, "help" command allows seeing names of the packages as well as of individual program files; the latter are operations that can be executed within MatLab. Example: Command "help" shows a bunch of packages, "matlab\datafun" among them; command "help datafun" displays a number of operations such as "max— largest component"; command "help max" explains the operation in detail.

A4.2 Loading and Storing Files

A numeric data file should be organized as an entity-to-feature data table: rows correspond to entities, columns to features (see studn.dat and studn.var). Such a data structure, with all entries numerical, is referred to as a 2D array, corresponding to a matrix in mathematics; 1D arrays correspond to solitary entities or columns (features) or rows (entity records). Array is a most important MatLab data format to hold numeric values. It works on the principle of a chess-board: its (i,k)-th entry arr(i,k) is the element in i-th row and k-th column. An Excel file has a similar structure but it is interlaced with strings. A 2D array's defining feature is that every row has the same number of digits.

To load such a file, one may use a command from package "iofun". A simple one is "load" to load a numeric array, organized as described, into the current MatLab processor memory:

>>arr=load('Data\stud.dat');
% symbol "%" is used for comments:
% MatLab interpreter doesn't read lines beginning with "%".
% "arr" is a place in computer's memory to put the data (variable);
% semicolon ";" should stand at the end of an instruction;
% if it does not, then the result will be printed to the screen,
% which can be very useful for the user for checking the process of computation
% studn.dat is a 100x8 file of 100 part-time students with 8 features:
% 3 binary for Occupation categories; then Age, NumberChildren,
% and scores over three disciplines.
% All feature names are in file studn.var stored in Data folder.

An 1D array can be put into the workspace with a command like
≫a=[3 4 7 0]
which is a 4×1 array, which can be transposed into a 1×4 array with a "transpose" command
≫b=a'
Since no semicolon is put in the end, b will be displayed on screen as

$$3$$
$$4$$
$$7$$
$$0$$

To get its 2D entry, a command
≫c=b(2)
can be utilized. Similarly, command
≫d=arr(7,8)
puts the value in arr's 7th row and 8th column into workspace as variable d.

If a numeric array in working memory is to be stored, one may use MatLab command "save" which admits a number of storage formats including internal .mat format (see more with "help save"). To store array X into file Result\good.res in ASCII format (which is a text format covering characters in a standard keypad set), one may use command
≫save Result\good.res X −ascii

If you need to check, before saving, what files and variables are currently in the workspace, you may use the upper-left part of the MatLab window or command
≫whos
that produces the list on the screen.

Names are handled as *strings*, with ' ' symbol. The entity/feature name sizes may vary, thus cannot be handled in the array format.

To do this, another data format is used: the *cell*. Cells involve curly braces rather than round brackets (parentheses) utilized for arrays. See the difference: arr(i) is 1D array arr's i-th element, whereas brr{i} is cell brr's i-th element, which can be not only a number or character, as in arrays, but also a string, an array, or even another cell.

There are other data structures as well in MatLab (video, audio, internet) which are not covered here.

MatLab supports several data formats, including Excel which is popular among scientists and practitioners alike (see more in help iofun). An Excel file with extension .xls can be dealt with in MatLab by using commands xlsread and xlswrite. Straightforward as they are, the user should not expect a comfortable switch between Excel and MatLab with these commands. Take a look, for example, onto Excel data file of several students in the Table A.4.

The xlsread command produces three data structures from an xls file: one for numeric part, the other for text part, and the third for all data in the file. Specifically, if the table above is stored in Data subfolder as student.xls file, this works as follows:

≫[nn,tt,rr]=xlsread('Data\student.xls');
% nn is array of numeric values, tt – is cell of text,
% and rr is cell covering all the data in file

to produce a numeric 5×4 array nn:

Table A.4 An Excel spreadsheet with data of five students over four features (Age in years, Number of Children, Occupation [Information Technology IT or Business Administration BA or Other AN], and Mark over a range of 0–100%)

Student	Feature			
	Age	#Children	Occup	CI_Mark
John	35	0	IT	94
Peggy	28	2	BA	67
Fred	27	1	BA	85
Chris	28	0	OT	48
Liz	25	0	IT	87

$$
nn = \begin{array}{cccc}
35 & 0 & \text{NaN} & 94 \\
28 & 2 & \text{NaN} & 67 \\
27 & 1 & \text{NaN} & 85 \\
28 & 0 & \text{NaN} & 48 \\
25 & 0 & \text{NaN} & 87
\end{array}
$$

and a text 8×5 cell tt:

$$
tt = \begin{array}{ccccc}
\text{'Feature'} & \text{'Age'} & \text{'\#Children'} & \text{'Occup'} & \text{'ML\_Mark'} \\
\text{'Student'} & '' & '' & '' & '' \\
'' & '' & '' & '' & '' \\
\text{'John'} & '' & '' & \text{'IT'} & '' \\
\text{'Peggy'} & '' & '' & \text{'BA'} & '' \\
\text{'Fred'} & '' & '' & \text{'BA'} & '' \\
\text{'Chris'} & '' & '' & \text{'OT'} & '' \\
\text{'Liz'} & '' & '' & \text{'IT'} & ''
\end{array}
$$

The full dataset is in 8×5 cell rr:

$$
rr = \begin{array}{ccccc}
\text{'Feature'} & \text{'Age'} & \text{'\#Children'} & \text{'Occup'} & \text{'CI\_Mark'} \\
\text{'Student'} & [\text{NaN}] & [\text{NaN}] & [\text{NaN}] & [\text{NaN}] \\
[\text{NaN}] & [\text{NaN}] & [\text{NaN}] & [\text{NaN}] & [\text{NaN}] \\
\text{'John'} & [35] & [0] & \text{'IT'} & [94] \\
\text{'Peggy'} & [28] & [2] & \text{'BA'} & [67] \\
\text{'Fred'} & [27] & [1] & \text{'BA'} & [85] \\
\text{'Chris'} & [28] & [0] & \text{'OT'} & [48] \\
\text{'Liz'} & [25] & [0] & \text{'IT'} & [87]
\end{array}
$$

The NaN symbol applies in MatLab to undefined numeric values emerging from division by zero and the like.

As one can see, these are not exactly clean-cut structures to work with. The numerical array nn contains an incomprehensible column of NaN values, and the text file tt mixes up names of students and features.

A4.3 Using Subsets of Entity/Feature Sets

If one wants working with only three of the six features, say "Age", "Children" and "OOProgramming_Score", one must put together their indices into a named 1D array:

≫ii=[457]
% no semicolon in the end to display ii on screen as a row;

Then commands to reduce the dataset and the set of feature names over ii columns can be like these:

≫newa=arr(:,ii); %new data array
≫newb=b(ii);
% newb is new feature set: to set it, one uses round braces rather than curly ones,
% in spite of the fact that cells are involved here, not arrays

A similar command makes it to a subset of entities. If, for instance, we want to limit our attention to only those students who received 60 or more at "OOProgramming", we first find their indices with command "find":

≫jj=find(arr(:,7)>=60);
% jj is the set of the students defined in find( )
% arr(:,7) is the seventh column of arr

Now we can apply "arr" to "jj":

≫al=arr(jj,:); % partial data of better off students

The size of the data file al can be found with command

≫size(al)
% note: no semicolon to see the size on the screen

to produce a screen output:

$$ans =$$

$$55 \quad 8$$

meaning that al consists of 55 rows and 8 columns. If one needs to maintain these in the workspace, use command

≫[n,m]=size(al)

that will put 55 into n and 8 into m.

Now we are ready to discern meaningful data from numerical array nn and text cell tt in workspace for Table A.4 on p. 501. The array nn's meaningless column is 3. Thus we can remove it like this:

≫[rnn,cnn]=size(nn);

% thus, the number of columns is cnn

≫vv=setdiff([1:cnn],3);
% operation setdiff(x,y) removes from x all elements of array y occurring in x
% [1:cnn] is an array of all integers from 1 to cnn inclusive, e.g., [1:4] is [1 2 3 4]
% thus, vv consists of all indices but 3

≫nnr=nn(:,vv);
% this puts all nn, except for column 3, into nnr:

$$nnr = \begin{matrix} 35 & 0 & 94 \\ 28 & 2 & 67 \\ 27 & 1 & 85 \\ 28 & 0 & 48 \\ 25 & 0 & 87 \end{matrix}$$

To create a cell containing the corresponding feature set, we need first to have a cell with all features. These constitute the final fragment of the first row of cell tt, without the very first string, "Feature", as can be seen from the tt contents shown above. Thus command

≫fe=tt(1,2:5);
% only first row in tt concerning its four columns, 2 to 5, goes to cell fe

leads to cell fe of size 1×4 containing of four features. To remove feature 3, we apply the array vv produced above:

≫fer=fe(vv); .

Cell fer contains strings 'Age', '#Children', 'ML_Mark' indexed by 1, 2 and 3 and corresponding, in respect, to columns of array nnr.

Many additional operations of MatLab are introduced and utilized in projects, worked examples and case studies within the text throughout.

A5 MatLab Program Codes

A5.1. Minkowski's Center: Evolutionary Algorithm

```
%cm.m, computing Minkowski p-distance central point c of a series x
%along with the average distance and its proportion in the sum

function [c,d,pe]=cm(x,p)

n=length(x);
lb=min(x);
rb=max(x); %-----------------lb, rb are boundaries of the area (i)---
de=0;
for ik=1:n
   de=de+(abs(x(ik)))^p;
end
de=de/n;%--------------------------average p-th power of the data


%-------------population setting (ii), setting the limit, iter, to iterations
pp=15; %population size
feas=(rb-lb)*rand(pp,1)+lb; %  generated population of p c values within the
range
flag=1;
count=0;
iter=5000;
%---------- evaluation of the initially generated population (iii) ----
funp=0;
for ii=1:pp
   vv(ii)=mink(p,x,feas(ii));
end
[funi, ini]=min(vv);
soli=feas(ini) %initial best c value
funi %initial error
si=1;%0.5; %step of change
%-------------evolution of the population (iv) -----------------
while flag==1
   count=count+1;
   feas=feas+si*randn(pp,1); % Gaussian mutation added with step si
   for ii=1:pp
```

```
      feas(ii)=max(lb, feas(ii));
      feas(ii,:)=min(rb,feas(ii));% keeping the population within the range
      vec(ii)=mink(p,x,feas(ii)); %evaluation
   end
%------------- elite maintenance (v) ----------------
   [fun, in]=min(vec); %best distance value
   sol=feas(in,:);%corresponding c value
   [wf,wi]=max(vec);
   wun=feas(wi); %worst c
   if wf>funi
      feas(wi)=soli;
      vec(wi)=funi; % changing the worst for the elite
   end
   if fun < funi
      soli=sol;
      funi=fun;
   end
   if (count>=iter)
      flag=0;
   end
   pe=funi/de;
%------------ screen the results of every 1000th iteration
   if rem(count,1000)==0
      %funp=funi;
      disp([soli pe]);
   end
end
c=soli;
d=funi;
pe=d/de;

return


%--------computing the quality of ce, the average deviation in p-th power
function dis=mink(p,x,ce)

nn=length(x);
dis=0;
for ik=1:nn
   dis=dis+(abs(x(ik)-ce))^p;
end
dis=dis/nn;

return
```

A5.2 Fitting Power Law: Non-linear Evolutionary and Linearization

```
% plan.m, power law analysis assuming the predictor x and target y are
% available as variables in matlab
% the power law is a function:    y=ax^b              (1)
% its linearized form:            log(y)=log(a)+b*log(x)  (2)


function plan(x,y)

%-----linear regression analysis of log(x) and log(y)

for ii=1:length(x);xc(ii)=max(.05,x(ii));yc(ii)=max(0.05,y(ii));end;
%0.05 instead of 0 to make logarithms possible
xll=log(xc);
yll=log(yc);
[all,bll,cll, rvll]=lr(xll,yll);

%all the slope, bll the intercept of the linear regression
%cll the correlation coefficient, rvll the residual variance of the linear regression
yle=all*xll+bll;% linear-regression estimated yll
cd=cll^2;%determinacy coefficient, it should be cd=1-rvll
cd
rvll
%figure(1);plot(xll,yll,'k.',xll,yle,'rp');

%-----linearized: fitting equation (1) by first fitting equation(2)

[al,bl, rl]=llr(x,y);
% al the estimate of a, bl the estimate of b and
% rl the proportion of the residual variance in the variance of y

% ylr - the linearized rule estimate for the power law
for ii=1:length(x);ylr(ii)=al*x(ii)^bl;end;

%-------as is: fitting equation (1) by straightforwardly minimizing the
%-------residual variance with an evolutionary algorithm

[an,bn,f, rn]=nlr(x,y);
% an the estimate of a, bn the estimate of b and
% rn the proportion of the residual variance in the variance of y
for ii=1:length(x);yn(ii)=an*x(ii)^bn;end; %estimated power law

%-----------output: two-plot figure, real on the left, log on the right
%figure(2);
subplot(1,2,1);
plot(x,y,'k.',x,ylr,'b.',x,yn,'r.');%data scatter with two estimated power laws,
```

```
% blue-linearized, red- as is
subplot(1,2,2);plot(xll,yll,'k.',xll,yle,'rp');

%-----------output: text file of the results
saveplan('rep', cll, al, bl, rl, an, bn, rn,cd);

return

% llr.m, fitting a nonlinear regression function y=ax^b
% using linearization
% x is predictor, y is target, a,b -regression parameters to be fitted

function [a,b, residvar]=llr(xt,yt);

% regression is power law y=a*x^b as reflected in the procedure
% residvar is the average square error's proportion to the variance of y;
% xt, yt are predictor and target

%-----an elementary check of length compatibility--------
ll=length(xt);
if ll~=length(yt)
   disp('Something wrong is with the data');
   pause;
end

%--------- calculating a and b using the linearization
for ii=1:ll;xc(ii)=max(.05,xt(ii));yc(ii)=max(0.05,yt(ii));end;
%putting 0.05 instead of zero to make possible logarithms of the data
xl=log(xc); %taking log of x and y
yl=log(yc);

[al,bl,dl]=lr(xl,yl);
b=al;
a=exp(bl);
ab=[a b];
residvar=delta(ab,xt,yt)/var(yt,1);
return

%-------- computing the quality of the approximation y=a*(x^b)
%which is the residual variance

function esq=delta(tt,x,y)%tt=[a, b]; x predictor, y target
a=tt(1);
b=tt(2);
esq=0;
for ii=1:length(x)
```

```
   yp(ii)=a*(x(ii)^b); %this power law function can be changed
   esq=esq+(y(ii)-yp(ii))^2;
end
esq=esq/length(x);
return;
```

```
% nlr.m, evolutionary fitting of a nonlinear regression function y=f(x,a,b)
% x is predictor, y is target, a,b -regression prameters to be fitted
```

```
function [a,b, funi,residvar]=nlr(xt,yt);
```

```
% in this version the regression equation is power law y=a*x^b which is
% reflected only in the subroutine 'delta' in the bottom for computing the
% value of the average error squared;
% funi is the average square error's best value;
% residvar is its proportion to the variance of y;
% xt, yt are predictor and target
```

```
%-----an elementary check of length compatibility--------
ll=length(xt);
if ll~=length(yt)
   disp('Something is wrong with the data');
   pause;
end
%-------------- determine rectangle at which (a,b)-populations fluctuate
[ab,bb]=ddr(xt,yt);
```

```
lb=[ab(1) bb(1)];
rb=[ab(2) bb(2)];
lb
rb
disp('Hit ENTER if you wish to proceed. ');
pause;
%-------------organisation of the iterations, iter the limit to their number
p=15; %population size
for ii=1:p;feas(ii,:)=(rb-lb).*rand(1,2)+lb;end; %  generated population of p pairs
coefficients within the range
flag=1;
count=0;
iter=10000;%5000;
%---------- evaluation of the initially generated population
funp=0;
for ii=1:p
   vv(ii)=delta(feas(ii,:),xt,yt);
end
[funi, ini]=min(vv);
```

```
      soli=feas(ini,:) %initial coeffts
      funi %initial error
      si=1;%0.5; %step of change
      %-------------evolution of the population
      while flag==1
        count=count+1;
        feas=feas+si*randn(p,2); %mutation added with step si
        for ii=1:p
          feas(ii,:)=max([lb;feas(ii,:)]);
          feas(ii,:)=min([rb;feas(ii,:)]);% keeping the population within the range
          vec(ii)=delta(feas(ii,:),xt,yt); %evaluation
        end

        [fun, in]=min(vec); %best approximation value
        sol=feas(in,:);%corresponding  parameters
        [wf,wi]=max(vec);
        wun=feas(wi,:); %worst case
        if wf>funi
          feas(wi,:)=soli;
          vec(wi)=funi;
  %changing the worst for the best of the previous generation
        end
        if fun < funi
          soli=sol;
          funi=fun;
        end
        if (count>=iter)
          flag=0;
        end
      residvar=funi/var(yt,1);
      %------------ screen the results of every 500th iteration
        if rem(count,500)==0
          %funp=funi;
          disp([soli residvar]);
        end
      end
      a=soli(1);
      b=soli(2);
      return

      %-------- computing the quality of the approximation y==a*(x^b)
      function esq=delta(tt,x,y)%tt=[a, b]; x predictor, y target
      a=tt(1);
      b=tt(2);
      esq=0;
      for ii=1:length(x)
```

```
    yp(ii)=a*(x(ii)^b); %this is a power law function
    esq=esq+(y(ii)-yp(ii))^2;
end
esq=esq/length(x); %the average difference squared
return;



% ddr.m, determinacy of the domain for power law y=a*x^b with b
% restricted
function [ab,bb]=ddr(x,y)
n=length(x);
bm=(log(y(1))-log(y(2)))/(log(x(1))-log(x(2)));
am=y(1)/(x(1)^bm);
ab=[am am];
bb=[bm bm];
%-------------finding extreme values for a and b using pairwise equations
bs=0;as=0; bsq=0;asq=0;
count=0;
for ii=1:(n-1);
   if min(x(ii),y(ii))>.25
   for jj=(ii+1):n
     if min(x(jj),y(jj))>.25
     if (x(ii)/x(jj)<0.75)|(x(ii)/x(jj)>1.25)
        count=count+1;
        bt=(log(y(ii))-log(y(jj)))/(log(x(ii))-log(x(jj)));
        aij=y(ii)/(x(ii)^bt);
        aij=min(aij,100);%restriction
        %if (aij>100)
        %   disp([ii jj]); aij
        %end;
        bs=bs+bt;
        bsq=bsq+bt*bt;
        as=as+aij;
        asq=asq+aij*aij;
        if bt>bb(2)
           bb(2)=bt;
        end;
        if bt<bb(1)
           bb(1)=bt;
        end;
        if aij>ab(2)
           ab(2)=aij;
        end;
        if aij<ab(1)
           ab(1)=aij;
        end;
```

```
        end;
        end;
      end;
      end;
    end;
  as=as/count
  asq=asq/count;
  sas=sqrt(asq-as^2)
  bs=bs/count
  bsq=bsq/count;
  sbs=sqrt(bsq-bs^2)
  ab(1)=as-4*sas;ab(2)=as+4*sas;
  bb(1)=bs-4*sbs;bb(2)=bs+4*sbs;
  count
  return
```

% saveplan.m, saving results of the power-law analysis in plan.m

```
function saveplan(file, cc, al, bl, rl, an, bn, rn,cd);

ct =num2str(cc);
first=['Results of the power-law analysis y=ax^b' ];
alla=[ 'On the level of logarithms, the correlation is  ' num2str(cc)];
alex=['Explained proportion of log(y)-variance is ' num2str(100*cd) '%'];
nt=[ ];
lt1=['Linearized estimate parameter values are a= ' num2str(al) ', b= ' num2str(bl)];
lt2=['Explained proportion of y-variance is r= ' num2str(100*(1-rl)) '%' ];
nt1=['"As is" estimate parameter values are a= ' num2str(an) ', b= ' num2str(bn)];
nt2=['Explained proportion of y-variance is r= ' num2str(100*(1-rn)) '%'];
alltext=strvcat(alla, lt1,lt2,nt1,nt2);

oul=[' These are visualized on the Figure produced:']
our=[' The power-law estimates on the left, the logarithms, on the right'];
  alltt=strvcat(alltext, oul, our);
  alltt
Filename=[ file '.out'];
fid= fopen(Filename, 'at');
if fid~=-1
   fprintf(fid, '%s\n', first);
   fprintf(fid, '%s\n', '    ');
   fprintf(fid, '%s\n', alla);
   fprintf(fid, '%s\n', alex);
   fprintf(fid, '%s\n', '    ');
   fprintf(fid, '%s\n', lt1);
   fprintf(fid, '%s\n', lt2);
   fprintf(fid, '%s\n', '    ');
```

```
        fprintf(fid, '%s\n', nt1);
        fprintf(fid, '%s\n', nt2);
        fprintf(fid, '%s\n', '    ');
        fprintf(fid, '%s\n', oul);
        fprintf(fid, '%s\n', our);
        fprintf(fid, '%s\n', '    ');
        fprintf(fid, '%s\n', '    ');
        fclose(fid);
    end;
    return
```

A5.3 Training Neural Network with One Hidden Layer

```
% nnn.m for learning a set of features from a data set
% with a neural net with a single hidden layer
% with the symmetric sigmoid (hyperbolic tangent) in the hidden layer
% and data normalisation to [-10,10] interval

function [V,W, mede]=nnn(hiddenn,muin)

% hiddenn - number of neurons in the hidden layer
% muin - the learning rate, should be of order of 0.0001 or less
% V, W - wiring coefficients learnt
% mede - vector of absolute values of errors in output features

%--------------1.loading data ---------------------
da=load('Data\studn.dat'); %this is where the data file is put!!!
% da=load('Data\iris.dat'); %this will be for iris data
[n,m]=size(da);

%-------2.normalizing to [-10,10] scale---------------------
mr=max(da);
ml=min(da);
ra=mr-ml;
ba=mr+ml;
tda=2*da-ones(n,1)*ba;
dan=tda./(ones(n,1)*ra);
dan=10*dan;
%------------3. preparing input and output target)--------
ip=[1:5];  % here is list of indexes of input features!!!
%ip=[1:2];%only two input features in the case of iris
ic=length(ip);
op=[6:8];  % here is list of indexes of output features!!!
%op=[3:4];% output iris features
oc=length(op);
output=dan(:,op); %target features file
```

```
input=dan(:,ip);  %input features file
input(:,ic+1)=10;     %bias component
%----------------4.initialising the network --------------------
h=hiddenn;      %the number of hidden neurons!!!
W=randn(ic+1,h); %initialising w weights
V=randn(h,oc);  %initialising v weights
W0=W;
V0=V;
count=0; %counter of epochs
stopp=0; %stop-condition to change
%pause(3);


while(stopp==0)
mede=zeros(1,oc); % mean errors after an epoch
%----------------5. cycling over entities in a random order
  ror=randperm(n);
  for ii=1:n
    x=input(ror(ii),:); %current instance's input
    u=output(ror(ii),:);% current instance's output
%---------------6. forward pass (to calculate response ru)------
    ow=x*W;
    o1=1+exp(-ow);
    oow=ones(1,h)./o1;
    oow=2*oow-1;% symmetric sigmoid output of the hidden layer
    ov=oow*V; %output of the output layer
    err=u-ov; %the error
    mede=mede+abs(err)/n;
%----------- 7. error back-propagation-------------------------
    gV=-oow'*err;    % gradient vector for matrix V
    t1=V*err'; % error propagated to the hidden layer
    t2=(1-oow).*(1+oow)/2; %the derivative
    t3=t2.*t1';% error multiplied by the th's derivative
    gW=-x'*t3;  % gradient vector for matrix W
%----------------8. weights update-----------------------
    mu=muin; %the learning rate from the input!!!
    V=V-mu*gV;
    W=W-mu*gW;
  end;
%-----------------. stop-condition ------------------------
  count=count+1;
  ss=mean(mede);
  if ss<0.01|count>=10000
    stopp=1;
  end;
  mede;
  if rem(count,500)==0
      count
      mede
    end
  end;
```

A5.4 Building Classification Trees

```
% clatree.m a program for building a decision tree over quantitative data,
% according to a method, 'gini', 'chi' or 'ing' in 3.5
% and specified stopping conditions: (a) number of entities,
% (b) prevailing feature; Inputs: data matrix X, partition as cell s, method
% variables untouched

function Clusters=clatree(X, s, method)

[n,mm]=size(X)
TS=10; %cluster size threshold
ee=0.8;%threshold to an s-class contents in a cluster
tin=0; %threshold on the scoring function to be set
switch method
        case 'gini'
            tin=0.03;
        case 'chi'
            tin=0.08;
        case 'ing'
            tin=0.15;
        otherwise
            disp('The method is wrong ');
            pause(10);
     end
for ik=1:length(s);
   ds(ik)=length(s{ik});
end
ds=ds/sum(ds);
%distribution of s
ss=1; %cluster counter
bb=ss; %the last cluster's index
Clusters{ss,1}=[1:n];%entity set to cluster
%Clusters{ss,2}=[1:m];%features to be used
if max(ds)<ee
   Clusters{ss,2}=1; %should be split further
else
   Clusters{ss,2}=0; %should not be split further
end
Clusters{ss,3}=[0]; %parent's index
Clusters{ss,4}=[]; %characteristics
```

```
Clusters{ss,5}=ds;%distribution of s
tt=0;%counter of clusters to split taking into account added clusters
while ~(tt==ss),
  for uu=(tt+1):ss
    uu
    realnum=Clusters{uu,1}; %cluster to be split
    flag=Clusters{uu,2};
    if (flag==1)·
      ma=-1;%starting gain value
      vv=0;%starting feature
      yy=-1000;%starting value
      for v0=1:mm
        xs=X(realnum,v0); %variable to be used
        [g,res,y]=msplit(xs,s,method);%producing split
      disp(['var ' num2str(v0) ' val ' num2str(y) ' ' num2str(res)])
      % this line is to see action of each feature at each cluster
        if res>=ma
          ma=res;
          yy=y;
          vv=v0;
        end;
      end
      if ma>tin
        xt=X(realnum,vv);
        g{1}=realnum(find(xt<=yy));
        g{2}=realnum(find(xt>yy));
        l1=length(g{1});
        l2=length(g{2});
        if (l1*l2)==0
          Clusters{uu,2}=0;
        else
          if (l1>TS & l2>TS)
            cc=clfil(g{1},s,ee,vv,uu,-1,yy,ma);
            for il=1:5
              Clusters{bb+1,il}=cc{il};
            end
            cc=clfil(g{2},s,ee,vv,uu,1,yy,ma);
            for il=1:5
              Clusters{bb+2,il}=cc{il};
            end
            bb=bb+2;
          elseif l1>TS
            cc=clfil(g{1},s,ee,vv,uu,-1,yy,ma);
            for il=1:5
              Clusters{bb+1,il}=cc{il};
            end
```

```
                    bb=bb+1;
                elseif l2>TS
                    cc=clfil(g{2},s,ee,vv,uu,1,yy,ma);
                    for il=1:5
                        Clusters{bb+1,il}=cc{il};
                    end
                    bb=bb+1;
                end;
                Clusters{uu,2}=0;
            end
        end;
      end;
    end;
  tt=ss;
  ss=bb;
end;
%savrdnew(file,Clusters,CC,B,yent);
return
%-------------- assigning a cluster object
function cc=clfil(gg,s,ee,vv,uu,t,y,ma)
%t=-1 for 1-split, 1 for 2-split
 cc{1}=gg;
 for ik=1:length(s)
    ds(ik)=length(intersect(s{ik},gg));
 end;
 ds=ds/sum(ds);%distribution of s in gg
 if (max(ds)>ee)
    cc{2}=0;
 else
    cc{2}=1;
 end
 cc{3}=uu; %parent
 cc{4}=[vv t y ma];
 % variable, less/more than, split y, gain
 cc{5}=ds;
 return
```

A6 Two Random Samples

A6.1 Short.dat is a Dataset of Random Samples from Three Different Distributions in Table A.5

Table A.5 Three columns from three different distributions

8	20	1512
12	21	50
11	23	48
10	21	206
9	9	12
7	20	199
10	22	51
12	18	50
9	20	198
13	21	843
9	5	12
13	13	8
10	10	7
11	14	9
9	18	39
9	13	12
7	21	51
11	20	46
11	21	50
9	18	54
8	20	1391
10	19	49
10	19	41
13	24	35
12	23	45
10	13	11
12	9	9
10	21	49
7	10	10
8	17	52
12	8	8
11	20	48
12	17	199
8	11	9
8	11	13

(continued)

Table A.5 (continued)

9	20	978
12	17	51
9	20	6233
13	19	23
10	21	47
11	11	8
11	20	973
11	7	43
13	20	201
9	18	200
10	19	49
9	10	7
14	20	36
9	10	8
11	21	203

A6.2 A Sample of 280 N(0,10) Values

Here is a sample from the Gaussian distribution N(0,10). The sample has been sorted in the ascending order (Table A.6).

Table A.6 Sample of 280 values from N(0,1) sorted in the ascending order

−30.29	−12.48	−7.01	−2.99	1.76	5.58	10.35
−25.57	−12.29	−6.94	−2.91	1.98	5.59	10.5
−25.34	−12.27	−6.83	−2.83	1.98	5.63	10.94
−23.79	−11.89	−6.79	−2.78	2.07	5.65	10.98
−23.34	−11.61	−6.65	−2.75	2.08	5.65	11.08
−22.38	−11.5	−6.64	−2.66	2.14	5.74	11.13
−22.37	−11.33	−6.11	−2.66	2.14	5.74	11.64
−21.78	−11.1	−6.02	−2.58	2.18	5.81	12.28
−21.05	−10.78	−5.98	−2.52	2.21	5.82	12.33
−20.89	−10.57	−5.87	−2.23	2.27	5.89	12.59
−20.65	−10.52	−5.53	−2.07	2.28	6.13	12.79
−19.1	−10.44	−5.35	−2.06	2.29	6.26	12.93
−18.16	−10.13	−5.33	−1.91	2.36	6.29	13.15
−17.95	−10.09	−5.22	−1.9	2.37	6.51	13.24
−17.79	−10.08	−5.17	−1.74	2.56	6.55	13.42
−17.58	−10.06	−4.91	−1.6	2.71	6.59	13.44
−16.47	−9.79	−4.82	−1.51	2.79	6.59	13.48
−16.43	−9.11	−4.62	−1.44	2.85	6.65	13.56
−16.31	−9.08	−4.58	−1.42	2.91	7	13.99
−16.19	−9.01	−4.53	−1.28	2.94	7.09	14.27
−16.15	−8.95	−4.43	−1.26	2.98	7.16	14.69
−16.14	−8.93	−4.26	−0.8	3.16	7.3	14.95
−15.9	−8.71	−4.18	−0.79	3.21	7.58	15.35
−15.89	−8.53	−4.17	−0.73	3.27	7.99	15.74
−15.67	−8.49	−4.08	−0.5	3.27	8.34	15.82
−15.56	−8.01	−4.01	−0.49	3.46	8.57	15.84
−15.5	−7.98	−3.98	−0.23	3.66	8.58	15.99
−15.04	−7.97	−3.95	−0.21	3.74	8.7	16.03
−15	−7.75	−3.84	−0.08	3.8	8.85	16.84
−14.91	−7.67	−3.78	−0.02	4.29	8.87	16.87
−14.16	−7.48	−3.74	0.03	4.39	8.97	17.29
−14.14	−7.46	−3.65	0.33	4.41	9.02	17.62
−14.04	−7.44	−3.61	0.65	4.42	9.08	18.43
−13.88	−7.37	−3.59	0.7	4.48	9.12	19.57
−13.84	−7.37	−3.47	0.78	4.6	9.39	19.58
−13.72	−7.35	−3.46	0.8	4.78	9.57	20.8
−13.58	−7.27	−3.39	1.1	4.94	9.83	22.38
−13.33	−7.24	−3.14	1.2	5.28	10.02	22.66
−12.98	−7.2	−3.02	1.38	5.41	10.08	29.5
−12.68	−7.03	−3.01	1.58	5.54	10.09	32.03

Index

A

Absolute deviation, 90
Absolute Quetelet index, 225
Accuracy, 287
Activation function, 204, 207
Additive clustering, 465
AddRemAdd algorithm, 471
AddRem algorithm, 449
Adequate statement, 85
Admissible transformation, 84
ADO algorithm, 473
Affinity data, 24
Agglomerative clustering, 368, 409
Alternating minimization, 491
Analytical Hierarchy Process, 159
Anomalous pattern, 327
Anti-cluster, 469
Area Under the Curve (AUC), 289
Artificial neuron, 204
Attraction, 447

B

Background similarity, 446
Bag-of-words model, 279
Baire distance, 407
Batch K-means, 297
Bayes decision rule, 277, 278
Bernoulli model, 111, 282
Betweenness, 245
Binary feature, 107
Binary tree, 254
Bisecting K-means, 411
Bootstrap, 102, 173, 353
Build algorithm, 320

C

Canonical correlation, 232, 497
Categorical feature, 33, 105
Center of cluster, 293
Chi-squared contingency coefficient, 222
Chi-square distance, 239
Classification tree, 253
Clustering, 294
Cluster intensity, 457
Cluster representative, 345
Coefficient of determination, 192
Comparing means, 353
Complementary clustering criterion, 324
Conceptual association, 218
Conceptual clustering, 411
Conceptual description of clusters, 351
Conditional probability, 222
Confusion table, 260, 285
Connected component, 432, 434
Consensus partition, 387
Contingency table, 217
Conventional PCA formulation, 128
Contribution of PCA model to data scatter, 126
Correlation coefficient, 176
Correlation ratio, 265
Correspondence analysis, 233, 234
Covariance matrix, 127, 279
Cross-over, 315
Cross-validation, 203

D

Data analytics, 45
Data recovery approach, 7
Data scatter, 82, 114

© Springer Nature Switzerland AG 2019
B. Mirkin, *Core Data Analysis: Summarization, Correlation, and Visualization*, Undergraduate Topics in Computer Science,
https://doi.org/10.1007/978-3-030-00271-8